8/71

D0163573

ORGANIZATION AND CONTROL IN PROKARYOTIC AND EUKARYOTIC CELLS

Other Publications of the
*Society for General Microbiology**

THE JOURNAL OF GENERAL MICROBIOLOGY
THE JOURNAL OF GENERAL VIROLOGY

SYMPOSIA

* Published by the Cambridge University Press, except for the first Symposium, which was published by Blackwell's Scientific Publications Limited.

ORGANIZATION AND CONTROL IN PROKARYOTIC AND EUKARYOTIC CELLS

TWENTIETH SYMPOSIUM OF THE
SOCIETY FOR GENERAL MICROBIOLOGY
HELD AT
IMPERIAL COLLEGE LONDON
APRIL 1970

CAMBRIDGE
Published for the Society for General Microbiology
AT THE UNIVERSITY PRESS
1970

Published by the Syndics of the Cambridge University Press
Bentley House, 200 Euston Road, London, N.W.1
American Branch: 32 East 57th Street, New York, N.Y.10022

Library of Congress Catalogue Card Number: 70–111131

Standard Book Number: 521 07815 6

Printed in Great Britain
at the University Printing House, Cambridge
(Brooke Crutchley, University Printer)

CONTRIBUTORS

Bodmer, W. F., Department of Genetics, Stanford School of Medicine, Stanford University.

Brightwell, R., B.B.C., P.O. Box 1 AA, Broadcasting House, London, W. 1.

Echlin, P., Department of Botany, University of Cambridge.

Ellar, D., Sub-department of Chemical Microbiology, Department of Biochemistry, University of Cambridge.

Evans, M. C. W., Department of Botany, King's College, London.

Flexer, A. S., Department of Molecular, Cellular and Developmental Biology, University of Colorado, Boulder, Colorado 80302, U.S.A.

Holliday, R., National Institute for Medical Research, Mill Hill, London.

Hughes, D. E., Department of Microbiology, University College, Cardiff.

Lloyd, D., Department of Microbiology, University College, Cardiff.

Loening, U. E., Department of Zoology, University of Edinburgh.

Mitchell, P., Glynn Research Laboratories, Bodmin, Cornwall.

Newton, Alison, A., Department of Biochemistry, University of Cambridge.

Raper, J. R., Department of Biology, Harvard University.

Richmond, M. H., Department of Bacteriology, University of Bristol.

Shockman, G. D., Department of Pathology, College of Physicians and Surgeons, Columbia University, New York.

Stanier, R. Y., Department of Bacteriology and Immunology, University of California.

Thompson, J. S., Department of Biological Chemistry, University of Manchester.

Vogel, H. J., Department of Pathology, College of Physicians and Surgeons, Columbia University, New York.

WHATLEY, F. R., Department of Botany, King's College, London.

WILKIE, D., Department of Botany and Microbiology, University College London.

WITTMANN, H. G., Max-Planck-Institut für Molekulare Genetik, Berlin-Dahlem, Germany.

WOESE, C. R., Department of Microbiology, University of Illinois.

CONTENTS

EDITORS' PREFACE

Since Darwin a major generalization of biology has been the theory of evolution. A subsequent generalization is that of the unity of bio-chemistry at the cellular level, and now there is the generalization of the universality of the genetic code. The latter perhaps takes precedence among biological generalizations because it implies both the theory of evolution and the theory that all living organisms had a common origin. The use of the electron microscope has brought, however, acceptance of a major morphological *discontinuity* among the cells of living organisms, namely as between prokaryotes—the bacteria and blue-green algae—on the one hand, and the eukaryotic cells of other living forms. The prokaryotes do not have nuclear membranes and 'true' chromosomes, nor do they have mitochondria, chloroplasts and other organelles characteristic of the cells of other organisms. It may, however, be questioned whether the morphological discontinuity between pro-karyotic cells and eukaryotic cells is as great as we now think; what first appear to be sharp distinctions in science often become less sharp on further research.

If living organisms did indeed have a common origin then eukaryotic cells presumably evolved from prokaryotic cells, because the latter have the less complex organization; in which case there are gaps in knowledge about the way in which this evolution took place. It is possible that some or all of the intracellular organelles of eukaryotic cells may have evolved from prokaryotes dwelling as symbionts of cells which then evolved into eukaryotes (see article by R. Y. Stanier).

Wide generalizations have been drawn from the study of relatively few micro-organisms; but deeper understanding of the apparent morphological discontinuity between prokaryotes and eukaryotes, and of the evolution of eukaryotes from prokaryotes will require much greater knowledge of the less studied groups of micro-organisms.

The purpose of this Symposium has been to gather some knowledge about prokaryotic and eukaryotic cells so that the two kinds may be seen in relation to each other. The Symposium cannot be restricted to the consideration only of micro-organisms because in some areas, as in the study of chromosomes, knowledge has hitherto come mostly from the study of cells of higher plants and animals rather than from those of micro-organisms.

The general problem of the interrelations between prokaryotes and eukaryotes is discussed by R. Y. Stanier in the first article of this Symposium. C. R. Woese discusses the problem of the universality of the genetic code. Related to the problem of the genetic code is the nature of ribosomes; the articles by H. G. Wittmann and by U. E. Loening deal with selected topics about these. Another aspect of comparative biochemistry is considered by H. J. Vogel, J. S. Thompson & G. D. Shockman; their article shows the value of the study of lysine biosynthesis for the understanding of microbial relationships.

Photosynthesis is of special significance in the comparative study of prokaryotes and eukaryotes. Some of the processes are common to both kinds of organism, but chloroplasts are found only in eukaryotes. The ability to evolve oxygen from water is a property shared by one peculiar group of prokaryotes, the blue-green algae, with all photosynthetic eukaryotes. The comparative study of photosynthesis therefore reveals the blue-green algae as uniquely and puzzlingly related to eukaryotes. The chemical and morphological aspects of photosynthesis are dealt with by M. C. W. Evans & F. R. Whatley, and by P. Echlin, respectively.

A striking difference between prokaryotes and eukaryotes is that the latter have mitochondria, chloroplasts and other organelles. We are now aware that these organelles contain DNA, which confirms that they have some degree of autonomy. The significance of this DNA is discussed by D. Wilkie. Other characters of organelles are discussed by D. E. Hughes, D. Lloyd & R. Brightwell, and by R. Y. Stanier. Whilst bacteria do not have mitochondria, etc., they do have extrachromosomal DNA; this is discussed by M. H. Richmond. Both prokaryotes and eukaryotes have mechanisms whereby the DNA from one cell line is brought in contact with the DNA from another; aspects of this are discussed by M. H. Richmond, by W. F. Bodmer and by J. R. Raper & A. S. Flexer.

All cells have membranes: aspects of these are dealt with by P. Mitchell, by D. E. Hughes, D. Lloyd & R. Brightwell in relation to cell organelles, and by D. Ellar in relation to protective surface structures.

The morphology of prokaryote 'chromosomes' is relatively simple and understood, but that of eukaryote chromosomes is proving a very difficult problem, as also is the significance of the high DNA content of eukaryotes. These problems are discussed by R. Holliday, chiefly in relation to work with higher organisms. A related problem is why the diploid phase has become dominant in both higher plants and higher animals; this is discussed by J. R. Raper & A. S. Flexer in relation to nuclear phenomena in fungi.

A common feature of prokaryotes and eukaryotes is that they are parasitized by viruses. In view of the differences between the two kinds of cells, questions arise about the mechanism of virus infection and multiplication; this is discussed by Alison A. Newton.

In attempting to bring together discussion of prokaryotes and eukaryotes, problems of terminology arise. For example, terms such as nucleus, chromosome and flagellum are used for structures which are rather different in prokaryotes and eukaryotes. This problem is raised by R. Y. Stanier in his article, and we would draw attention to his glossary (page 31) in which he defines his use of terms.

We have not attempted to impose uniformity of terminology in the articles of this Symposium, except to spell the words prokaryote and eukaryote with k and not with c. The terms prokaryotic and eukaryotic were introduced by the late E. C. Dougherty in a short article (1957) which is reprinted on p. 433. The root words which he devised were prokaryon, as a name for the nucleus of bacteria and blue-green algae, and eukaryon, for the nucleus of all other cells. Strictly, therefore, it is incorrect to use the expression 'prokaryotic nucleus' and 'eukaryotic nucleus', but the terms prokaryon and eukaryon have not yet become generally used.

We thank Mrs Patricia H. C. Cooper and Mrs Elizabeth W. Tisley for excellent secretarial work.

And for ourselves we would quote Moses Maimonides (1135–1204)*: 'Today he can discover his errors of yesterday and tomorrow he may obtain light on what he thinks himself sure of today.'

H. P. CHARLES

B. C. J. G. KNIGHT

Department of Microbiology
University of Reading

---

* From *Physics, Psychology and Medicine*, by J. H. Woodger, 1956. Cambridge University Press.

SOME ASPECTS OF THE BIOLOGY OF CELLS AND THEIR POSSIBLE EVOLUTIONARY SIGNIFICANCE

R. Y. STANIER

Department of Bacteriology and Immunology,
University of California, Berkeley, California, U.S.A.

ASPECTS OF THE BIOLOGY OF CELLS

All organisms except viruses can be assigned to one of two primary groups, readily distinguishable by differences in cellular organization. The more complex eukaryotic cell is the unit of structure in metazoans, vascular plants, bryophytes, protozoa, fungi and all except one of the groups traditionally assigned to the algae. The less complex prokaryotic cell is the unit of structure in bacteria *sensu lato* and blue-green algae. This fundamental dichotomy at the cellular level is not well correlated with the conventional division of organisms into major groups in terms of organismal characters. The eukaryotes represent an extremely large and diverse assemblage of organisms, which includes the overwhelming majority of the biological species that now exist on earth. All prokaryotes are micro-organisms. Their relatively narrow range of organismal diversity parallels to a considerable degree that characteristic of microbial eukaryotes (protozoa, fungi, eukaryotic algae). Indeed, at the microbial level it is not difficult to find many striking specific examples of organismal analogy between eukaryotes and prokaryotes: cellular slime moulds and fruiting myxobacteria; mycelial fungi and actinomycetes; the green alga *Hydrodictyon* and the green bacterium *Pelodictyon*, to mention only a few.

The fact that bacteria and blue-green algae are in some sense atypical cellular organisms has been recognized for a century. Ferdinand Cohn (1875) showed a clear awareness of it in his discussion of the possible natural relationships of these two microbial groups: 'Perhaps the designation of *Schizophytae* may recommend itself for this first and simplest division of living beings, which appears to me to be naturally segregated from the higher plant groups, even though its distinguishing characters are negative rather than positive.'*

---

* Vielleicht möchte sich die Bezeichnung *Schizophytae* für diese erste und einfachste Abtheilung lebender Wesen empfehlen, die mir, den höheren Pflanzengruppen gegenüber, natürlich abgegrenzt erscheint, wenn auch die Merkmale, durch welche sie charakterisirt ist, mehr negativer als positiver Art sind (Cohn (1875), p. 201).

Even as recently as 30 years ago, Stanier & van Niel (1941) could do little better in an attempt to define these two groups. The definition of bacteria and blue-green algae in terms of positive rather than negative characters had to await the revolution in our knowledge of cell structure which followed the introduction of the electron microscope as a tool in biological research. Light microscopy cannot reveal many of the most significant structural properties of cells; and the limitations of the light microscope are particularly restrictive in the context of the prokaryotic cell, which is built on a smaller scale than the eukaryotic one. Once cell structure could be visualized at magnifications many times those achievable with a light microscope, it became easy to define in precise terms the principal differences between the two kinds of cells (Stanier, 1961; Murray, 1962; Stanier & van Niel, 1962).

These differences are now so widely recognized that descriptions of them can be found in the better textbooks of general biology (e.g. Curtis, 1968), a sure indication that they have acquired the status of truisms. It therefore seems unnecessary to start with a simple recapitulation. Instead, I shall devote this introductory essay to the consideration of two questions. Firstly, has work of recent years in the domain of cell biology turned up any new evidence for *least common denominators*, either structural or functional, characteristic of either eukaryotic or pro-karyotic cells? I think that there have been some significant new insights, even though the major differences were already evident ten years ago. Secondly, what are the possible evolutionary connections between pro-karyotes and eukaryotes? Here, too, recent work in the domain of cell and molecular biology has yielded some suggestive information.

The problem of recognizing least common denominators

For all practical purposes, research on the cytology of prokaryotic cells began only 25 years ago. Thanks to the comparative simplicity and uniformity of this type of cell, knowledge has accumulated very rapidly, so that it is now rather easy to recognize the least common denominators of prokaryotic structure and function. To make valid generalizations about the eukaryotic cell is more difficult for several reasons. As a result of its much greater inherent complexity, progress has been relatively slow; one has only to compare the state of knowledge about eukaryotic chromosomes (cf. Du Praw, 1968) with our present delight-fully complete and simple picture of the structure and replication of the bacterial *genophore* (Cairns, 1963), to become aware of this. A second difficulty in attempting to discern least common denominators of structure and function in the eukaryotic cell arises from its extraordinary diversity, a consequence both of its evolutionary specializations in the many divergent branches of the eukaryotic world, and of the ontogenic modifications that can be imposed upon it among the more highly differentiated multicellular eukaryotes. Even an attempt to generalize about such a central phenomenon as nuclear division must appear foolhardy, when this process can display the extreme variations exemplified in dinoflagellates (Kubai & Ris, 1969) and in fungi (Robinow & Bakerspiegel, 1965). For these reasons, I embarked on the search for new common denominators of eukaryotic cellular structure and function with considerable diffidence, all the more justified because my knowledge of this vast field is not first-hand.

A preliminary note on terminology

Terminological redundancies and ambiguities now present a significant and wholly unnecessary obstacle to the comprehension of cell structure. An international commission on cellular nomenclature is badly needed ! Much of the redundancy occurs in the naming of eukaryotic cell components; it can be largely explained (but not excused) by the long separation between botanical and zoological cytology. Much of the ambiguity has arisen from the use of the same term for isofunctional components of eukaryotic and prokaryotic cells, before their profound organizational differences had become evident. Ambiguity is particularly acute in nuclear cytology, where the word 'chromosome' has different meanings in eukaryotic and prokaryotic contexts. This specific ambiguity can be avoided only by the introduction of a new term, *genophore*, as proposed by Ris (1961). The glossary (p. 31) gives the definitions and synonyms of the principal cytological terms that will appear in the following pages.

SOME COMMON DENOMINATORS OF THE EUKARYOTIC CELL

Dispersion of the genome

The nucleus is almost never the sole repository of DNA in a eukaryotic cell. (Possible exceptions are strictly anaerobic, non-photosynthetic, eukaryotic protists.) A fraction of the genome—quantitatively minor, but of great importance for cell function—is carried by genophores located in mitochondria and chloroplasts. This recent discovery had been foreshadowed by genetic findings, which showed that some traits expressed at the level of chloroplasts and mitochondria are subject to non-Mendelian (so-called 'cytoplasmic') inheritance (Rhoades, 1946; Mitchell & Mitchell, 1952; Ephrussi, 1953). However, the material bases of such anomalous genetic behaviour remained obscure until it was shown by cytological techniques (e.g. Ris & Plaut, 1962; Nass & Nass, 1963) and subsequently by direct isolation of DNA from purified organelles (e.g. Wolstenholme & Gross, 1968; Avers, Billheimer, Hoffmann & Pauli, 1968) that both mitochondria and chloroplasts contain DNA. In some eukaryotes (though not all), this organellar DNA can be distinguished physically from the nuclear DNA of the cell as a result of its markedly different buoyant density, the reflection of a substantial difference in mean DNA base composition (data summarized by Iwamura, 1966 and Nass, 1969).

The organellar genophores also differ from the nuclear genophores in structural respects. They can be isolated in the form of circular double helical molecules which resemble in structure (but are much smaller than) the bacterial genophore (e.g. Avers *et al.* 1968). There is no evidence to suggest that organellar genophores contain histones, as do the chromosomes of all eukaryotes except dinoflagellates.

This discovery immediately raised a further question: how much of the machinery of replication, translation and transcription essential for the maintenance and phenotypic expression of organellar genophores is also located in mitochondria and chloroplasts? This problem is now being actively explored in a variety of cellular systems; and there is already good evidence for the presence in each class of organelle of DNA polymerase, DNA-dependent RNA polymerase, activating enzymes, transfer RNAs and ribosomes, some of which can be readily distinguished in functional terms from the corresponding extra-organellar components of the cell (see Nass, 1969, for a summary of the evidence in mitochondrial systems). A very elegant autoradiographic demonstration of RNA synthesis in both the mitochondria and the chloroplasts of an algal flagellate, *Ochromonas*, has been provided recently by Gibbs (1968).

It cannot be stated that the entire machinery of translation in each class of organelle is distinct from that in the cytoplasm; there may well be some common components. However, the organellar specificity of one major component, the ribosome, is evident from its physical properties. In all chloroplasts and mitochondria so far examined in this respect, the ribosomes are significantly smaller than the 80s ribosomes that occur in the cytoplasm; they are in the size range of prokaryotic (70s) ribosomes. Their resemblance to prokaryotic ribosomes has been further confirmed in a few cases by the demonstration that the major species of RNA isolated from organellar ribosomes have sedimentation constants of 16s and 22s, significantly different from those of the major RNA species in 80s ribosomes, and closely similar to those of the major

RNA species in prokaryotic ribosomes. A detailed comparison between cytoplasmic and chloroplastic ribosomes of the green alga *Chlamydomonas* has been made by Hoober & Blobel (1969).

The total amount of DNA in the genophore of mitochondria and chloroplasts is relatively small. Most estimates for mitochondria are of the order of 10^7 daltons, which is approximately 1 % of the amount of DNA in the genophore of the bacterium *Escherichia coli* (Nass, 1969). Since there is evidence that at least part of the mitochondrial translation system is coded for by mitochondrial DNA, the total information content of any mitochondrion is almost certainly insufficient to code for all the enzymes and structural proteins which confer respiratory function on this organelle.

The most extensive functional analysis of an organellar genetic system has been performed with the mitochondria of yeast (*Saccharomyces cerevisiae*). The question of the degree of organellar autonomy can therefore at present be discussed most profitably in this specific context. Yeast is a particularly favourable object for the analysis of this problem, because mitochondrial function is dispensable; glycolysis can provide the cell with an alternative source of ATP. Normal mitochondrial development in yeast can be irreversibly abolished by treatment of the cell with acridine dyes (Ephrussi, Hottinguer & Tavlitski, 1949). Such treatment leaves vestigial mitochondria in the cell (Yotsuyanagi, 1962) and has been shown to produce major changes in the nucleotide sequence of the mitochondrial DNA (Mounoulou, Jakob & Slonimski, 1966). The stable, respiration-deficient, so-called *petite* (*colonie*) mutants produced by acridine treatment therefore probably retain very few (if any) intact mitochondrial genes. It is also possible to block reversibly the *normal phenotypic expression* of the mitochondrial genophore by growing yeast in the presence of chloramphenicol, an antibiotic which selectively inhibits protein synthesis on ribosomes of the 70s type (Clark-Walker & Linnane, 1967).

These two treatments, one genetic and one physiological, provoke the synthesis of defective organelles which lack many normal mitochondrial components. A functional analysis can therefore reveal those constituents of the mitochondrion which are endogenous, and those which are synthesized in the cytoplasm, presumably under the control of chromosomal genes. Both in *petite* mutants and in chloramphenicol-treated cells of the wild type, the synthesis of cytochromes a, a_3, b and c_1 is abolished; these mitochondrial components therefore appear to be of endogenous origin. Synthesis of two c cytochromes is unaffected, and these components of the electron transport system are probably synthesized in the cytoplasm and determined by chromosomal genes. Clark-Walker & Linnane (1967) suggested that the soluble enzymes of the TCA cycle are also of extra-mitochondrial origin. The autonomy of the yeast mitochondrion is therefore far from absolute.

Petite mutants are, strictly speaking, genetic mutilates. However, a specific heritable mutation which affects a particular mitochondrial gene has been recently reported. At the cellular level, the mutation is expressed by increased resistance to erythromycin, an antibiotic which inhibits protein synthesis on 70s ribosomes. The mitochondrial location of this mutation is shown by its non-Mendelian inheritance (Linnane, Saunders, Gingold & Lukins, 1968), as well as by the fact that it affects specifically mitochondrial protein synthesis (Linnane, Lamb, Christodoulou & Lukins, 1968).

The ability of both mitochondria and chloroplasts to multiply by division is now well established. The large size and small number of chloroplasts in some algae provide favourable material for studying the replication of these organelles; their increase by division had been established by continuous microscopic observation in the nineteenth century (see Granick, 1961, for a summary). An unambiguous microscopic demonstration of mitochondrial division is more difficult, because of their smallness and abundance in most eukaryotic cells. However, there is now much evidence, direct and indirect, that this is the mode of mitochondrial multiplication (see Granick & Gibor, 1967, for a summary). Some particularly convincing micrographs showing division stages in the mitochondria of a myxomycete, *Physarum*, have been published by Guttes, Guttes & Devi (1969).

The very rapid recent extension of knowledge about mitochondria and chloroplasts has led to a major change in the interpretation of their nature. These organelles are evidently not just complex membrane-bounded multi-enzyme systems responsible for respiratory and photosynthetic energy generation in eukaryotic cells; they contain many structural elements and functional attributes hitherto considered distinctive of cells of the prokaryotic type (Nass, 1969). Only their partial dependence on the genetic and translational machinery of the enclosing eukaryotic cell—a dependence of which the extent cannot yet be fully assessed—distinguishes them formally from prokaryotic endosymbionts. At the operational level, it may not always be easy to make this distinction, as shown by some specific cases that I shall discuss presently. The evolutionary implications are obvious and important; they will be explored in the last part of this essay.

In the context of the question now being considered—the differences between eukaryotes and prokaryotes—the recent work on chloroplasts and mitochondria has revealed a major new common denominator of eukaryotic cells, which can be stated as follows. The eukaryotic cell contains two (or more) genetic systems, each housed and replicated in a separate intracellular structure, and each associated with a machinery of replication, transcription and translation, some elements of which are specific to that genetic system. The cellular phenotype is the expression of the sum of these genetic systems.

The functions of cytoplasmic unit membrane systems
in eukaryotic cells

The extra-organellar cytoplasmic region of the eukaryotic cells is always traversed by a complex system of unit membranes, the endoplasmic reticulum. For the most part, this system takes the form of irregular, flattened vesicles, the outer surfaces of which may be lined with adherent 80s ribosomes. In one or more regions, these membranes are more densely and regularly packed, forming a complex membranous organelle of fairly well-defined shape, which is never lined with ribosomes. This structure, the Golgi apparatus, was first seen at the end of the nineteenth century in stained sections of certain animal tissues. However, it is not easy to make visible by the methods of classical cytology; and its very existence, even in the cells where it had first been detected, remained controversial for many decades (Beams & Kessel, 1968). Electron microscopy revealed both the physical reality of the Golgi apparatus, and its presence in all types of eukaryotic cells. It is clearly a major element of structure in the eukaryotic cell; the work of the past decade suggests that it plays a variety of roles in cellular function, all of great importance.

The structure of the Golgi apparatus and its topological relations to other cell-components can best be discussed in the context of an interpretative diagram (Fig. 1), borrowed from Du Praw (1968). The unit membranes of the Golgi apparatus are always smooth, and devoid of adherent ribosomes. Secondly, the organelle has a well-defined polarity. One pole, which I shall term the basal pole, is deep in the cell; it sometimes lies near the nuclear membrane, as shown in Fig. 1. Its vesicles are elongated and flattened, and are arrayed in a more or less regular stack, with their long axes parallel. The shape of the vesicles changes gradually towards the opposite, apical pole; they become less flattened, and often give place, close to the apical pole, to large numbers of irregular but more or less isodiametric vesicles. The whole system may be of considerable size, and the apical pole is usually located fairly close to the cell surface. The polarity of this cellular element immediately suggests

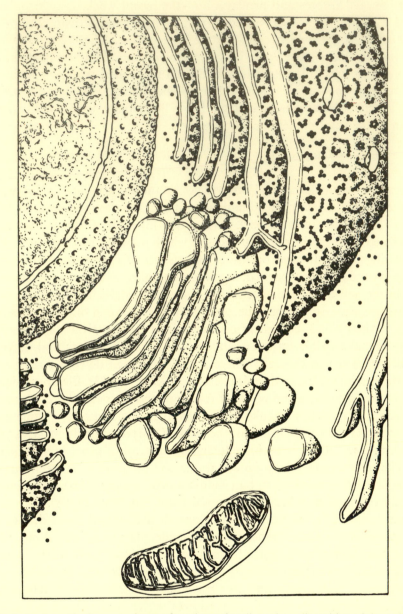

Fig. 1. A schematic diagram of part of a eukaryotic cell, to show disposition of the vesicles of the Golgi apparatus (centre) relative to the ribosome-coated membranes of the endoplasmic reticulum. At top left, part of the interphase nucleus; at bottom centre, a mitochondrion. Reproduced with permission from E. J. Du Praw (1968), *Cell and Molecular Biology*. London and New York: Academic Press.

that it is in a state of dynamic structural equilibrium, a view well expressed by Grimstone (1961): 'It is not a static organelle, therefore, but a steady-state system, through which there is a constant flow of membranes.' The flow proceeds, as we shall presently see, from the basal to the apical pole; the membrane stack is presumably replenished by the addition of vesicles derived from the endoplasmic reticulum.

In vertebrates, specialized cells that possess a secretory function have a very well-developed Golgi apparatus; studies of such cell types in mammals provided the first indications of the function (or, more correctly, of some of the functions) of this organelle. Extensive observations and autoradiographic experiments by Palade and his coworkers (Siekevitz & Palade, 1958; Palade 1961; Caro & Palade, 1964) on exocrine pancreatic cells, which secrete the digestive enzymes of pancreatic juice, have established the role of the Golgi apparatus in this particular secretory process. The secretory products are packaged within so-called 'zymogen granules', which arise from the Golgi apparatus. Radioautographic evidence suggests that the contents are synthesized on ribosomes adherent to the endoplasmic reticulum, and thereafter transferred to the Golgi apparatus. The discharge of zymogen granules from the cell involves their fusion with the cytoplasmic membrane at the apical end of the cell, a process for which Caro & Palade (1964) used the significant term 'reversed pinocytosis'.

By using analogous methods, Porter (1964) showed that the secretion of collagen by vertebrate fibroblasts involves an essentially comparable sequence of events. Collagen accumulates within the vesicles of the Golgi apparatus; these vesicles subsequently migrate towards the cell surface and discharge their contents into the intercellular space.

Sano & Knoop (1959) provided evidence that yet another secretory process, the secretion of the hormone norepinephrine by the medullary cells of the adrenal gland, occurs through the formation of secretory granules in the Golgi apparatus, and their discharge into blood capillaries from the axon terminal of secreting cells (which are modified neural cells).

These studies accordingly suggest that, in several different highly specialized vertebrate cell lines, the Golgi apparatus acts as a vehicle for the packaging and transport to the cell surface of the specific secretory products formed by each type of cell. However, the universality of the Golgi apparatus as a eukaryotic cell component implies that it must have more general functions. There is a certain amount of evidence from work with vertebrate cells which suggests that lysosomes are formed from the Golgi apparatus; but the mode (or modes) of formation of these structures is still not clear (see Novikoff, Essner & Quintana, 1964, for a discussion). If the Golgi apparatus is indeed the source of all or some types of lysosomes, it would be in this context an *endosecretory* organelle, since lysosomes are the vehicles of the enzymes responsible for the intracellular digestion of food materials brought into the cell by phagocytosis or pinocytosis (de Duve, 1964; Hirsch & Cohn, 1964). This is an essential cellular function in all phagotrophic eukaryotes.

Perhaps the most interesting observations concerning the function of the Golgi apparatus have been made by Manton (1966, 1967*a*, *b*) on the flagellate algae *Prymnesium* and *Chrysochromulina*. The structurally specialized cells of these organisms are covered, over the entire body surface, by small scales with an elaborate and characteristically asymmetric fine structure, which permits a ready discrimination between the inner scale surface (namely, that facing towards the cell membrane) and the outer one. A chemical analysis of the scales of *Chrysochromulina* (Green & Jennings, 1967) has shown that they consist largely, if not entirely, of carbohydrate, the principal constituent sugars being ribose and galactose. The scaly investment of these algae can therefore be interpreted as a highly specialized, discontinuous cell wall, comparable to the walls composed of many individual plates that are characteristic of the armoured dinoflagellates.

Manton has shown beyond any question that the scales are formed *de novo* in individual vesicles of the Golgi apparatus. Furthermore, it is possible to infer the temporal sequence of their synthesis, since the profiles of the scales become increasingly mature, the closer the vesicles containing them lie to the apical end of the Golgi apparatus. A second remarkable feature is the intravesicular orientation of the scales: what is destined to become the outer surface of the scale *in situ* always faces towards the apical end of the organelle. Precisely how the scales make their way from Golgi vesicles to their eventual positions on the cell surface is unclear; but Manton (1966) suggested that they may be liberated in the area of a deep pit near the base of the flagellum. This fundamental discovery has been followed up by Brown (1969), who has shown that the continuous multilaminate wall of a coccoid chrysophyte alga, *Pleurochrysis*, is likewise synthesized in Golgi vesicles. Elements of the future wall, of considerable area, develop in flattened vesicles near the apical pole, which lies close to the cell surface; the vesicles appear to discharge their contents by fusion with the cell membrane. In this alga, there is no evidence for a localized site of extrusion; and Brown suggests that the external deposition of wall material may occur in an orderly sequence, mediated by slow translocation of the Golgi apparatus through protoplasmic streaming.

These studies on wall formation in algae throw a new and somewhat different light on Golgi function. In animal cells where the products transported in Golgi vesicles appear to be in the main proteins, the organelle has been interpreted as a centre for the packaging and transport of material synthesized at another site in the cell. This conclusion seems irrefutable, since ribosomes never occur either on the surface of or within the Golgi vesicles. However, the *de novo* formation of polysaccharidic wall constituents within vesicles of the Golgi system in algae very strongly suggests that it is not (or perhaps not always) a metabolically passive structure.

From the work described above it will be evident that we are just beginning to understand the roles in cellular function, both general and special, which can be played by the Golgi apparatus. But I think it is already possible to draw some general conclusions that are relevant to the definition of the eukaryotic cell.

The Golgi apparatus is a device which permits eukaryotic cells to perform the operation characterized by Caro & Palade (1964) as 'reversed pinocytosis', but for which I prefer the more general designation *exocytosis*, as suggested by Du Praw (1968). Relatively large accumulations of biosynthetic products, either soluble (e.g. extracellular enzymes) or solid (e.g. scales) can be extruded from the cell, through coalescence of Golgi vesicles with the cytoplasmic membrane. Precisely what happens at the moment of extrusion is unclear. If the extrusion process is to be construed in the strictest sense as a reversal of pinocytosis or phagocytosis, it must be assumed that the unit membrane surrounding the Golgi vesicle fuses with the cytoplasmic membrane, the inner surface of the vesicle membrane then becoming a part of the outer surface of the cytoplasmic membrane. The much more familiar inverse processes of phagocytosis and pinocytosis (Holter, 1965), which can be collectively termed *endocytosis*, are the devices by which all eukaryotes not completely enclosed by walls take complex organic foodstuffs, either soluble or particulate, into the cell. These foodstuffs subsequently undergo intravacuolar digestion, mediated by fusion of lysosomes with the food vacuoles. All these endocytotic and exocytotic activities are made possible by the extraordinary plasticity of the eukaryotic cytoplasmic membrane. They allow the passage across the cell boundary, in both directions, of objects large enough, at the upper limit, to be visible with the light microscope.

The prokaryotic cytoplasmic membrane, on the other hand, poses a much more selective barrier to the passage of materials into and out of the cell. The largest objects that appear to traverse this barrier are pieces of DNA (e.g. transforming DNA fragments) or relatively small proteins (extracellular enzymes, if indeed they cross the membrane, rather than being synthesized at its surface). In this context, the differences between the known mechanisms of prokaryotic and eukaryotic wall formation are highly instructive. Such algae as *Prymnesium, Chrysochromulina* and *Pleurochrysis* synthesize relatively large pieces of wall material in Golgi vesicles and then extrude them for final positioning and integration on the cell surface. In bacteria, the extension of two major wall components, the peptidoglycans (Strominger & Tipper, 1965) and the polysaccharide moiety of lipopolysaccharides (Robbins, Bray, Dankert & Wright, 1967), involves *molecular* extension. The formation of the molecular sub-units of each polymer class is completed on a specific polyisoprenoid carrier built into the cytoplasmic membrane, and these sub-units are then directly transferred to molecular growing points in the wall.

The capacity to harbour cellular endosymbionts

Cellular endosymbionts, either eukaryotic or prokaryotic, occur in all major groups of eukaryotes. Among vascular plants, a classical example is the *Rhizobium* endosymbiosis in the root nodules of legumes (Nutman, 1963). Cellular endosymbionts occur in some members of almost every metazoan phylum; the vast literature on this subject has been surveyed by Buchner (1965). Because it is particularly relevant to the evolutionary problem analysed in the last part of this essay, I shall discuss here in detail the nature of certain endosymbioses in which the hosts are uni-cellular eukaryotes.

Endosymbiotic algae, which occur frequently in Coelenterata and Mollusca (see Buchner, 1965), are also harboured, either constantly or intermittently, by certain protozoa. A well studied case is *Paramecium bursaria*, which harbours an endosymbiotic unicellular green alga (Pringsheim, 1928). Provided that the photosynthetic endosymbiont is a eukaryotic alga (and most of these endosymbionts appear to be dino-flagellates or green algae) the recognition of its nature is not difficult, even if its organismal independence cannot be shown by its isolation and cultivation, since the endosymbiont (like its host) bears the characteristic structural stigmata of a eukaryotic cell. The problem of recognition becomes much more delicate when the presumed photosynthetic endo-symbiont is a prokaryote (namely, a blue-green alga). A thecate amoeba containing blue-green inclusions was first described by Lauterborn (1895); other examples of amoeboid or flagellate eukaryotes which contain similar objects were discovered by Korschikoff (1924) and Pascher (1929). Pascher (1929) systematically interpreted such intracellular inclusions as endosymbiotic blue-green algae ('cyanelles'). The subject has been reviewed more recently by Pringsheim (1958) and Geitler (1959), both of whom likewise interpret the cyanelles as endosymbionts. However, in the light of the discoveries of the past decade about the properties of chloroplasts, the criteria used by Pascher, Geitler and Pringsheim in deciding that cyanelles are organisms rather than organelles begin to appear somewhat inadequate. Pringsheim (1958) listed these criteria as follows:

(1) The cyanelles sometimes have forms not found in the chloroplasts of any lower organisms, despite the great diversity in form of these cellular elements. (2) The cyanelles commonly show an internal light region, similar to the colourless centroplasm of blue-green algae. (3) Cyanelles are found in organisms the relatives of which never possess chloroplasts with a blue-green colour.*

Geitler (1959) mentioned an additional criterion for the recognition of the organismal nature of the cyanelles: 'furthermore, when cyanelles are squeezed out of the protoplasts, they do not become disorganized, as chloroplasts would' (ausserdem lassen sich die Cyanellen aus den Protoplasten ausdrücken und desorganisieren dabei nicht, wie dies Chromatophoren im Sinne Plastiden täten).

With all due respect to these two great protistologists (and although I think it probable that cyanelles really are unicellular blue-green algae), I do not find their proposed criteria formally sufficient, either individually or collectively, to establish the organismal nature of the structures in question. We must then ask: what *would* be a sufficient proof that cyanelles are blue-green algae? A decisive proof would be their isolation and cultivation outside the host; but so far, this has not been achieved in any case. Almost as conclusive would be the demonstration that cyanelles are enclosed by walls that have a peptidoglycan layer; this is characteristic of all free-living blue-green algae so far examined, and peptidoglycans are a class of biopolymers completely absent from eukaryotic organisms. However, this possible proof appears to be excluded by ultrastructural studies, which show that neither the cyanelles of *Glaucocystis nostochinearum* (Lefort, 1965; Hall & Claus, 1967) nor those of *Cyanophora paradoxa* (Hall & Claus, 1963) are enclosed by structural elements interpretable as cell walls. If the cyanelles are endosymbionts, they appear to have lost the ability to make walls, which

---

* Die Cyanellen haben zuweilen Gestalten, wie sie, trotz der grossen Mannigfaltigkeit der Form, bei Chromatophoren keines niederen Organismus vorkommen; 2. Sie weisen gewöhnlich im Inneren eine helle Region auf, die dem farblosen Centroplasma der Cyanophyceen gleicht; 3. Cyanellen werden bei Lebewesen gefunden, deren Verwandte nie blaugrüne Chromatophoren besitzen (Pringsheim (1958), p. 167).

suggests the demonstration of their organismal independence by cultivation outside the host might offer considerable technical difficulties. In the last analysis, accordingly, the claim that cyanelles are organisms, not organelles, really rests on two highly subjective value-judgements: they have an internal structure which looks more like that of a blue-green alga than that of objects definitely known to be chloroplasts and they occur in hosts with cellular properties which would not normally lead a protistologist to predict the presence of chloroplasts. I have dwelt on this particular problem at some length, because it shows how difficult the formal differentiation between a chloroplast and a prokaryotic endosymbiont may become.

Many instances of bacterial endosymbioses in protozoa have been recorded. Information about such endosymbionts is mostly marginal, since as a rule they have been detected incidentally, in the course of studies on the structure of the host organism. A good deal is known, however, about the bacteria harboured by so-called 'killer stocks' of *Paramecium aurelia* (Sonneborn, 1959). These objects (kappa and related particles) were initially interpreted as plasmids, and their inheritance was extensively analysed in terms of this assumption. The bacterial nature of lambda particles was established by van Wagtendonk, Clark & Godoy (1963), who succeeded in growing them outside the host; and a recent careful ultrastructural study (Jurand & Preer, 1969) has shown that lambda particles are equipped with typical peritrichously inserted bacterial flagella. The work on these agents again reveals the difficulty of making an operational distinction between an organelle and a prokaryotic endosymbiont.

To the best of my knowledge, *a stable endosymbiosis in which the host is a prokaryote has never been described. The only known stable associations involving two prokaryotic partners are ectosymbioses.* These may be very close, as exemplified by the *chlorochromatium* association, which consists of a relatively large, non-photosynthetic, polarly-flagellate rod, coated with a monolayer of smaller, rod-shaped green bacteria arranged in longitudinal rows, like the staves of a barrel (Lauterborn, 1915).

One special case, which at first appeared interpretable as an unstable endosymbiosis, is the relationship between small predatory bacteria of the *Bdellovibrio* group and their larger eubacterial prey, such as *Pseudomonas* and *Escherichia* (Stolp & Petzold, 1962; Stolp & Starr, 1963; Shilo & Bruff, 1965). The attack of the predator is initiated by a leech-like attachment to the host cell surface, followed by penetration of the predator through the wall of the host. The host cell subsequently lyses, generally after rounding up, and the predator and its progeny eventually emerge from the cellular carcass. However, several analyses of predator-prey interaction by electron microscopy (e.g. Burnham, Hashimoto & Conti, 1968) have in fact shown that *Bdellovibrio* does not enter the cytoplasm of its victim. Growth and multiplication of the predator occur in the space between the wall and the partly contracted protoplast of the host. Even this relationship is not, therefore, an endosymbiotic one in the strict sense.

A necessary condition (though not of course a sufficient one) for the primary establishment of an endosymbiosis is the passage of the future endosymbiont through the surface barriers of the future host cell. It might be argued that the cell wall, a well-nigh universal structure in prokaryotic cells, presents an insurmountable barrier to the establishment of endosymbioses in these organisms. However, the case of *Bdellovibrio* shows that, even when this barrier is penetrated, true endosymbiosis does not ensue. Furthermore, a rigid enclosing wall does not preclude the establishment of endosymbioses in plant cells; the *Rhizobium* endosymbiosis, mentioned earlier, is only the most familiar and carefully studied example. Hence it seems more probable that the real barrier to the establishment of endosymbioses in prokaryotes lies at the level of the cytoplasmic membrane. The ability of eukaryotes to internalize objects of cellular dimensions by endocytosis provides an easy and natural mode of passage for foreign cells across the cytoplasmic membrane; and provided that such cells are not immediately subjected to intracellular digestion the basic precondition for the establishment of an endosymbiotic relationship has been met. In prokaryotes, the cytoplasmic membrane may undergo deep invaginations (as in mesosomes, or the membranous intrusions which bear the photosynthetic pigments of purple bacteria); but it apparently lacks the plasticity to undergo a complete involution, necessary to internalize a relatively large external object. The impenetrability of the prokaryotic cytoplasmic membrane by any object of supramolecular dimensions effectively precludes the acquisition of endosymbionts. This may well be considered a fundamental biological difference between eukaryotes and prokaryotes.

Directed intracellular translocations in the eukaryotic cell

When one considers the recent information on the functioning of the Golgi apparatus, it becomes clear that there must be a mechanism which determines the orientation and movement of the Golgi vesicles, as they pass from the basal to the apical pole, and thereafter to the cell surface. This is just one aspect of a very old problem in cellular dynamics: how do eukaryotic cells effect directed intracellular translocations? The light-

induced movements of chloroplasts, the localized concentration of
mitochondria, and the movement of the chromosomes during mitosis
are a few of the other phenomena which reveal the precision with which
the relative positions of structural components within the eukaryotic
cell can be controlled. On a larger scale, directed cytoplasmic streaming
in walled eukaryotic cells and the extension of pseudopodia in cells
without walls constitute manifestations of the same mysterious property.
Such phenomena have no counterparts at the prokaryotic level. Although
a really satisfactory explanation of the mechanism (or mechanisms) is
still not in sight, recent cytological research may perhaps have revealed
part of the machinery.

Microtubular systems

A microtubule can be defined as a rigid hollow filament of considerable
length, 150 to 250 Å wide and with a wall thickness of 45 to 70 Å.
Microtubules are now recognized to be ubiquitous elements in the cytoplasm of eukaryotic cells. Some microtubular systems were revealed
early in the era of biological electron microscopy. Only much later,
however, did their ultrastructure, their very wide distribution and their
varied roles in cellular function begin to be appreciated. The full extent
of their contributions to the structure and operation of eukaryotic cells
can almost certainly not yet be assessed.

Two members of the class of microtubules have been recognized for
some time, because of their conspicuous association with major structural components of the eukaryotic cell. One consists of the microtubules in the mitotic apparatus, which provide its material framework
and are at the same time an important part of its machinery (Mazia,
1961). The other consists of the longitudinally arranged filaments in the
core of the eukaryotic cilium (under which term I include eukaryotic
flagella and sperm tails, as well as cilia *sensu stricto*; see Glossary). In
this location, the microtubules have a fixed number and arrangement:
the familiar circle of nine outer pairs surrounding an inner pair of
different origin and slightly different structure (see Fawcett, 1961, for a
summary). Since the tubular fine structure of these two classes of filaments was not at first recognized, they have been (and often still are)
referred to as 'fibres' or 'fibrils'.

Ciliary microtubules are extremely sturdy structures, able to survive
much mishandling during preparative manipulations; their arrangement
and nature were thus easily determined. The microtubules of the mitotic
apparatus are more delicate. Although the filamentous organization of
this organelle was firmly established through Inoué's (1953) observations
on living cells with the polarizing microscope, and through the almost

simultaneous isolation in a cell-free state of the mitotic apparatus by Mazia & Dan (1952), the visualization of the microtubules in thin sections of the spindle region of fixed cells proved more difficult. In 1962, Harris showed that the presence of divalent cations during fixation is essential for the preservation of spindle microtubules. The systematic demonstration of other classes of microtubules which are still more labile followed the general adoption of glutaraldehyde fixation, after the introduction of this technique by Sabatini, Bensch & Barnett (1963). It then very quickly became apparent that all types of eukaryotic cells contain microtubules in other cytoplasmic locations; these are often locations where the possible presence of a microtubular system is not betrayed by grosser structural features, as it is in cilia and in the dividing nucleus (Porter, 1966). In recent years, microtubular systems have also been discovered in certain non-ciliar filiform appendages of eukaryotes: the suctorian tentacle (Rudzinska, 1965) and the heliozoan axopodium, a highly specialized pseudopodium characteristic of this class of rhizopods (Kitching, 1964). Both tentacles and axopodia (unlike typical pseudopodia) have a regular microtubular endoskeleton. They cannot, however, be interpreted as modified cilia, since both the number and the arrangement of the microtubules, characteristic for each kind of organelle, are quite different from those of all cilia.

Intimately associated with certain microtubular systems in eukaryotes is an enigmatic organelle, the centriole (Stubblefield & Brinkley, 1968). It is a cylinder some 1200 to 1500 Å wide, and of variable length. The cylinder wall is composed of nine microtubular triplets, arranged parallel to the long axis. Centrioles serve invariably as the infra-cytoplasmic anchoring structures of cilia and can be plausibly construed as organizing centres for the outgrowth of ciliar microtubular systems, since the nine outer pairs of ciliar microtubules originate in, and are continuations of, two microtubules in each of the nine triplets in the wall of the centriole.

The mode of formation of centrioles, long a mystery, is beginning to become clearer (Randall & Disbrey, 1965; Dirksen & Crocker, 1966; Smith-Sonneborn & Plaut, 1967; Dippell, 1968). Some (possibly all) centrioles contain DNA. However, their replication does not involve division of the organelle (as in the case of chloroplasts and mitochondria), but rather neoformation of additional organelles around a proliferative element (DNA?) derived from a pre-existing one. Centriolar synthesis thus appears analogous to the spontaneous assembly of a virion, rather than to the duplication of a cell.

In some eukaryotic groups which possess centrioles as organizing centres for the synthesis of cilia, centrioles have a second function: they serve as organizing centres for the synthesis of the mitotic apparatus. In metazoans, a pair of centrioles defines the polarity of the mitotic spindle, and establishes its areas of termination (Mazia, 1961). In eukaryotic groups which have undergone an evolutionary loss of the ability to form flagella, and hence do not contain centrioles (e.g. some amoebae, flowering plants, higher fungi), the mitotic spindle, even though it may have a 'classical' form (e.g. in flowering plants) appears to be centred in two immaterial poles!

Among eukaryotic protists which retain ciliary movement, centrioles sometimes intervene in karyokinesis, but more commonly do not. The ciliates provide a striking example of a protistan group in which the cell typically contains hundreds of centrioles (located in the cortical region underlying the cilia) although karyokinesis, both macronuclear and micronuclear, is non-centriolar. The use of centrioles as part of the machinery of mitosis is evidently a biological luxury, not a necessity.

Centrioles do not appear to play a role in the formation of non-ciliar eukaryotic filiform appendages containing an internal system of microtubules, such as the suctorian tentacle and the heliozoan axopodium. However, in some cells microtubular systems other than those of the mitotic apparatus and the cilium are associated with centrioles. This has long been evident for the complex and regular cortical microtubular system of ciliates, the so-called infraciliature. Gibbins, Tilney & Porter (1969) have recently observed that many elements of the much more diffuse cytoplasmic microtubular system in mesenchymal cells of sea urchin embryos converge on centrioles. The only general conclusion which can be derived from the known distribution and location of centrioles is that they are associated with the formation of some types of microtubular systems in some eukaryotes, but are by no means necessary for the formation of such systems, with the single exception of ciliar microtubules.

Some kinds of microtubules—for example, those of the mitotic apparatus (Mazia, Chaffee & Iverson, 1961) and of cilia (Stephens, 1968)—have been isolated and subjected to chemical study. They are composed largely if not entirely of protein, and can be disaggregated by appropriate chemical treatment to yield homogeneous preparations of sub-units having a relatively low molecular weight ($\sim 60,000$). Since high resolution electron microscopy reveals globular sub-units with an approximate diameter of 40 Å in the intact microtubules (e.g. Stephens, 1968), microtubules provide one of the increasingly numerous examples of complex biological structures which owe their special form to 'self-assembly': viz. an ordered aggregation of like protein sub-units.

Mazia & Ruby (1968) drew attention to the resemblances (particularly striking when amino acid compositions are compared) between microtubular proteins and other eukaryotic structural proteins: those of the erythrocyte membrane and of the mitochondrion, and the two major proteins of muscle, actin and myosin. They suggest that all these proteins, which participate in the assembly of microtubular, filamentous or membranous elements of cell structure, belong to a single class, for which they propose the name *tektins*. The monomeric constituents of each tektin undergo sensitive and specific associations with other molecules of their own kind, each self-assembly process yielding a specific structural element of the eukaryotic cell. Proteins which possess this associative property are not confined to eukaryotes; the bacterial flagellum is constructed on the same principle (Kerridge, Horne & Glauert, 1962; Abram & Koffler, 1964). However, as Mazia & Ruby (1968) point out, the flagellins (that is, the monomeric elements of bacterial flagella) do not share the pattern of amino acid composition characteristic of the tektins, and are thus members of a different class of proteins.

Do microtubules constitute a single homologous class of cellular microstructures? This is by no means evident: the diversity of their special locations and functions, the differences in their degree of permanence, and the apparent variations of their dimensions, all perhaps argue against such a conclusion. However, there is one very interesting piece of evidence which does suggest a substantial degree of homology: the response of microtubular systems to treatment of cells with colchicine. This extremely toxic alkaloid first entered the frame of reference of cell biologists as an antimitotic agent. Studies on its mode of action in this context suggest that colchicine inhibits mitosis by binding to the sub-units of the spindle microtubules (Taylor, 1965). Dramatic colchicine effects have recently been shown with other cellular microtubular systems. Although it does not affect the mature ciliary apparatus, this alkaloid completely inhibits ciliar regeneration by de-ciliated (but viable) cells of *Tetrahymena* (Rosenbaum & Carlson, 1969). In developing sea urchin gastrulae, colchicine treatment causes a disappearance of the microtubular system that is widely dispersed through the cytoplasm of the embryonic cells: the cells at the same time lose their distinctive shapes, and development of the primary mesenchyme ceases (Tilney & Gibbins, 1969). Colchicine has a remarkable effect on the heliozoan *Actinosphaerium* (Tilney, 1968). In common with other members of this protozoan group, *Actinosphaerium* has a spherical cell from which there project in all directions long, thin and relatively stable axopodia, the form of which is determined, as already mentioned, by their cores of microtubules (the so-called *axoneme*). Between the axoneme and the enclosing cytoplasmic membrane is a cortical layer of cytoplasm, in a state of active streaming movement. Exposure to colchicine provokes a very rapid disorganization of axonemal structure; as a result, the axopodia retract into the main body of the cell within thirty minutes. This effect is reversible: when colchicine is removed, the axopodia reform through the outgrowth of new axonemes.

Much of our recent knowledge about microtubules has been contributed by the observations and experiments of K. R. Porter and his group. I cannot attempt a detailed presentation of this extensive, complex and extremely important body of work; its earlier phases have been summarized by Porter (1966). The most firmly based conclusion from it all is that microtubules, in addition to their specific functions in mitosis and flagellar movement, provide a transient cytoplasmic endoskeleton which is the primary determinant of cell shape (at least in eukaryotic cells without walls). This is suggested with particular force by the correlations between intracytoplasmic microtubule distribution and the development of complex cell form in the differentiating mesenchyme of the sea urchin embryo (Gibbins *et al.* 1969). Porter's group has emphasized a semi-static architectural role for microtubules, but has been much more cautious about proposing shorter term dynamic functions for these structures. Of course, a conclusive experimental demonstration of the part played by microtubules in more rapid changes of cell form is far more difficult, for technical reasons. Nevertheless, the literature of the past few years is replete with suggestive indications of the involvement of microtubules in directed intracellular translocations, cytoplasmic streaming and pseudopodial movement (of certain types, at least). Since their involvement in one special kind of directed intracellular translocation (namely, the movement of chromosomes during mitosis) has long been evident (see Mazia, 1961, for a discussion), such an extrapolation is not unreasonable, now that the pervasiveness of these structures in eukaryotic cells has been demonstrated.

Colchicine treatment causes a rapid retraction of the heliozoan axopodium. In contrast, it does not affect preformed cilia, but prevents their synthesis *de novo*. I find this contrast profoundly suggestive. A cilium is a static structure: its microtubular system, once formed, is there for the lifetime of the organelle, and has a fixed role to play in the mediation of ciliar movement. A pseudopodium, almost by definition, is a highly transient structure. In the special modification represented by the axopodium, the pseudopodium is given a seemingly more permanent form by its microtubular core, which could be said to convert it into a structural analogue of a cilium. However, the extraordinarily rapid disaggregation of the axostyle provoked by colchicine treatment suggests that the microtubular core of the axopodium is not a static structural element, like the ciliar microtubules, but a dynamic one; its seeming permanence reflects a steady-state condition, in which sub-units flow continuously through the visible structure. This flow might well be causally related to the streaming that is so conspicuous in the thin cortical layer of cytoplasm surrounding the axostyle. Just because of its extreme specialization, the axopodium may provide the best model system for the study of cytoplasmic streaming and pseudopodial loco-motion. Of course, it is still an unverified assumption that all the phenomena of intracellular movement are mechanistically homologous; even the various kinds of pseudopodial locomotion may not share a common mechanism. Nevertheless, I hope this brief exposé justifies to some degree my earlier statement that if we do not yet understand the mechanisms of directed cytoplasmic movements, some parts of the machinery have now been glimpsed.

SOME NEWLY RECOGNIZED STRUCTURAL FEATURES OF PROKARYOTIC CELLS

The generalization that the machinery of respiration and photosynthesis in prokaryotic organisms is carried in the cytoplasmic membrane, or its intrusions into the cytoplasm, was proposed some years ago (Stanier, 1963, 1966). It still appears to be valid for aerobic bacteria and one group of photosynthetic bacteria, the purple bacteria; but was certainly too sweeping. The blue-green algae represent one possible exception. Direct connections between the cytoplasmic membrane and the flattened lamellae or thylakoids which bear the lipid-soluble photopigments in this group are very rarely detectable in electron micrographs of thin sections (Germaine Cohen-Bazire, personal communication). Hence in these prokaryotes the unit membrane system which bears the photo-synthetic apparatus may be physically separate from the cytoplasmic membrane, even though this does not necessarily preclude a primary derivation of the internal membrane system from the cytoplasmic membrane. Further work on this particular problem is required before firm conclusions can be reached. A clear exception to the generalization, discussed below, has been found in a second group of photosynthetic bacteria.

Chlorobium vesicles

In green bacteria (genera *Chlorobium* and *Pelodictyon*) the photo-synthetic apparatus has a unique structural basis, different from that in any other photosynthetic group, prokaryotic or eukaryotic (Cohen-Bazire, Pfennig & Kunisawa, 1964; Pfennig & Cohen-Bazire, 1967). These are the only known organisms in which the chlorophylls and carotenoids are not incorporated into unit membranes. Instead, the photopigments are housed in cigar-shaped vesicles, about 300 to 500 Å wide and 1000 to 1500 Å long, each surrounded by a non-unit membrane some 30 Å thick. These so-called *chlorobium vesicles* are located in the cortical region of the cytoplasm, immediately underlying, but physically distinct from, the cytoplasmic membrane (Pl. 1). Although there are indications of a fine structure within the vesicles, the seeming internal regularities are on an extremely small scale, and have not so far been clearly resolved. The green bacteria also contain a well-developed unit membrane system; rather complex mesosomal intrusions apparently derived from the cytoplasmic membrane are common, particularly at sites of transverse wall formation.

Cruden (1968) has succeeded in physically separating the chlorobium vesicles from the fragments of the unit membrane system in extracts of *Chlorobium* cells (Pl. 2). Her analyses confirm that both molecular species of chlorophyll found in green bacteria (bacteriochlorophyll *a* and either bacteriochlorophyll *b* or *c*, depending on the strain) are located in the chlorobium vesicles, not in the membrane fraction. Succinoxidase activity, on the other hand, is restricted to the membrane fraction. Like the internal lamellae of blue-green algae and of eukaryotic chloroplasts, chlorobium vesicles (but not the chlorobium membrane fraction) contain galactolipids. This class of lipids does not occur at all in the other group of photosynthetic bacteria, the purple bacteria (Gorcheim, 1968).

The green photosynthetic bacteria appear to have the most highly differentiated type of cell so far encountered among prokaryotes. However, even though the photosynthetic apparatus of green bacteria is segregated from other cell components in membrane-bounded organelles, these organelles cannot be homologized with chloroplasts: they are not enclosed by, and do not contain, any unit membranes. With respect to other structural features, the green bacteria are typically prokaryotic. These organisms do not accordingly offer any promise of providing a missing link between the two levels of cellular organization; they can be more plausibly construed as lying at the end of a terminal branch of prokaryotic cellular evolution.

Gas vacuoles and gas vesicles

A cellular inclusion of irregular form and peculiar optical properties known as a *gas vacuole* occurs widely but sporadically among prokaryotes. Gas vacuoles are relatively common in blue-green algae (Fogg, 1941), and also occur in a few bacteria: some purple and green bacteria (Pfennig, 1967); extreme halophiles of the genus *Halobacterium* (Petter, 1931; Houwink, 1956); and some recently discovered non-photosynthetic prosthecate freshwater bacteria (Staley, 1968). Most of the fundamental observations on gas vacuoles were made with blue-green algae (Klebahn, 1929), and their existence as bacterial cell inclusions has been largely overlooked, even by authors of textbooks who should have known better (e.g. Stanier, Doudoroff & Adelberg, 1963). These bodies are filled with gas (Klebahn, 1929) and serve as organelles of flotation, which permit the aquatic organisms that possess them to regulate their position in a vertical water gradient. In blue-green algae, gas vacuoles usually make the cells totally buoyant. The ecological observations of Pfennig (1967) show this is not necessarily true in gas-vacuole-containing purple and green bacteria, which can adopt a closely fixed position within the water gradient, behaving like internally regulated Cartesian divers.

Ultrastructural studies on the gas vacuoles of blue-green algae (e.g. Bowen & Jensen, 1965; Smith & Peat, 1967) show that they are compound objects, composed of a variable number of *gas vesicles* (Pl. 3). Each gas vesicle is a cylinder about 80 Å wide and of variable length, with conical ends. The wall of the vesicle is composed of a non-unit membrane some 20 to 30 Å thick, with a very regular banded structure, consisting of transverse rows of beads (Jost, 1965). The gas vesicles of *Halobacterium* (Stockenius & Kunau, 1968) and of purple and green bacteria (Pfennig & Cohen-Bazire, 1967; and in preparation) show identical ultrastructural features, leaving no doubt that the gas vacuoles in all prokaryotic groups are homologous cellular elements.

Gas vesicles and chlorobium vesicles represent, accordingly, two classes of prokaryotic organelles which are enclosed by non-unit membranes quite distinct from the cytoplasmic membrane, and which have no counterparts among eukaryotes.

HOW IT MIGHT HAVE HAPPENED

Oh! let us never, never doubt
What nobody is sure about! *H. Belloc* (1898)

Advancing knowledge in the domain of cell biology has done nothing to diminish the apparent magnitude of the differences between eukaryotic and prokaryotic cells that could be descried some ten years ago; if anything, the differences now seem greater. The only major links which have emerged from recent work are the many significant parallelisms between the *entire* prokaryotic cell and two *component parts* of the eukaryotic cell, its mitochondria and chloroplasts. The organizers of this Symposium imprudently requested a speculative introduction, and I shall now meet their request with some speculations about cellular evolution.

Some evolutionary premises

Leaving aside for the moment the special question of the origin of chloroplasts and mitochondria, do we really have good grounds for believing that eukaryotic and prokaryotic cells are branches from a common stem of *cellular* evolution? In other words, might they not have arisen independently from pre-cellular forms of living matter? Mazia (1965) raised this question in a specific context: namely, that of chromosome structure. I think completely separate origins are improbable, even though most of the evidences of homology are to be found only at the deepest level (i.e. the molecular one). They include the possession of the same genetic code; common mechanisms for the replication, transcription and translation of the genetic message; and largely common mechanisms for the biosynthesis of major classes of cell-constituents. The nexus of shared properties is sufficiently complex to suggest an origin from a common ancestor which could already be described as 'cellular'.

Is the comparative structural simplicity of prokaryotic organisms really indicative of great evolutionary antiquity? In view of their similarities to mitochondria and chloroplasts, it could be argued that they are relatively late products of cellular evolution, which arose through the occasional escape from eukaryotes of organelles which had acquired sufficient autonomy to face life on their own. This is a far-fetched assumption; but I do not think one can afford to dismiss it out of hand. The principal counter-arguments are derived from considerations of comparative biochemistry and physiology.

It seems likely that biological photosynthesis as we know it today had a single evolutionary origin, since much of its complex mechanism and machinery is shared by all existing phototrophs. The ultimate version of this kind of energy-generating metabolism is photosynthesis with oxygen production (the use of water as a photosynthetic electron donor), which requires the presence in the photosynthetic apparatus of two different kinds of photochemical reaction centres. This is the only mode of photosynthesis among eukaryotes. Among prokaryotes it occurs in blue-green algae, in parallel with a mechanistically simpler type of photosynthesis, which requires the presence of only one kind of reaction centre in the photosynthetic apparatus, and is characteristic of purple and green bacteria. This strongly implies that the mechanism and machinery of photosynthesis evolved in organisms with prokaryotic cells, only the final version becoming implanted in eukaryotic cell lines.

Geochemical evidence indicates that the early earth had an anaerobic atmosphere, which became aerobic, probably in large measure as a result of the metabolic activity of oxygen-producing photosynthetic organisms, between 1·2 and 2·1 thousand million years ago (Cloud, 1965). Early forms of life were accordingly dependent on anaerobic modes of energy-yielding metabolism. The range and variety of such modes of metabolism among bacteria are remarkable, whereas eukaryotes are confined to the use of one, glycolysis. Among bacteria, possession of one of these diverse modes of anaerobic energy generation is frequently associated with the physiological property of obligate anaerobiosis. This is characteristic of all the members of many specialized and taxonomically isolated bacterial groups: for example, the green bacteria, the desulfovibrios and the methane-producing bacteria. It is, on the other hand, a rare physiological property among eukaryotic micro-organisms. Only one strictly anaerobic fungus, *Aqualinderella*, has been discovered to date (Emerson & Held, 1969); and there are scattered examples of obligate anaerobes among flagellate and ciliate protozoa. All anaerobic eukaryotes can be construed as organisms which have undergone comparatively recent adaptations to highly specialized anaerobic ecological niches; for example, the rumen (Hungate, 1967). It is accordingly the prokaryotic lines, and not the eukaryotic ones, which appear to retain today the most numerous biochemical and physiological vestiges of the anaerobic phase of terrestrial evolution. For these reasons I shall assume that the prokaryotic cell has existed for as long as the eukaryotic one, if not longer.

Many biochemical attributes that are nearly universal or widely distributed among prokaryotes are not found at all among eukaryotes. They include: the peptidoglycan cell wall; poly-β-hydroxybutyrate as a reserve cell material; fixation of nitrogen; use of reduced inorganic compounds as energy sources; use of non-glycolytic mechanisms for anaerobic energy generation.

Until very recently, one nearly universal biochemical attribute of eukaryotes has appeared to be absent from prokaryotes: the ability to synthesize sterols. However, two recent reports (de Souza & Nes, 1968; Reitz & Hamilton, 1968) have clearly established the presence of sterols in blue-green algae. The cellular concentrations of these substances are between one and two orders of magnitude lower than those commonly found in eukaryotes, which suggests that the negative outcome of previous searches for sterols in prokaryotes may have been caused by insufficiently sensitive methods of detection. If so, the difference between eukaryotes and prokaryotes in this particular respect may be quantitative, not qualitative.

Some evolutionary hypotheses

In terms of the premises outlined above, only two general classes of hypotheses about the evolutionary links between eukaryotic and prokaryotic cells appear to be possible.

(1) Direct filiation: all the structural elements of the eukaryotic cell are derived from a prokaryotic ancestry (Nass, 1969).

(2) Parallel evolution from a common early cellular ancestor and subsequent infection of eukaryotic cell lines by prokaryotes, necessary to account for the origin of chloroplasts and mitochondria (Ris, 1961; Sagan, 1967; Margulis, 1968). This class of hypothesis has also had earlier proponents; the historical background was reviewed by Sagan (1967).

Nass (1969) has attempted to develop systematically a hypothesis of the first class. Although he accepts specific prokaryotic cellular origins for chloroplasts and mitochondria, he tries to maintain the hypothesis of complete evolutionary filiation by postulating that the eukaryotic cell arose from a colonial rather than from a unicellular prokaryote. This hypothetical colonial ancestor underwent extreme specialization of cellular structure and function, so that the various cell components characteristic of the eukaryotic cell evolved separately in different individuals of the colony. The eukaryotic cell then arose through cellular fusion, causing reversion from a colonial to a quasi-unicellular state. I do not find this hypothesis attractive: it has a distinct flavour of science fiction. Nass was led to formulate it, I suspect, because he could find no escape from a paradox seemingly inherent in all hypotheses of the second class.

The paradox can be stated as follows. Without chloroplasts or mitochondria, a cell of the eukaryotic type is dependent on glycolysis for its energy supply. All hypotheses of the second class therefore assume that this mechanistically primitive and inefficient mechanism of energy generation was the only one available to the evolving eukaryotic cell line until the capacities for respiration and photosynthesis were implanted in it by the acquisition of prokaryotic endosymbionts. This implies that most of the characteristic complexities of eukaryotic cellular design arose in a metabolically primitive cell line, while the main thrust of progressive biochemical evolution was being expressed in parallel prokaryotic cell lines, accompanied by a far smaller degree of structural evolution. Hypotheses of the second class can be made plausible only if one can propose a new kind of selective force, of sufficient strength to provoke a rapid tempo of progressive structural evolution in the eukaryotic line. Neither Ris (1961) nor Sagan (1967) really came to grips with this difficulty. More recently, Sagan (under another name: Margulis, 1968) has attempted to circumvent it by proposing a very early implantation of mitochondria in the evolving eukaryotic line, considerably before the acquisition of chloroplasts. This sub-hypothesis, which does not seem plausible to me, will be discussed later.

I should now like to suggest a possible solution of the paradox: namely, that the progressive structural evolution of the eukaryotic cell received its initial impetus from the acquisition of a novel cellular property, *the capacity to perform endocytosis*. As discussed in an earlier section, this capacity does not seem to exist in any contemporary prokaryotic lines, even those with large cells; it is an important differential character of cells of the eukaryotic type. The triggering event in conferring endocytotic ability in the eukaryotic cell line was no doubt a change in the properties of the surface membrane of the cell. The capacity for endocytosis would have conferred on its early possessors a new biological means for obtaining nutrients: predation on other cells. Since the cell materials of the prey provided substrates for glycolysis, the members of this predator cell line were no longer subject to selection for the evolution of new modes of energy-yielding metabolism. Such selection appears to have been a major factor in the evolution of prokaryotes, as shown by their unrivalled contemporary diversity with respect to mechanisms of energy generation.

If this hypothesis be correct, selection in the emerging eukaryotic cell line would have centred on the improvement of the efficiency of predation. It would have therefore tended to receive expression primarily in a structural complexification of the cell, leading to an increase of cell size, and to innovations that affected active movement, food capture and intracellular digestion. I shall not attempt a detailed exposition of this hypothesis, beyond pointing out that it can perhaps provide an evolutionary rationale for the evolution of the Golgi apparatus and—even more important—of microtubular systems, such a pervasive element of cytoplasmic structure in eukaryotes today. This hypothesis implies that microtubules arose in the context of the acquisition of novel mechanisms for active cellular movement (cilia, pseudopodia), so necessary for predation, and of means for directed intracellular translocation. At the same time, microtubular systems, once developed, provided a precondition for the eventual evolution of the machinery of mitosis, when increase in the total size of the genome and its consequent dispersion over numerous genophores made imperative a special mechanism for the equipartition of chromosomes. The dinoflagellates, which in all other respects possess typical eukaryotic cellular characters, have preserved a nucleus which perhaps exemplifies an early stage in the evolution of the mitotic apparatus. The chromosomes of dinoflagellates do not contain histones (Dodge, 1964), and never undergo supercoiling during the divisional cycle, remaining visible in the interphase nucleus (Giesbrecht, 1962). The first really complete analysis of nuclear division in a flagellate of this group, *Gyrodinium* (Kubai & Ris, 1969), has shown that the nuclear membrane (which never dissolves) invaginates in a complex manner, to become traversed by a number of parallel cylindrical channels, within the cytoplasm of which bundles of microtubules develop. There is never a direct association between the chromosomes (segregated within the nuclear membrane) and these microtubules. Kubai & Ris (1969) suggest that the chromosomes have attachment points to the nuclear membrane, and that the separation of daughter sets occurs through membrane growth or translocation, perhaps mediated or directed by the microtubular system. Apart from the involvement of microtubules, the mechanism of genophore separation proposed is analogous to that which has been suggested to account for the separation of genophores in bacteria (see Ryter, 1968).

The terminal step in the development of the eukaryotic cell—acquisition of the capacity to perform oxygen-producing photosynthesis and respiration in a world that was now becoming aerobic—had to await the evolution of these modes of metabolism in prokaryotic cell lines. It seems to have occurred through the capture and intracellular maintenance of prokaryotic cells which possessed these metabolic capacities. The evolutionary shift from predation pure and simple, to predation combined with (or replaced by) endosymbiosis, probably took place repeatedly, in many different lines of eukaryotes, and over a long evolutionary period. A vestige of this stage of cellular evolution may perhaps be evident in the various contemporary groups of eukaryotic algae, as suggested by Sagan (1967). Each major algal group is distinguished in part by the distinctive properties of its chloroplasts, particularly their complement of photopigments. Only one type of algal chloroplast, that characteristic of the green algae, was carried into the major evolutionary line which gave rise to vascular plants. This diversity of the algal plastid could reflect the establishment of a stable endosymbiotic partnership by the eukaryotic progenitor of each major algal group with a different kind of oxygen-evolving photosynthetic prokaryote, which thereby conferred its specific photopigment system on the host cell. In this event, the present-day diversity of plastid structure and composition among eukaryotic algae is really a reflection of an ancient evolutionary diversity at the prokaryotic level, not of cellular evolution at the eukaryotic one.

The contemporary diversity of photopigment systems is of great ecological significance, particularly at the prokaryotic level, where the differences are greatest (Stanier & Cohen-Bazire, 1957). It permits the co-existence of different kinds of photosynthetic organisms in one habitat, since each can absorb preferentially certain regions of the solar emission spectrum, thanks to the specific absorptive properties of its light-gathering pigments. Even though the hypothesis of Sagan (1967) concerning the polyphyletic origin of algal chloroplasts is thus an attractive one in ecological terms, it receives scant support from the properties of photopigment systems in contemporary oxygen-producing prokaryotes (blue-green algae). All these organisms have a remarkably uniform photopigment system, similar to but not completely identical with the photopigment system of one eukaryotic algal group, the *Rhodophyta*. Hence Sagan's hypothesis implies that many lines of oxygen-producing photosynthetic prokaryotes have become extinct as free-living organisms, although their photopigment systems have been preserved in the photosynthetic organelles of such eukaryotic algal groups as the *Chlorophyta*, the *Pyrrophyta*, the *Phaeophyta* and the *Euglenophyta*.

An analogous diversity with respect to functional respiratory components is not characteristic of mitochondria: the mitochondrial electron transport systems of eukaryotes are remarkably uniform, particularly in contrast to the highly diverse electron transport systems of contemporary aerobic bacteria (Smith, 1961). This could imply that the implantation of respiratory endosymbionts in eukaryotes took place over a short evolutionary time span, soon after the first appearance of this new piece of metabolic machinery among prokaryotes. A less plausible alternative explanation (Margulis, 1968) is that the ancestor of the mitochondrion made a very early entry into the eukaryotic cell line, from which it was carried into all the divergent branches of the contemporary eukaryotic world. On geochemical grounds, I think we must assume that aerobic respiration appeared on earth after (and probably long after) the evolutionary emergence of oxygen-producing photosynthesis. Cloud (1965) interpreted the geochemical data as indicative of a slow build-up of atmospheric oxygen; and it is obvious that the net accumulation of free oxygen on earth could not have begun at all until all the readily auto-oxidizable constituents of the crust had been converted to the oxidized state. Hence, it seems probable that mitochondria were the last components established in the eukaryotic cell. This event brought to a close the purely cellular phase of evolution. Subsequently, the thrust of biological evolution received its primary expression at the organismal level.

One element of the cell of a free-living prokaryote becomes immediately dispensable when it adopts a permanent intracellular habitat: the cell wall. Loss of walls was presumably an early event in the intracellular evolution of the prokaryotic precursors of mitochondria and chloroplasts; and, as a result, the ability to synthesize peptidoglycans, acquired by eukaryotes together with their endosymbionts, was soon lost again. At a later period of evolution, when cell walls came to possess adaptive value in some eukaryotic lines (e.g. algae and fungi), the evolutionary problem of their synthesis had to be resolved in new ways. A number of different molecular solutions were discovered, as evidenced by the present wide diversity of wall structure in these eukaryotic groups. Furthermore, if we may safely generalize from the work of Manton (1966, 1967a, b) and Brown (1969), the Golgi apparatus was put to use in effecting a solution.

An evolutionary change in the structure of the eukaryotic genetic system assured the permanence of the captivity of photosynthetic and respiratory endosymbionts. This was the abstraction of some determinants of endosymbiotic function from the genophores of the endosymbionts, and their incorporation in the nuclear genophores of the host. Thereafter, the evolutionary paths of the partners were no longer free to diverge; the endosymbionts lost the status of organisms, and acquired that of organelles.

It might have happened thus; but we shall surely never know with certainty. Evolutionary speculation constitutes a kind of metascience, which has the same intellectual fascination for some biologists that metaphysical speculation possessed for some mediaeval scholastics. It can be considered a relatively harmless habit, like eating peanuts, unless it assumes the form of an obsession; then it becomes a vice. The most appropriate response to such speculations (if they are plausible and logically consistent) is an Italian rejoinder, of which the amiable cynicism cannot be adequately translated:

Se non è vero, è ben trovato.

Part of the reading and thinking that underlies this essay was done during the tenure of a fellowship from the J. S. Guggenheim Memorial Foundation in 1967–8. My fellowship proposal had contained a rash promise to start the preparation of a book on the prokaryotic cell, viewed in the light of general cell theory. However, the flesh proved weak; and I fear that the present analysis will remain the sole return from this intellectual venture-capital.

Many useful comments and criticisms were made by Daniel Branton, Germaine Cohen-Bazire, Daniel Mazia, Murdoch Mitchison and Peter Satir, who were kind enough to read a first draft. I am most grateful for their help; but this acknowledgement should not be construed to mean that they are responsible for any of the opinions expressed.

GLOSSARY

Cytological terms used in this essay, with some of the synonyms frequently found in the cytological literature

Term	Synonyms	Definition
Unit membrane	Double membrane, triple membrane	Any membrane, irrespective of cellular location, which can be resolved in sections of osmium-fixed material as a triple-layered structure 60 to 100 Å wide, consisting of two electron-dense outer layers and a less dense central layer
Cytoplasmic membrane	Cell membrane, plasmalemma	The unit membrane which constitutes the external boundary of the cytoplasm in both eukaryotic and prokaryotic cells
Endoplasmic reticulum	Ergastoplasm	The irregular network of unit membranes which traverses the cytoplasmic region of a eukaryotic cell, often bearing ribosomes on its surface
Golgi apparatus	Dictyosome	A compound membranous eukaryotic organelle, consisting of numerous ribosome-free vesicles, in part flattened and elongated, and arrayed in a more or less regular stack
Vacuole*	—	A closed structure completely surrounded by a unit membrane, the contents of which are in the liquid state. Found only in eukaryotic cells.

* *See* note on next page.

Term	Synonyms	Definition
Vesicle*	Cisterna	A closed structure completely surrounded by a unit membrane, similar to a vacuole, but of which the contents are not (or are not known to be) in the liquid state. Found only in eukaryotic cells
Non-unit membrane	—	Any membrane less than 60 Å wide, and appearing in section as a single, electron-dense layer
Microtubules	Fibres, fibrils (e.g. 'spindle fibres')	Rigid, hollow filaments with diameters ranging from 150 to 250 Å, which occur in many regions of the cytoplasm of the eukaryotic cell. They are particularly conspicuous (on account of their regular arrangements) in cilia and in the mitotic apparatus
Centriole	Basal body, kinetosome	A hollow cylinder with microtubular walls, 1200 to 1500 Å in diameter and of variable length, found in the cytoplasm of some eukaryotic cells, and associated with the production of certain microtubular systems. The two synonyms are partial ones, referring specifically to centrioles at the base of the microtubular system of cilia
Cilium	Eukaryotic flagellum, sperm tail	A specialized eukaryotic locomotor organelle which consists of a filiform extrusion of the cell surface. It is bounded by an extension of the cytoplasmic membrane, and contains a regular longitudinal array of microtubules, anchored basally in a centriole. 'Sperm tail' is a partial synonym, referring specifically to organelles of this type borne by specialized male gametes
Flagellum	Prokaryotic flagellum, bacterial flagellum	A specialized prokaryotic locomotor organelle, which consists of a filiform extension through the cell surface, rarely bounded by an extension of the cytoplasmic membrane, and composed of a single, helically-wound fibril, ∼ 140 Å wide
Nucleus	—	A localized cellular region which contains most (or all) of the DNA. In eukaryotes, the nuclear DNA is carried on more than one genophore (chromosome); in prokaryotes, it is generally contained in a single genophore.
Genophore	—	A discrete element of genetic material: the physical entity that corresponds to a linkage group
Chromosome	—	One of the constituent genophores in a eukaryotic nucleus

* The terms 'vesicle' and 'vacuole' will also be employed, but always with qualifying adjectives, for specialized prokaryotic cell structures which do not conform to the definitions presented here.

REFERENCES

ABRAM, D. & KOFFLER, H. (1964). *In vitro* formation of flagella-like filaments and other structures from flagellin. *J. molec. Biol.* **9**, 168.

AVERS, C. J., BILLHEIMER, F. E., HOFFMANN, H. P. & PAULI, R. M. (1968). Circularity of yeast mitochondrial DNA. *Proc. natn. Acad. Sci. U.S.A.* **61**, 90.

BEAMS, H. W. & KESSEL, R. G. (1968). The Golgi apparatus: structure and function. *Int. Rev. Cytol.* **23**, 209.

BELLOC, H. (1898). *More Beasts for Worse Children.* London: Duckworth.

BOWEN, C. C. & JENSEN, T. E. (1965). Blue-green algae: fine structure of the gas vacuoles. *Science, N.Y.* **147**, 1460.

BROWN, Jr., R. M. (1969). Observations on the relationship of the Golgi apparatus to wall formation in the marine chrysophycean alga, *Pleurochrysis scherffelii* Pringsheim. *J. Cell Biol.* **41**, 109.

BUCHNER, P. (1965). *Endosymbiosis of Animals with Plant Micro-organisms.* New York: Wiley (Interscience).

BURNHAM, J. C., HASHIMOTO, T. & CONTI, S. F. (1968). Electron microscopic observations on the penetration of *Bdellovibrio bacteriovorus* into Gram-negative bacterial hosts. *J. Bact.* **96**, 1366.

CAIRNS, J. (1963). The chromosome of *Escherichia coli. Cold Spring Harb. Symp. quant. Biol.* **28**, 43.

CARO, L. G. & PALADE, G. E. (1964). Protein synthesis, storage, and discharge in the pancreatic exocrine cell. An autoradiographic study. *J. Cell Biol.* **20**, 473.

CLARK-WALKER, G. D. & LINNANE, A. W. (1967). The biogenesis of mitochondria in *Saccharomyces cerevisiae.* A comparison between cytoplasmic respiratory-deficient mutant yeast and chloramphenicol-inhibited wild type cells. *J. Cell Biol.* **34**, 1.

CLOUD, Jr., P. E. (1965). Significance of the gunflint (precambrian) microflora. *Science, N.Y.* **148**, 27.

COHEN-BAZIRE, G., PFENNIG, N. & KUNISAWA, R. (1964). The fine structure of green bacteria. *J. Cell Biol.* **22**, 207.

COHN, F. (1875). Untersuchungen über Bacterien. II. *Beitr. Biol. Pfl.* **1**, 141.

CRUDEN, D. L. (1968). Structure and function in photosynthetic procaryotic organisms. *Thesis: University of California, Berkeley.*

CURTIS, H. (1968). *Biology.* New York: Worth Publishers.

DIPPELL, R. (1968). The development of basal bodies in *Paramecium. Proc. natn. Acad. Sci., U.S.A.* **61**, 461.

DIRKSEN, E. R. & CROCKER, T. T. (1966). Centriole replication in differentiating ciliated cells of mammalian respiratory epithelium. *J. Microscopie*, **5**, 629.

DODGE, J. D. (1964). Cytochemical staining of sections from plastic-embedded flagellates. *Stain Technol.* **39**, 381.

DE DUVE, C. (1964). From cytases to lysosomes. *Fedn Proc. Fedn Am. Socs exp. Biol.* **23**, 1045.

DU PRAW, E. J. (1968). *Cell and Molecular Biology.* London and New York: Academic Press.

EMERSON, R. & HELD, A. A. (1969). *Aqualinderella fermentans* gen. et sp. nov., a phycomycete adapted to stagnant waters. II. Isolation, culture and gas relationships. *Am. J. Bot.* **56**. (In Press.)

EPHRUSSI, B. (1953). *Nucleo-cytoplasmic Relations in Micro-organisms.* Oxford: Clarendon Press.

EPHRUSSI, B., HOTTINGUER, H. & TAVLITSKI, J. (1949). Action de l'acriflavine sur les levures. II. Étude génétique du mutant 'petite colonie'. *Annls Inst. Pasteur, Paris* **76**, 419.

FAWCETT, D. (1961). Cilia and flagella. In *The Cell*, vol. 2, p. 217. Eds. J. Brachet and A. E. Mirsky. London and New York: Academic Press.

FOGG, G. E. (1941). The gas-vacuoles of the Myxophyceae (Cyanophyceae). *Biol. Rev.* **16**, 205.

GEITLER, L. (1959). Syncyanosen. In *Handbuch der Pflanzenphysiologie*, vol. 11, p. 530. Ed. W. Ruhland.

GIBBINS, J. R., TILNEY, L. G. & PORTER, K. R. (1969). Microtubules in the formation and development of the primary mesenchyme in *Arbacia punctulata.* I. The distribution of microtubules. *J. Cell Biol.* **41**, 201.

GIBBS, S. P. (1968). Autoradiographic evidence for the *in situ* synthesis of chloroplast and mitochondrial RNA. *J. Cell Sci.* **3**, 327.

GIESBRECHT, P. (1962). Vergleichende Untersuchungen an den Chromosomen des Dinoflagellaten *Amphidinium elegans* und denen der Bakterien. *Zent. Bakt. ParasitKde* (1. Abt. Orig.) **187**, 452.

GORCHEIM, A. (1968). The separation and identification of the lipids of *Rhodopseudomonas spheroides*. *Proc. Roy. Soc. Lond.* B **170**, 279.

GRANICK, S. (1961). The chloroplasts: inheritance, structure and function. *The Cell*, vol. 1, p. 489. Ed. J. Brachet and A. E. Mirsky. London and New York: Academic Press.

GRANICK, S. & GIBOR, A. (1967). The DNA of chloroplasts, mitochondria, and centrioles. In *Progress in Nucleic Acid Research and Molecular Biology*, vol. 6, p. 143. Eds. J. N. Davidson and W. E. Cohn. London and New York: Academic Press.

GREEN, J. C. & JENNINGS, P. H. (1967). A physical and chemical investigation of the scales produced by the Golgi apparatus within and found on the surface of the cells of *Chrysochromulina chiton* Parke et Manton. *J. exp. Bot.* **18**, 359.

GRIMSTONE, A. V. (1961). Fine structure and morphogenesis in protozoa. *Biol. Rev.* **36**, 97.

GUTTES, E., GUTTES, S. & DEVI, R. V. (1969). Division stages of the mitochondria in normal and actinomycin-treated plasmodia of *Physarum polycephalum*. *Experientia* **25**, 66.

HALL, W. T. & CLAUS, G. J. (1963). Ultrastructural studies on the blue-green algal symbiont in *Cyanophora paradoxa* Korschikoff. *J. Cell Biol.* **19**, 55.

HALL, W. T. & CLAUS, G. J. (1967). Ultrastructural studies on the cyanelles of *Glaucocystis nostochinearum* Itzigsohn. *J. Phycol.* **3**, 37.

HARRIS, P. (1962). Some structural and functional aspects of the mitotic apparatus in sea urchin embryos. *J. Cell Biol.* **14**, 475.

HIRSCH, J. G. & COHN, Z. A. (1964). Digestive and autolytic functions of lysosomes in phagocytic cells. *Fedn Proc. Fedn Am. Socs exp. Biol.* **23**, 1023.

HOLTER, H. (1965). Passage of particles and macromolecules through cell membranes. In *Function and Structure in Micro-organisms*, p. 89. Eds. M. R. Pollock and M. H. Richmond. Cambridge University Press.

HOOBER, J. K. & BLOBEL, G. (1969). Characterization of the chloroplastic and cytoplasmic ribosomes of *Chlamydomonas reinhardii*. *J. molec. Biol.* **41**, 121.

HOUWINK, A. L. (1956). Flagella, gas vacuoles and cell-wall structure in *Halobacterium halobium*: an electron microscope study. *J. gen. Microbiol.* **15**, 146.

HUNGATE, R. E. (1967). *The Rumen and its Microbes*. London and New York: Academic Press.

INOUÉ, S. (1953). Polarization optically studied of the mitotic spindle. I. The demonstration of spindle fibres in the living cells. *Chromosoma* **5**, 487.

IWAMURA, T. (1966). Nucleic acids in chloroplasts and metabolic DNA. In *Progress in Nucleic Acid Research and Molecular Biology*, vol. 5, p. 153. Eds. J. N. Davidson and W. E. Cohn. London and New York: Academic Press.

JURAND, A. & PREER, L. B. (1969). Ultrastructure of flagellated lambda symbionts in *Paramecium aurelia*. *J. gen. Microbiol.* **54**, 359.

JOST, M. (1965). Die Ultrastruktur von *Oscillatoria rubescens*. *Arch. Mikrobiol.* **50**, 211.

KERRIDGE, D., HORNE, R. W. & GLAUERT, A. (1962). Structural components of flagella from *Salmonella typhimurium*. *J. molec Biol.* **4**, 227.

KITCHING, J. A. (1964). The axopods of the sun animalcule *Actinophrys sol* (Heliozoa). In *Primitive Motile Systems in Cell Biology*, p. 445. Eds. R. D. Allen and N. Kamiya. London and New York: Academic Press.

KLEBAHN, H. (1929). Über die Gasvacuolen der Cyanophyceen. *Ver. int. Verein. theor. angew. Limnol.* **4**, 408.

KORSCHIKOFF, A. A. (1924). Protistologische Beobachtungen. I. *Cyanophora paradoxa. Arch. Russ. Protistenk.* **3**, 57.

KUBAI, D. F. & RIS, H. (1969). Division in the dinoflagellate *Gyrodinium cohnii* (Schiller). A new type of nuclear reproduction. *J. Cell Biol.* **40**, 508.

LAUTERBORN, R. (1895). Protozoen—Studien. III. *Paulinella chromatophora. Z. wiss. Zool.* **59**, 537.

LAUTERBORN, R. (1915). *Die sapropelische Lebewelt. Verh. naturht.-med. Ver. Heidelb.* **13**, 395.

LEFORT, M. (1965). Sur le chromatoplasma d'une Cyanophycée endosymbiotique: *Glaucocystis nostochinearum* Itzigs. *Cr. hebd. Séanc. Acad. Sci. Paris* **261**, 233.

LINNANE, A. W., LAMB, A. J., CHRISTODOULOU, C. & LUKINS, H. B. (1968). The biogenesis of mitochondria. VI. The biochemical basis of the resistance of *Saccharomyces cerevisiae* towards antibiotics which inhibit mitochondrial protein synthesis. *Proc. natn. Acad. Sci. U.S.A.* **59**, 1288.

LINNANE, A. W., SAUNDERS, G. W., GINGOLD, E. B. & LUKINS, H. B. (1968). The biogenesis of mitochondria. V. Cytoplasmic inheritance of erythromycin resistance in *Saccharomyces cerevisiae. Proc. natn. Acad. Sci. U.S.A.* **59**, 903.

MANTON, I. (1966). Observations on scale production in *Prymnesium parvum. J. Cell Sci.* **1**, 375.

MANTON, I. (1967a). Further observations on the fine structure of *Chrysochromulina chiton* with special reference to the haptonema, 'peculiar' Golgi structure and scale production. *J. Cell Sci.* **2**, 265.

MANTON, I. (1967b). Further observations on scale formation in *Chrysochromulina chiton. J. Cell Sci.* **2**, 411.

MARGULIS, L. (1968). Evolutionary criteria in thallophytes: a radical alternative. *Science, N.Y.* **161**, 1020.

MAZIA, D. (1961). Mitosis and the physiology of cell division. In *The Cell*, vol. 3, p. 77. Eds. J. Brachet and A. E. Mirsky. London and New York: Academic Press.

MAZIA, D. (1965). The partitioning of genomes. In *Function and Structure in Microorganisms*, p. 379. Eds. M. R. Pollock and M. H. Richmond. Cambridge University Press.

MAZIA, D., CHAFFEE, R. R. & IVERSON, R. M. (1961). Adenosine triphosphatase in the mitotic apparatus. *Proc. natn. Acad. Sci. U.S.A.* **47**, 788.

MAZIA, D. & DAN, K. (1952). The isolation and biochemical characterization of the mitotic apparatus of dividing cells. *Proc. natn. Acad. Sci. U.S.A.* **38**, 826.

MAZIA, D. & RUBY, A. (1968). Dissolution of erythrocyte membranes in water and comparison of the membrane protein with other structural proteins. *Proc. natn. Acad. Sci. U.S.A.* **61**, 1005.

MITCHELL, M. B. & MITCHELL, H. K. (1952). A case of 'maternal inheritance'. *Proc. natn. Acad. Sci. U.S.A.* **38**, 442.

MOUNOULOU, J. C., JAKOB, H. & SLONIMSKI, P. P. (1966). Mitochondrial DNA from yeast 'petite' mutants: specific changes of buoyant density corresponding to different cytoplasmic mutations. *Biochem. biophys. Res. Commun.* **24**, 218.

MURRAY, R. G. E. (1962). Fine structure and taxonomy of bacteria. In *Microbial Classification*, p. 119. Eds. G. C. Ainsworth and P. H. A. Sneath. Cambridge University Press.

NASS, S. (1969). The significance of the structural and functional similarities of bacteria and mitochondria. *Int. Rev. Cytol.* **25**, 55.

NASS, M. M. K. & NASS, S. (1963). Intramitochondrial fibers with DNA characteristics. *J. Cell Biol.* **19**, 593.

NOVIKOFF, A. B., ESSNER, E. & QUINTANA, N. (1964). Golgi apparatus and lysosomes. *Fedn Proc. Fedn Am. Socs exp. Biol.* **23**, 1010.

NUTMAN, P. S. (1963). Factors influencing the balance of mutual advantage in legume symbiosis. In *Sybiotic Associations*, p. 51. Eds. P. S. Nutman and B. Mosse. Cambridge University Press.

PALADE, G. (1961). The secretory process of the pancreatic exocrine cell. In *Electron Microscopy in Anatomy*, p. 176. Eds. J. Boyd, F. Johnson and J. Levers. Baltimore: Williams and Wilkins.

PASCHER, A. (1929). Über einige Endosymbiosen von Blaualgen in Einzellern. *Jb. wiss. Bot.* **71**, 386.

PETTER, H. F. M. (1931). On bacteria of salted fish. *Proc. Acad. Sci. Amst.* **34**, 1417.

PFENNIG, N. (1967). Photosynthetic bacteria. *A. Rev. Microbiol.* **21**, 285.

PFENNIG, N. & COHEN-BAZIRE, G. (1967). Some properties of the green bacterium *Pelodictyon clathratiforme*. *Arch. Mikrobiol.* **59**, 226.

PORTER, K. R. (1964). Cell fine structure and biosynthesis of inter-cellular macromolecules. *Biophys. J.* **4**, 167.

PORTER, K. R. (1966). Cytoplasmic microtubules and their functions. In *Ciba Foundation Symposium on Principles of Biomolecular Organization*, p. 308. Eds. G. E. W. Wolstenholme and M. O'Connor. London: Churchill.

PRINGSHEIM, E. G. (1928). Physiologische Untersuchungen an *Paramecium bursaria*. Ein Beitrag zur Symbioseforschung. *Arch. Protistenk.* **64**, 289.

PRINGSHEIM, E. G. (1958). Organismen mit blaugrünen Assimilatoren. In *Studies in Plant Physiology*. Ed. S. Prát. Praha: Československa Akademie Věd.

RANDALL, J. & DISBREY, C. (1965). Evidence of the presence of DNA at basal body sites in *Tetrahymena pyriformis*. *Proc. Roy. Soc. Lond.* B **162**, 473.

REITZ, R. C. & HAMILTON, J. G. (1968). The isolation and identification of two sterols from two species of blue-green algae. *Comp. Biochem. Physiol.* **25**, 401.

RHOADES, M. M. (1946). Plastid mutations. *Cold Spring Harb. Symp. quant. Biol.* **11**, 202.

RIS, H. (1961). Ultrastructure and molecular organization of genetic systems. *Can. J. Cytol.* **3**, 95.

RIS, H. & PLAUT, W. (1962). Ultrastructure of DNA-containing areas in the chloroplast of *Chlamydomonas*. *J. Cell Biol.* **13**, 383.

ROBBINS, P. W., BRAY, D., DANKERT, M. & WRIGHT, A. (1967). Direction of chain growth in polysaccharide synthesis. *Science, N.Y.* **158**, 1536.

ROBINOW, C. F. & BAKERSPIEGEL, A. (1965). Somatic nuclei and forms of mitosis in fungi. In *The Fungi, an Advanced Treatise*, vol. 1, p. 119. Eds. G. C. Ainsworth and A. S. Sussman. London and New York: Academic Press.

ROSENBAUM, J. L. & CARLSON, K. (1969). Cilia regeneration in *Tetrahymena* and its inhibition by colchicine. *J. Cell Biol.* **40**, 415.

RUDZINSKA, M. A. (1965). The fine structure and function of the tentacle in *Tokophrya infusionum*. *J. Cell Biol.* **25**, 459.

RYTER, A. (1968). Association of the nucleus and the membrane of bacteria: a morphological study. *Bact. Rev.* **32**, 39.

SABATINI, D. D., BENSCH, K. & BARNETT, R. J. (1963). Cytochemistry and electron microscopy—the preservation of cellular ultrastructure and enzymatic activity by aldehyde fixation. *J. Cell Biol.* **17**, 19.

SAGAN, L. (1967). On the origin of mitosing cells. *J. theor. Biol.* **14**, 225.

SANO, Y. & KNOOP, A. (1959). Elektronenmikroskopische Untersuchungen am kaudalen neurosekretorischen system von *Tinca vulgaris*. *Z. Zellforsch. mikrosk. Anat.* **49**, 464.

SHILO, M. & BRUFF, B. (1965). Lysis of Gram-negative bacteria by host-independent ectoparasitic *Bdellovibrio bacteriovorus* isolates. *J. gen. Microbiol.* **40**, 317.

SIEKEVITZ, P. & PALADE, G. E. (1958). A cytochemical study on the pancreas of the guinea pig. I. Isolation and enzymatic activities of all fractions. *J. biophys. biochem. Cytol.* **4**, 203.

SMITH, L. (1961). Cytochrome systems in aerobic electron transport. In *The Bacteria*, vol. 2, p. 365. Eds. I. C. Gunsalus and R. Y. Stanier. London and New York: Academic Press.

SMITH, R. V. & PEAT, A. (1967). Comparative structure of the gas-vacuoles of blue-green algae. *Arch. Mikrobiol.* **57**, 111.

SMITH-SONNEBORN, J. & PLAUT, W. (1967). Evidence for the presence of DNA in the pellicle of *Paramecium. J. Cell Sci.* **2**, 225.

SONNEBORN, T. M. (1959). Kappa and related particles in *Paramecium aurelia. Adv. Virus Res.* **6**, 229.

DE SOUZA, N. J. & NES, W. R. (1968). Sterols: isolation from a blue-green alga. *Science, N.Y.* **162**, 363.

STALEY, J. T. (1968). *Prosthecomicrobium* and *Ancalomicrobium*, new prosthecate freshwater bacteria. *J. Bact.* **95**, 1921.

STANIER, R. Y. (1961). La place des bactéries dans le monde vivant. *Annls Inst. Pasteur, Paris* **101**, 297.

STANIER, R. Y. (1963). The organization of the photosynthetic apparatus in purple bacteria. In *The General Physiology of Cell Specialization*, p. 242. Eds. D. Mazia and A. Tyler. New York: McGraw-Hill.

STANIER, R. Y. (1966). The organization of respiratory and photosynthetic function in procaryotic organisms. In *Problemas Actuales de Biologia*. Eds. J. R. Villanueva and J. L. Rodriguez-Candela. Madrid: Consejo Superior de Investigaciones Científicas.

STANIER, R. Y. & COHEN-BAZIRE, G. (1957). The role of light in the microbial world: some facts and speculations. In *Microbial Ecology*, p. 56. Eds. R. E. O. Williams and C. C. Spicer. Cambridge University Press.

STANIER, R. Y., DOUDOROFF, M. & ADELBERG, E. A. (1963). *The Microbial World*. Second edition. Englewood Cliffs: Prentice-Hall.

STANIER, R. Y. & VAN NIEL, C. B. (1941). The main outlines of bacterial classification. *J. Bact.* **42**, 437.

STANIER, R. Y. & VAN NIEL, C. B. (1962). The concept of a bacterium. *Arch. Mikrobiol.* **42**, 17.

STEPHENS, R. E. (1968). On the structural protein of flagellar outer fibers. *J. molec. Biol.* **32**, 277.

STOECKENIUS, W. & KUNAU, W. H. (1968). Further characterization of particulate fractions from lysed cell envelopes of *Halobacterium halobium* and isolation of gas vacuole membranes. *J. Cell Biol.* **38**, 337.

STOLP, H. & PETZOLD, H. (1962). Untersuchungen über einen obligat parasitischen Mikroorganismus mit lytischer Aktivität für Pseudomonas-Bakterien. *Phytopath. Z.* **45**, 364.

STOLP, H. & STARR, M. P. (1963). *Bdellovibrio bacteriovorus* gen. et sp. n., a predatory, ectoparasitic, and bacteriolytic microorganism. *Antonie van Leeuwenhoek J. Microbiol. Serol.* **29**, 217.

STROMINGER, J. L. & TIPPER, D. J. (1965). Bacterial cell wall synthesis and structure in relation to the mechanism of action of penicillins and other antibacterial agents. *Am. J. Med.* **39**, 708.

STUBBLEFIELD, E. & BRINKLEY, B. R. (1968). Architecture and function of the mammalian centriole. *Symp. Int. Soc. Cell Biol.* **6**, 175. London and New York: Academic Press.

TAYLOR, E. W. (1965). The mechanism of colchicine inhibition of mitosis. I. Kinetics of inhibition and the binding of H^3-colchicine. *J. Cell Biol.* **25**, 145.

TILNEY, L. G. (1968). Studies on the microtubules in Heliozoa. IV. The effect of colchicine on the formation and maintenance of the axopodia and the redevelopment of pattern in *Actinosphaerium nucleofilum* (Barrett). *J. Cell Sci.* **3**, 549.

TILNEY, L. G. & GIBBINS, J. R. (1969). Microtubules in the formation and development of the primary mesenchyme in *Arbacia punctulata*. II. An experimental analysis of their role in development and maintenance of cell shape. *J. Cell Biol.* **41**, 227.

VAN WAGTENDONK, W. J., CLARK, J. A. D. & GODOY, G. A. (1963). The biological status of lambda and related particles in *Paramecium aurelia*. *Proc. natn. Acad. Sci. U.S.A.* **50**, 835.

WOLSTENHOLME, D. R. & GROSS, N. J. (1968). The form and size of mitochondrial DNA of the red bean, *Phaseolus vulgaris*. *Proc. natn. Acad. Sci. U.S.A.* **61**, 245.

YOTSUYANAGI, Y. (1962). Études sur le chondriome de la levure. II. Chondriomes des mutants à déficience respiratoire. *J. Ultrastruct. Res.* **7**, 141.

EXPLANATION OF PLATES

Plate 1. Electron micrograph of a thin section of the green bacterium, *Pelodictyon clathratiforme*. × 75,000. This organism contains two different types of organelles bounded by non-unit membranes: chlorobium vesicles (cv), containing the photopigment system, which are arrayed in a cortical layer immediately beneath the cytoplasmic membrane; and gas vesicles (gv), which comprise the gas vacuoles characteristic of this species. Fairly complex mesosomal intrusions (m) of the cytoplasmic membrane also occur. (Courtesy of Dr Germaine Cohen-Bazire.)

Plate 2. Electron micrographs of negatively-stained preparations of partly purified chlorobium vesicles and unit membrane fragments, isolated from a cell-free extract of *Chlorobium thiosulfatophilum*. Fig. 1: vesicle fraction, × 224,000. Fig. 2: membrane fraction, × 100,000. (Courtesy of Dr Diana Loeb Cruden.)

Plate 3. Electron micrograph of a negatively-stained preparation of purified gas vesicles isolated from a filamentous blue-green alga, *Oscillatoria agardhii* var. *suspensa*. Most of the vesicles are still inflated; one (arrow) is partly collapsed, × 164,000. (Courtesy of Dr Germaine Cohen-Bazire.)

PLATE 1

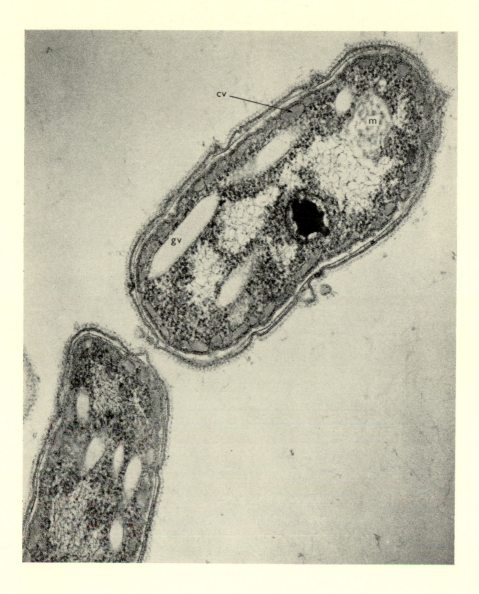

PLATE 2

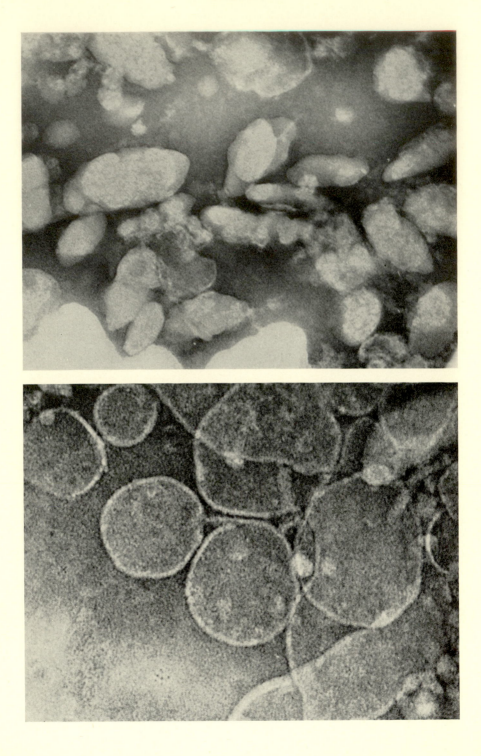

PLATE 3

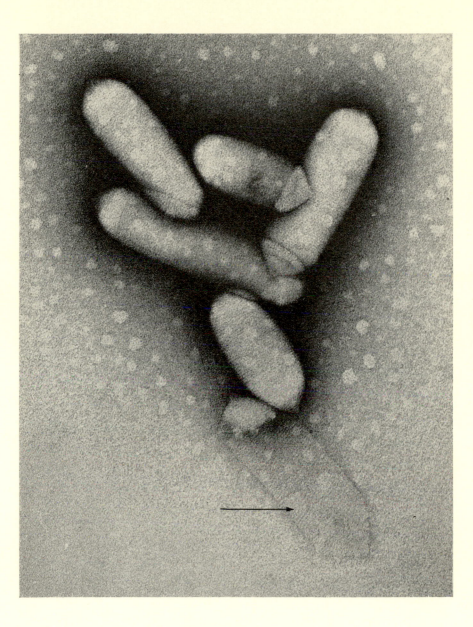

THE GENETIC CODE IN PROKARYOTES AND EUKARYOTES

C. R. WOESE

Department of Microbiology, University of Illinois,
Urbana, Illinois, 61801, U.S.A.

INTRODUCTION

'The genetic code' is at best unfortunate, at worst deceptive, terminology as applied to the phenomenon of structural gene expression. To the average reader 'genetic code' merely calls to mind a table—such as Table 1—in which the symbols for the 20 encoded amino acids are paired in some ordered array with the symbols for the 64 possible nucleotide triplet codons. To some, genetic code may also mean certain 'black box' experiments in which polynucleotides of more or less defined compositions were added to an '*in vitro* system' (black box) with the result that polypeptides of more or less defined compositions were produced—these results being used to construct Table 1.

The problem with the terminology is that it causes one to overlook what may be the central issue in gene expression; the terminology

TABLE 1. *The genetic code*

The table shows the complete set of 64 triplets, and their corresponding amino acids. Three triplets do not code for amino acids; two of these triplets (CT-1, CT-2) terminate the polypeptide chain and the third (CT-3?) probably does so.

1st position	2nd position				1st position
	U	C	A	G	
U	Phe	Ser	Tyr	Cys	U
	Phe	Ser	Tyr	Cys	C
	Leu	Ser	CT-1	CT-3?	A
	Leu	Ser	CT-2	Typ	G
C	Leu	Pro	His	Arg	U
	Leu	Pro	His	Arg	C
	Leu	Pro	Gln	Arg	A
	Leu	Pro	Gln	Arg	G
A	Ile	Thr	Asn	Ser	U
	Ile	Thr	Asn	Ser	C
	Ile	Thr	Lys	Arg	A
	Met	Thr	Lys	Arg	G
G	Val	Ala	Asp	Gly	U
	Val	Ala	Asp	Gly	C
	Val	Ala	Glu	Gly	A
	Val	Ala	Glu	Gly	G

artificially fractures a monolith. The code, the table of formal relationships between amino acids and codons, does not exist in a vacuum, it
exists in the context of a machine that does the decoding. There is good
reason to believe that the form of the set of codon assignments is not
necessarily independent of the nature of the decoding, translating,
machine. In fact, it is quite reasonable that the evolution of codon assignments and that of the decoding machine were interconnected happenings.
Once we realize that a set of codon assignments (a codon catalogue) and
a biological decoding machine are facets of the same monolith, then we
can define 'genetic code' accordingly, and not impede our understanding of the molecular mechanisms involved in gene expression.

   The matter before us is the comparison of the genetic code of eukaryotes with that of prokaryotes. The central issue is the extent to which the
genetic code is universal. Since we have defined the code in its broad
sense, the question of universality is not confined to the codon catalogue
alone. We must also enquire whether the translation machine is universal
and even whether the way in which it is biosynthesized is universal. We
have to make distinctions between 'essential' and 'trivial' non-universality. It has become apparent that a host of primary structures are consistent with any given function of a molecule—cytochrome c being an
excellent example (Fitch & Margoliash, 1966). If true for cytochrome c,
a small molecule, this principle should hold to a greater extent—perhaps
even on a higher level—for the translation machine. Therefore, to
show that the primary structures of *Escherichia coli* ribosomal RNAs
differ from those of *Bacillus subtilis*, for example, is certainly a demonstration of non-universality. But if such heterology is not reflected in
functional (or gross structural) heterology, the case is one of 'trivial non-
universality', for it reflects only the 'evolutionary wanderings' of the
non-essential portions of the rRNA primary structures.

   At this point we must distinguish a more subtle kind of meaningless
non-universality, which I will call 'incidental non-universality'. It is
conceivable that 'evolutionary wanderings' of non-essential portions of
various primary structures of components of the translation apparatus
change the overall geometry of the component in question, without
altering the critical sites of the molecules—i.e. the determinants of the
molecule's function. One test for universality of translation has been the
construction of a system in which different components of the apparatus
have been isolated from unrelated sources. If such a chimaera system
does not function, it could conceivably reflect merely these 'incidental'
changes in the geometry of the molecule(s)—i.e. changes in one or both
components of the chimaera system prevent the critical sites of the

molecules from assuming the correct juxtaposition, even though these sites are otherwise perfectly 'compatible' with one another. Obviously such 'incidental' non-universality is of little or no biological significance.

The reason for concern with universality of the genetic code is obvious, but should, nevertheless, be detailed here. If the genetic code proves to be universal, in all but trivial or incidental ways, the question whether all forms living today share a common ancestor would for all intents and purposes be settled. This is not the problem of whether life arose only once, but a more sophisticated one, in that the ancestor cell would be in no sense the 'first cell', but a cell that had already come to the point of possessing a fully evolved genetic code. Thus, the problem would be to explain why one cell-line, one species, survived to inherit the earth, while all its presumed contemporaries did not.

The more intriguing possibility would be that the genetic code is not completely universal in non-trivial ways. In this case we would entertain two possibilities. The first is that any common ancestor of eukaryotes and prokaryotes was so far back in the evolutionary scheme that the genetic code was not fully evolved at the time the eukaryote–prokaryote divergence occurred, making it almost certain that any further evolution of the code would occur in different ways in the different lines. In other words, traces of the evolution of the cell itself would be revealed in eukaryote–prokaryote comparisons.

The second possibility is the less interesting one that since the time when they shared a common ancestor (an ancestor with a fully evolved genetic code), the prokaryotes and eukaryotes altered their genetic coding systems. In this latter case, the differences would be of very little evolutionary significance.

Evolution of the translation apparatus

It is first necessary to understand the nature of the problem of the evolution of the genetic code, because the importance of the eukaryote–prokaryote comparison lies in its significance for evolution. Discussion of this is necessarily confined to 'reasonable' argumentation and intuitive pronouncements. What follows represents largely my own prejudices on the matter; it is intended to be thought-provoking rather than necessarily correct.

The starting point for an understanding of the genetic code's evolution lies in three considerations. (1) In the cell today the matching of amino acid to codon is unambiguous and very precise—mistranslation of a codon occurs no more than a few times in every 10,000 (Loftfield, 1963;

Szer & Ochoa, 1964). (2) In the case of nucleic acids, a single strong interaction, the base-pairing and stacking interaction, determines the structure, mode of replication and aspects of the evolution of these entities and their functions in the cell today. The analogous situation does *not* hold for the genetic code, i.e. it is almost certain that no strong and highly specific interactions between amino acids (or derivatives thereof) and simple oligonucleotides underlie the amino acid:codon relationship and its evolution (Zubay & Doty, 1958; Woese, 1967). (3) The translation apparatus is extremely complex; the complexity is evident not only in the number and kinds of parts which the apparatus contains, and in their inter-relationships, but also in the behaviour of the machine as a whole under various perturbations. In general, a simple machine, when perturbed sufficiently will cease functioning altogether, whereas a more complex machine will often show characteristic and intricate types of aberrant behaviour. The obvious reason for this is that a complex machine is generally composed of a simple basic machine(s) upon which has been superimposed a variety of interrelated devices which detect and correct errors, etc. Perturbations which affect these devices will not necessarily cause the machine to stop altogether. Instead, one may observe either characteristic aberrant responses or atavistic, simple or less accurate response. The idiosyncrasies of television sets provide an apt analogy here. The behaviour of the cell's translation apparatus when operating in too high a Mg^{2+} concentration, or in the presence of streptomycin, or when translating a message such as polybromuracil, certainly places this machine in the class of complex machines (Friedman & Weinstein, 1964; Davies, Gilbert & Gorini, 1964; Grunberg-Manago & Michelson, 1964).

Translation, then, seems to have evolved through many successive stages, in all likelihood starting from an initial process that was far simpler and far less accurate than its present counterpart. I have given the argument for this conjecture in more detail elsewhere (Woese, 1967). The essential points are as follows. The accuracy of translation, as we now find it, depends to a large extent on the characteristics of the activating enzymes which recognize the amino acids; these enzymes are proteins whose functions are obviously highly evolved. However, to have evolved these functions the cell needed an accurate translation apparatus from the first. But, in the absence of strong and specific interactions between amino acids and oligonucleotides, it is impossible for the cell to have such a translation apparatus. The only way out of this paradox is to conclude that initially translation was not an accurate process and that accuracy was evolved through cycles whereby the degree of ambiguity and error of the system were gradually decreased to the present acceptable values.

I picture the evolution of translation as occurring in three major stages. The first stage would be the establishment of the rudimentary mechanism, a machine which 'reads' nucleic acid 'tape' in modulo three, and which can position and polymerize amino acids in a rudimentary way (presumably through proto-*t*RNAs), etc. The machine probably contains no components that distinguish clearly among amino acids—i.e. like an activating enzyme, etc. 'Recognition' of amino acids by this machine is taken to be *almost* non-existent at this stage; any distinctions that *are* made being only between crudely defined *groups* of amino acids. A second stage would follow in which the actual codon assignments became established. This would involve a refinement of the original 'assignments' (which are large groups of amino acids assigned as a whole to corresponding *groups* of codons) into progressively smaller and therefore more numerous subgroup assignments, a process that is imagined to continue until any amino acid group assignment comprises no more than one amino acid. During this second stage improvements in the precision of translation will undoubtedly be occurring. The third and final stage would be a logical extension of the second. Once actual ambiguity in codon assignments (i.e. the 'group assignments') has been eliminated, the residuum, translation error, has to be decreased to an acceptable degree, e.g. a point where in 80 to 90 % of the cases the translation is perfect. In this final stage the 'fine control' regulation, including the error detection and correction devices, would evolve. For example, several considerations suggest that the interaction of codon and anticodon alone is not precise enough to account for the accuracy of protein synthesis. To account for this accuracy, an auxiliary mechanism has been postulated in which the ribosome is capable to some extent of reading the codon. On this basis the ribosome will assume one of perhaps four different allosteric configurations. The *t*RNAs, for their part, are divided into four (or thereabouts) comparable classes on the basis of the configurations of their anticodon loops. Each ribosome configuration permits only certain of the *t*RNA classes access to the codon. The classes are arranged so that the *t*RNAs which tend to misread a codon, i.e. produce a translation error, are not permitted access to that codon (Woese, 1969).

Universality of the set of codon assignments

Let us now review the evidence bearing upon universality of the genetic code and see whether there is any significant non-universality.

On the basis of the present evidence, codon assignments themselves are probably completely universal (Speyer *et al.* 1963; Sager, Weinstein & Ashkenazi, 1963; Marshall, Caskey & Nirenberg, 1967). An occasional 'exception' is reported, but in view of the overwhelming evidence which favours universality, such reports should remain suspect until they are established beyond doubt (Rifkin, Hirsch, Rifkin & Konigsberg, 1966).

The phenomena of non-sense and mis-sense suppression (via *t*RNA alterations) might be taken as examples of non-universality in codon assignments (Carbon, Berg & Yanofsky, 1966; Brenner & Beckwith, 1965; Brenner, Stretton & Kaplan, 1965; Brenner, Barnett, Katz & Crick, 1967; Garen, Garen & Wilhelm, 1965; Weigert, Lanka & Garen, 1967*a*, *b*). However, from present evidence these phenomena are evolutionary dead ends, i.e. they never lead to permanent codon assignment changes. At best they can be classed as ambiguous codon assignments; examples of some (perhaps all) of these types of suppression occur in eukaryotes and prokaryotes alike, and so the 'exceptions' themselves are universal.

A point of interest, related to the above, is the relative frequency of use of a codon. For reasons which are not clear certain of the synonymous codons for some of the amino acids tend to be used in preference to others. To give several examples: in *Escherichia coli* there appears to be preferential usage of the AAA and GAA codons for lysine and glutamic acid, respectively, over their AAG and GAG synonyms (Weigert, Gallucci, Lanka & Garen, 1966). This organism also seems to use the arginine AGA, AGG codons quite sparingly (Carbon *et al.* 1966; P. Berg, personal communication). On the other hand mammals, if not all eukaryotes, may use the arginine GGU, CGC, CGA and CGG codons sparingly (Josse, Kaiser & Kornberg, 1961; Woese, 1967; Subak-Sharpe, 1968). In some of these instances at least, the relative concentrations of the corresponding *t*RNAs reflect the frequency of codon usage (Marshall *et al.* 1967; Söll, Cherayil & Bock, 1967). As yet there is no certainty that synonym codon usage falls into different patterns in prokaryotes and eukaryotes.

Universality and the translation apparatus

The question of the degree of generality of the translation apparatus is not clear-cut. Certainly, what I have called 'trivial' and 'incidental' non-universality (above) are encountered, but there is still a question as to whether all the non-universality falls into these meaningless categories. Consider the *t*RNAs and the *t*RNA 'charging' reactions, in which the amino acids are combined with their respective *t*RNAs.

Primary structure analyses of eukaryotic and prokaryotic *t*RNAs reveal many striking similarities. The general secondary structure of the molecule (suggested by the common base-pairing patterns) appears to be approximately the same for all *t*RNAs, regardless of their source or their codon response patterns. Also, all *t*RNAs have CCA at their 3' ends; all have a GTψCG segment (ψ = pseudouridine) in the same place in the 'loop' (Zamir, Holley & Marquissee, 1965); all have the same general structure for their anticodon loop; and the base adjacent to the anticodon on the 3' side seems to be modified in nearly the same way in eukaryotes and prokaryotes in the few cases where this comparison has been made (Madison, Everett & Kung, 1966; RajBhandary *et al.* 1967; Burrows *et al.* 1968; Goodman *et al.* 1968; Zachau, Dutting & Feldman, 1966; Doctor, Doebel, Sodd & Winter, 1969). Some *t*RNAs such as the *t*RNA for serine (*t*RNA$_{ser}$) from yeast contain an additional small 'vestigial' loop that others do not (Zachau *et al.* 1968).

In two respects, however, one can begin to see what are probably real eukaryote-prokaryote differences in *t*RNAs. In the typical eukaryotic *t*RNA one of the loop regions is (so far) composed solely of purines and dihydrouridine (Holley *et al.* 1965; Zachau *et al.* 1966; Madison *et al.* 1966; RajBhandary *et al.* 1967; Bayev *et al.* 1967). The corresponding loop in the prokaryotes so far comprises almost exclusively purine and cytidylate bases (Goodman *et al.* 1968; Dube, Marcker, Clark & Cory, 1968; Doctor *et al.* 1969). If reports that the anticodon plays no detectable role in *t*RNA charging are correct, the loop in question is probably concerned with the interaction between *t*RNA and the activating enzyme (Stulberg & Isham, 1967; Chuguev, Axelrod & Bayev, 1969).

A second major difference between eukaryotic and prokaryotic *t*RNAs resides in the patterns of degeneracy in codon recognition. Eukaryotic *t*RNAs often possess an inosine residue in the III' position of the anticodon. Such *t*RNAs will consequently recognize three codons, i.e. those of the form XYU, XYC and XYA (Crick, 1966; Söll *et al.* 1966). In these cases one generally also finds a second *t*RNA that recognizes the remaining XYG codon for the amino acid in question (Crick, 1966; Söll *et al.* 1966). The prokaryotic pattern of codon recognition, on the other hand, tends toward having one *t*RNA species that recognizes XYU and XYC (presumably by virtue of having a G residue in the III' position of the anticodon), with a second *t*RNA recognizing the XYA and XYG codons (a U in the III' anticodon position). A *t*RNA of the type which recognizes XYG alone also occurs in prokaryotes (Söll *et al.* 1966, 1967; Kellogg, Doctor, Loebel & Nirenberg, 1966).

With regard to the *t*RNA charging reaction, the eukaryote:prokaryote comparison presents a confused picture. The universality of the *t*RNA charging reaction has been studied through the use of 'chimaera systems', reaction mixtures in which the activating enzyme preparation derives from one source, the *t*RNA preparation from another. When the sources are either both prokaryotic or both eukaryotic, one encounters few, if any, cases indicating non-universality (Sueoka, 1965). On the other hand, some sort of non-universality appears to be as much the rule as the exception when these chimaera systems are constructed with one component of eukaryotic, the other of prokaryotic origin (Sueoka, 1965). Almost all of the results indicating non-universality in these latter cases are negative evidence, i.e. failure of a chimaera system to perform a particular charging reaction. Such results may be no more than the 'incidental' non-universality as defined above. That this may be the case is indicated by unpublished studies of Barnett and coworkers, which show that the concentrations and ratios of small molecules such as Mg^{2+} and ATP are critical in chimaera systems involving eukaryotic and prokaryotic components, much more than in a normal homologous charging-system. In some cases, where proper concentrations of small molecules have been determined, these workers found that *t*RNAs not formerly considered to be chargeable (or only very slightly chargeable) in a eukaryote prokaryote chimaera system, could be charged perfectly well. Charging in these cases could not be attributed to contamination with mitochondrial enzymes or mitochondrial *t*RNAs (W. E. Barnett, personal communication).

Far more puzzling than the failure to charge in a chimaera system are the cases of what can be called 'false-positive' charging; these are instances (few in number) where the wrong amino acid is combined with a *t*RNA in a chimaera system. The best known case is that of an activating enzyme preparation from Neurospora that places phenylalanine on at least three species of *t*RNA from *Escherichia coli*: $tRNA_{phe}$, $tRNA_{val}$, $tRNA_{ala}$ (Barnett & Jacobson, 1964; Barnett & Epler, 1966). The two false-positive results can be attributed to phenylalanine-activating enzymes of the cytoplasmic fraction, while the correct charging is due to a phenylalanine activating enzyme of mitochondrial origin (Barnett, 1965; Barnett & Epler, 1966; Barnett, Brown & Epler, 1967). This result may be significant, given the probable origin of mitochondria. Mitochondria have recently been shown to contain *t*RNAs and activating enzymes for at least 18 of the amino acids. The *t*RNAs of mitochondria are distinguishable from normal cytoplasmic *t*RNAs, and the enzymes seem to be as distinct from prokaryotic enzymes as they are from the cytoplasmic enzymes (J. L. Epler, personal communication). It is possible that cases of false-positive charging are coincidences, in which the enzyme in one species has by chance evolved its *t*RNA-recognition site to fit the wrong *t*RNA in some other species. Such an explanation would suggest that the number of *t*RNA configurations *not* recognized by activating enzymes is not enormous, i.e. the number of possible configurations which the site on the *t*RNA molecule can assume is sufficiently small so that coincidences of the sort leading to false-positive charging can occur with appreciable frequency. A more intriguing possibility is that the cases of false-positive charging reflect some sort of atavistic recognition, involving parts of the molecule that may have been important in the past, but now do not function in the same way or do not function at all normally.

To consider the ribosome itself. One initially notices a rather high degree of non-universality in the primary structures of the various components within the prokaryotes and eukaryotes. However, the demonstration that a functional 30s ribosome can be reconstituted by using a 16s *r*RNA from one prokaryote and 30s ribosomal proteins from an entirely different prokaryote, makes it highly likely that most if not all of this sort of non-universality is of the trivial sort (Nomura *et al.* 1968). Attempts to reconstitute eukaryote + prokaryote ribosomal chimaeras of this sort have so far failed (Nomura, Traub & Bechman, 1968). Perhaps the apparent non-universality encountered in this latter instance is not entirely trivial.

The RNA : protein ratios and the *r*RNA sizes differ in prokaryotic and eukaryotic ribosomes. Prokaryotic *r*RNAs all seem to be of the 16s and 23s varieties, while eukaryotic *r*RNAs are (almost) always 18s and 25 to 28s (Osawa, 1965, 1968; Loening, 1968). The prokaryotic *r*RNA composition seems fixed at about 53 mole % GC, while the eukaryotic *r*RNA composition seems more variable (Midgley, 1962). Within the prokaryotes it appears possible to show some degree of primary structural homology between the *r*RNAs of most if not all species (Moore & McCarthy, 1967); but the eukaryote : prokaryote comparison reveals little or no such homology (Moore & McCarthy, 1967).

Differences in antibiotic sensitivity as between eukaryote and prokaryote (including mitochondrial, etc.) ribosomes are well known. In spite of the above differences, ribosomes from the one kind of cell will utilize charged *t*RNAs from the other kind in the translation process, indicating some very basic similarity (Weisblum, Benzer & Holley, 1962, 1965; Gonano, 1967). Work with this type of chimaera system has not been extensive, therefore one cannot be confident that the phenomenon is perfectly general. Also, whether such chimaera translation systems translate with anything like normal speed, and particularly normal accuracy, is still an open question.

Another line of evidence that suggests significant differences between the eukaryotic and prokaryotic translation systems comes from a comparison of the patterns of ribosome biosynthesis in the two cases. I assume it possible, and likely, that the biosynthetic patterns may to some degree reflect the evolution of the ribosome. In both prokaryotes and eukaryotes the genome contains multiple copies of the cistrons which code for the rRNA (Yankofsky & Spiegelman, 1962a, b, 1963; Perry, 1967). This multiplicity is generally assumed to be the cell's solution to the problem of making large quantities of these RNAs quickly. But here the prokaryote : eukaryote similarity ends. The functional organization of the multiple rRNA cistrons and the fate of the nascent rRNAs is strikingly different in the two instances. The eukaryote produces a 38 to 45s high molecular weight rRNA precursor as the apparent primary transcription product (Brown & Gurdon, 1964; Taber & Vincent, 1969). The genes producing these precursors may be grouped together into still larger operons however (Quagliarotti & Ritossa, 1968). The 38 to 45s precursor form of the rRNA is subsequently processed to yield the 18s and 25 to 28s mature forms of rRNA, and in some cases additional RNA of unknown significance (Weinberg, Loening, Willems & Penman, 1967). The 5s rRNA cistrons in eukaryotes are not located in the vicinity of the cistrons for the larger rRNAs (Brown & Weber, 1968). In the prokaryote *no* large molecular weight precursor is produced, although precursors of the 16s and 23s rRNAs which are very close to their mature counterparts in size, etc., do exist (Hecht & Woese, 1968). Although 16s and 23s rRNA cistrons appear to be clustered in the prokaryote genome, all evidence suggests that polycistronic organization of these cistrons does not occur (Oishi & Sueoka, 1965; Smith, Dubnau, Morrell & Marmur, 1968; Bleyman, Kondo, Hecht & Woese, 1969). On the other hand, each 5s rRNA cistron in the prokaryote appears to be a part of an operon containing also one 23s rRNA cistron. It is possible that the 5s rRNA and the mature (final) form of the 23s rRNA derive from the same precursor rRNA, a molecule slightly larger than the mature form of the 23s rRNA (Hecht, Bleyman & Woese, 1968; Bleyman, Kondo, Hecht & Woese, 1969).

CONCLUSIONS

From the above discussion it seems safe to conclude that the genetic code is universal with respect to its most basic features, e.g. codon assignments, the essential workings of the translation machine, etc. Nevertheless, the facts cause one to doubt whether universality extends to all the fine points of the genetic code. Admittedly, given our present knowledge, one cannot interpret the prokaryote : eukaryote differences unequivocally. Thus, we fall far short of the goal of using such differences to reveal features of the evolution of the cell. However, I feel it useful to comment on the existing data on the assumption that they do reflect cellular evolution. I will assume that the presumed divergence of a primitive cell-line into proto-prokaryote and proto-eukaryote lines might have occurred before the final steps in the evolution of the genetic code. In this context the differences in the patterns for ribosome biosynthesis in the two lines suggests that each line encountered the need for multiple copies of the rRNA cistrons independently, and each solved the problem in its own way. In other words, the prokaryote:eukaryote divergence occurred at a stage when the cell was simple (small, or slow) enough to function with one rRNA cistron per genome. The fact that the 5s rRNA cistrons are differently organized and expressed in the two lines may merely reflect the way in which multiplicity of rRNA cistrons evolved. Alternatively the two different organizations may reflect something deeper, namely a separate evolution of the 5s rRNA molecule in each line, suggesting that 5s rRNA was a relatively late arrival on the evolutionary scene. This alternative is consistent with there being no noticeable primary structural similarity between eukaryote and prokaryote 5s rRNAs (Brownlee, Sanger & Barrell, 1968; Forget & Weissman, 1967).

Further, the striking size differences between functionally equivalent rRNAs in the two lines, differences in general ribosome composition, in antibiotic sensitivity, all suggest that the final stages in the evolution of the genetic code may have occurred independently in the two lines. It is noteworthy here that the precursor form of the prokaryotic 16s rRNA actually manifests the physical characteristics of its eukaryotic counterpart, i.e. it is an 18s molecule (Osawa, 1965; Hecht & Woese, 1968). These final evolutionary stages should be those concerned with the development of the refinements which give translation its great precision, speed, etc. For example, it is well known that streptomycin affects the accuracy of translation, and that streptomycin resistance (if not its actual binding) resides in 30s protein (Davies, 1964; Cox, White & Flaks, 1964). However, streptomycin action is confined to prokaryotes, showing that the structures on the ribosome affected by streptomycin are different in kind, and so perhaps of independent evolutionary origin.

One of the features of the translation machine that differs most as between prokaryotes and eukaryotes is a particular 'loop' in the *t*RNA molecule—the loop mentioned above as probably concerned with *t*RNA-activating enzyme interaction. This difference may then underlie some of the negative or false-positive results obtained with eukaryote + prokaryote chimaera *t*RNA charging systems. If implications of such a difference are pushed to the limit, they suggest that activating enzymes (as we know them) were a relatively recent evolutionary development, and that to some extent their evolution occurred independently in the two cell lines. It is interesting that mitochondria—which seem to have arisen from an 'invasion' of a proto-eukaryote by some sort of pro-karyote—possess activating enzymes that as a group are no more like prokaryotic activating enzymes than they are like eukaryote cytoplasmic activating enzymes (see above). If true, this would support the thesis that the final stages in the evolution of activating enzymes occurred after the prokaryote:eukaryote divergence.

Some consideration should be given to the fact that the set of codon assignments (codon catalogue) is universal. The prime question is whether or not to attribute any real significance to this fact, i.e. whether this universality is an essential property of the genetic code, or reflects merely genetic drift, strong selection for one particular characteristic at a critical point in evolution, etc. Several models have been proposed to account for the origin of the codon assignments (Sonneborn, 1965; Woese, 1965; Crick, 1967, 1968). These fall into two classes, the stochastic and the deterministic models (Woese, 1967). Stochastic models of necessity assume that the final codon catalogue might be any one of a large number, and that during its evolution many versions of the codon catalogue existed, even simultaneously. Stochastic models, if required to account for an essential universality of the codon catalogue, are then forced to make special *ad hoc* assumptions, some of which border on the miraculous (Woese, 1967, 1969). On the basis of a deter-ministic model, however, a single *a priori* predictable set of codon assignments is possible, i.e. all cell lines would head for the same final codon catalogue. The existence of a universal codon catalogue certainly does not, however, argue for a deterministic evolution to the codon catalogue, given that we are not sure at present that universality is an essential property of this catalogue.

The above comprises a highly biased view of the significance of the differences (or lack thereof) in the genetic code as between prokaryotes and eukaryotes. While this degree of idiosyncratic interpretation is not justifiable on purely scientific grounds, I feel it justifiable on didactic grounds: it is an attempt to outline some of the basic problems of evolution that science will have to face. The alternative to such an approach would be a bland recital of largely uninterpretable facts and useless generalities. It is unlikely that the above interpretations will turn out to be correct in all or even most instances. They may, nevertheless, together with other discussions in this Symposium, serve as an impetus for interest and experimentation about the evolution of the cell.

REFERENCES

BARNETT, W. E. (1965). Interspecies aminoacyl-sRNA formation: Fractionation of Neurospora enzymes involved in anomalous aminoacylation. *Proc. natn. Acad. Sci. U.S.A.* **53**, 1462.

BARNETT, W. E., BROWN, D. H. & EPLER, J. L. (1967). Mitochondrial-specific aminoacyl-RNA synthetases. *Proc. natn. Acad. Sci. U.S.A.* **57**, 1775.

BARNETT, W. E. & EPLER, J. L. (1966). Fractionation and specificities of two aspartyl-ribonucleic acid and two phenylalanyl-ribonucleic acid synthetases. *Proc. natn. Acad. Sci. U.S.A.* **55**, 184.

BARNETT, W. E. & JACOBSON, K. B. (1964). Evidence for degeneracy and ambiguity in interspecies aminoacyl-sRNA formation. *Proc. natn. Acad. Sci. U.S.A.* **51**, 642.

BAYEV, A. A., VENKSTERN, T. V., MIRSABEKOV, A. D., KRUTILINA, A. I., LI, L. & AXELROD, V. D. (1967). *Mol. Biol. USSR* **1**, 754.

BLEYMAN, M. A., KONDO, M., HECHT, N. B. & WOESE, C. (1969). Transcriptional mapping: regarding the functional organization of the ribosomal and transfer ribonucleic acid cistrons in the *Bacillus subtilis* genome. *J. Bact.* (In press.)

BRENNER, S., BARNETT, L., KATZ, E. R. & CRICK, F. H. (1967). UGA: a third nonsense triplet in the genetic code. *Nature, Lond.* **213**, 449.

BRENNER, S. & BECKWITH, J. R. (1965). Ochre mutants, a new class of suppressible nonsense mutants. *J. molec. Biol.* **13**, 1499.

BRENNER, S., STRETTON, A. O. W. & KAPLAN, S. (1965). Genetic code: The 'nonsense' triplets for chain termination and their suppression. *Nature, Lond.* **206**, 994.

BROWN, D. D. & GURDON, J. B. (1964). Absence of ribosomal RNA synthesis in the anucleate mutant of *Xenopus laevis*. *Proc. natn. Acad. Sci. U.S.A.* **51**, 139.

BROWN, D. D. & WEBER, C. S. (1968). Gene linkage by RNA–DNA hybridization. I. Unique DNA sequences homologous to 4s RNA, 5s RNA, and ribosomal RNA. *J. molec. biol.* **34**, 661.

BROWNLEE, G. G., SANGER, F. & BARRELL, B. G. (1968). The sequence of 5s ribosomal ribonucleic acid. *J. molec. Biol.* **34**, 379.

BURROWS, J., ARMSTRONG, D. J., SKOOG, F., HECHE, S. M., BOYLE, J. T. A., LEONARD, H. J. & OCCOLOWITZ, J. (1968). *Science, N.Y.* **161**, 691.

CARBON, J., BERG, P. & YANOFSKY, C. (1966). Studies of missence suppression of the tryptophan synthetase A-protein mutant A36. *Proc. natn. Acad. Sci. U.S.A.* **56**, 764.

CHUGUEV, I. I., AXELROD, V. D. & BAYEV, A. A. (1969). The role of anticodon in the acceptor function of *t*RNA. *Biochem. biophys. Res. Commun.* **34**, 348.

Cox, E. C., White, J. R. & Flaks, J. G. (1964). Streptomycin action and the ribosome. *Proc. natn. Acad. Sci. U.S.A.* **51**, 703.

Crick, F. H. C. (1966). Codon-anticodon pairing: the wobble hypothesis. *J. molec. Biol.* **19**, 548.

Crick, F. H. C. (1967). Origin of the genetic code. *Nature, Lond.* **213**, 119.

Crick, F. H. C. (1968). The origin of the genetic code. *J. molec. Biol.* **38**, 367.

Davies, J. E. (1964). Studies on the ribosomes of streptomycin-sensitive and resistant strains of *Escherichia coli. Proc. natn. Acad. Sci. U.S.A.* **51**, 659.

Davies, J., Gilbert, W. & Gorini, L. (1964). Streptomycin suppression and the code. *Proc. natn. Acad. Sci. U.S.A.* **51**, 883.

Doctor, B. P., Doebel, J. E., Sodd, M. A. & Winter, D. B. (1969). Nucleotide sequence of *Escherichia coli* tyrosine transfer ribonucleic acid. *Science, N.Y.* **163**, 693.

Dube, S. K., Marcker, K. A., Clark, B. F. C. & Cory, S. (1968). Nucleotide sequence of N-formyl-methionine-transfer RNA. *Nature, Lond.* **218**, 232.

Fitch, W. M. & Margoliash, E. (1966). Construction of phylogenetic trees. *Science, N.Y.* **155**, 279.

Forget, B. G. & Weissman, S. B. (1967). Nucleotide sequence of KB cell 5s RNA. *Science, N.Y.* **158**, 1695.

Friedman, M. & Weinstein, I. B. (1964). Lack of fidelity in the translation of synthetic polyribonucleotides. *Proc. natn. Acad. Sci. U.S.A.* **52**, 988.

Garen, A., Garen, S. & Wilhelm, R. C. (1965). Suppressor genes for nonsense mutations. I. Su-1, Su-2, Su-3 genes of *Escherichia coli. J. molec. Biol.* **14**, 167.

Gonano, F. (1967). Specificity of serine transfer ribonucleic acids in the synthesis of haemoglobin. *Biochemistry, N.Y.* **6**, 977.

Goodman, H. M., Abelson, J., Landy, S., Brenner, S. & Smith, J. S. (1968). Amber suppression; a nucleotide change in the anticodon of a tyrosine transfer RNA. *Nature, Lond.* **217**, 1019.

Grunberg-Manago, M. & Michelson, A. (1964). Polynucleotide analogues. II. Stimulation of amino acid incorporation by polynucleotide analogues. *Biochim. biophys. Acta* **80**, 431.

Hecht, N. B., Bleyman, M. & Woese, C. R. (1968). The formation of 5s ribosomal ribonucleic acid in *Bacillus subtilis* by posttranscriptional modification. *Proc. natn. Acad. Sci. U.S.A.* **59**, 1278.

Hecht, N. B. & Woese, C. R. (1968). Separation of bacterial ribosomal ribonucleic acid from its macromolecular precursors by polyacrylamide gel electrophoresis. *J. Bact.* **95**, 986.

Holley, R. W., Apgar, J., Everett, G. A., Madison, J. T., Marquissee, M., Merrill, S. H., Penswick, J. R. & Zamir, A. (1965). Structures of a ribonucleic acid. *Science, N.Y.* **147**, 1462.

Josse, J., Kaiser, A. D. & Kornberg, A. (1961). Enzymatic synthesis of deoxyribonucleic acid VIII. Frequencies of nearest neighbour base sequences in deoxyribonucleic acid.

Kellogg, D. A., Doctor, B. P., Loebel, J. E. & Nirenberg, M. W. (1966). RNA codons and protein synthesis. 1s. synonym codon recognition by multiple species of valine-, alanine-, and methionine-sRNA. *Proc. natn. Acad. Sci. U.S.A.* **55**, 912.

Loening, U. E. (1968). Molecular weights of ribosomal RNA in relation to evolution. *J. molec. Biol.* **38**, 365.

Loftfield, R. B. (1963). Errors in protein synthesis. *Biochem. J.* **89**, 82.

Madison, J. T., Everett, G. A. & Kung, H. (1966). Nucleotide sequence of a yeast tyrosine transfer RNA. *Science, N.Y.* **153**, 531.

MARSHALL, R. E., CASKEY, C. T. & NIRENBERG, M. (1967). Fine structure of RNA codewords recognized by bacterial, amphibian, and mammalian transfer RNA. *Science, N.Y.* **155**, 820.

MIDGLEY, J. T. A. (1962). The base composition of RNA from several microbial species. *Biochim. biophys. Acta* **61**, 513.

MOORE, R. L. & McCARTHY, B. J. (1967). Comparative study of ribosomal ribonucleic acid cistrons in enterobacteria and mycobacteria. *J. Bact.* **94**, 1066.

NOMURA, M., TRAUB, P. & BECHMAN, H. (1968). Hybrid 30s ribosomal particles reconstituted from components of different bacterial origins. *Nature, Lond.* **219**, 793.

OISHI, M. & SUEOKA, N. (1965). Location of genetic loci of ribosomal RNA on *Bacillus subtilis* chromosome. *Proc. natn. Acad. Sci. U.S.A.* **54**, 483.

OSAWA, S. (1965). Biosynthesis of ribosomes in bacterial cells. *Prog. Nuc. Acid Res. Molec. Biol.* **4**, 161.

OSAWA, S. (1968). Ribosome formation and structure. *Ann. Rev. Biochem.* **37**, 109.

PERRY, R. P. (1967). The nucleolus and the synthesis of ribosomes. *Prog. Nuc. Acid Res. Molec. Biol.* **6**, 219.

QUAGLIAROTTI, G. & RITOSSA, F. M. (1968). On the arrangement of genes for 28s and 18s ribosomal RNAs in *Drosophila melanogaster*. *J. molec. Biol.* **36**, 57.

RAJBHANDARY, U. L., CHANG, S. H., STUART, A., FAULKNER, R. D., HOSKINSON, M. & KHORANA, H. G. (1967). Studies on polynucleotides LXVIII. The primary structure of yeast phenylalanine transfer RNA. *Proc. natn. Acad. Sci. U.S.A.* **57**, 751.

RIFKIN, D. B., HIRSCH, D. I., RIFKIN, M. R. & KONIGSBERG, W. (1966). A possible ambiguity in the coding of mouse haemoglobin. *Cold Spring Harb. Symp. quant. Biol.* **31**, 715.

SAGER, R., WEINSTEIN, I. B. & ASHKENAZI, Y. (1963). Coding ambiguity in cell-free extracts of *Chlamydomonas*. *Science, N.Y.* **140**, 304.

SMITH, I., DUBNAU, D., MORRELL, P. & MARMUR, J. (1968). Chromosomal location of DNA complementary to transfer RNA, and to 5s, 16s, and 23s ribosomal RNA in *Bacillus subtilis*. *J. molec. Biol.* **33**, 123.

SÖLL, D., CHERAYIL, J. D. & BOCK, R. M. (1967). Studies on polynucleotides. LXXV. Specificity of *t*RNA for codon recognition as studied by amino acid incorporation. *J. molec. Biol.* **29**, 97.

SÖLL, D., JONES, D. S., OHTAKA, E., FAULKNER, R. D., LOHRMANN, R., HAYATSU, H. & KHORANA, H. G. (1966). Specificity of *s*RNA recognition of codons as studied by the ribosomal binding technique. *J. molec. Biol.* **19**, 556.

SÖLL, D. & RAJBHANDARY, U. L. (1967). Studies on polynucleotides. LXXVI. Specificity of transfer RNA for codon recognition as studied by amino acid incorporation. *J. molec. Biol.* **29**, 113.

SONNEBORN, T. M. (1965). Degeneracy of the genetic code: Extent, nature, and genetic implications. In *Evolving Genes and Proteins*, p. 377. Eds. B. Bryson and H. Vogel. London and New York: Academic Press.

SPEYER, J., LENGYEL, P., BASILIO, C., WAHBA, A., GARDNER, R. & OCHOA, S. (1963). Synthetic polynucleotides and the amino acid code. *Cold Spring Harb. Symp. quant. Biol.* **28**, 559.

STULBERG, M. P. & ISHARM, K. R. (1967). Studies on the locus of the enzyme recognition site in phenylalanine transfer RNA. *Proc. natn. Acad. Sci. U.S.A.* **57**, 1310.

SUBAK-SHARPE, H. (1968). Virus-induced changes in translation mechanisms. *Symp. Soc. gen. Microbiol.* **18**, 47.

SUEOKA, N. (1965). In *Evolving Genes and Proteins*, p. 479. Eds. V. Bryson and H. Vogel. London and New York: Academic Press.

Szer, W. & Ochoa, S. (1964). Complexing ability and coding properties of synthetic polynucleotides. *J. molec. Biol.* **8**, 823.

Taber, R. L., Jr. & Vincent, W. S. (1969). Effects of cycloheximide on ribosomal RNA synthesis in yeast. *Biochem. biophys. Res. Commun.* **34**, 488.

Weigert, M. G., Gallucci, E., Lanka, E. & Garen, A. (1966). Characteristics of the genetic code *in vivo*. *Cold Spring Harb. Symp. quant. Biol.* **31**, 145.

Weigert, M. G., Lanka, A. & Garen, A. (1967a). Base composition of nonsense codons in *Escherichia coli*. II. The N2 codon, UAA. *J. molec. Biol.* **23**, 391.

Weigert, M. G., Lanka, A. & Garen, A. (1967b). Amino acid substitutions resulting from suppression of nonsense mutations. III. Tyrosine insertion by the Su-4 gene. *J. molec. Biol.* **23**, 401.

Weinberg, R. A., Loening, U., Willems, M. & Penman, S. (1967). Acrylamide gel electrophoresis of hela cell nucleolar RNA. *Proc. natn. Acad. Sci. U.S.A.* **58**, 1088.

Weisblum, B., Benzer, S. & Holley, R. W. (1962). A physical basis for degeneracy in the amino acid code. *Proc. natn. Acad. Sci. U.S.A.* **48**, 1449.

Weisblum, B. S., Gonano, F., Ehrenstein, von G. & Benzer, S. (1965). A demonstration of coding degeneracy for leucine in the synthesis of protein. *Proc. natn. Acad. Sci. U.S.A.* **53**, 329.

Woese, C. (1965). On the evolution of the genetic code. *Proc. natn. Acad. Sci. U.S.A.* **54**, 1546.

Woese, C. (1967). *The Genetic Code: The Molecular Basis of Genetic Expression.* New York: Harper and Row.

Woese, C. (1969a). Models for the evolution of codon assignments. *J. molec. Biol.* **43**, 235.

Woese, C. (1969b). Concerning the accuracy of the translation process. *J. theor. Biol.* **25**.

Yankofsky, S. A. & Spiegelman, S. (1962a). The identification of the ribosomal RNA cistron by sequence complementarity I. Specificity of complex formation. *Proc. natn. Acad. Sci. U.S.A.* **48**, 1069.

Yankofsky, S. A. & Spiegelman, S. (1962b). The identification of the ribosomal RNA cistron by sequence complementarity II. Saturation of and competitive interaction at the RNA cistron. *Proc. natn. Acad. Sci. U.S.A.* **48**, 1466.

Yankofsky, S. A. & Spiegelman, S. (1963). Distinct cistrons for the two ribosomal RNA components. *Proc. natn. Acad. Sci. U.S.A.* **49**, 539.

Zachau, von H. G., Dutting, D. & Feldman, H. (1966). *Angew. Chem.* **78**, 392.

Zamir, A., Holley, R. W. & Marquissee, M. (1965). Evidence for the occurrence of a common pentanucleotide sequence in the structures of transfer RNAs. *J. biol. Chem.* **240**, 1267.

Zubay, G. & Doty, P. (1958). Nucleic acid interactions with metal ions and amino acids. *Biochim. biophys. Acta* **29**, 47.

A COMPARISON OF RIBOSOMES FROM PROKARYOTES AND EUKARYOTES

H. G. WITTMANN

Max-Planck-Institut für Molekulare Genetik, Berlin-Dahlem, Germany

Ribosomes play a central role in the biosynthesis of proteins which is an essential and complicated process in all organisms. Because of the importance of ribosomes and their complex mode of action many studies have been made of their structural and functional properties. In most studies ribosomes from *Escherichia coli* have been used. Differences and similarities have been found by comparing the structure and function of ribosomes from different sources. The differences are especially distinct between ribosomes from prokaryotes (bacteria, blue-green algae), which do not have a nucleus, and eukaryotes, i.e. organisms with a true nucleus. In this article a comparison of the chemical, physical, serological and functional properties of ribosomes and their components from different organisms will be given.

RIBOSOMAL PARTICLES

The ribosomes of all organisms studied to date can be divided according to their size into two main groups: (1) those of prokaryotes which have a sedimentation value of about 70s; (2) those of eukaryotes which have a sedimentation value of 70s in certain cell organelles (mitochondria, chloroplasts) and 80s in the cytoplasm. In all organisms studied the ribosomes consist of two unequal sub-units. 70s ribosomes have sub-units with sedimentation coefficients of about 50s and 30s, and the sub-units of 80s ribosomes sediment at about 60s and 40s. This suggests that two sub-units differing in size are a necessary requirement for the function of ribosomes. Taylor & Storck (1964) compared the sedimentation behaviour of ribosomes from 25 bacterial species, two species of blue-green algae and 26 species of fungi, under the same experimental conditions. They found 68·4s as the mean value for the '70s' ribosomes, and 81·3s for the '80s' ribosomes. The difference between the mean values is statistically significant. The cytoplasmic ribosomes of other eukaryotes, e.g. protozoa, algae, higher plants, invertebrates, vertebrates, also belong to the class of 80s ribosomes (data collected by Peterman, 1964).

The cytoplasmic ribosomes of the flagellate *Euglena gracilis* were first considered to be an exception to the rule that ribosomes from the eukaryotic cytoplasm are 80s. It was reported by Brawerman & Eisenstadt (1964) and by Gnanam & Kahn (1967) that they belonged to the 70s class, but it was later shown by Rawson & Stutz (1968) that these ribosomes were 80s, as expected. A report that cytoplasmic ribosomes of the alga *Acetabularia* sedimented at 82s (Janowski, 1966) was later retracted (Janowski, Bonotto & Boloukhere, 1969). The new sedimentation value is 70s and identical with that of the chloroplast ribosomes of this organism. This case is interesting and further confirmation is desirable before *Acetabularia* is accepted as an exception to the rule. Although the sedimentation values of cytoplasmic ribosomes are the same for all eukaryotes, these ribosomes can be further divided into two subgroups according to the size of the ribosomal RNAs: the sum of the molecular weights of the *r*RNAs is higher for the cytoplasmic ribosomes of animals than for those of plants. Therefore, three classes of ribosomes have been found: of prokaryotes, of plants (including fungi) and of animals.

In spite of the difference in size of 70s and 80s ribosomes, the X-ray diffraction patterns of ribosomes from prokaryotes and eukaryotes (*Drosophila*, rat, rabbit) have certain similarities which indicate common substructural features. Langridge (1969) suggested that part of the RNA in all of these ribosomes is in the form of 4 to 5 parallel double helices, 45 to 50 Å apart.

Whilst there are considerable differences in the immunologically detectable surface structures of ribosomes from different organisms (e.g. various mammals), no tissue specificity (e.g. in liver, kidney) of the antigenic determinants has been found (Noll & Bielka, 1969). This was confirmed by comparative electrophoresis of ribosomal proteins from different tissues and species (Bielka & Welfle, 1968).

Mitochondria and chloroplasts are interesting because in these organelles ribosomes of the 70s type occur. The first evidence for the occurrence of 70s ribosomes in certain organelles of eukaryotes was the finding by Lyttleton (1962) that two kinds of ribosomes were present in plants, namely the 70s type in the chloroplasts and the 80s type in the cytoplasm. This result has been extended by several groups of workers to higher plants and algae.

Not only the size of the undissociated particles but also that of their sub-units and of their *r*RNA molecules is the same in bacterial and chloroplast ribosomes (Stutz & Noll, 1967; Loening & Ingle, 1967). From this concurrence, a phylogenetic relationship between these two types of ribosomes can be inferred. This assumption has been strengthened by further findings, namely that the same Mg^{2+} ion concentration is required both for reversible dissociation into sub-units and for optimal amino acid incorporation in a cell-free system (Boardman, Francki & Wildman, 1966) and that they have a similar degree of sensitivity to certain antibiotics, e.g. chloramphenicol and cycloheximide which inhibit the functions of 70s and 80s ribosomes, respectively.

Serological tests showed that there is no detectable relationship between ribosomes from chloroplasts of higher plants, and those of bacteria and blue-green algae. Furthermore, although chloroplast ribosomes are related to the cytoplasmic ribosomes of the same plant and related species (e.g. bean and pea), they are only weakly related to chloroplast ribosomes of other plant families. Moreover chloroplast ribosomes are much more weakly related to each other than are the cytoplasmic ribosomes of these plants. Cytoplasmic ribosomes of all higher plants examined are serologically related to each other and have at least one band in common on gel electrophoretograms. Also, the cytoplasmic ribosomes of plants belonging to different families are immunologically more closely related than ribosomes of bacteria belonging to different families. There is no immunological relationship between cytoplasmic ribosomes of higher plant on the one hand and those of yeast or rabbit reticulocytes on the other (Janda & Wittmann, 1968; G. Stöffler, H.-G. Janda & H. G. Wittmann, to be published).

In addition to chloroplasts, mitochondria have ribosomes which differ in size and other properties from the corresponding cytoplasmic ribosomes. The mitochondrial ribosomes which have been most studied are those of fungi and mammalian tissues. Rifkin, Wood & Luck (1967) found values of 81s, 61s and 47s for the sedimentation values of mitochondrial ribosomes and their sub-units from *Neurospora crassa*; Küntzel & Noll (1967) reported 73s for these ribosomes and Küntzel (1969) reported the sedimentation coefficients for the sub-units to be 50s and 37s. The small sub-units of both the mitochondrial and cytoplasmic ribosomes have a sedimentation value of 37s.

The mitochondrial ribosomes require a higher Mg^{2+} concentration for their structural integrity than the cytoplasmic ribosomes (Rifkin *et al.* 1967; Küntzel, 1969). As discussed later, the mitochondrial ribosomes differ from the corresponding cytoplasmic ribosomes both in their RNA and protein structures and in their response to certain antibiotics.

Size estimations of the mitochondrial ribosomes from mammalian tissues by different workers (Truman, 1963; Elaev, 1964; Rabinowitz, *et al.* 1966; O'Brien & Kalf, 1967) disagree even more than the estimations for the mitochondrial ribosomes of fungi. Because of the experimental difficulties in isolating ribosomes from mammalian mitochondria it is possible, as suggested by Rifkin *et al.* (1967) and Georgatsos & Papasarantopoulou (1968), that the 'mitochondrial' ribosomes may be cytoplasmic ribosomes or their large sub-units. However, there is certainly a difference between the ribosomes of the cytoplasm and those of the mitochondria of mammals (see below).

RIBOSOMAL RNA

Size

The interesting question as to whether the size of ribosomal RNAs from various organisms differ in size is difficult to answer by the comparison of data obtained in various laboratories under dissimilar conditions. However, comparisons have been made under identical experimental conditions in some laboratories, either by sedimentation analyses (Stutz & Noll, 1967; Taylor, Glasgow & Storck, 1967; Click & Tint, 1967) or by electrophoresis in polyacrylamide gel (Loening, 1968). According to these analyses the organisms investigated can be divided into the following three classes.

(1) Prokaryotes (bacteria, actinomycetes, blue-green algae) and certain organelles (chloroplasts, mitochondria) of eukaryotes have *r*RNAs with molecular weights of $5 \cdot 6 \times 10^5$ and $1 \cdot 1 \times 10^6$ in the small and large sub-unit respectively. Ribosomal RNAs of these sizes have been found in all bacterial species so far investigated (Stanley & Bock, 1965; Taylor *et al.* 1967). The report that there is no 23s component in ribosomes of *Rhodopseudomonas spheroides* (Lessie, 1965) is invalid when the extraction conditions used inhibit RNase activity (Szilagyi, 1968). Furthermore, the report that the 16s component but not the 23s RNA can be isolated from ribosomes of chloroplasts of higher plants (Spencer & Whitfeld, 1966) was shown to be erroneous by Stutz & Noll (1967), Ruppel (1968) and Loening (1969). Ribosomal RNAs isolated from mitochondria of fungi (Wintersberger, 1967; Rogers, Preston, Titchener & Linnane, 1967; Küntzel & Noll, 1967; Rifkin *et al.* 1967; Dure, Epler & Barnett, 1967; Küntzel, 1969) and mammals (Dubnin & Brown, 1967; Kroon, 1968) differ distinctly in size from the RNAs of the corresponding cytoplasmic ribosomes but cannot be distinguished from the ribosomal RNAs of bacteria.

(2) Cytoplasmic ribosomes of organisms belonging to the plant kingdom (fungi, algae, ferns, higher plants, some protozoa) contain

rRNAs with molecular weights of 7.0×10^5 and 1.3×10^6, values which are 20 % higher than those for the prokaryotes (Loening, 1968). There is no doubt that the rRNA of the large sub-unit (25s) of plants is bigger than that of bacteria (23s). However, doubt still persists about the relative sizes of the small sub-units: whereas Stutz & Noll (1967) and Click & Tint (1967) reported the same size for the RNA of the small sub-units of both prokaryotes and eukaryotes, Taylor *et al.* (1967) and Loening (1968) found distinct size differences. This disagreement however may be due to the use of different methods (sucrose gradients, analytical boundary centrifugation, gel electrophoresis). By using the size of the rRNA as a criterion, protozoa are not all alike: the rRNA of *Tetrahymena* corresponds in size to that of plants whereas the rRNA of *Euglena* and *Amoeba* is markedly larger, with molecular weights of 9×10^5 and 1.5×10^6 for the rRNA of the sub-units (Loening, 1968; Rawson & Stutz, 1968). For *Paramecium* the reports disagree: according to Loening (1969) both rRNAs are as large as in plants, whereas Reisner, Rowe & MacIndoe (1968) considered that this was true only for the large rRNA and that the small one is as large as in mammals.

(3) In animals the small rRNA corresponds in molecular weight (7×10^5) to that of all other eukaryotes, with the exception of some protozoa. The large rRNA increases in size from the invertebrates (e.g. 1.4×10^6 in *Drosophila*) and the amphibians (1.5×10^6 in *Xenopus*) to the mammals (1.75×10^6), according to Loening (1969). A larger RNA in animals than in plants was detected by Click & Tint (1967).

It is interesting that there has been strong conservation of the size of rRNA within bacteria and plants, whereas in animals the size of the larger RNA increased during evolution; the reason for this is not yet understood.

Nucleotide composition and sequence

Analyses of the base compositions of rRNAs (Miura, 1962; Midgley, 1962) and of total RNA of bacteria (Belozersky & Spirin, 1960) reveal a small variation range for rRNAs compared with the wide variation in the composition of bacterial DNAs. Distinct differences in the rRNA base composition can in general only be found between bacterial groups which are taxonomically unrelated. The base composition of rRNA from bacteria differing greatly in the GC content of their DNAs shows only a very slight correlation with that of the DNA (Midgley, 1962). Amaldi (1969) concluded from the literature on the base compositions of rRNAs of the small and large sub-units isolated from various phyla that there is a definite correlation between the base compositions of the two rRNA components: the $G + C : A + U$ ratio which differs greatly in the various groups of organisms is always the same for both sub-units. The reasons for this are not understood.

Distinct differences between the base compositions of rRNAs from various classes of organisms are inferred from studies on the rRNAs of bacteria, fungi, higher plants, invertebrates and mammals. Also distinct differences in the base composition of rRNAs from cytoplasmic and mitochondrial ribosomes of the same organism have been found, e.g. *Neurospora crassa* (Pollard, Stemler & Blaydes, 1966; Rifkin *et al.* 1967; Küntzel & Noll, 1967).

Comparison of the terminal nucleotide sequences of rRNAs from ribosomal sub-units of various groups of organisms (bacteria, fungi, higher plants, mammals) reveals four facts (Sugiura & Takanami, 1967; Madison, 1968): (1) All the rRNAs are phosphorylated at their 5'-terminal end. (2) All four nucleoside diphosphates are present at the 5'-terminus of the large RNA of various organisms, but only adenine and uracil have been found at the 5'-end of the small rRNAs. (3) In contrast to the 5'-terminus there is far less variation of base sequence at the 3'-end in the rRNAs of the investigated species. However, these results have been obtained from a too few species to justify generalization. (4) The 5'-terminal oligonucleotides of the small and large rRNAs are identical within three species of *Bacillus* but they vary between different bacterial families.

Studies of the partial nuclease digestion products of rRNAs from various groups of organisms reveal a surprisingly strong conservation of the structure of the rRNAs during evolution (Gould, Bonanou & Kanagalingham, 1966; Pinder, Gould & Smith, 1969). There is a general similarity in the gel electrophoresis patterns between bacterial species (especially for those belonging to the same family), between different fungi and between various mammals. Few similarities were found between the gel electrophoresis patterns from bacteria, fungi and animals.

Nomura, Traub & Bechman (1968) concluded from 30s sub-unit reconstitution experiments on different bacterial ribosomes that some functional property of the small rRNA has been conserved in evolution, namely its ability to interact successfully with sets of 30s proteins derived from a fairly wide range of bacterial species. This is compatible with the known conservation of part of the base sequence among various bacterial species. It is possible that only certain small regions, namely the conserved regions of the small rRNA, are directly involved in the specific interaction with ribosomal proteins in the assembly of ribosomal particles. Non-conserved regions may not be involved in the direct protein interaction necessary for physical assembly, yet may have important functions in the completed ribosomal particle.

As already mentioned, ribosomes contain a 5s RNA component with 120 nucleotides besides the two other rRNAs. The complete nucleotide sequence of this RNA from *Escherichia coli* (Brownlee, Sanger & Barrell, 1967) and *E. coli* KB (Forget & Weissmann, 1967) has been determined. The two RNAs differ in length by one nucleotide. There are few obvious similarities, except for a tetranucleotide near the 3′-end. Whether other sequence homologies are significant or random remains uncertain.

Hybridization

The extent of sequence homology between rRNAs and DNA of the same and other species was tested by DNA + RNA-hybridization techniques for several organisms from various phyla. Most of these studies have been done with bacteria, especially those which differ in the GC content of their DNA (Yankofsky & Spiegelman, 1963; Doi & Igarashi, 1965; Dubnau, Smith, Morrel & Marmur, 1965; Attardi, Huang & Kabat, 1965; Moore & McCarthy, 1967; Takahashi, Saito & Ikeda, 1967; Nomura *et al.* 1968; Pinder *et al.* 1969). Although the results of these investigations do not agree in all details the following two conclusions can be drawn. (1) The cistrons for rRNAs have a common primary structure leading to cross-hybridization between DNA and rRNA from different bacterial species. This conservation of rRNA sites during the evolution of bacteria is probably due to the basic importance of ribosomes in protein biosynthesis. (2) The nucleotide sequence within the cistrons for rRNAs is conserved more than that of other cistrons, e.g. in mRNAs. This finding also accords with the similarities in the base compositions and the constancy of the molecular weights of rRNAs in bacteria.

It follows from the studies on the reconstitution of 30s particles from RNA and proteins of various bacterial species (Nomura *et al.* 1968) that a large part of the base sequences in the small rRNAs can be different in various species. This has a marked effect on the reconstitution process, i.e. the ability of the RNA to recognize ribosomal proteins and to interact with them in a specific manner. The regions which interact with the proteins should have certain base sequences either alike or very similar even in rRNAs of different families. They should therefore be conserved in evolution, whereas other regions not essential for the assembly or function of ribosomes might vary between species to give the differences in hybridization and gel electrophoresis patterns after partial nuclease digestion, referred to above.

However, the conservation of base sequences is not absolute. Sequence differences between rRNAs of even closely related bacterial species are revealed by the heterologous DNA + rRNA cross-hybridization. These differences are large between remotely related species. In general there is relatively good agreement between the taxonomical relationship, or the GC content of the DNA, on one hand and the results obtained by the DNA + DNA hybridization technique on the other. But there is only a poor agreement between the results of hybridization experiments with DNA + DNA on one hand, and DNA + rRNA on the other as shown for many bacterial species.

Not only the extent of DNA + rRNA cross-hybridization but also the stability of the hybrids at various temperatures is important for determining the degree of relationship. The stability of the hybrids depends very much on the partners. In general when there is a high degree of hybridization the rRNA hybrids are very stable. However, taxonomically closely related species sometimes give hybrids with a low degree of cross-reaction but a very high stability. This indicates that some cistrons for rRNA are very similar in base sequences between related bacteria whereas other cistrons are quite different (Moore & McCarthy, 1967).

The question whether there are similar sequences in the rRNAs from the small and large sub-units has been investigated by using hybridization techniques; the results depend on the bacterial species. Extensive cross-hybridization between the 16s and 23s rRNAs of *Escherichia coli* has been found (Attardi *et al.* 1965), indicating an evolutionary relationship between them, but no hybridization between the rRNA sub-units was observed with some *Bacillus* species (Yankofsky & Spiegelman, 1963; Doi & Igarashi, 1966). These apparently contradictory results were confirmed for *E. coli* and *B. subtilis* by Mangiarotti, Apirion, Schlessinger & Silengo (1968). Further experiments by using nucleic acids from eukaryotes are discussed below.

The intergeneric DNA + rRNA hybridization tests with bacteria always show a relatively strong cross-reaction as compared to the very low hybridization values obtained between bacteria on one hand and fungi, higher plants or animals on the other. Only the DNA of fungi seems to give a degree of hybridization with rRNA of *E. coli* which is above the background (Attardi *et al.* 1965).

The number of cistrons for rRNA is distinctly higher in eukaryotes than in prokaryotes. 130 to 140 cistrons have been found for the rRNA of yeast (Retel & Planta, 1968; Schweizer, Mackechnie & Halvorson, 1969), 100 to 200 for pea (Chipchase & Birnstiel, 1963; Trewavas & Gibson, 1968), 130 for *Drosophila* (Ritossa, Catwood, Lindsley &

Spiegelman, 1966), 600 to 800 for *Xenopus* tadpoles (Wallace & Birnstiel, 1966), 100 to 200 for chicken (Merits, Schulze & Overby, 1966) and 130 to 300 for human cells (Attardi *et al.* 1965; McConkey & Hopkins, 1964). The highest numbers were reported by Tewari & Wildman (1968) for cytoplasmic ribosomes of tobacco (about 2000) and by Wallace & Birnstiel (1966) for reticulocytes (1600 to 2400).

The number of cistrons is the same for the small and large *r*RNA (Ritossa *et al.* 1966; Birnstiel *et al.* 1968; Schweizer, Mackechnie & Halvorson, 1969). There is evidence from several eukaryotes that both *r*RNAs derive from a common 45s precursor. The arrangement of the cistrons for the *r*RNA on the genome in operons for the 45s RNA would guarantee a stoichiometric synthesis of both *r*RNA components. The arrangement of the cistrons for both *r*RNAs has been recently discussed by Birnstiel *et al.* (1968) and Schweizer, Mackechnie & Halvorson (1969).

As mentioned earlier the extent of homology of base sequences between small and large bacterial *r*RNAs depends on the species. There is no cross-hybridization for various species of *Bacillus* whereas extensive cross-reaction has been found for *Escherichia coli*. Hybridization studies with eukaryotes, namely with *Saccharomyces cerevisiae* (Schweizer, Mackechnie & Halvorson, 1969) *Drosophila* (Ritossa *et al.* 1966) and rabbit (Girolamo, Buziello & Girolamo, 1969) show that there are separate cistrons for both *r*RNAs, with no or only a low cross-reaction. This is also true for the small and large *r*RNA of mitochondrial ribosomes of *Neurospora crassa* (Wood & Luck, 1969). On the other hand, Retel & Planta (1968) reported considerable cross-hybridization for *Sacchoromyces carlsbergensis*. It remains to be seen whether there are species differences among eukaryotes. No cross-reaction between ribosomal and transfer RNA has yet been found.

Hybridization studies with mitochondrial *r*RNA (Wintersberger, 1967; Fukuhara, 1967; Suyama, 1967; Wood & Luck, 1969) of fungi and protozoa gave the following results. There was no detectable hybridization of mitochondrial *r*RNA with nuclear DNA, or of cytoplasmic *r*RNA with mitochondrial DNA. However, because of the insensitivity of the method, these results do not allow one to exclude the presence of cistrons in the nucleus for the mitochondrial *r*RNA. From the extensive hybridization of mitochondrial *r*RNA with mitochondrial DNA it can be calculated that there are several (at least four) cistrons in the mitochondrial DNA for each of the small and large *r*RNAs of mitochondrial ribosomes. The competition for DNA cistrons between the small and large mitochondrial *r*RNAs is, if any, very weak.

It is interesting to compare the hybridization studies of mitochondria with those of chloroplasts, the other system with autonomous ribosomes in eukaryotes. As with mitochondria, chloroplast DNA contains cistrons for chloroplast rRNA (Scott & Smillie, 1967; Tewari & Wildman, 1968). There are about four cistrons for each of the two rRNAs. No hybridization between chloroplast DNA and cytoplasmic rRNA occurs, but an extensive hybridization between nuclear DNA and chloroplast rRNA has been found. The cistrons in nuclear DNA for cytoplasmic and chloroplast rRNA are separated on the genome. In each cell there is about three times as much information coding for chloroplast rRNA in the nuclear DNA as there is in the chloroplast DNA. For this reason it has been suggested (Tewari & Wildman, 1968) that the ribosome population in chloroplasts may be heterogeneous, and a small number of highly specific ribosomes may be coded for by chloroplast DNA whereas the majority are coded for by nuclear DNA.

RIBOSOMAL PROTEINS

The studies of Waller & Harris (1961) and Waller (1964) showed a surprisingly complex electrophoretic pattern for ribosomal proteins of *Escherichia coli*. Isolation of single ribosomal proteins and their chemical and physical characterization (Traut *et al.* 1967; Kaltschmidt, Dzionara, Donner & Wittmann, 1967; Moore *et al.* 1968; Fogel & Sypherd, 1968; Hardy, Kurland, Voynow & Mora, 1969; Craven, Voynow, Hardy & Kurland, 1969; Wittmann *et al.* 1969) clearly showed that these were different ribosomal proteins which differed in their molecular weights and amino acid compositions. The best evidence for differences in the structure of ribosomal proteins is: (*a*) comparison of tryptic peptide maps of numerous proteins (Craven *et al.* 1969); (*b*) isolation of tryptic peptides of pure proteins and determination of their amino acid compositions (B. Wittmann-Liebold, to be published); (*c*) immunological studies with antibodies against about 30 individual ribosomal proteins (G. Stöffler & H. G. Wittmann, to be published).

The K and B strains of *Escherichia coli* show a difference in the rate of migration of two proteins of the 30s sub-unit in gel electrophoresis and chromatography (Leboy, Cox & Flaks, 1964; Otaka, Itoh & Osawa, 1968) and it is possible that there are more protein differences between these two strains. Chromatographic comparison of the ribosomal proteins of several bacteria revealed a strong relationship between *E. coli* and *Salmonella abony*, both of which belong to the same family, Enterobacteriaceae. On the other hand there are almost no similarities in the chromatographic elution patterns of ribosomal proteins from several *Bacillus* species indicating a heterogeneity within Bacillaceae at least with respect to the ribosomal proteins (Otaka *et al.* 1968).

The preparation of antisera against proteins isolated from ribosomes of *Escherichia coli* made it possible to test immunologically whether related proteins are present in ribosomes from *E. coli* and from bacteria of other genera and families (G. Stöffler & H. G. Wittmann, to be published). It was of special interest to determine in which bacterial species protein structures homologous to those in ribosomes of *E. coli* occur. Antibodies against ten isolated proteins of *E. coli* ribosomes reacted almost exclusively with ribosomes of bacteria belonging to the Enterobacteriaceae. This result shows that the Enterobacteriaceae are a homogeneous group with respect to the proteins of their ribosomes. This conclusion is confirmed by the experiments now to be described. Antisera against ribosomes and their sub-units of *E. coli*, *Bacillus stearothermophilus* and *B. subtilis* were tested by three different methods (Ouchterlony gel diffusion technique, immunoelectrophoresis, quantitative immunoprecipitation) against ribosomes and ribosomal proteins of 19 species of bacteria belonging to various families. These studies gave the following results (G. Stöffler & H. G. Wittmann, to be published). Antisera against *Escherichia coli* ribosomes, their sub-units and other ribosomal components (core particles, split proteins) reacted as strongly as in the homologous system only with ribosomes from members of Enterobacteriaceae (*Proteus*, *Salmonella*, *Serratia*); they reacted less strongly with ribosomes from *Azotobacter*; intermediately with ribosomes from *Hydrogenomonas* and *Rhodopseudomonas*; weakly with ribosomes from *Lactobacillus* and *Sarcina*; very weakly with ribosomes from *Micrococcus*, *Propionibacterium*, *Streptococcus*, *Bacillus*, *Anacystis*, and a yeast; not at all with ribosomes from higher plants and rabbit reticulocytes.

The immunological cross-reaction of the ribosomal proteins was weaker and more heterogeneous between members of the Bacillaceae than between those of the Enterobacteriaceae. This result shows that the Bacillaceae are, at least with respect to the structure of their ribosomes, a much more heterogeneous family than Enterobacteriaceae. As mentioned above, this conclusion was also reached by Otaka *et al.* (1968), who used different methods.

There was only a weak immunological reaction between the 30s sub-unit of *Escherichia coli* and the antiserum against 30s sub-unit of *Bacillus stearothermophilus*, and *vice versa*. This finding is remarkable in view of the complete reconstitution between the RNA of the 30s sub-unit of one organism and the 30s ribosomal proteins of the other (Nomura *et al.* 1968) and can be explained by postulating conserved RNA and protein regions as discussed below.

From reconstitution experiments with rRNAs and ribosomal proteins from various bacterial species (Nomura *et al.* 1968) a similar conservation among different bacteria of specific parts of ribosomal proteins interacting with rRNAs can be accepted as discussed earlier for the conservation of specific parts of rRNAs. It is an open question whether for conservation of a specific function during evolution, e.g. the process of ribosome assembly, the same structures have to be conserved, or whether the same function can be achieved by several different structures.

Since the studies of Waller & Harris (1961) it is well known that there is a non-random distribution of the *N*-terminal groups among ribosomal proteins: About 50 % of the ribosomal proteins of *Escherichia coli* start with methionine, 25 % with alanine and the rest with threonine or a few other amino acids. The *N*-terminal amino acids of several proteins are formylated (Hauschild-Rogat, 1968). The occurrence of such a high percentage of methionine is in general correlated with chain initiation by formylmethionine in prokaryotes. Therefore it is interesting to identify *N*-terminal groups of ribosomal proteins in other bacterial species. In *Bacillus subtilis* about 90 % of the end groups are alanine and only 5 % methionine (Horikoshi & Doi, 1968). Unfortunately no further determinations of *N*-terminal amino acids in ribosomal proteins of other bacteria are so far known.

A complexity in ribosomal proteins similar to that found in bacteria has also been found in other organisms whose ribosomal proteins have been examined by gel electrophoresis. The protein patterns of ribosomes isolated from various organs (e.g. liver, pancreas, kidney) of the same mammalian species did not show differences. On the other hand, distinct differences are detectable between different mammalian species although they are only weak among taxonomically closely related species, e.g. mouse and rat (Low & Wool, 1966; Bielka & Welfle, 1968; Keller, Cohen & Hollinshead Beeley, 1968).

Gel electrophoresis studies with *Neurospora crassa* showed that ten wild strains which were isolated in various parts of the world were remarkably similar in the protein patterns of their ribosomes. The various stages of development from conidia to mature hyphae also yielded the same electrophoretic patterns, although it has to be admitted that only relatively big differences can be detected by this technique. Furthermore no antigenic differences were found among ribosomes or ribosomal proteins of the ten wild strains or conidial and hyphal samples of one strain (Rothschild, Itikawa & Suskind, 1967). Similar results were obtained by comparing the gel electrophoresis patterns of ribosomal proteins isolated from numerous wild strains of *Saccharomyces cerevisiae* (H. G. Wittmann, unpublished).

Ribosomal proteins from *Neurospora crassa* possess antigenic determinants present also in *Escherichia coli* ribosomal proteins (Alberghina & Suskind, 1967). Antiserum against *E. coli* ribosomes reacted very weakly with ribosomes from the blue-green alga *Anacystis nidulans* and from a yeast, but not at all with ribosomes from higher plants and rabbit reticulocytes (G. Stöffler & H. G. Wittmann, to be published).

Most of the ribosomal proteins from *Neurospora crassa* mitochondria differ from the proteins of cytoplasmic ribosomes when analysed on a carboxymethylcellulose column (Küntzel, 1969). Similarly the proteins of cytoplasmic and chloroplast ribosomes of the same plant distinctly differ in their gel electrophoresis patterns (Lyttleton, 1968; Janda & Wittmann, 1968; Odintsova & Yurina, 1969) whereas the ribosomal proteins from different plant organs, as for mammals, do not differ detectably in their protein patterns. It would appear therefore that control of differentiation is not affected by the synthesis of different classes of ribosomes.

The isolation and characterization of proteins of eukaryotic ribosomes has been done for a yeast (Chersi, Dzionara, Donner & Wittmann, 1968; Horstmann, 1968; Schmidt & Reid, 1968; H. G. Janda, M. Cech & H. G. Wittmann, to be published) and for mammals (Hamilton & Ruth, 1967; Westermann, Bielka & Böttger, 1969; Welfle, Bielka & Böttger, 1969). The results are similar to those obtained with *Escherichia coli* with respect to the range of molecular weights (10 to 30,000) and the variations in amino acid compositions among the isolated ribosomal proteins. The number of isolated ribosomal proteins from eukaryotes is too low and their chemical and physical characterization insufficient to enable more than tentative conclusions to be made about similarities and differences in the structure of prokaryotic and eukaryotic ribosomes.

FUNCTION OF RIBOSOMES

It is interesting to examine whether the structural division of ribosomes into two groups, namely 70s and 80s, is also reflected in their non-interchangeability for protein synthesis. Component specificity in protein synthesizing systems is due at least to the following incompatibility (1) between *t*RNA and aminoacyl-*t*RNA-synthetases; (2) between the ribosomes and supernatant enzymes from organisms of different groups. Only the latter interaction, known as ribosome specificity, is considered in this review.

By extending earlier studies on species specificities in protein bio-synthesis (Nathans & Lipmann, 1961; Rendi & Ochoa, 1962; Griffin, Holland & Canning, 1965; Allende & Bravo, 1966; Heredia & Halvor-son, 1966; Parisi & Ciferri, 1966; Herrlich, Schweiger, Zillig & Lang, 1967; Parisi *et al.* 1967), Schweiger, Herrlich & Zillig (1969) tested the compatibility of *m*RNAs and ribosomes from numerous prokaryotes and eukaryotes in cell-free systems of *Escherichia coli* and rat liver. Two groups (bacterial and non-bacterial) were differentiated according to the specificity of the interaction between their *m*RNAs and ribosomes. Bacterial *m*RNA directs protein synthesis on bacterial ribosomes but not on other ribosomes and *vice versa*. The observed specificity relates to the attachment of the *m*RNA to the ribosome since the supernatant factors can be exchanged in these experiments without affecting the results. Preliminary experiments (M. Schweiger & P. Herrlich, to be published) indicate that it is the binding of *m*RNA to the 30s ribosomal sub-unit which is the group-specific step. The two groups (bacterial and non-bacterial) do not coincide simply with prokaryotes and eukaryotes: *Anacystis nidulans*, which belongs to the prokaryotic blue-green algae and has 70s ribosomes, behaves like eukaryotes in protein synthesis, whereas the nucleated alga *Cyanidium caldarium* functions like bacteria in protein synthesis. This group-specificity finds some supports from experiments with certain antibiotics (Spencer, 1965).

In attempting to resolve why bacterial ribosomes cannot replace eukaryotic ribosomes (and *vice versa*) in cell-free systems of translation, Parisi *et al.* (1967) concluded that the specificity resides in the inter-action between ribosomes and enzymes from the supernatant. It was then found (Ciferri, Parisi, Perani & Grandi, 1968) that of the transfer factors (G, T_s, T_u) required for chain elongation in *Escherichia coli*, only T_s is ribosome-specific. This factors is thought to interact with a site on the ribosome which is different in 70s and 80s ribosomes. The other two factors (G, T_u) do not interact directly with ribosomes or do so with identical sites in both types of ribosomes.

Whereas the process of chain initiation involving N-formyl-methionyl-*t*RNA is well known in bacteria, little information is available about chain initiation in eukaryotes. From the existence of temperature-sensitive yeast mutants, which are unable to initiate the synthesis of new protein chains under restrictive temperatures, it can be concluded that in eukaryotes, as in bacteria, there are processes unique to the initiation of protein chains (Hartwell & McLaughlin, 1969).

The experiments of Hunt (1969) with mRNA-directed globin synthesis indicated that globin can be synthesized in a bacterial cell-free system supplemented with N-acetylvalyl-tRNA. Studying the unfinished fragments of the α-chain of globin from rabbit reticulocytes, Wilson (1969) concluded that a modified amino acid which cannot react with dansyl chloride or phenylisothiocyanate is first incorporated at the N-terminus during initiation and that the blocking group is removed before the globin chain is finished on the ribosome, resulting in a free valine as the first amino acid.

In contrast to chain initiation, the mechanism of chain elongation is functionally analogous in prokaryotes and eukaryotes (Skogerson & Moldave, 1968; Felicetti & Lipmann, 1968). Two mammalian transfer factors (TF_1, TF_2) have been purified (Arlinghaus, Shaeffer, Bishop & Schweet, 1968; Hardesty, 1969; Moldave, 1969) which seem to correspond to the transfer factors $T_u + T_s$ and G respectively from *Escherichia coli*. The specific inhibition of the bacterial G factor and the mammalian factor TF_2 by the antibiotic fusidic acid strengthens this belief (Malkin & Lipmann, 1969). Another functional step during the biosynthesis of proteins which is similar in prokaryote and eukaryotes is the exchange of ribosomal sub-units. Dissociation of ribosomes and the association of sub-units to ribosomes between successive rounds of mRNA reading is well documented for *Escherichia coli*; this mechanism has recently been shown to occur also with cytoplasmic ribosomes of a yeast (Kaempfer, 1969).

It has been known for several years that numerous factors (temperature, magnesium concentration, polyamines, organic solvents, streptomycin and other aminoglycoside antibiotics, concentration of tRNA or amino acids) induce miscoding in bacterial cell-free systems such as those from *Bacillus stearothermophilus* and *Escherichia coli*. In contrast to these findings, these factors have no or only a very weak effect on miscoding in subcellular systems of eukaryotes (Weinstein, Friedman & Ochoa, 1966). It was shown that the factors mentioned above exert their miscoding effects in bacteria at the ribosomal level and that the high fidelity of mammalian systems can also be localized to the ribosomes (Friedman, Berezney & Weinstein, 1968). Furthermore, the finding that the ribosome species has a strong influence on the specificity of the translation process clearly illustrates the active role of the ribosome itself in the process of codon recognition. During evolution ribosomes apparently appeared which became more resistant to environmental changes than the more primitive bacterial ribosomes, and which translate the genetic information into the programmed protein structure without disturbance by unphysiological influences.

The existence of 70s ribosomes in mitochondria and chloroplasts of eukaryotes raises the question whether there are functional differences between these ribosomes and those of prokaryotes. The occurrence of N-formylmethionyl-*t*RNA in yeast and rat liver is exclusively mito-chondrial and not cytoplasmic in origin (Smith & Marcker, 1968). This suggests that there is a process of chain initiation in mitochondria com-parable to that in bacteria. The analogy between the mechanisms of protein synthesis in mitochondria and bacteria is further illustrated by the finding of the same reaction of both systems to numerous antibiotics known to inhibit bacterial protein biosynthesis (Clark-Walker & Linnane, 1966). Similar results have been obtained with chloroplasts, the protein synthesis of which is affected by antibiotic inhibitors of bacterial protein synthesis, e.g. chloramphenicol, but not by inhibitors of 80s ribosomes, e.g. cycloheximide (Brawerman & Eisenstadt, 1964; Ellis, 1969). As mentioned earlier, several other properties are identical for bacterial and chloroplast ribosomes and different for cytoplasmic and chloroplast ribosomes of the same plant, e.g. Mg^{2+} concentration for optimal amino acid incorporation and reversible dissociation into sub-units.

All results so far obtained point to a strong similarity in the function of ribosomes of mitochondria and chloroplasts on the one hand and of prokaryotes on the other, whereas distinct functional differences exist between the ribosomes of these organelles and their corresponding cytoplasmic ribosomes.

I would like to thank Dr R. A. Garrett for help in translating this article.

REFERENCES

ALBERGHINA, F. A. M. & SUSKIND, S. R. (1967). Ribosomes and ribosomal protein from *Neurospora crassa. J. Bact.* **94**, 630.

ALLENDE, J. E. & BRAVO, M. (1966). Amino acid incorporation and aminoacyl transfer in a wheat embryo system. *J. biol. Chem.* **241**, 5813.

AMALDI, F. (1969). Non-random variability in evolution of base compositions of ribosomal RNA. *Nature, Lond.* **221**, 95.

ARLINGHAUS, R., SHAEFFER, J., BISHOP, J. & SCHWEET, R. (1968). Purification of the transfer enzymes from reticulocytes and properties of the transfer reaction. *Archs Biochem. Biophys.* **125**, 604.

ARONSON, A. I. & HOLOWCZYK, M. A. (1965). Composition of bacterial ribosomal RNA. *Biochim. biophys. Acta* **95**, 217.

ATTARDI, G., HUANG, P. C. & KABAT, S. (1965). Recognition of ribosomal RNA sites in DNA. *Proc. natn. Acad. Sci. U.S.A.* **53**, 1490.

BELOZERSKY, A. N. & SPIRIN, A. S. (1960). Chemistry of the nucleic acids of micro-organisms. In *The Nucleic Acids*, vol. 3, p. 147. Eds. E. Chargaff and J. N. Davidson. London and New York: Academic Press.

BIELKA, H. & WELFLE, H. (1968). Characterization of ribosomal proteins from different tissues and species of animals by electrophoresis on polyacrylamide gel. *Molec. Gen. Genetics* **102**, 128.

BIRNSTIEL, M., SPEIRS, J., PURDOM, I., JONES, K. & LOENING, U. E. (1968). Properties and composition of the isolated ribosomal DNA satellite of *Xenopus laevis*. *Nature, Lond.* **219**, 454.

BOARDMAN, N. K., FRANCKI, R. I. B. & WILDMAN, S. G. (1966). Protein synthesis by cell-free extracts of tobacco leaves. *J. molec. Biol.* **17**, 470.

BRAWERMAN, G. & EISENSTADT, J. M. (1964). Template and ribosomal ribonucleic acids associated with the chloroplasts and the cytoplasm of Euglena gracilis. *J. molec. Biol.* **10**, 403.

BROWNLEE, G. G., SANGER, F. & BARRELL, B. G. (1967). Nucleotide sequence of 5s-ribosomal RNA from *Escherichia coli. Nature, Lond.* **215**, 735.

CHERSI, A., DZIONARA, M., DONNER, D. & WITTMANN, H. G. (1968). Ribosomal Proteins. IV. Isolation, amino acid composition, peptide maps and molecular weights of yeast ribosomal proteins. *Molec. Gen. Genetics* **101**, 82.

CHIPCHASE, M. I. H. & BIRNSTIEL, M. L. (1963). On the nature of nucleolar RNA. *Proc. natn. Acad. Sci. U.S.A.* **50**, 1101.

CIFERRI, O., PARISI, B., PERANI, A. & GRANDI, M. (1968). Different specificity of yeast transfer enzymes for *Escherichia coli* ribosomes. *J. molec. Biol.* **37**, 529.

CLARK-WALKER, G. D. & LINNANE, A. W. (1966). In vivo differentiation of yeast cytoplasmic and mitochondrial protein synthesis with antibiotics. *Biochem. biophys. Res. Commun.* **25**, 8.

CLICK, R. E. & TINT, B. L. (1967). Comparative sedimentation rates of plant, bacterial and animal ribosomal RNA. *J. molec. Biol.* **25**, 111.

CRAVEN, G. R., VOYNOW, P., HARDY, S. J. S. & KURLAND, C. G. (1969). Ribosomal proteins of *Escherichia coli*. II. Chemical and physical characterization of the 30s ribosomal proteins. *Biochemistry* **8**, 2906.

DOI, R. H. & IGARASHI, R. T. (1965). Conservation of ribosomal and messenger RNA cistrons in *Bacillus* species. *J. Bact.* **90**, 384.

DUBIN, D. T. & BROWN, R. E. (1967). A novel ribosomal RNA in hamster cell mitochondria. *Biochim. biophys. Acta* **145**, 538.

DUBNAU, D., SMITH, I. MORELL, P. & MARMUR, J. (1965). Gene conservation in *Bacillus* species. *Proc. natn. Acad. Sci. U.S.A.* **54**, 491.

DURE, L. S., EPLER, J. L. & BARNETT, W. E. (1967). Sedimentation properties of mitochondrial and cytoplasmic ribosomal RNAs from *Neurospora. Proc. natn. Acad. Sci. U.S.A.* **58**, 1883.

ELAEV, N. R. (1964). An investigation into the metabolic activity of cytoplasmic, nuclear and mitochondrial ribosomes. *Biokhimiya* **29**, 359.

ELLIS, R. J. (1969). Chloroplast ribosomes: stereospecificity of inhibition by chloramphenicol. *Science, N.Y.* **163**, 477.

FELICETTI, L. & LIPMANN, F. (1968). Comparison of amino acid polymerization factors isolated from rat liver and rabbit reticulocytes. *Archs Biochem. Biophys.* **125**, 548.

FOGEL, S. & SYPHERD, P. S. (1968). Chemical basis for heterogeneity of ribosomal proteins. *Proc. natn. Acad. Sci. U.S.A.* **59**, 1329.

FORGET, B. G. & WEISSMANN, S. M. (1967). Nucleotide sequence of KB cell 5s RNA. *Science, N.Y.* **158**, 1695.

FRIEDMAN, S. M., BEREZNEY, R. & WEINSTEIN, I. B. (1968). Fidelity in protein synthesis. *J. biol. Chem.* **243**, 5044.

FUKUHARA, H. (1967). Informational role of mitochondrial DNA studied by hybridization with different classes of RNA in yeast. *Proc. natn. Acad. Sci. U.S.A.* **58**, 1065.

GEORGATSOS, J. G. & PAPASARANTOPOULOU, N. (1968). Evidence for the cytoplasmic origin of 78s ribosomes of mouse liver mitochondria. *Archs biochem. Biophys.* **126**, 771.

GIROLAMO, A. D., BUSIELLO, E. & GIROLAMO, M. D. (1969). Hybridization properties of ribosomal RNA from rabbit tissues. *Biochim. Biophys. Acta* **182**, 169.

GNANAM, A. & KAHN, J. S. (1967). Biochemical studies on the induction of chloroplast development in *Euglena gracilis*. III. Ribosome metabolism associated with chloroplast development. *Biochim. biophys. Acta* **142**, 493.

GOULD, H., BONANOU, S. & KANAGALINGHAM, K. (1966). Structural characterization of ribosomal RNA's from various species by a new 'fingerprinting' technique. *J. molec. Biol.* **22**, 397.

GRIFFIN, A. C., HOLLAND, B. H. & CANNING, L. (1965). Comparison of protein-synthesizing systems from normal and tumor tissues. In *Developmental and Metabolic Control Mechanisms and Neoplasia*, p. 475. *19th Annual Symposium at the University of Texas M.D. Anderson Hospital and Tumor Institute.* Baltimore: The Williams and Wilkie Company.

HAMILTON, M. G. & RUTH, M. E. (1967). Characterization of some of the proteins of the large sub-unit of rat liver ribosomes. *Biochemistry* **6**, 2585.

HARDESTY, B. (1969). Paper given at Cold Spring Harb. Symp. cited in *Nature, Lond.* **223**, 133.

HARDY, S. J. S., KURLAND, C. G., VOYNOW, P. & MORA, G. (1969). Ribosomal proteins of *Escherichia coli*. I. Purification of the 30 s proteins. *Biochemistry* **8**, 2897.

HARTWELL, L. H. & MCLAUGHLIN, C. S. (1969). A mutant of yeast apparently defective in the initiation of protein synthesis. *Proc. natn. Acad. Sci. U.S.A.* **62**, 468.

HAUSCHILD-ROGAT, P. (1968). N-Formylmethionine as a N-terminal group of *Escherichia coli* ribosomal protein. *Molec. Gen. Genetics* **102**, 95.

HEREDIA, C. F. & HALVORSON, H. O. (1966). Transfer of amino acids from aminoacyl soluble ribonucleic acid to protein by cell-free extracts from yeast. *Biochemistry* **5**, 946.

HERRLICH, P., SCHWEIGER, M., ZILLIG, W. & LANG, N. (1967). Klassenspezifität bei der Bildung von aktiven Polysomen aus Ribosomen und Matrizen-Ribonukleinsäuren. *Hoppe-Seyler's Z. physiol Chem.* **348**, 1207.

HORIKOSHI, K. & DOI, R. H. (1968). The NH_2-terminal residues of *Bacillus subtilis* proteins. *J. biol. Chem.* **243**, 2381.

HORSTMANN, H.-J. (1968). Untersuchungen über ribosomale Proteine. *Hoppe-Seyler's Z. physiol. Chem.* **349**, 405.

HUNT, J. A. (1969). Paper given at Cold Spring Harb. Symp. cited in *Nature, Lond.* **223**, 133.

JANDA, H.-G. & WITTMANN, H. G. (1968). Ribosomal proteins. V. Comparison of protein patterns of 70 s and 80 s ribosomes from various plants by polyacrylamide gel electrophoresis. *Molec. Gen. Genetics* **103**, 238.

JANOWSKI, M. (1966). Detection of ribosomes and polysomes in Acetabularia mediterranea. *Life Sci.* **5**, 2113.

JANOWSKI, M., BONOTTO, S. & BOLOUKHERE, M. (1969). Ribosomes of *Acetabularia mediterranea*. *Biochim. biophys. Acta* **174**, 525.

KAEMPFER, R. (1969). Ribosomal sub-unit exchange in the cytoplasm of a eukaryote. *Nature, Lond.* **222**, 950.

KALTSCHMIDT, E., DZIONARA, M., DONNER, D. & WITTMANN, H. G. (1967). Ribosomal proteins. I. Isolation, amino acid composition, molecular weights and peptide mapping of proteins from *Escherichia coli* ribosomes. *Molec. Gen. Genetics* **100**, 364.

KELLER, P. J., COHEN, E. & HOLLINSHEAD BEELEY, J. A. (1968). Canine pancreatic ribosomes. II. The protein moiety. *J. biol. Chem.* **243**, 1271.

KROON, A. M. (1968), RNA of rat liver mitochondrial ribosomes. *Abstracts 5th Meeting Fed. Europ. Biochem. Soc. (Prague)*, p. 51.

KÜNTZEL, H. (1969). Mitochondrial and cytoplasmic ribosomes from *Neurospora crassa*: characterization of their sub-units. *J. molec. Biol.* **40**. 315.

KÜNTZEL, H. & NOLL, H. (1967). Mitochondrial and cytoplasmic polysomes from *Neurospora crassa*. *Nature, Lond.* **215**, 1340.

LANGRIDGE, R. (1969). Ribosomes: A common structural feature. *Science, N.Y.* **140**, 1000.

LEBOY, P. S., COX, E. C. & FLAKS, J. G. (1964). The chromosomal site specifying a ribosomal protein in *Escherichia coli*. *Proc. natn. Acad. Sci. U.S.A.* **52**, 1367.

LESSIE, T. G. (1965). The atypical ribosomal RNA complement of *Rhodopseudomonas spheroides*. *J. gen. Microbiol.* **39**, 311.

LOENING, U. E. (1968). Molecular weights of ribosomal RNA in relation to evolution. *J. molec. Biol.* **38**, 355.

LOENING, U. E. & INGLE, J. (1967). Diversity of RNA components in green plant tissues. *Nature, Lond.* **215**, 363.

LOW, R. B. & WOOL, I. G. (1967). Mammalian ribosomal protein: Analysis by electrophoresis on polyacrylamide gel. *Science, N.Y.* **155**, 330.

LYTTLETON, J. W. (1962). Isolation of ribosomes from spinach chloroplasts. *Expl Cell Res.* **26**, 312.

LYTTLETON, J. W. (1968). Protein constituents of plant ribosomes. *Biochim. biophys. Acta* **154**, 145.

MADISON, J. T. (1968). Primary structure of RNA. *Ann. Rev. Biochem.* **37**, 131.

MALKIN, M. & LIPMANN, F. (1969). Fusidic acid: inhibition of factor T_2 in reticulocyte protein synthetis. *Science, N.Y.* **164**, 71.

MANGIAROTTI, G., APIRION, D., SCHLESSINGER, D. & SILENGO, L. (1968). Biosynthetic precursors of 30 s and 50 s ribosomal particles in *Escherichia coli*. *Biochemistry* **7**, 456.

McCONKEY, E. H. & HOPKINS, J. W. (1964). The relationship of the nucleolus to the synthesis of ribosomal RNA in HeLa cells. *Proc. natn. Acad. Sci. U.S.A.* **51**, 1197.

MERITS, I., SCHULZE, W. & OVERBY, L. R. (1966). Ribosomal ribonucleic acid sites in chicken DNA. *Archs Biochem. Biophys.* **115**, 197.

MIDGLEY, J. E. M. (1962). The nucleotide base composition of RNA from several microbial species. *Biochim. biophys. Acta* **61**, 513.

MIURA, K. I. (1962). The nucleotide composition of RNAs of soluble and particle fractions in several species of bacteria. *Biochim. biophys. Acta* **55**, 62.

MOLDAVE, K. (1969). Paper given at Cold Spring Harb. Symp., cited in *Nature, Lond.* **223**, 133.

MOORE, P. B., TRAUT, R. R., NOLLER, H., PEARSON, P. & DELIUS, H. (1968). Ribosomal proteins of *Escherichia coli*. II. Proteins of the 30 s sub-unit. *J. molec. Biol.* **31**, 441.

MOORE, R. L. & McCARTHY, B. J. (1967). Comparative study of ribosomal RNA cistrons in Enterobacteria and Myxobacteria. *J. Bact.* **94**, 1066.

NATHANS, D. & LIPMANN, F. (1961). Amino acid transfer from aminoacyl-RNAs to protein on ribosomes of *Escherichia coli*. *Proc. natn. Acad. Sci. U.S.A.* **47**, 497.

NOLL, F. & BIELKA, H. (1969). Studies on proteins of animal ribosomes. III. Immunochemical analyses of whole ribosomes from different tissues and species of animals. *Molec. Gen. Genetics*. (In Press.)

NOMURA, M., TRAUB, P. & BECHMANN, H. (1968). Hybrid 30 s ribosomal particles reconstituted from components of different bacterial origins. *Nature, Lond.* **219**, 793.

O'BRIEN, T. W. & KALF, G. F. (1967). Ribosomes from rat liver mitochondria. *J. biol. Chem.* **242**, 2180.

ODINTSOVA, M. S. & YURINA, N. P. (1969). Proteins of chloroplast and cytoplasmic ribosomes. *J. molec. Biol.* **40**, 503.

OTAKA, E., ITOH, T. & OSAWA, S. (1968). Ribosomal proteins of bacterial cells: Strain- and species-specificity. *J. molec. Biol.* **33**, 93.

PARISI, B. & CIFERRI, O. (1966). Protein synthesis by cell-free extracts from castor bean seedlings. *Biochemistry* **5**, 1638.

PARISI, B., MILANESI, G., VAN ETTEN, J. L., PERANI, A. & CIFERRI, O. (1967). Species specificity in protein synthesis. *J. molec. Biol.* **28**, 295.

PETERMAN, M. L. (1964). *The Physical and Chemical Properties of Ribosomes.* Amsterdam: Elsevier Publishing Co.

PINDER, J. C., GOULD, H. J. & SMITH, J. (1969). Conservation of the structure of ribosomal RNA during evolution. *J. molec. Biol.* **40**, 289.

POLLARD, C. J., STEMLER, A. & BLAYDES, D. F. (1966). Ribosomal ribonucleic acids of chloroplastic and mitochondrial preparations. *Pl. Physiol., Lancaster* **41**, 1323.

RABINOWITZ, M., DESALLE, L., SINCLAIR, J., STIREWALT, R. & SWIFT, H. (1966). Ribosomes isolated from rat liver mitochondrial preparations. *Fedn Proc. Fedn Am. Socs exp. Biol.* **25**, 581.

RAWSON, J. R. & STUTZ, E. (1968). Characterization of *Euglena* cytoplasmic ribosomes and ribosomal RNA by zone velocity sedimentation in sucrose gradients. *J. molec. Biol.* **33**, 309.

REISNER, A. H., ROWE, J. & MACINDOE, H. M. (1968). Structural studies on the ribosomes of paramecium: Evidence for a 'primitive' animal ribosome. *J. molec. Biol.* **32**, 587.

RENDI, R. & OCHOA, S. (1962). Species specificity in activation and transfer of leucine from carrier RNA to ribosomes. *J. biol. Chem.* **237**, 3707.

RETEL, J. & PLANTA, R. J. (1968). The investigation of the ribosomal RNA sites in yeast DNA by the hybridization technique. *Biochim. biophys. Acta* **169**, 416.

RIFKIN, M. R., WOOD, D. D. & LUCK, D. J. L. (1967). Ribosomal DNA and ribosomes from mitochondria of *Neurospora crassa. Proc. natn. Acad. Sci. U.S.A.* **58**, 1025.

RITOSSA, F. J., CATWOOD, K. C., LINDSLEY, D. L. & SPIEGELMAN, S. (1966). On the chromosomal distribution of DNA complementary to ribosomal and soluble RNA. *Natn. Cancer Inst. Monogr.* **23**, 449.

ROGERS, P. J., PRESTON, B. N., TITCHENER, E. B. & LINNANE, A. W. (1967). Differences between the sedimentation characteristics of the ribonucleic acid prepared from yeast cytoplasmic ribosomes and mitochondria. *Biochem. biophys. Res. Commun.* **27**, 405.

ROTHSCHILD, H., ITIKAWA, H. & SUSKIND, S. R. (1967). Ribosomes and ribosomal proteins from *Neurospora crassa.* II. Ribosomal proteins in different wild-type strains and during various stages of development. *J. Bact.* **94**, 1800.

RUPPEL, H. G. (1968). Nukleinsäuren in Chloroplasten. II. Isolierung und Charakterisierung der Chloroplasten-Ribosomen von *Antirrhinum majus. Z. Naturf.* **23**b, 997.

SCHMIDT, J. & REID, B. R. (1968). Fractionation of yeast ribosomal proteins. *Biochem. biophys. Res. Commun.* **31**, 654.

SCHWEIGER, M., HERRLICH, P. & ZILLIG, W. (1969). Group specificity in protein synthesis. III. *Hoppe-Seyler's Z. physiol. Chem.* **350**, 775.

SCHWEIZER, E., MACKECHNIE, C. & HALVORSON, H. O. (1969). The redundancy of ribosomal and transfer RNA genes in *Saccharomyces cerevisiae. J. molec. Biol.* **40**, 261.

SCOTT, N. S. & SMILLIE, R. (1967). Evidence for the direction of chloroplast ribosomal RNA synthesis by chloroplast DNA. *Biochem. Biophys. Res. Commun.* **28**, 598.

SKOGERSON, L. & MOLDAVE, K. (1968). Characterization of the interaction of aminoacyltransferase II with ribosomes. *J. biol. Chem.* **243**, 5354.

SMITH, A. E. & MARCKER, K. A. (1968). N-Formylmethionyl-*t*RNA in mitochondria from yeast and rat liver. *J. molec. Biol.* **38**, 241.

SPENCER, D. (1965). Protein synthesis by isolated spinach chloroplasts. *Arch. Biochem. Biophys.* **111**, 381.

SPENCER, D. & WHITFELD, P. R. (1966). The nature of the RNA of isolated chloroplasts. *Archs Biochem. Biophys.* **117**, 337.

STANLEY, W. M. & BOCK, R. M. (1965). Isolation and physical properties of the ribosomal RNA of *Escherichia coli. Biochemistry* **4**, 1302.

STUTZ, E. & NOLL, H. (1967). Characterization of cytoplasmic and chloroplast polysomes in plants. *Proc. natn. Acad. Sci. U.S.A.* **57**, 774.

SUGIURA, M. & TAKANAMI, M. (1967). Analysis of the 5′-terminal nucleotide sequences of RNAs. *Proc. natn. Acad. Sci. U.S.A.* **58**, 1595.

SUYAMA, Y. (1967). The origins of mitochondrial ribonucleic acids in *Tetrahymena pyriformis. Biochemistry* **6**, 2829.

SZILAGYI, J. F. (1968). 16 s and 23 s components in the ribosomal RNA of *Rhodopseudomonas spheroides. Biochem. J.* **109**, 191.

TAKAHASHI, H., SAITO, H. & IKEDA, Y. (1967). Species specificity of the ribosomal RNA cistrons in bacteria. *Biochim. Biophys. Acta* **134**, 124.

TAYLOR, M. M. & STORCK, R. (1964). Uniqueness of bacterial ribosomes. *Proc. natn. Acad. Sci. U.S.A.* **52**, 958.

TAYLOR, M. M., GLASGOW, J. E. & STORCK, R. (1967). Sedimentation coefficients of RNA from 70 s and 80 s ribosomes. *Proc. natn. Acad. Sci. U.S.A.* **57**, 164.

TEWARI, K. K. & WILDMAN, S. G. (1968). Function of chloroplast DNA. *Proc. natn. Acad. Sci. U.S.A.* **59**, 569.

TRAUT, R. R., MOORE, R. B., DELIUS, H., NOLLER, H. & TISSIERES, A. (1967). Ribosomal proteins of *Escherichia coli.* I. Demonstration of different primary structures. *Proc. natn. Acad. Sci. U.S.A.* **57**, 1294.

TREWAVAS, A. J. & GIBSON, I. (1968). Ribosomal nucleotide sequence homologies in plants. *Pl. Physiol., Lancaster* **43**, 445.

TRUMAN, D. E. S. (1963). Incorporation of amino acids into the protein of submitochondrial particles. *Expl Cell Res.* **31**, 313.

WALLACE, H. & BIRNSTIEL, M. L. (1966). Ribosomal cistrons and the nuclear organizer. *Biochim. biophys. Acta* **114**, 226.

WALLER, J. P. (1964). Fractionation of the ribosomal protein from *Escherichia coli. J. molec. Biol.* **10**, 319.

WALLER, J. P. & HARRIS, J. I. (1961). Studies on the composition of the protein from *Escherichia coli* ribosomes. *Proc. natn. Acad. Sci. U.S.A.* **47**, 18.

WEINSTEIN, I. B., FRIEDMAN, S. M. & OCHOA, M. (1966). Fidelity during translation of the genetic code. *Cold Spring Harb. Symp. quant. Biol.* **31**, 671.

WELFLE, H., BIELKA, H. & BÖTTGER, M. (1969). Studies on proteins of animal ribosomes. *Molec. Gen. Genetics* **104**, 165.

WESTERMAN, P., BIELKA, H. & BÖTTGER, M. (1969). Studies on proteins of animal ribosomes. *Molec. Gen. Genetics* **104**, 157.

WILSON, D. B. (1969). Paper given at Cold Spring Harb. Symp. cited in *Nature, Lond.* **223**, 133.

WINTERSBERGER, E. (1967). A distinct class of ribosomal RNA components in yeast mitochondria as revealed by gradient centrifugation and by DNA-RNA-hybridization. *Hoppe-Seyler's Z. physiol. Chem.* **348**, 1701.

WITTMANN, H. G., STÖFFLER, G., KALTSCHMIDT, E., RUDLOFF, V., JANDA, H.-G., DZIONARA, M., DONNER, D., NIERHAUS, K., CECH, M., HINDENNACH, I. & WITTMANN, B. (1969). Proteinchemical and serological studies on ribosomes of bacteria, yeast and plants. *Proc. Fed. Europ. Biochem. Soc. Meeting, Madrid.* (In Press.)

WOOD, D. D. & LUCK, D. J. L. (1969). Hybridization of mitochondrial ribosomal RNA. *J. molec. Biol.* **41**, 211.

YANKOFSKY, S. A. & SPIEGELMAN, S. (1963). Distinct cistrons for the two ribosomal RNA components. *Proc. natn. Acad. Sci. U.S.A.* **49**, 538.

THE MECHANISM OF SYNTHESIS OF RIBOSOMAL RNA

ULRICH E. LOENING*

Department of Botany, University of Edinburgh

Ribosomes contain three distinct types of ribonucleic acid (*r*RNA), conventionally known by their sedimentation values as 28s, 18s and 5s in higher organisms and 23s, 16s and 5s in bacteria. There are about 10^4 ribosomes in one bacterial cell and between 10^6 and 10^7 in one plant or animal cell; the high molecular weight *r*RNA comprises most of the RNA of the cell. A bacterial cell which divides every 30 min. must synthesize about 5 ribosomes/sec. A rapidly growing eukaryote may divide once in 24 hr; this requires a rate of synthesis of 10 to 100 new ribosomes/sec. Each pair of 18s and 28s RNA components, which are synthesized together as a single RNA chain, takes about 2 to 3 min. to complete. There must therefore be some 15,000 simultaneously growing chains to maintain the required rate. Multiple sites of synthesis are thus a necessity and are provided partly by multiple copies of the genes coding for *r*RNA and partly by the synthesis of several chains on one gene at a time. (The term 'redundant' has frequently been used for the repeated copies of the ribosomal genes; this is correct only in the genetic sense; 'reiterated' is to be preferred as a descriptive term without functional implications. 'Redundant' could also be used for the excess RNA in the ribosomal precursor.)

The synthesis of *r*RNA (and of transfer RNA, *t*RNA) is a special case in which the molecule finally required is synthesized directly on the gene. There is therefore no possibility of the intermediate amplification which is provided by messenger RNA (*m*RNA) for the synthesis of proteins. The required multiplicity of the genes for *r*RNA and the large amount of immediate gene product leads to the possibility of fundamental experiments apart from the interest in the mechanism of ribosome synthesis. This is the only case in which a gene of known function has been isolated (Birnstiel, Wallace, Sirlin & Fischberg, 1966) and the only case in which the primary gene product is present in sufficient amount to be characterized. This system may therefore be an experimentally suitable model for the study of the mechanism of gene action and the control of transcription. Also, the primary gene product, the ribosomal

* Present address: Department of Zoology, University of Edinburgh.

precursor-RNA, contains in addition some non-ribosomal RNA which is later discarded. Such synthesis of apparently excess RNA may be a general phenomenon and apply also to the synthesis of *m*RNA. A study of the synthesis and assembly of ribosomes, which is experimentally feasible, may be the best approach to an understanding of the control of RNA synthesis during cell growth and development. This paper will discuss the synthesis and properties of the ribosomal precursor-RNA and its fate during maturation to *r*RNA. Since most attention has so far been focused on mammalian cells, and especially on the HeLa cell, stress is laid here on results obtained with other species. A brief comparison is also made with *r*RNA synthesis in bacteria, which lack a nucleolus and synthesize the two ribosomal components separately. The 5s RNA is not discussed here, since it is synthesized on a different part of the genome and not made in the nucleolus.

Comparison of the size of ribosomal RNA of different organisms

The sedimentation coefficients of *r*RNA vary in different species and even apparently in different laboratories for the same species (see for example summary by Click & Tint, 1967). A reasonably exact knowledge of the molecular weights is however essential for a comparison of the *r*RNA with its precursors and the determination of the amounts of precursor-RNA which are discarded. The high resolution obtained by gel electrophoresis of RNA provides a rapid means of comparing the RNAs of different species. The method provides a relative, and perhaps an absolute, measure of molecular weight which is not much dependent on the configuration of the RNA (Loening, 1969).

It is generally accepted that the sedimentation coefficient of bacterial *r*RNA (23s and 16s) is less than that of higher cells. The difference may be correlated with differences in the properties of the ribosome, namely: ease of dissociation into sub-units at low magnesium ion concentrations, optimum magnesium concentrations for *in vitro* amino acid incorporation, different sensitivity to antibiotics like chloramphenicol for the prokaryotes and cycloheximide for the eukaryotes, the probable differences in chain initiation requirements. While the compositions of the *r*RNA of different bacteria vary somewhat, the molecular weights of all *r*RNAs of prokaryotes seems to be the same (Table 1). These weights are more than 20 % lower than the weights of the *r*RNAs of eukaryotes. Among eukaryotes, the 18s *r*RNA seems to have been conserved throughout evolution without any change in weight, although there are large differences in composition. The 28s *r*RNAs of plants, fungi, algae, ferns and many protozoa seem to have identical molecular weights. The

28s *r*RNA of animals varies; the results so far suggest that the molecular weight is greater for the more recently evolved animals (see Table 1). Exceptions to these striking generalizations have so far been found only in Protozoa; the weights for the *r*RNA of *Amoeba* and *Euglena* are higher than those for other plants or lower animals. In both cases, however, the 28s RNA is very unstable; the properties of these *r*RNA species require further investigation. These results emphasize the sharp distinction between prokaryotic and eukaryotic ribosomes, which is reflected in the different method of *r*RNA synthesis and in the presence of the nucleolus in eukaryotes. In this paper, specific RNA molecules will be referred to by their molecular weights determined by gel electrophoresis. The sedimentation values 28s and 18s will be used as names for the larger and smaller *r*RNA components.

Table 1. *Molecular weights of ribosomal RNA of different organisms*

RNA was fractionated by polyacrylamide gel electrophoresis and the molecular weights of the ribosomal components determined with those of HeLa cells and *Escherichia coli* as standards (Loening, 1968).

Species	Mol. wt $\times 10^{-6}$	
Prokaryotes		
Bacteria	1·08	0·56
Actinomycetes	1·12	0·56
Blue-green algae	1·07	0·56
Chloroplasts	1·07 to 1·11	0·56
Plants and protozoa		
Higher plants	1·27 to 1·31	0·70 to 0·72
Algae, ferns	1·28 to 1·34	0·68 to 0·73
Fungi	1·28 to 1·3	0·68 to 0·73
Tetrahymena	1·30	0·69
Amoeba, Euglena	1·3 ?, 1·5 (unstable)	0·85 to 0·89
Animals		
Arbacia	1·40	0·68
Drosophila	1·40	0·73
Xenopus	1·5	0·70
Chick	1·6	0·7
Mouse	1·71	0·70
HeLa cell	1·75	0·70

The site of ribosome formation in eukaryotes

Many different approaches, summarized in recent reviews (Perry, 1966, 1967; Birnstiel, 1967; Birnstiel *et al*. 1966; Maden, 1968), have shown that the principal and perhaps only function of the nucleolus is the production of ribosomes. The older correlation between the size or density of the nucleolus and the basophilia or number of ribosomes has been strengthened by the finding that inhibition of nucleolar RNA synthesis results in inhibition of ribosome synthesis (Perry, 1962; Roberts & Newman,

1966; Penman, Vesco & Penman, 1968). The finding that the mutant anucleolate embryos of *Xenopus laevis* are unable to produce ribosomes is the most extreme example of such 'inhibition' and proved the function of the nucleolus in *r*RNA synthesis (Brown & Gurdon, 1964). Hybridization of *r*RNA to DNA showed that the anucleolate mutant lacked the DNA complementary to *r*RNA. The deletion of the nucleolar organizer region in the mutant was thus correlated with loss of the nucleolus and loss of the DNA coding for *r*RNA (Birnstiel *et al.* 1966). Similarly, different numbers of nucleolar organizers in the diploid genome in Drosophila were proportional to the amount of DNA complementary to *r*RNA (Ritossa & Spiegelman, 1965). In passing, it is interesting that the correlation between the nucleolar organizer region of the genome and the formation of the nucleolus has long been known (McClintock, 1934) and that mutants of maize suitable for hybridization of *r*RNA to DNA have long been available. Finally, the RNA precursors of *r*RNA have been identified in preparations of isolated nucleoli. The synthesis and processing of the precursor and the assembly of the ribosomal particles with protein takes place in the nucleolus (Perry, 1962; Penman, 1966; Muramatsu, Hodnett & Busch, 1966; Warner, Girard, Latham & Darnell, 1966). Some of these experiments are described below.

The size and composition of the ribosomal RNA precursor

The bacterial precursor RNA

There have been many studies on the nascent ribosomal particles in bacteria (reviewed by Osawa, 1968) but little is known about the first RNA molecule made. It is probable that the *r*RNA precursor molecule has a higher molecular weight than the final mature *r*RNA. Osawa (1968) and colleagues showed that the precursor to 16s RNA is about 18s, and probably therefore of higher weight. Hecht & Woese (1968) and M. Adesnick & C. Levinthal (personal communication) used polyacrylamide gel electrophoresis to analyse rapidly-labelled bacterial RNA; Fig. 1 shows the fractionation, as compared with long-labelled RNA. The label near the 23s has a slightly lower mobility than the 23s RNA, and there is a prominent peak apparently heavier than the 16s RNA. Both these rapidly-labelled components become coincident with the mature *r*RNA after a few minutes of labelling. The low mobility of the rapidly-labelled components was stable to melting of the RNA, suggesting that these components have a higher molecular weight and not a more open structure. The excess RNA, which appears to be discarded a few minutes after synthesis, is of molecular weight about 0.05×10^6 for the 23s RNA, and about 0.15×10^6 for the 16s component. Thus bacteria, which have no nucleolus, do not synthesize a polycistronic *r*RNA precursor. These experiments suggest however that the synthesis of a larger molecule, from which excess RNA is later discarded, is not confined only to eukaryotes. It would be interesting to see what happens in other prokaryotes such as the blue-green algae, and in chloroplasts.

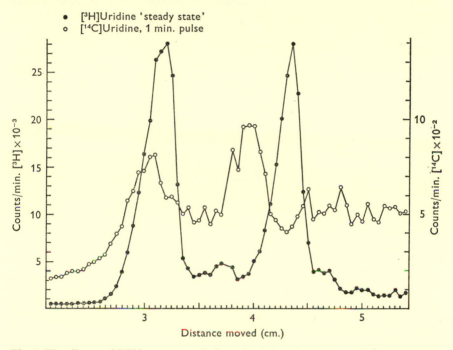

Fig. 1. The ribosomal RNA precursors in bacteria. A log-phase culture of *Bacillus subtilis* was labelled with [³H]uridine for two cell generations and with [¹⁴C]uridine for 1 min. The RNA was extracted and analysed by electrophoresis on a 2·4 % polyacrylamide gel. The [³H]radioactivity (●—●—●—) shows the positions of mature 23 s and 16 s *r*RNA; the [¹⁴C]radioactivity (—O—O—O—) shows the rapidly-labelled precursors to the *r*RNA and some polydispersed label (from Hecht & Woese, 1968).

The polycistronic precursor in higher organisms

A rapidly-labelled high molecular weight RNA with a ribosomal type base composition and a sedimentation value of 45 s was identified in mammalian cells by Perry (1962) and by Scherrer, Latham & Darnell, (1963). This RNA was not readily extracted from cell homogenates; either heat or strong detergents were required, which may explain why it had not been found earlier. The composition and kinetics of labelling suggested that it is the precursor of *r*RNA; this has been confirmed by experiments described below.

The molecular weight of the 45s RNA in HeLa cells has been determined by five independent methods: (1) the sedimentation coefficient in comparison to *r*RNA; (2) equilibrium centrifugation (McConkey & Hopkins, 1969); (3) gel electrophoresis (Weinberg, Loening, Willems & Penman, 1967); (4) content of 2′-O-methylribose as compared to uridine, described below (Weinberg *et al.* 1967); (5) length of shadowed preparations seen in the electron microscope (Granboulan & Scherrer, 1969). All these methods agree in giving molecular weights of about

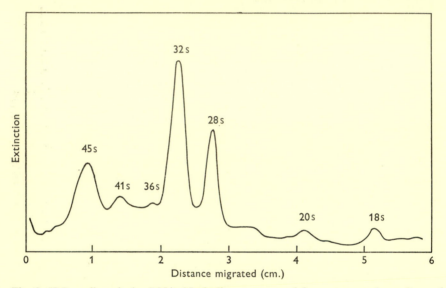

Fig. 2. HeLa cell nucleolar RNA. Nucleoli were prepared from a suspension culture of HeLa cells and the RNA extracted with phenol at 55° (Penman, 1966). The RNA was analysed on a 2·4 % polyacrylamide gel; electrophoresis for 4 hr at 50 V. The gel was scanned at 265 nm in a Joyce Loebl Chromoscan. This was one of the first scans obtained for the experiments described by Weinberg *et al.* (1967). The approximate equivalents of sedimentation coefficients are indicated.

$4·5 \times 10^6$. This is 2×10^6 greater than the sum of the molecular weights of the *r*RNA components; this excess RNA is presumed to be discarded during the processing of precursor (described later). A gel electrophoretic separation of rapidly-labelled HeLa cell RNA is shown later in Fig. 10 and of nucleolar RNA in Fig. 2.

There has been much discussion whether this rapidly-labelled 45s precursor RNA is one molecule, whether it is an aggregate or perhaps held together by carbohydrate or protein, and even whether the high sedimentation coefficient is due only to a compact structure. Bramwell & Harris (1967) and Parish & Kirby (1966) described experiments which suggested that the sedimentation coefficients of all such rapidly-labelled apparently high-molecular-weight RNA can be decreased by salt or solvent treatments which minimize aggregation. It is true that aggregation can occur during RNA isolation, especially at high RNA concentrations such as are found in the nucleolus and at the high salt concentrations and high temperatures which are required to extract the 45s RNA (Wagner, Katz & Penman, 1967). The conditions must be controlled, as the latter authors showed, to avoid such aggregation. It is also clear that RNA, and especially rapidly-labelled RNA, is easily degraded in tissue homogenates by ribonucleases, although a 'nicked' molecule can be held together by its secondary structure. Such hidden breaks would become apparent only when the molecule is partially unfolded, as in the experiments of Bramwell & Harris (1967) and Parish & Kirby (1966). Experiments which purport to show that the 45s RNA is not a continuous chain must therefore show that there has been no scission of the molecule during extraction. It is not possible at present to define conditions for the extraction of RNA such that neither aggregation nor chain-breakage occurs.

The evidence that the 45s RNA is a continuous chain is provided by the five independent molecular weight determinations, by the electron microscopy of the molecule when unfolded in urea (Granboulan & Scherrer, 1969), and by the fact that it remains intact under many different conditions and enzyme treatments (Fujisawa & Muramatsu, 1969) including extraction with hot phenol (Penman, 1966) as well as with cold detergent as used by Parish & Kirby (1966) as in Fig. 10.

Since the molecular weight of the 45s RNA is greater than that of the sum of the two main rRNA components (neglecting the 5s RNA which is made separately) it is necessary to show that this precursor molecule does contain one unit of each of the rRNA components, and to account for the excess molecular weight. The base compositions, fingerprints of partial digests and methylation patterns all provide consistent evidence. The true base composition of the 45s RNA was difficult to determine because of contaminating DNA-like RNA which contained about 43 mole % GC. 45s RNA from specially purified nucleoli of HeLa cells was nearly free from such contamination, and contained 70 % GC. This may be compared to the combined composition of the rRNA of 64 mole % GC. The difference in composition and in molecular weight between precursor and rRNA was used to calculate the apparent composition of the excess non-ribosomal RNA component; this was about 77 mole % GC (Willems, Wagner, Laing & Penman, 1968). All oligonucleotides found in partial digests of rRNA were also found in 45s

RNA; the 45s RNA also contained sequences not found in rRNA (Jeantur, Amaldi & Attardi, 1968).

About 1 % of the 2-OH of ribose in rRNA is methylated; this confers alkali resistance to the adjacent phosphodiester bond. Characteristic alkali-stable dinucleotides are found in rRNA after alkaline hydrolysis. The frequencies of such dinucleotides in the 28s and 18s rRNA of HeLa cells were also found in the 45s precursor. No other dinucleotides were found in the precursor, indicating that the non-ribosomal portions were not methylated (Wagner, Penman & Ingram, 1968). The 32s precursor to 28s rRHA (see below) contained only the dinucleotides characteristic of 28s RNA. These experiments show that 45s RNA contained the two rRNA components plus about 40 % of its weight of non-methylated non-ribosomal RNA with a high GC content. This excess RNA was apparently lost during conversion of precursor to mature rRNA; this processing of the precursor is described as 'non-conservative'.

In organisms other than mammals a polycistronic precursor is also synthesized; this has been less studied. A 40s precursor, presumably of lower molecular weight than the 45s of mammals, has been described in the newt *Triturus* (Gall, 1966) and in the toad *Xenopus* (Landesman & Gross, 1969). The 40s *Xenopus* RNA is methylated during synthesis, as in HeLa cells, and the methyl-labelled RNA is transferred to rRNA during a 'chase' in actinomycin D which prevents further RNA synthesis. This provides convincing evidence that the 40s RNA is the precursor to rRNA. The labelling experiments were done with dissociated *Xenopus* embryos so that label could enter the cells: the synthesis of the 40s RNA, as of rRNA, started after gastrulation.

We have examined by gel electrophoresis the *Xenopus* precursor in cultured kidney cells (see Fig. 3) (Loening, Jones & Birnstiel, 1969). The RNA of the rapidly-labelled peak material (Fig. 3a) had a molecular weight of 2·5 to $2·6 \times 10^6$. This is about $0·35 \times 10^6$ greater than the molecular weight of the sum of the rRNA (Table 1). The base composition corrected for the underlying DNA-like RNA was about 63 mole % GC, and that of the combined rRNA components was 59 mole % GC. This small difference in composition suggests that the composition of the $0·35 \times 10^6$ excess RNA must be at least 80 mole % GC. Until recently a similar high molecular weight precursor had not been found in plant cells. Experience suggested that this was due to the lability of rapidly-labelled RNA in cell homogenates. Homogenization direct in phenol + detergent mixtures showed that rapidly-labelled high molecular weight components was found; suitable choice of species and of homogenization media showed that this RNA was present in crude

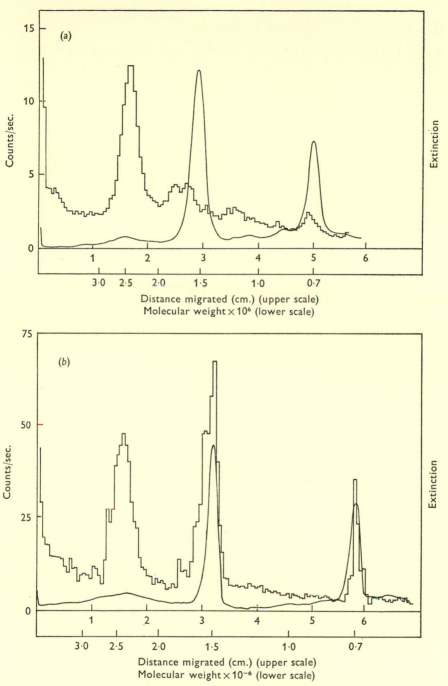

Fig. 3. Rapidly-labelled RNA from *Xenopus laevis*. Monolayer cultures of *Xenopus* kidney cells were labelled (*a*) for 1 hr and (*b*) for 3 hr with [^{32}P]. Nucleic acids were extracted in the cold, essentially as described by Parish & Kirby (1966), and the DNA digested with deoxyribonuclease. One-fifth to one-half of the RNA from a dish of cells was applied to a 2·0 % polyacrylamide gel and electrophoresis continued for 3 hr in (*a*) and 3·75 hr in (*b*), at 50 V. Gels were scanned on a Joyce Loebl Chromoscan and sliced on a Mickle gel slicer. ————————, extinction at 265 nm; ⌐⌐⌐, histogram, [^{32}P] labelled RNA. The log molecular weight scale is indicated, assuming that the weight of the *r*RNA is 1·5 × 10^6 and 0·7 × 10^6, as in Table 1 (Loening *et al.* 1969).

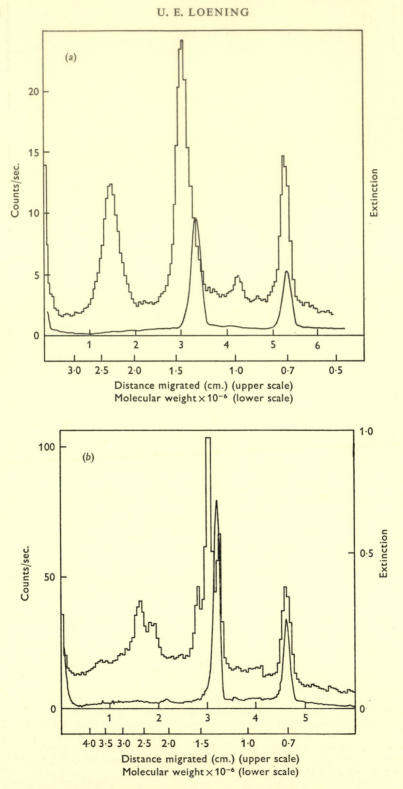

Distance migrated (cm.) (upper scale)
Molecular weight × 10⁻⁶ (lower scale)

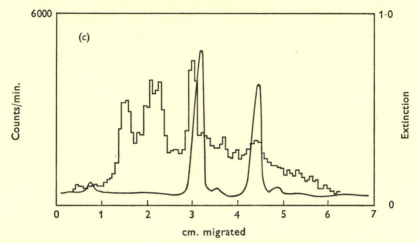

Fig. 4. Rapidly-labelled RNA in various plant species. (*a*) RNA from artichoke tuber explants was labelled by 45 min. incubation in [^{32}P]. The labelling was done after the explants had been in culture for about 30 hr; this is just after one cell division when the rate of ribosome synthesis is high. DNA was removed with deoxyribonuclease; electrophoresis was in a 2·4 % gel for 12·5 hr at 19 V. (Fraser & Loening, 1970). (*b*) RNA from the fission yeast, *Schizosaccharomyces pombe*, was labelled by 7·5 min. incubation in [^{32}P]. DNA was removed; electrophoresis was in a 2·2 % gel for 3 hr at 50 V. The extinction profile indicates a minor component with a molecular weight of 2×10^6; this is exactly the sum of the weights of the *r*RNA components and is probably an aggregate of them. The precursors were clearly heavier than this aggregate and the difference gives a direct visualization of the amount of excess RNA. (*c*) RNA from carrot slices, labelled for 40 min. in [^{32}P]. The slices had previously been washed in water for 6 hr, by which time the rate of ribosome synthesis was high (Leaver & Key, 1970). ————, extinction at 265 nm; ⌐⎺⌐ histogram, radioactivity.

nuclear fractions. (The choice of the HeLa cell for the earliest studies was fortunate; the precursor in cultured mouse cells is much less stable; S. Penman, personal communication).

Gel-electrophoretic fractionations of pulse-labelled RNA from several plant species are shown in Figs. 4, 6, 9. The apparent molecular weight of the precursors was different in different species and in some two distinct components were seen. RNA from the artichoke tuber tissue-culture showed a single heavy component, with an apparent molecular weight of $2·3 \times 10^6$. Pea precursor (Fig. 9) had a similar molecular weight but showed evidence of an additional slightly smaller component by the broadness of the labelled RNA peak. That there were two components was confirmed from an extinction scan of pea nuclear RNA (Rogers, Loening & Fraser, 1970). Yeasts are eukaryotic micro-organisms with a nucleus and nucleolus and showing mitotic division; the rapidly-labelled RNA has a high molecular weight like the precursor in higher plants (Fig. 4*b*). The time-course of labelling (Sassella & Loening, 1970) and competitive hybridization of the RNA to yeast DNA (Tabor & Vincent, 1969) suggested that this high molecular weight component is the precursor of rRNA. Its base composition could not be accurately determined because of the large amount of DNA-like RNA; it clearly consists of two components, with molecular weights of 2·2 and 2·6 × 10⁶ (Fig. 4b). Rapidly-labelled RNA from carrot discs (Leaver & Key, 1970) shows two precursor components like those in yeast (Fig. 4c).

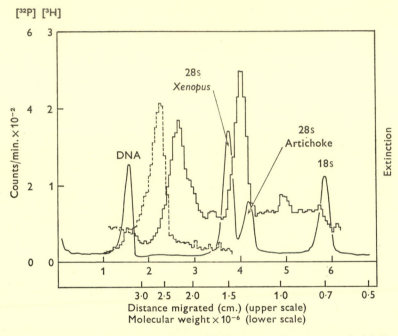

Fig. 5. Nuclear rapidly-labelled RNA from artichoke tissue culture compared with *Xenopus* RNA. *Xenopus* kidney cells were labelled with [³H]uridine for 1 hr and RNA extracted. Artichoke explants were labelled for 45 min. with [³²P] as in Fig. 4a. The tissue was homogenized in a buffered sucrose medium and the broken nuclei and cell walls sedimented at 1000 g. RNA was extracted from the sediment (Rogers, Loening & Fraser 1970). DNA was not removed from either preparation. Portions of the *Xenopus* and artichoke preparations were mixed and applied to a 2·2 % gel; electrophoresis was for 3·5 hr at 50 V. ————, extinction; ‿⌐ㄴ‐, [³H]; ‿⌐ㄴ , [³²P]radioactivity.

The molecular weights of these plant rRNA precursors were confirmed by comparison with the *Xenopus* precursor in double-labelled experiments. Fig. 5 shows that the artichoke rRNA precursor was about 0·2 × 10⁶ smaller than that of *Xenopus*, as is the '28s' RNA. Therefore the amount of excess non-ribosomal RNA is about the same in both species, 0·35 × 10⁶. Fig. 5 also shows that the artichoke precursor was nuclear, and that the nuclear preparation contained little labelled 18s (0·7 × 10⁶) rRNA (Rogers, Loening & Fraser 1970). Fig. 6 shows a comparison of mung bean (*Phaseolus aureus*) leaf RNA with RNA of *Xenopus*. The precursors were almost identical on gel electrophoresis; the bean precursor RNA therefore included nearly 0.6×10^6 of non-ribosomal excess. Also, the bean precursor showed a trail on the lighter side; the extent of this varied with age of leaf. A trail suggests a range of molecular sizes rather than a distinct second component. The compositions of the precursor

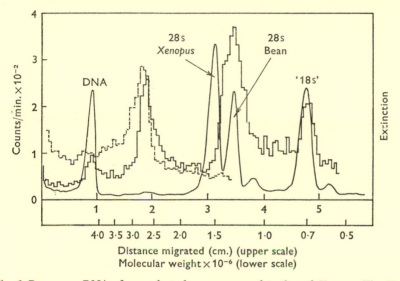

Fig. 6. Precursor rRNA of mung bean leaves compared to that of *Xenopus*. The *Xenopus* RNA preparation was the same as in Fig. 5. The leaves of mung bean seedlings (*Phaseolus aureus*, 4 days old) were labelled by direct application of [32P]. The RNA was extracted by the Parish & Kirby procedure (1966) (D. Grierson, unpublished). The RNA equivalent to that in one leaf was mixed with the *Xenopus* RNA and applied to a 2·2 % gel; electrophoresis was for 3 hr at 50 V. ———, extinction at 256 nm; ⌐⌐ , [32P] labelled RNA; ⌐⌐- , [3H] labelled RNA. The extinction scan shows the DNA of *Xenopus* and the bean as a single peak, the separation of the '28 s' rRNAs which had molecular weights of 1.5×10^6 and 1.3×10^6, the '18 s' rRNA which was the same for both species, and the chloroplast RNA from the bean leaves which had molecular weights of 1.1×10^6 and 0.56×10^6 (compare Table 1). The [32P] labelled bean RNA shows that the precursor was coincident with *Xenopus* RNA; the broad peak near the '28 s' RNA includes the 1.4×10^6 component, the 1.3×10^6 rRNA and some label in the 1.1×10^6 chloroplast RNA.

RNAs from the higher plants described above was 28 to 30 mole % G, which is a slightly lower G content than that of the combined rRNA. The determinations are not yet sufficiently exact to deduce the composition of the non-ribosomal component.

These experiments suggest that a rRNA precursor which includes both ribosomal components and some excess RNA occurs in all eukaryotes. The presence of a very high molecular weight precursor which contains a large excess of non-ribosomal RNA seems to be a peculiarity of mammals; it would be interesting to examine birds and reptiles. If there are differences in the molecular weights of precursor as between closely related species, and a spread of weights within a species, as suggested by the pea and bean experiments, then different tissues of the same organisms may be worth investigating. In any case, the properties of the plant precursor suggest a more complex situation than a single precursor molecule.

The stages of processing of the precursor of rRNA

Mammals. The sequence through which the precursor molecule is cleaved to rRNA and by which excess RNA is discarded has been most thoroughly investigated in HeLa cells by pulse and chase incubations. Low concentrations of actinomycin D added for the chase incubations were used to stop further rRNA synthesis while processing continued at slightly decreased rate. Labelling with ([^{14}C]methyl)-methionine, which labels only the 2′-O-methyl on ribose, also allowed effective chase incubations, since protein synthesis rapidly uses up methionine. Also, methionine does not label the heterogeneous RNA, so that the early stages of synthesis of the precursor could be studied. In these ways it was shown that label in the 45s precursor was transferred through the nucleolar RNA components in Fig. 2 to rRNA (Penman, 1966; Greenberg & Penman, 1966; Weinberg *et al.* 1967). The 45s precursor was labelled within a few minutes; the growth of the RNA chains was seen in electrophoresis of methionine-labelled RNA as a trail on the light side of the precursor peak. This indicated that the time of synthesis of precursor was about 2·5 min. The life time of the precursor was about 15 min.; it was then cleaved, in stages, to 32s RNA and 20s RNA. The 32s RNA had a lifetime of about 1 hr being cleaved to 28s rRNA which was transported to the cytoplasm. The 20s RNA seen in nucleolar preparations is probably the immediate precursor of 18s rRNA; this is rapidly transported to the cytoplasm (Girard, Penman & Darnell, 1964; McConkey & Hopkins, 1965; Edstrom & Daneholt, 1967; Gall, 1966; Rogers, 1968; Weinberg *et al.* 1967). In steady-state growth, the lifetimes of all the precursor molecules should be proportional to the molar amounts present in the nucleolar preparation. The two minor components seen between the 45s and 32s RNAs in Fig. 2 would have lifetimes of about 2 min. if they were intermediate cleavage products. This is too short to be directly measured, but evidence that they are intermediates in processing comes from their accumulation after polio virus infection discussed later (Weinberg *et al.* 1967).

Like heterogeneous RNA, the non-ribosomal RNA in the precursor is not methylated. The conservation of methyl-labelled RNA and loss of non-methylated RNA during processing was shown in an experiment suggested by Weinberg. In a long incubation of HeLa cells in a mixture of ([^{3}H]methyl)-methionine and [^{14}C]uridine, the ratio of [^{14}C] to [^{3}H] decreased with each successive stage of processing (Weinberg *et al.* 1967). This ratio should be proportional to the molecular weights of the various precursor molecules, assuming that the [^{3}H]methyl label is completely conserved in the *r*RNA and that the non-ribosomal RNA has the same average uridine content as *r*RNA. The molecular weight of the 45s RNA obtained in this way agreed with that obtained by the other methods. All the experiments showing non-conservative processing of precursor were therefore quantitatively consistent.

The non-ribosomal RNA which is discarded from the precursor has not been detected separately; it must be assumed to be broken down very rapidly after it is cleaved. Nucleoli incubated *in vitro* in suitable media continue to process the 45s precursor to some extent (Liau, Craig & Perry, 1969; Vesco & Penman, 1969). Under these conditions a non-methylated RNA component, of about 26s, was found; this may be one of the cleaved products which is not broken down, due perhaps to loss of the required nuclease during the preparation.

Other eukaryotes. The amount of excess non-ribosomal RNA in the precursors of other species is much less than in mammals; it is likely that there are then fewer stages in the processing. Only RNA from whole cells or from crude nuclear preparations have, at present, been analysed, so that less detail is observed than with purified nucleoli from HeLa cells or mammalian liver. In higher plants the *r*RNA precursor becomes labelled after about 10 min. of incubation, followed after 30 min. by the 18s (0.7×10^6) *r*RNA and a component which is slightly heavier than the 28s (1.3×10^6) *r*RNA and similar to it in composition. This component is the immediate precursor of 1.3×10^6 *r*RNA and thus is analogous to the 32s RNA of mammals; however, its molecular weight is only 1.4×10^6, whereas the 32s RNA is about 2.2×10^6, which is 0.45×10^6 greater than the *r*RNA. Fig. 4 shows the 2.3 to 2.6×10^6 and the 1.4×10^6 precursors from several plants; Fig. 5 shows that these occurred in the nuclear preparation. As in mammalian nuclei, there was almost no labelled 0.7×10^6 *r*RNA in the artichoke nuclear preparation, since this component was rapidly transported into the cytoplasm. Chase-incubations after short pulse-incubations with [^{32}P]phosphate were effective in this artichoke culture without the use of actinomycin D; the amount of label in the 2.3×10^6 precursor decreased during the chase, while the

$1\cdot4 \times 10^6$ RNA and the rRNA label increased (Fraser & Loening, 1970). In most other plant tissues such a chase-incubation was less effective, but timed incubations indicated that the $2\cdot3 \times 10^6$ component was the first precursor, with a short life, and that the $1\cdot4 \times 10^6$ RNA had a life up to 1 hr in the nucleus (Leaver & Key, 1970). In yeast and carrot, where there are two high molecular weight precursors, it is not yet known whether the heavier is a precursor of the lighter, or whether they are synthesized independently.

The components seen in pulse-labelled RNA of *Xenopus* cells are in general similar to those of plants (Fig. 3). The immediate precursor to the '28s' ($1\cdot5 \times 10^6$) rRNA has a molecular weight of about $1\cdot6 \times 10^6$; there may also be a short-lived component of about $1\cdot75 \times 10^6$ (Fig. 3a) but its existence was not established in these preparations of total cell RNA.

The proposed schemes of processing of the precursors in the different species is summarized in the following scheme.

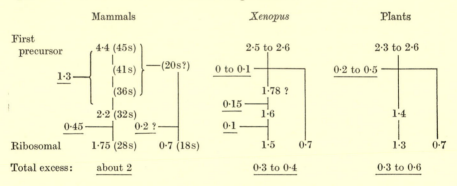

The diagram of the mammalian scheme is adapted from Weinberg *et al.* (1967) and gives apparent sedimentation coefficients and molecular weights as determined by gel electrophoresis. The scheme for *Xenopus* and the plants is the simplest consistent with the results of the labelling experiments described above. The amounts of excess RNA discarded at each stage are indicated underlined.

Common to all species is a 'polycistronic' precursor which is cleaved into at least two components: the larger of these remained for a relatively long period in the nucleolus and was then cleaved to '28s' rRNA; the smaller component was the 20s RNA, which was rapidly cleaved to 18s rRNA and transported into the cytoplasm. The existence of 20s RNA in *Xenopus* and in plants is not yet proved. A nucleolar particle containing 20s RNA has been found in amphibians (Rogers, 1968) but was not apparent in the RNA preparations illustrated in Fig. 3. Many plant tissues have an RNA component with a molecular weight of about 0.9×10^6 which might be like the 20s RNA of mammals. This can be seen in the nuclear preparation of the artichoke RNA in Fig. 5, and is prominent in pea root total cell RNA when processing is inhibited (see Fig. 8). If such a component is the precursor to 0.7×10^6 rRNA, then no further intermediates are possible: the 0.9 and 1.4×10^6 components together account for the molecular weight of the precursor of 2.3×10^6. Thus the exact details of the stages by which the non-ribosomal excess RNA is lost are not known. Even in the HeLa cell, it is not clear whether the 20s RNA is the first molecule to be cleaved from the precursor or whether a large piece of excess RNA is first. These uncertainties are indicated in the diagram above. The amount of detail is however sufficient to suggest two further questions. One question is what is the structure of the genes which code for the rRNA; the other question is whether or not there are any differences in the stages of processing which may be correlated with the stage of growth or metabolism of the tissue.

Correlation with the structure of the ribosomal genes

The proportion of the genome which codes for rRNA varies in different species from less than 0.1% to 2%, according to the total amount of DNA/nucleus and to the number of reiterated copies of ribosomal genes. In yeast 140 copies account for about 2% of the total genome (Schweizer, Mackechnie & Halvorson, 1969). In *Xenopus* the diploid genome contains about 1500 copies, which is about 0.08% of the DNA (Birnstiel *et al.* 1966; Brown, 1967). The synthesis of a polycistronic rRNA precursor on these genes suggests that the DNA sequences coding for 18s and 28s RNA alternate with each other. It is improbable that the DNA which codes for each rRNA component exists as a long block of repeating sequences. This question was investigated by molecular hybridization of rRNA with the isolated rDNA satellite of *Xenopus* (Birnstiel *et al.* 1968; Brown & Weber, 1968). The rDNA was isolated and studied by caesium chloride gradient centrifugation; it had a higher content of GC and thus a higher buoyant density than the bulk DNA. The rDNA was sheared or ultrasonically treated to determine how low the molecular weight must be before the regions coding for the 1.5 and 0.7×10^6 rRNA can become separated. The 0.7×10^6 rRNA had a lower GC content than the 1.5×10^6, so that the DNA coding for it should have a lighter buoyant density. This should be observed when the molecular weight of the DNA is low enough to allow the regions to separate. The technical difficulty which arises is that the band width on caesium chloride gradient centrifugation increases at low molecular weights of DNA, making the determinations less precise. One would expect half of any given DNA segment to have been liberated when the molecular weight of the DNA is half that of the segment under scrutiny.

The buoyant density of *Xenopus* rDNA is 1·731 g./cm^3. When the DNA had a molecular weight of about 4×10^6, both the 0·7 and the $1·5 \times 10^6$ rRNA hybridized to DNA of this density. When the molecular weight was decreased to $1·6 \times 10^6$, and the DNA fractionated on caesium chloride gradient centrifugation, the $0·7 \times 10^6$ rRNA hybridized to DNA of slightly lower density. When the molecular weight was further decreased to 0·3 to $0·5 \times 10^6$, the $0·7 \times 10^6$ rRNA hybridized to DNA fractions with an average density of 1·712 to 1·714, and the $1·5 \times 10^6$ rRNA to DNA at 1·720 to 1·722 density. Some separation had therefore been achieved; these densities were near those calculated from the GC contents of the rRNA. It appears therefore that it is necessary to shear the DNA to below the length of the RNA to achieve significant separation of the cistrons. Therefore the repeated stretches which code for each rRNA component do not occur in long blocks but are closely integrated with each other.

If the two ribosomal regions of the DNA alternate, then hybridization of saturating amounts of $0·7 \times 10^6$ rRNA should increase the apparent density of DNA complementary to the $1·5 \times 10^6$ rRNA since a hybrid molecule is considerably denser than the denatured DNA, and the hybridized $0·7 \times 10^6$ rRNA region will carry the non-hybridized DNA with it. Additional hybridization with $1·5 \times 10^6$ rRNA will then further increase the apparent density of both regions if they are integrated. The DNA must be sheared as before if hybridization of one RNA component is not to increase the density of the other hybrid. The increased density of RNA–DNA hybrids above denatured DNA is large, so that the method is very sensitive. When single-stranded rDNA with a molecular weight of 0·18 to $0·2 \times 10^6$ was hybridized to saturating amounts of $0·7 \times 10^6$ rRNA or to both rRNA components the density of the hybrids was high. This indicated that the ratio RNA:DNA was high and little non-hybridized DNA was attached to the hybrid; hybridization of one component did not affect that of the other. At this molecular weight therefore the two regions on the DNA behaved independently. When $0·7 \times 10^6$ rRNA was hybridized to DNA with a molecular weight of about 3×10^6, the hybrid had a lower density than before, showing that some non-hybridized DNA remained attached. Further hybridization with $1·5 \times 10^6$ rRNA increased the density of the $0·7 \times 10^6$ rRNA hybrid to a value close to that obtained with the 0·18 to $0·2 \times 10^6$ ultrasonically treated DNA (Fig. 7). Thus when the DNA molecular weight was about equivalent to one genome length of the two ribosomal components the sequences were joined. It follows that the sequences for *r*RNA alternate.

These experiments also showed that there was some DNA with a high GC content which did not hybridize with *r*RNA. This is readily

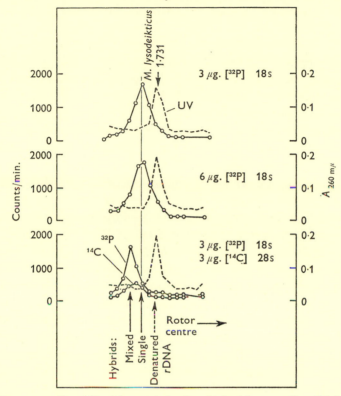

Fig. 7. The co-operative shift of buoyant density obtained by annealing '28s' and '18s' *r*RNA to *r*DNA. Denatured *r*DNA was prepared with a single-strand molecular weight of 3×10^6; this is slightly longer than the ribosomal RNA precursor. The DNA was hybridized in solution to [32P] labelled 18s *r*RNA (upper two figures) or to both [32P] labelled 18s and [14C] labelled 28s *r*RNA (lower figure). The buoyant densities of the hybrids formed were determined on caesium chloride gradients using *Micrococcus lysodeikticus* DNA (1·731 g./ cm³) as marker. The top curve shows the density of the single hybrid with 18s RNA and the second curve that a larger excess of 18s RNA made no difference. The lower curve shows that the hybrid was shifted to a higher density when 28s RNA also hybridized (Birnstiel *et al.* 1968).

seen, since the buoyant density of the whole *r*DNA is higher than that of either of the separated sequences complementary to *r*RNA. Some of this extra DNA must be that which codes for the excess RNA in the ribosomal precursor. This however can be only 12 % of the total, since that is the percentage of excess RNA in the precursor. The amount of extra DNA not hybridizable to *r*RNA is less than 60 %. It is very likely therefore that there are stretches of DNA in the *r*DNA which are not transcribed.

We may conclude then that the structure of the genes which code for rRNA consists of alternating regions coding for the 0·7 and the 1·5 × 10⁶ rRNA components, interspersed with small regions coding for the excess RNA. Each whole group is separated by non-transcribed DNA which may be up to one-third or one-half of the total length.

Electron microscopy of the nucleolar genes

The above results can be correlated with the electron microscope pictures obtained by Miller & Beatty (1969 a, b), showing the process of RNA synthesis in the nucleolar core. In preparations from the amplified nucleoli of *Xenopus* and *Triturus* oocytes, the pictures show the growing chains of the ribosomal precursor nucleoprotein on the polycistronic deoxyribonucleoprotein chain (Plate 1). Selective enzyme digestion showed that the central filament contained DNA, and the side chains, of increasing length in each region, contained RNA. There are about 100 growing ribonucleoprotein chains on each 'gene' or matrix unit. The length of the deoxyribonucleoprotein of these transcribed regions is constant and approximately $2·5\,\mu$m long. This corresponds closely to the length of the DNA double helix, when in the B-conformation, which should code for an RNA molecule with a molecular weight of $2·5 \times 10^6$. This is in exact agreement with the molecular weight of the *Xenopus* rRNA precursor. This agreement may only be apparent, because the deoxyribonucleoprotein strand can be stretched without breaking during preparation; whether such extension of the DNA helix could occur is not known (Dr O. L. Miller, personal communication). The exactness of the correlation with the weight of the precursor must therefore be further investigated.

Between each region of transcription, or matrix unit, there are regions which are apparently not transcribed. The lengths of these are variable and account for about one-third of the total DNA. This is also in approximate agreement with the hybridization experiments described above.

Inhibition and regulation of rRNA synthesis

This section describes some experiments with inhibitors and special growth conditions which represent the beginnings of an investigation into control mechanisms.

The synthesis of ribosomes must be matched to the requirements of the cell: a rapidly dividing cell will make more new ribosomes in unit time than a differentiated cell such as a liver cell or the older cells of a root tip. Synthesis could be controlled at transcription, thus limiting the supply of precursor, or by the rate of processing. The latter might result in some form of feed-back control when the concentration of nucleolar precursor particles increased. When the requirement for ribosome synthesis is extremely high, as in developing amphibian oocytes, the amount of rDNA is enormously increased by the synthesis of large numbers of extra-nucleoli per cell (Gall, 1968; Evans & Birnstiel, 1968). In general, however, it is unlikely that the rate of rRNA synthesis is directly controlled by the amount of rDNA. For example, the heterozygote *Xenopus* mutant which lacks one nucleolus and so has only the haploid amount of rDNA (Birnstiel *et al.* 1966) synthesizes the normal amount of ribosomes and grows into a phenotypically normal organism.

The processing of precursor must depend on the availability of the proteins required for the assembly of nucleolar particles. It may be expected that inhibitors of protein synthesis will inhibit processing and maturation of ribosomes, but may not have a direct effect on the transcription of precursor RNA. Inhibition of protein synthesis in HeLa cells results in a decreased rate of ribosome production (Warner *et al.* 1966). Since some ribosomes are still formed there must be a pool of ribosomal structural proteins. The formation of 18s RNA from the 45s precursor ceases soon after inhibition, although 32s and 28s are still formed (Ennis, 1966; Soeiro, Vaughan & Darnell, 1968). With puromycin inhibition (Soeiro *et al.* 1968) and cycloheximide inhibition (Willems, Penman & Penman, 1969) the lifetime of the 45s precursor is increased and its rate of synthesis decreased; the amount of 45s precursor in the cycloheximide-inhibited cell remains approximately constant. The slow processing of the precursor led to a quantitative production of 32s and some 28s RNA, but no 18s RNA was found. It must be concluded that the latter is selectively destroyed. The cleavage of the 32s RNA to 28s appears to be regulated so that the amount of 32s RNA in the nucleolus remains constant.

Cycloheximide also inhibits ribosome formation in yeast and probably results in accumulation of precursor (Tabor & Vincent, 1969). Vinblastine sulphate, which is used medically in tumour therapy, appears similarly to inhibit the transport of ribosomal particles to the cytoplasm (Wagner & Roizman, 1968). Inhibition of processing of precursor, without much inhibition of its synthesis nor of protein synthesis, was observed in HeLa cells grown at 42° instead of the normal 37° (Warocquier & Scherrer, 1969). Infection of HeLa cells with polio virus, even in the presence of guanidine when the virus cannot multiply but does inhibit host protein synthesis, inhibited processing at the 41s stage; this component accumulated as indicated by its extinction in nucleolar RNA after infection (Weinberg *et al.* 1967).

The cells of pea-root tip, in which the rate of r-RNA synthesis is high, can be caused to 'shift-down' by excising the tip or the region of rapidly expanding cells just behind the tip, and growing in 2 % sucrose. Under these conditions cell expansion continues and is in some way influenced by RNA metabolism (Heyes & Vaughan, 1967); but cell division stops and net accumulation of protein and RNA is inhibited; turnover of these continues, however, and some enzyme activities change. Cell division can be started again with auxins in suitable culture media. Fig. 8 shows that in the excised root tip the rRNA precursor was labelled by [^{32}P]. However, even after prolonged incubation, very little cleavage products were formed. After labelling for about 3 hr, the precursor was still the most prominent labelled component in the RNA of the excised root, whereas in the intact root the mature rRNA was labelled (Fig. 8) (Rogers, Loening & Fraser 1970). The older cells of the root tip showed a greatly decreased rate of rRNA synthesis; it was interesting to ask whether this was also correlated with a slower processing of precursor. Fig. 9 compares the RNA from the tip and older part of the root after a short incubation with [^{32}P]. The time of incubation was such that the radioactivity in the $1\cdot4 \times 10^6$ and $0\cdot7 \times 10^6$ RNA was increasing rapidly. The ratio of label in these components to that in the $2\cdot3 \times 10^6$ precursor then gives a rough measure of the rate of processing. The results in Fig. 9 suggest that the rate of cleavage of precursor was as high or even higher in the older cells as in the younger, since the proportion of label in the precursor of the older cells was lower. We have also found that the absolute amount of precursor per cell is less in older cells than in younger cells (Rogers, Loening & Fraser 1970). This shows that the average rate of synthesis of precursor was decreased in older cells, without a change in the rate of processing.

None of these experiments shows whether under normal conditions of growth and development the synthesis of ribosomes could be regulated by the rate of processing of precursor. Even in the 'shift-down' conditions protein synthesis was inhibited, so that the situation may be similar to that obtained by using inhibitors. Clearly the techniques are now suitable for a closer investigation of differentiating tissues. To illustrate the possibilities Fig. 10 shows a comparison of the HeLa cell and rat liver precursors (K. Shankar Narayan & U. E. Loening, unpublished). The HeLa preparation was of [^{3}H] labelled total cell RNA and the rat liver preparation of *in vivo* [^{14}C] labelled nucleolar RNA. The extinction profile shows the HeLa 28s and 18s components; all other components seen in the extinction profile are from liver nucleolar RNA. The 28s RNA of the HeLa cell and of rat liver cell have separated

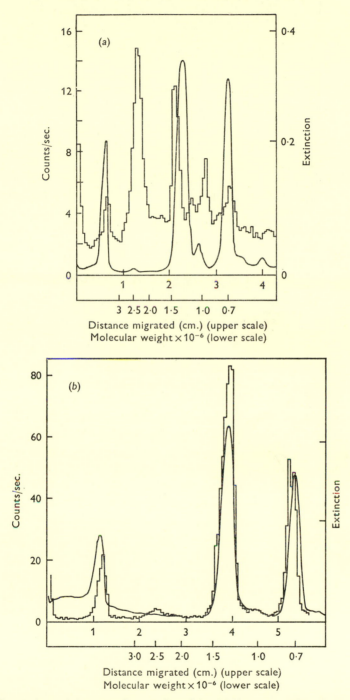

Fig. 8. Comparison of the rates of processing of the rRNA precursor in excised and intact pea root tips. (a) Excised root tips, 3 mm. long, were labelled in [³²P] for 1 hr followed by [³¹P]phosphate for 2 hr; the medium contained also 2 % sucrose. Electrophoresis of the nucleic acids was on a 2·2 % gel for 3 hr at 50 V. (b) Intact seedlings were labelled in [³²P] for 45 min. followed by incubation in water for 3 hr. Electrophoresis of the nucleic acids in a 2·2 % gel for 3·5 hr (Rogers, Loening & Fraser 1970). ————, extinction; ⌐‾⌐, [³²P]radioactivity.

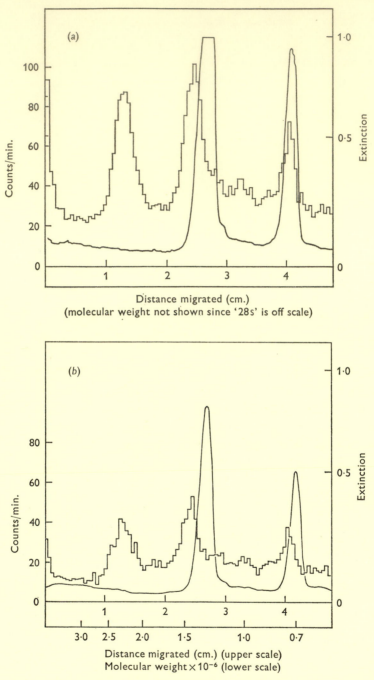

Distance migrated (cm.)
(molecular weight not shown since '28s' is off scale)

Distance migrated (cm.) (upper scale)
Molecular weight × 10⁻⁶ (lower scale)

Fig. 9. Comparison of *r*RNA synthesis in the meristem and older regions of pea root tips. Intact roots were labelled in [³²P] for 15 min. followed by incubation in 10^{-5} M-phosphate for 20 min. The roots were then cut into segments: (1) the tip 0 to 1·6 mm.; (2) 1·6 to 3·4 mm.; (3) 3·4 to 6·4 mm. Nucleic acids were extracted and DNA digested. RNA equivalent to about six root segments was applied to each polyacrylamide gel; electrophoresis was in 2·4 % gels for 3 hr at 50 V. The results from the first (*a*) and the third (*b*) segments are shown. The radioactivity profiles in this and similar experiments showed that the rate of formation of the $1·4 \times 10^6$ component was at least as great in the old as in the dividing parts of the root. ———, extinction; 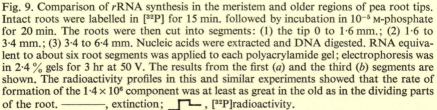, [³²P]radioactivity.

slightly, with molecular weights of about 1·75 and 1·70 × 10⁶. The time of labelling in both cases was just sufficient to label the precursor and the first cleavage products. It is seen that the precursors in the two prepara- tions were identical in molecular weight. However, the labelled HeLa 32s RNA did not compare with any of the rat liver components. The rat

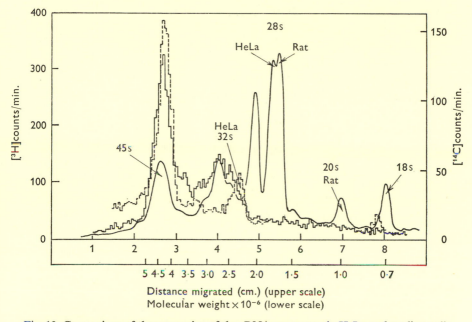

Fig. 10. Comparison of the processing of the rRNA precursors in HeLa and rat liver cells. A rat was labelled by intraperitoneal injection of [¹⁴C]orotic acid. After 10 min. the animal was killed and nucleolar particles prepared from the liver (K. Shankar Narayan, in preparation). RNA was extracted from the nucleolar preparation essentially as described by Parish & Kirby (1966), at about 5°. HeLa cells were labelled, in layer culture, by addition of [³H]uridine, for 1 hr. RNA was extracted from the whole cells by the same method and DNA digested. Samples of the rat and HeLa RNA were mixed and applied to a 2·0 % gel; electrophoresis was for 4 hr at 60 V. The gel was scanned at 265 nm in a Joyce Loebl Chro- moscan, and sliced on a Mickle gel slicer. The RNA in the slices was hydrolysed in 10 % piperidine and the two isotopes counted in a Packard scintillation spectrometer. ——————, extinction; ⌐⌐⌐, [³H] labelled HeLa cell RNA; ⌐⌐, [¹⁴C] labelled rat nucleolar RNA.

nucleolar RNA contained a complex labelled component which was heavier than the HeLa 32s, and an unlabelled component lighter than the 32s. This experiment makes three different comparisons: between species, between tissues and between RNA preparations from whole cells and from nucleoli. It serves to show the possibilities for further investigation of the stages of processing and to indicate that the scheme presented for the HeLa cell may not be common to all mammalian cells.

CONCLUSIONS

Ribosomal RNA is synthesized in bacteria and in higher organisms as molecules which are larger than the final product. The same is probably true of transfer RNA and of 5s RNA (Bernhardt & Darnell, 1969; Burdon & Clason, 1969; Hecht, Bleyman & Woese, 1968). Thus all the known homogeneous types of RNA are synthesized with some apparently redundant excess which is later discarded. This may therefore be a general phenomenon and may have wider implications. Messenger RNA is not at present sufficiently well defined or fractionated for its precursors to be identified. The amount of excess RNA in the ribosomal precursors cannot be correlated with any obvious property of the cells or with the evolutionary position of the species. One might, for example, have expected all the animals to be similar to each other and different from the plants; in fact of the species studied the presence of a large amount (40 %) of excess RNA is peculiar to the mammal. It is then unlikely that the excess RNA is a messenger RNA for the ribosomal proteins or for the enzymes required for processing, since the amounts in *Xenopus* and in plants are much too low. Similarly it is unlikely to have some structural function in the assembly of the relatively similar ribosomes. In addition to the apparently redundant RNA there is also some non-transcribed DNA. One might suggest that the difference between the *Xenopus* and HeLa cell rDNA is not in the length of the genome but in the positions on the DNA where transcription starts and stops. The synthesis of the high molecular weight precursor containing 40 % of excess RNA would then imply that the length of the non-transcribed regions in HeLa cell DNA is smaller.

The multiplicity of the ribosomal genes raises the problem of whether all the copies are identical and if so, how mutational drift is avoided. There is some evidence from the kinetics of hybridization of rRNA with rDNA that the sequences of ribosomal RNA are not exactly alike (Dr M. L. Birnstiel, personal communication). There is also evidence that rRNA precursor may not be quite homogeneous. The precursors from the different plant species (Fig. 4) suggest that at least two classes of molecule can be found. In the HeLa cell RNA (Fig. 2) and in *Xenopus* (Fig. 3) the precursor peak is broader than would be expected for a unique molecule. This apparent heterogeneity might be due to changes during synthesis and processing or to true differences in gene length. Further detailed analysis of this molecule is required.

All the eukaryotes seem to have in common the synthesis of the polycistronic precursor, the rapid transport of the smaller ribosomal sub-unit to the cytoplasm and the slow maturation of the larger sub-unit in the nucleolus. This system thus presumably has some common functional significance, such as with the assembly of ribosomal sub-units with messenger RNA (McConkey & Hopkins, 1965). Now that the intermediate molecules from gene product to rRNA can be reasonably clearly identified, one may hope that a detailed study of these and of the corresponding nucleolar particles will help to solve some of the mysteries of gene action.

REFERENCES

BERNHARDT, D. & DARNELL, J. E. (1969). t-RNA synthesis in HeLa cells: a precursor to t-RNA and the effects of methionine starvation on t-RNA synthesis. *J. molec. Biol.* **42**, 43.

BIRNSTIEL, M. L. (1967). The nucleolus in cell metabolism. *A. Rev. Pl. Physiol.* **18**, 25.

BIRNSTIEL, M. L., SPIERS, J., PURDOM, I., JONES, K. & LOENING, U. E. (1968). Properties and composition of the isolated ribosomal satellite of *Xenopus laevis. Nature, Lond.* **219**, 454.

BIRNSTIEL, M. L., WALLACE, H., SIRLIN, J. L. & FISCHBERG, M. (1966). Localization of the ribosomal DNA complements in the nucleolar organizer region of *Xenopus laevis. Natn. Cancer Inst. Monogr.* **23**, 431.

BRAMWELL, M. E. & HARRIS, H. (1967). The origin of the polydispercity in sedimentation patterns of rapidly labelled nuclear ribonucleic acid. *Biochem. J.* **103**, 816.

BROWN, D. D. (1967). The genes for ribosomal RNA and their transcription during amphibian development. In *Current Topics in Developmental Biology*, vol. 2, p. 47. Eds. A. Monroy and A. Moscona.

BROWN, D. D. & GURDON, J. B. (1964). Absence of ribosomal RNA synthesis in the anucleolate mutant of *Xenopus laevis. Proc. natn. Acad. Sci. U.S.A.* **51**, 139.

BROWN, D. D. & WEBER, C. S. (1968). Gene linkage by RNA-DNA hybridisation. II. Arrangement of the redundant gene sequences for 28 s and 18 s ribosomal RNA. *J. molec. Biol.* **34**, 681.

BURDON, R. H. & CLASON, A. E. (1969). Intracellular location and molecular characteristics of tumour cell transfer RNA precursors. *J. molec. Biol.* **39**, 113.

CLICK, R. E. & TINT, B. L. (1967). Sedimentation coefficients of ribosomal RNA. *J. molec. Biol.* **25**, 111.

EDSTROM, J.-E. & DANEHOLT, B. (1967). Sedimentation properties of the newly synthesized RNA from isolated nuclear components of *Chironomus tentans* salivary gland cells. *J. molec. Biol.* **28**, 331.

ENNIS, H. L. (1966). Synthesis of ribonucleic acid in L cells during inhibition of protein synthesis by cycloheximide. *Mol. Pharmacol.* **2**, 543.

EVANS, D. & BIRNSTIEL, M. L. (1968). Localization of amplified ribosomal DNA in the oocyte of *Xenopus laevis. Biochim. biophys. Acta* **166**, 274.

FUJISAWA, T. & MURAMATSU, M. (1969). Studies on the sedimentation properties of nucleolar and extra-nucleolar nuclear RNA. *Biochim. biophys. Acta* **169**, 175.

GALL, J. G. (1966). Nuclear RNA of the Salamander oocyte. *Natn. Cancer Inst. Monogr.* **23**, 475.

GALL, J. G. (1968). Differential synthesis of the genes for ribosomal RNA during amphibian oogenesis. *Proc. natn. Acad. Sci. U.S.A.* **60**, 553.

GIRARD, M., PENMAN, S. & DARNELL, J. E. (1964). The effect of actinomycin D on ribosome formation in HeLa cells. *Proc. natn. Acad. Sci. U.S.A.* **51**, 205.

GRANBOULAN, N. & SCHERRER, K. (1969). Visualization in the electron microscope and size of RNA from animal cells. *Europ. J. Biochem.* **9**, 1.

GREENBERG, K. & PENMAN, S. (1966). Methylation and processing of ribosomal RNA in HeLa cells. *J. molec. Biol.* **21**, 527.

HECHT, N. B., BLEYMAN, M. & WOESE, C. R. (1968). The formation of 5 s ribosomal ribonucleic acid in *Bacillus subtilis* by post-transcriptional modification. *Proc. natn. Acad. Sci. U.S.A.* **59**, 1278.

HECHT, N. B. & WOESE, C. R. (1968). Separation of bacterial ribosomal ribonucleic acid from its macromolecular precursors by polyacrylamide gel electrophoresis. *J. Bact.* **95**, 986.

HEYES, J. K. & VAUGHAN, D. (1967). The effects of thiouracil on the growth of excised pea root tips. *Proc. Roy. Soc. Lond.* B **169**, 77.

JEANTUR, PH., AMALDI, F. & ATTARDI, G. (1968). Partial sequence analysis of ribosomal RNA from HeLa cells. II. Evidence for sequences of non-ribosomal type in 45 and 32 s ribosomal RNA precursors. *J. molec. Biol.* **33**, 757.

LANDESMAN, R. & GROSS, P. R. (1969). Patterns of macromolecular synthesis during development of *Xenopus laevis*; II. Identification of the 40 s precursor to ribosomal RNA. *Devl Biol.* **19**, 244.

LEAVER, C. J. & KEY, J. L. (1970). rRNA synthesis in plants. (In Press.)

LIAU, M. C., CRAIG, N. C. & PERRY, R. P. (1969). The production of ribosomal RNA from high molecular weight precursors. I. Factors which influence the ability of isolated nucleoli to process 45 s RNA. *Biochim. biophys. Acta* **169**, 196.

LOENING, U. E. (1968). Molecular weights of ribosomal RNA in relation to evolution. *J. molec. Biol.* **38**, 355.

LOENING, U. E. (1969). The determination of the molecular weight of ribonucleic acid by polyacrylamide gel electrophoresis. The effects of changes in conformation. *Biochem. J.* **113**, 131.

LOENING, U. E., JONES, K. & BIRNSTIEL, M. L. (1969). The properties of the ribosomal RNA precursor in *Xenopus laevis*; Comparison to the precursor in mammals and in plants. *J. molec. Biol.* (In Press.)

MADEN, B. E. H. (1968). Ribosome formation in animal cells. *Nature, Lond.* **219**, 685.

MAYO, V. S., ANDREAN, B. A. G. & DE KLOET, S. R. (1969). The effects of cycloheximide and 5-fluorouracil on the synthesis of ribosomal nucleic acid in yeast. *Biochim. biophys. Acta* **169**, 297.

McCLINTOCK, B. (1934). The relationship of a particular chromosomal element to the development of the nucleoli in *Zea mays. Z. Zellforsch. Mikrosk. Anat.* **21**, 294.

McCONKEY, E. H. & HOPKINS, J. W. (1965). Subribosomal particles and the transport of messenger RNA in HeLa cells. *J. molec. Biol.* **14**, 253.

McCONKEY, E. H. & HOPKINS, J. W. (1969). Molecular weights of some HeLa ribosomal RNAs. *J. molec. Biol.* **39**, 545.

MILLER, O. L. & BEATTY, B. R. (1969a). Visualization of nucleolar genes. *Science, N.Y.* **164**, 955.

MILLER, O. L. & BEATTY, B. R. (1969b). Extrachromosomal nucleolar genes in amphibian oocytes. *Genetics, Princeton.* (In Press.)

MURAMATSU, M., HODNETT, J. & BUSCH, H. (1966). Base compositions of fractions of nuclear and nucleolar ribonucleic acid obtained by sedimentation and chromatography. *J. biol. Chem.* **241**, 1544.

OSAWA, S. (1968). Ribosome formation and structure. *A. Rev. Biochem.* **37**, 109.

PARISH, J. H. & KIRBY, K. S. (1966). Reagents which reduce interactions between ribosomal RNA and rapidly labelled RNA from rat liver. *Biochim. biophys. Acta* **129**, 554.

PENMAN, S. (1966). RNA metabolism in the HeLa cell nucleus. *J. molec. Biol.* **17**, 117.

PENMAN, S., VESCO, C. & PENMAN, M. (1968). Localization and kinetics of formation of nuclear heterodisperse RNA, cytoplasmic heterodisperse RNA and polyribosome-associated messenger RNA in HeLa cells. *J. molec. Biol.* **34**, 49.

PERRY, R. P. (1962). The cellular sites of synthesis of ribosomal and 4 s RNA, *Proc. natn. Acad. Sci. U.S.A.* **48**, 2179.

PERRY, R. P. (1966). On ribosome biogenesis. *Natn. Cancer Inst. Monogr.* **23**, 527.

PERRY, R. P. (1967). The nucleolus and the synthesis of ribosomes. *Progress in Nucleic Acid Research* **6**, 219.

RITOSSA, F. M. & SPIEGELMAN, S. (1965). Localization of DNA complementary to ribosomal RNA in the nucleolus organizer region of *Drosophila melanogaster*. *Proc. natn. Acad. Sci. U.S.A.* **53**, 737.

ROBERTS, W. K. & NEWMAN, J. (1966). Use of low concentrations of actinomycin D in the study of RNA synthesis in Ehrlich ascites cells. *J. molec. Biol.* **20**, 63.

ROGERS, M. E. (1968). Ribonucleoprotein particles in the amphibian oocyte nucleus. *J. Cell Biol.* **36**, 421.

ROGERS, M. E., LOENING, U. E. & FRASER, R. S. S. (1970). *r*RNA precursors in plants. *J. molec. Biol.* (In Press.)

SASSELLA, D. & LOENING, U. E. (1970). The *r*RNA precursor in yeast. *J. molec. Biol.* (In Press.)

SCHERRER, K., LATHAM, H. & DARNELL, J. E. (1963). Demonstration of an unstable RNA and of a precursor to ribosomal RNA in HeLa cells. *Proc. natn. Acad. Sci. U.S.A.* **49**, 240.

SCHWEIZER, E., MACKECHNIE, C. & HALVORSON, H. O. (1969). The redundancy of ribosomal and transfer RNA genes in *Saccharomyces cerevisiae*. *J. molec. Biol.* **40**, 261.

SHANKER NARAYAN, K. (1969). (In preparation.)

SOEIRO, R., VAUGHAN, M. & DARNELL, J. E. (1968). The effect of puromycin on intranuclear steps in ribosome formation. *J. Cell Biol.* **36**, 91.

TABOR, V. & VINCENT, W. S. (1969). The effects of cycloheximide on ribosomal RNA synthesis in *Schizosaccharomyces pombe*. *Biochem. biophys. Res. Commun.* **34**, 488.

VESCO, C. & PENMAN, S. (1969). The fractionation of nuclei and the integrity of purified nucleoli in HeLa cells. *Biochim. biophys. Acta* **169**, 188.

WAGNER, E., KATZ, L. & PENMAN, S. (1968). The possibility of aggregation of ribosomal RNA during hot phenol-SDS deproteinization. *Biochem. biophys. Res. Commun.* **28**, 152.

WAGNER, E., PENMAN, S. & INGRAM, V. (1968). Methylation patterns of HeLa cell ribosomal RNA and its nucleolar precursors. *J. molec. Biol.* **29**, 371.

WAGNER, E. K. & ROIZMAN, B. (1968). Effect of the Vinca alkaloids on RNA synthesis in human cells in vitro. *Science, N.Y.* **162**, 569.

WARNER, J., GIRARD, M., LATHAM, H. & DARNELL, J. (1966). Ribosome formation in HeLa cells in the absence of protein synthesis. *J. molec. Biol.* **19**, 373.

WAROCQUIER, R. & SCHERRER, K. (1969). Synthesis of ribosomal RNA in HeLa cells grown at elevated temperatures. *Europ. J. Biochem.* (In Press.)

WEINBERG, R. A., LOENING, U. E., WILLEMS, M. & PENMAN, S. (1967). Acrylamide gel electrophoresis of HeLa cell nucleolar RNA. *Proc. natn. Acad. Sci. U.S.A.* **58**, 1088.

WILLEMS, M., PENMAN, M. & PENMAN, S. (1969). The regulation of RNA synthesis and processing in the nucleolus during inhibition of protein synthesis. *J. Cell Biol.* **41**, 177.

WILLEMS, M., WAGNER, E., LAING, R. & PENMAN, S. (1968). Base composition of ribosomal RNA precursors in the HeLa cell nucleolus: further evidence of non-conservative processing. *J. molec. Biol.* **32**, 211.

ZIMMERMAN, E. E. & HOLLER, B. W. (1967). Methylation of 45 s ribosomal RNA precursor in HeLa cells. *J. molec. Biol.* **23**, 149.

EXPLANATION OF PLATE

Plate 1. Visualization of the nucleolar genes. Nucleoli from *Triturus viridescens* oocytes were isolated in distilled water at pH 8·5, centrifuged through sucrose and formalin onto carbon-coated grids and stained with phosphotungstic acid. The electron micrograph shows matrix units which are about 2·5 μm long; about 100 fibrils of ribonucleoprotein of increasing lengths are attached to each matrix unit axis. Presumptive RNA polymerase molecules can be seen at the bases of the RNP fibrils. The small particles at the tips of the longer fibrils have not been identified. × 22,300. Photograph kindly provided by Dr Miller, similar to those in Miller & Beatty, (1969 *a*, *b*).

PLATE 1

(*Facing p.* 106)

CHARACTERISTIC METABOLIC PATTERNS OF PROKARYOTES AND EUKARYOTES

H. J. VOGEL, J. S. THOMPSON AND G. D. SHOCKMAN

Department of Pathology, College of Physicians and Surgeons,
Columbia University, New York, N.Y.

and

Department of Microbiology, Temple University School of Medicine,
Philadelphia, Pa. U.S.A.

It probably is accurate to say that most of the characteristic metabolic patterns of the prokaryotic or the eukaryotic cell are those associated with metabolites that are peculiar to one or the other of these cell types. However, from certain evolutionary points of view, special interest attaches to cases of different pathways leading to a metabolite that is common to prokaryotes and eukaryotes. Perhaps the most striking of such cases, in its scope and consistency, is that of lysine biosynthesis (Vogel, 1965). The present paper, therefore, will be concerned mainly with the formation of lysine, and to some extent with the formation of ornithine, a lower homologue of lysine.

The lysine dichotomy

Two distinctive lysine pathways are known to occur in Nature. One of them has α,ϵ-diaminopimelic acid (DAP) as a key intermediate and the other, α-aminoadipic acid (AAA); see Fig. 1, 2. The distribution of the two lysine pathways was explored with the aid of certain specifically labelled radiocarbon tracers (Vogel, 1959 a–c; 1960–1965). The general procedure consisted in allowing the organism under study to utilize the desired tracer (in the presence of an unlabelled main carbon source), and measuring the relative specific radioactivity of the lysine and aspartic acid from the protein of the culture obtained. Particularly useful tracers included 3- and 4-labelled aspartic acids, as well as 1-labelled alanine and 2-labelled acetate. With these tracers, the mode of lysine synthesis is revealed through diagnostic labelling patterns (Vogel, 1964, 1965). For example, in the case of the DAP-lysine path, aspartate-[4-^{14}C] labels protein aspartic acid and lysine at approximately equal specific activity; in the case of the AAA-lysine path, aspartate-[4-^{14}C] tends to label protein aspartic acid but not lysine. In contrast, aspartate-

[3-^{14}C] labels protein aspartic acid and lysine at about the same specific activity in both kinds of pathway. The results of a survey of various forms of life are summarized in Tables 1 and 2. In some instances, the

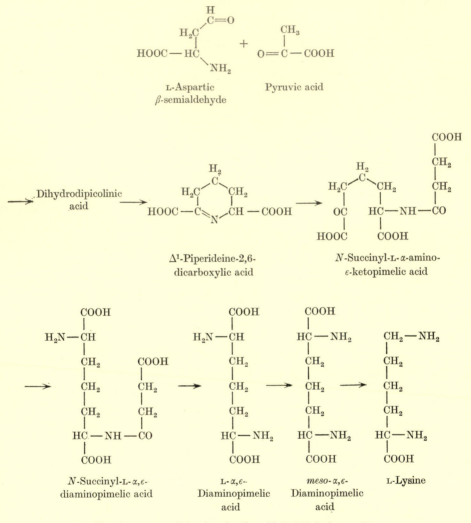

Fig. 1. *meso-α, ε*-Diaminopimelic acid (DAP)-lysine path.

tracer experiments were supplemented with enzyme studies (Shimura & Vogel, 1961, 1966).

As seen in Table 1, the DAP path occurs in prokaryotes as well as eukaryotes, and is in fact general in the plant kingdom: this path is indicated for bacteria, blue-green and green algae, mosses, liverworts,

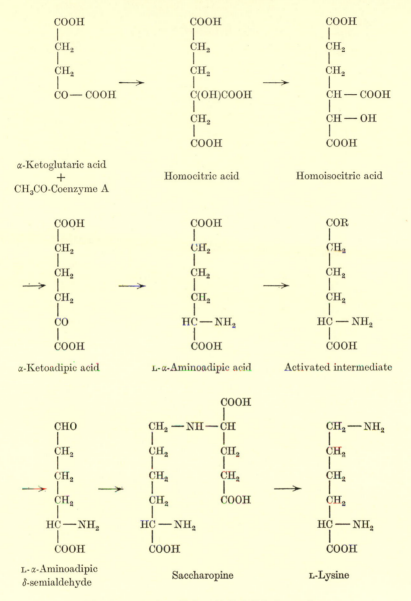

Fig. 2. L-α-Aminoadipic acid (AAA)-lysine path.

ferns, gymnosperms and angiosperms. Bacteria capable of anaerobic growth by photosynthesis, like other plants, use the DAP path.

The only photosynthesizing organisms thus far found to have the AAA path are the euglenoids, which have veered away from the plant kingdom in the direction of animality. The largest group in which the

Table 1. *Lysine synthesis in various organisms*: *DAP path*

From Vogel (1965) and unpublished observations. The phycomycetes listed produce either biflagellate or anteriorly uniflagellate spores. The bryophytes and higher organisms were tested as whole plants, except *Ginkgo biloba* and *Melilotus officinalis*, which were tested as pollen tissue and habituated root tissue, respectively.

Species	Higher taxon
Pseudomonads	
Rhodopseudomonas spheroides	Athiorhodaceae
Hydrogenomonas facilis	Methanomonadaceae
Pseudomonas fluorescens	Pseudomonadaceae
Eubacteria	
Azotobacter agilis	Azotobacteraceae
Agrobacterium radiobacter	Rhizobiaceae
Alcaligenes faecalis	Achromobacteraceae
Escherichia freundii	Enterobacteriaceae
Micrococcus lysodeikticus	Micrococcaceae
Streptococcus bovis	Lactobacillaceae
Arthrobacter globiformis	Corynebacteriaceae
Bacillus subtilis	Bacillaceae
Actinomycetes	
Mycobacterium smegmatis	Mycobacteriaceae
Streptomyces griseus	Streptomycetaceae
Actinoplanes philippinensis	Actinoplanaceae
Blue-green alga	
Plectonema boreanum	Nostocales
Phycomycetes	
Achlya bisexualis	Saprolegniales
Sapromyces elongatus	Leptomitales
Sirolpidium zoophthorum	Lagenidiales
Pythium ultimum	Peronosporales
Hyphochytrium catenoides	Hyphochytriales
Green alga	
Chlorella vulgaris	Chlorococcales
Bryophytes	
Riccia fluitans	Hepaticae
Physcomitrella patens	Musci
Tracheophytes	
Azolla caroliniana	Filicineae
Ginkgo biloba	Gymnospermae
Melilotus officinalis	Dicotyledoneae
Lemna minor	Monocotyledoneae

AAA path occurs is the higher fungi, including yeasts, moulds, morels, smuts, mushrooms and puffballs (Table 2). Interestingly, fungi have long been recognized to have certain animal-like features.

The lower fungi are cleanly split with respect to lysine synthesis. Those which produce non-motile spores or spores with a posterior flagellum use the AAA path, and those which make anteriorly flagellated or bi-flagellated spores, such as the so-called water moulds, have the DAP path. The finding that the morphological character, *spore flagellation*, correlates perfectly with the biochemical character, *lysine synthesis*, is noteworthy in indicating that both of these characters must have been handled very conservatively in evolution (Tables 1, 2).

Table 2. *Lysine synthesis in various organisms*: *AAA path*

From Vogel (1965). The phycomycetes listed produce either posteriorly uniflagellate or nonflagellate spores.

Species	Higher taxon
Euglenoid	
Euglena gracilis	Euglenida
Phycomycetes	
Rhizophylctis rosea	Chytridiales
Allomyces macrogynus	Blastocladiales
Monoblepharella laruei	Monoblepharidales
Rhizopus stolonifer	Mucorales
Ascomycetes	
Dipodascus uninucleatus	Endomycetales
Taphrina deformans	Taphrinales
Penicillium chrysogenum	Plectascales
Venturia inaequalis	Pseudosphaeriales
Neurospora crassa	Sphaeriales
Gibberella fujikuroi	Hypocreales
Morchella crassipes	Pezizales
Sclerotinia fructicola	Helotiales
Basidiomycetes	
Ustilago maydis	Ustilaginales
Polyporus tulipiferus	Polyporales
Coprinus radians	Agaricales
Calvatia gigantea	Lycoperdales

Higher animals and some protozoa have a nutritional requirement for lysine and do not seem to form this amino acid. Thus, the (other than sporadic) absence of lysine synthesis is a feature of animals; the DAP path is peculiar to plants—prokaryotic as well as eukaryotic; the AAA path is characteristic of groups of eukaryotes that have plantlike *and* animal-like traits.

A scheme of the possible evolutionary descent of the two lysine pathways is shown in Fig. 3. An important element of this scheme is the suggestion that neither path emerged in an organism possessing the other, since a partial appearance of either path in the presence of the other may well have been selected against, and an appearance *in toto* is thought unlikely in view of the number of enzymes involved in each path. The DAP path is pictured as the more ancient because it occurs in

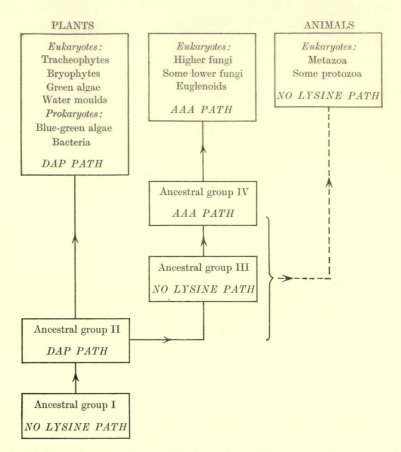

Fig. 3. Possible evolutionary descent of lysine pathways. Four ancestral groups may have been involved. Group I is thought to have depended on an external supply of lysine (primitive heterotrophy). Group II presumably developed the DAP-lysine path. This group is pictured as having given rise to modern plants as well as to ancestral Group III, which lost its lysine path and evolved in the direction of animality. The AAA-lysine path appears to have emerged in Group IV, which may have contained a common evolutionary precursor of modern forms having both plant-like and animal-like properties. In the more recent precursors of modern animals, the ability to synthesize lysine apparently remained lost (Vogel, 1965; Vogel & Vogel, 1967).

both prokaryotes and eukaryotes, whereas the AAA path seems to be present exclusively in eukaryotes. Since, at certain stages of evolution, forms having the DAP path may have been predominant in the biological world, it would be plausible to assume that the AAA path arose in an organism whose ancestors had lost the DAP path. In the following, some aspects of the loss of pathways or enzymes will be considered.

A latent pathway

Some bacteria contain DAP as a structural component of their cell wall, and some do not; nevertheless, DAP is indicated to be a lysine precursor in all known cases of bacterial lysine synthesis (Vogel, 1965). This finding is noteworthy in the light of the distribution of DAP as a structural component: this distribution splits the lactic acid bacteria (Work & Dewey, 1953; Cummins & Harris, 1956), which include some nutritionally exacting organisms. For example, *Streptococcus faecalis* (ATCC 9790) is a highly fastidious organism that can be used for the turbidimetric assay of lysine and other amino acids (Shockman, 1963). This organism does not utilize DAP as a source of lysine, and does not have DAP in its cell wall. In the assay of lysine with this organism, special precautions must be observed, since lysine-limited cultures, after the exponential growth phase, undergo rapid autolysis (Toennies & Gallant, 1949; Shockman *et al.* 1961). Following serial subculture (Shockman, 1963) of such autolysates and single-colony selection on an agar medium, a number of autolysis-resistant mutants were isolated (see Smith & Henderson, 1964). Some, but not all, of the resistant mutants were able to grow, although very slowly, in the absence of added lysine. Since preliminary studies with one such mutant (LNL I) showed that the organism is capable of lysine synthesis, it became a matter of considerable interest to determine the path of lysine formation used.

Accordingly, tracer experiments with 3-labelled and 4-labelled aspartic-[^{14}C] acids were performed in the general manner employed previously (Vogel, 1965). The results obtained (see Table 3) show that both of these tracers lead to the same specific radioactivity in the aspartic acid and the lysine from the protein of the organism, within the accuracy of the methods. This labelling pattern is characteristic of the DAP-lysine path, i.e. the same path that occurs in other bacteria. After

Table 3. *Incorporation of tracers into protein amino acids of Streptococcus faecalis strain LNL I as specific radioactivity relative to the respective protein aspartic acid values*

The general method used was that of Vogel (1959b). The medium was essentially that of Shockman (1963), with glucose as major carbon source, but without asparagine and lysine. The labelled aspartates were supplied at 0·25 mg. and at 2·5 μc/ml. Incubation was at 37° without agitation.

Tracer	Aspartic acid	Lysine
DL-[3-^{14}C]Aspartate	100	99
DL-[4-^{14}C]Aspartate	100	98

completion of our experiments, Gilboe, Friede & Henderson (1968) reported on a detailed investigation leading to similar conclusions. Earlier, McClure, Neuman & McCoy (1954), who used a different strain of *S. faecalis*, found that uniformly labelled aspartic-[^{14}C] acid was converted to lysine; this tracer, however, does not make it possible to differentiate between the DAP path and the AAA path.

A latent enzyme

Cases of dormant metabolic sequences, such as the latent path of lysine synthesis in *Streptococcus faecalis*, presumably reflect inadequate amounts or total absence of one or more, if not all, of the enzymes involved, although one must assume the presence of the relevant genetic information. An instance of a latent enzyme, which has been studied in some detail, is associated with the ornithine-arginine pathway (Fig. 4) of *Escherichia coli* (Vogel *et al.* 1967). The fourth enzyme of the path leading from glutamate to ornithine and to arginine is acetylornithine δ-transaminase (Albrecht & Vogel, 1964; Vogel & Jones, 1969), which catalyses the reaction of *N*-acetyl-L-glutamic-γ-semialdehyde and L-glutamate to yield N^α-acetyl-L-ornithine and α-ketoglutarate, with pyridoxal-5-phosphate as cofactor. In experiments with strain W of *E. coli*, it was possible to isolate arginine-requiring (*argD⁻*) mutants having no detectable acetylornithine δ-transaminase activity (Vogel, Bacon & Baich, 1963; Itikawa, Baumberg & Vogel, 1969). From such mutants, various kinds of (phenotypic) revertants could be obtained. For example, some of the revertants produce an acetylornithine δ-transaminase indistinguishable from the normal enzyme, which is characteristically repressible by arginine. Other revertants, however, form a transaminase which differs from the wild-type transaminase in various properties, including regulation behaviour, although the two enzymes catalyse the same reaction (Vogel *et al.* 1963; Bacon & Vogel, 1963; Vogel *et al.* 1967). Interestingly, the revertant transaminase, instead of being repressible, is actually inducible by arginine. The mutation giving rise to the appearance of the inducible transaminase maps at a locus (*argM*) which is not closely linked to any of the other arginine genes (Vogel & Bacon, 1966).

The revertants synthesizing the arginine-inducible transaminase maintain a certain level of the enzyme in the absence of added arginine, and, consequently, can grow on minimal medium. The inducible enzyme can thus function in the place of the normal enzyme. In the context of evolution, it seems intriguing that an *argD⁻* organism, in a one-step mutation, is able to achieve the formation of a substitute enzyme whose genetic information, therefore, must have been faithfully propagated without normally being expressed to any detectable extent.

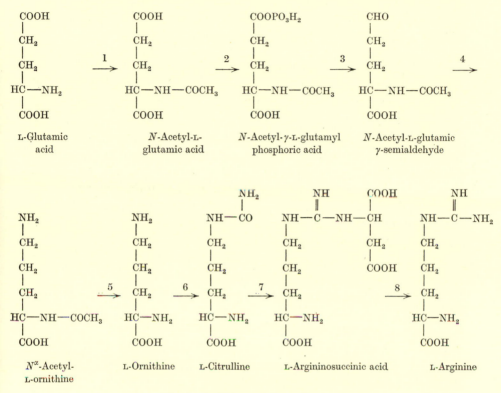

Fig. 4. Ornithine-arginine path.

Ornithine biosynthesis: minor variations

In the ornithine path of *Escherichia coli*, the step catalysed by ornithine δ-transaminase yields N^α-acetylornithine which, in this organism, is converted to ornithine by hydrolytic removal of the acetyl group, catalysed by acetylornithinase (Vogel, 1953; Vogel & Bonner, 1956). In certain other organisms, however, including *Micrococcus glutamicus* (Udaka & Kinoshita, 1958) and strains of *Neurospora* (Vogel & Vogel, 1963*a*), *Saccharomyces* (DeDeken, 1963) and *Chlamydomonas* (Dénes, 1969), the conversion of N^α-acetylornithine to ornithine is catalysed by an ornithine-glutamate acetyltransferase. The differences in ornithine synthesis between these organisms and *E. coli* thus are rather minor and affect a pair of enzymic mechanisms in pathways which utilize the same intermediates. The distribution of these enzymic mechanisms, though, does not appear to be highly consistent, since *Bacillus subtilis*, which like *M. glutamicus* is a Gram-positive bacterium, forms ornithine in the manner of *E. coli* (Vogel & Vogel, 1963*b*). The limited and somewhat sporadically occurring differences in ornithine synthesis, therefore, stand in contrast to the sharp and consistently distributed differences in the synthesis of the ornithine homologue, lysine.

Pathways in evolution

The emergence of the lysine-synthesizing system in strain LNL I of *Streptococcus faecalis* and the appearance of the arginine-inducible transaminase in an *arg*D⁻ revertant of *Escherichia coli* have something in common. We see that, in both cases, a (normally unutilized) metabolic potential can readily be expressed as a consequence of genetic change, which can reasonably be assumed in one case, and which has been demonstrated in the other. The organisms with the new metabolic function then have an advantage under suitable selective conditions (i.e. lysine-limited medium for strain LNL I, and minimal medium for the *arg*D⁻ revertant).

In the case of the two acetylornithine δ-transaminases in *Escherichia coli*, the repressible enzyme is produced in the wild type, whereas the inducible enzyme is formed only after the occurrence of a mutation at *arg*M (Bacon & Vogel, 1963). A related case, but with a significant difference, is that of the two ornithine transcarbamylases of *Bacillus licheniformis*, of which one is arginine-inducible and the other is arginine-repressible (Bernlohr, 1966). Here, both enzymes can be synthesized in the wild type, subject to physiological control; no genetic change is needed for the formation of either enzyme. The existence of such arginine-inducible enzymes suggests the possibility of catabolic sequences, although, in the instance of the *E. coli* transaminase, a conceivable catabolic function would have to be regarded as vestigial. A further variation on the theme is presented by strain K 12 of *E. coli*, which is indicated to have two genetic loci specifying ornithine transcarbamylase, both species of enzyme being arginine-repressible (Glansdorff, Sand & Verhoef, 1967). The presence of these two loci would explain the previously encountered difficulty in obtaining ornithine transcarbamylase mutants of this strain. Such redundancy is quite exceptional in organisms like *E. coli*, in which deficiency mutations can readily be demonstrated for almost any step of the type of pathway under discussion. The inferred two transcarbamylases of *E. coli* strain K 12 could be thought to furnish protection against mutational loss of biosynthetic capacity, since either of the enzymic species (especially in view of possible de-repression) could sustain the needed rate of ornithine → citrulline conversion. However the rarity of such redundancy indicates that this hypothetical kind of protection is not generally selected for; rather, it would seem that a selective advantage could accrue to the organism from being able to drop, mutationally, the formation of this or that enzyme, under particular circumstances. On the other hand, the propagation of the genetic information for the latent lysine path of *Streptococcus faecalis* or the latent transaminase of *E. coli* appears to carry no appreciable penalty and to have been positively selected for, possibly in terms of the flexibility mechanism provided. Such genetically conditioned latency is of special interest in relation to the metabolic regulatory processes of enzyme repression and induction, which tend to exert appropriate control over cellular enzyme levels. Since, in repression as well as in induction, the enzymes involved do not decrease below 'basal levels' the latency phenomenon would seem to go further than do these metabolic regulatory processes in decreasing the cellular levels of functional enzymes.

One can then contemplate a hierarchy of mechanisms affecting biosynthetic function: *feedback inhibition*, which regulates the flow of metabolites in a pathway, through curtailing enzyme action; *enzyme repression* (*or induction*), which regulates enzyme level through influencing the rate of enzyme synthesis; and *latency*, which represents a genetically alterable lack of functional-enzyme formation. Latency can be viewed as involving a reversible loss, since the latent function presumably was expressed at some stage in evolution. In addition, we would contemplate an *irreversible loss* in biosynthetic function. Indeed, it is the irreversible loss of the DAP-lysine path that is believed to have preceded the evolution of the AAA-lysine path, because a reversible loss of the DAP path probably would have resulted in the re-establishment of this path rather than the emergence of a new one.

Finally, having thus returned to the amphikaryotic DAP path and the eukaryotic AAA path, we should like to point out briefly some phylogenetic implications suggested by the dichotomous distribution found, although caution is indicated in relying on the single character, *lysine synthesis*. We would propose that:

> bacteria are no fission fungi;
> fungi are no plants;
> water moulds are no moulds; and
> euglenoids are no algae.

This work was aided by the National Science Foundation and the U.S. Public Health Service. The portion of the research dealing with

Streptococcus faecalis was completed in 1966, when one of the authors (H.J.V.) was at the Institute of Microbiology, Rutgers University, and another (J.S.T.) was at the Department of Microbiology, Temple University, School of Medicine.

REFERENCES

ALBRECHT, A. M. & VOGEL, H. J. (1964). Acetylornithine δ-transaminase: partial purification and repression behaviour. *J. biol. Chem.* **239**, 1872.

BACON, D. F. & VOGEL, H. J. (1963). A regulatory gene simultaneously involved in repression and induction. *Cold Spring Harb. Symp. quant. Biol.* **28**, 437.

BERNLOHR, R. W. (1966). Ornithine transcarbamylase enzymes: occurrence in *Bacillus licheniformis*. *Science, N.Y.* **152**, 87.

CUMMINS, C. S. & HARRIS, H. (1956). The chemical composition of the cell wall in some Gram-positive bacteria and its possible value as a taxonomic character. *J. gen. Microbiol.* **14**, 583.

DEDEKEN, R. H. (1963). Biosynthèse de l'arginine chez la levure. I. Le sort de la Nα-acétylornithine. *Biochim. biophys. Acta* **78**, 606.

DÉNES, G. (1969). Ornithine acetyltransferase of *Chlamydomonas*. In *Methods in Enzymology*. Eds. H. Tabor and C. W. Tabor, vol. 17. (In Press.)

GILBOE, D. P., FRIEDE, J. D. & HENDERSON, L. M. (1968). Effect of hydroxylysine on the biosynthesis of lysine in *Streptococcus faecalis*. *J. Bact.* **95**, 856.

GLANSDORFF, N., SAND, G. & VERHOEF, C. (1967). The dual genetic control of ornithine transcarbamylase synthesis in *Escherichia coli* K12. *Mutation Res.* **4**, 743.

ITIKAWA, H., BAUMBERG, S. & VOGEL, H. J. (1968). Enzymic basis for a genetic suppression: accumulation and deacylation of N-acetylglutamic γ-semialdehyde in enterobacterial mutants. *Biochim. biophys. Acta* **159**, 547.

McCLURE, L. E., NEUMAN, R. E. & McCOY, T. A. (1954). Amino acid metabolic studies. VI. Aspartic-lysine interrelations in *Streptococcus faecalis*. *Arch. Biochem.* **53**, 50.

SHIMURA, Y. & VOGEL, H. J. (1961). Lysine synthesis and biochemical evolution: diaminopimelic decarboxylase in higher plants. *Fedn Proc. Fedn Am. Socs exp. Biol.* **20**, 10.

SHIMURA, Y. & VOGEL, H. J. (1966). Diaminopimelic decarboxylase of *Lemna perpusilla*: partial purification and some properties. *Biochim. biophys. Acta* **118**, 396.

SHOCKMAN, G. D. (1963). Amino-acids. In *Analytical Microbiology*, p. 567. Ed. F. Kavanagh. London and New York: Academic Press.

SHOCKMAN, G. D., CONOVER, M. J., KOLB, J. J., PHILLIPS, P. M., RILEY, L. S. & TOENNIES, G. (1961). Lysis of *Streptococcus faecalis*. *J. Bact.* **81**, 36.

SMITH, W. G. & HENDERSON, L. M. (1964). Relationships of lysine and hydroxylysine in *Streptococcus faecalis* and *Leuconostoc mesenteroides*. *J. biol. Chem.* **239**, 1867.

TOENNIES, G. & GALLANT, D. L. (1949). Bacterimetric studies. II. The role of lysine in bacterial maintenance. *J. biol. Chem.* **177**, 831.

UDAKA, S. & KINOSHITA, S. (1958). Studies on L-ornithine fermentation. I. The biosynthetic pathway of L-ornithine in *Micrococcus glutamicus*. *J. gen. appl. Microbiol.* **4**, 272.

VOGEL, H. J. (1953). Path of ornithine synthesis in *Escherichia coli*. *Proc. natn. Acad. Sci. U.S.A.* **39**, 578.

VOGEL, H. J. (1959a). On biochemical evolution: a dichotomy in microbial lysine synthesis. *Fedn Proc. Fedn Am. Socs exp. Biol.* **18**, 345.

VOGEL, H. J. (1959b). Lysine biosynthesis in *Chlorella* and *Euglena*: phylogenetic significance. *Biochim. biophys. Acta* **34**, 282.

VOGEL, H. J. (1959c). On biochemical evolution: lysine formation in higher plants. *Proc. natn. Acad. Sci. U.S.A.* **45**, 1717.

VOGEL, H. J. (1960). Two modes of lysine synthesis among lower fungi: evolutionary significance. *Biochim biophys. Acta* **41**, 172.

VOGEL, H. J. (1961). Lysine synthesis and phylogeny of lower fungi: some chytrids versus *Hyphochytrium*. *Nature, Lond.* **189**, 1026.

VOGEL, H. J. (1962). Lysine synthesis in some 'algal fungi' and blue-green algae: phylogenetic implications. *Genetics, Princeton* **47**, 992.

VOGEL, H. J. (1963). Lysine pathways as 'biochemical fossils'. In *Evolutionary Biochemistry*, vol. 3, p. 341. Ed. A. I. Oparin. *Proc. 5th Int. Congr. Biochem.*

VOGEL, H. J. (1964). Distribution of lysine pathways among fungi: evolutionary implications. *Am. Nat.* **98**, 435.

VOGEL, H. J. (1965). Lysine biosynthesis and evolution. In *Evolving Genes and Proteins*, p. 25. Eds. V. Bryson and H. J. Vogel. London and New York: Academic Press.

VOGEL, H. J. & BACON, D. F. (1966). Gene aggregation: evidence for a coming together of functionally related, not closely linked genes. *Proc. natn. Acad. Sci. U.S.A.* **55**, 1456.

VOGEL, H. J., BACON, D. F. & BAICH, A. (1963). Induction of acetylornithine δ-transaminase during pathway-wide repression. In *Informational Macromolecules*, p. 293. Eds. H. J. Vogel, V. Bryson and J. O. Lampen. London and New York: Academic Press.

VOGEL, H. J., BAUMBERG, S., BACON, D. F., JONES, E. E., UNGER, L. & VOGEL, R. H. (1967). In *Organizational Biosynthesis*, p. 223. Eds. H. J. Vogel, J. O. Lampen and V. Bryson. London and New York: Academic Press.

VOGEL, H. J. & BONNER, D. M. (1956). Acetylornithinase of *Escherichia coli*: partial purification and some properties. *J. biol. Chem.* **218**, 97.

VOGEL, H. J. & JONES, E. E. (1969). Acetylornithine δ-transaminase. In *Methods in Enzymology*. Eds. H. Tabor and C. W. Tabor, vol. 17. (In Press.)

VOGEL, H. J. & VOGEL, R. H. (1967). Some chemical glimpses of evolution. *Chem. Engng News* **45** (52), 90.

VOGEL, R. H. & VOGEL, H. J. (1963a). Evidence for acetylated intermediates of arginine synthesis in *Neurospora crassa*. *Genetics, Princeton* **48**, 914.

VOGEL, R. H. & VOGEL, H. J. (1963b). Acetylated intermediates of arginine synthesis in *Bacillus subtilis*. *Biochim. biophys. Acta* **69**, 174.

WORK, E. & DEWEY, D. L. (1953). The distribution of α,ε-diaminopimelic acid among various micro-organisms. *J. gen. Microbiol.* **9**, 394.

MEMBRANES OF CELLS AND ORGANELLES: MORPHOLOGY, TRANSPORT AND METABOLISM

P. MITCHELL

Glynn Research Laboratories, Bodmin, Cornwall

The natural membranes occupy a special position in the organization of living cells because they define, structurally and functionally, the boundary between the cell and its environment and the boundaries between intracell compartments. There are four main attributes of the boundaries defined by the natural membranes. (1) They represent regions of mechanical stress. Generally there are differences of pressure across the thickness of the membrane and corresponding tensions in the plane of the membrane. (2) They represent regions of chemical anisotropy between phases or compartments of different chemical composition. This anisotropy is generally related to an intrinsic sidedness of the membrane. (3) They act as osmotic barriers, impeding the diffusion of certain solutes between the phases on either side. (4) They act as osmotic links, permitting or catalysing the translocation of certain solutes or chemical groups between the phases on either side. The phases on either side of the natural membranes are generally aqueous solutions, while the membranes themselves are composed of lipid, protein (polypeptide) and polysaccharide components in the form of a laminar complex.

Some natural membranes are relatively porous and act as molecular sieves for solutes of small and intermediate molecular weight (e.g. the cell wall of bacteria and plants and the outer membrane of mitochondria); but in the case of most so-called lipid membranes the membrane structure includes a non-aqueous osmotic barrier layer (or M phase) of low permeability to hydrophilic solutes. The translocation of certain components may be specifically catalysed through lipid membranes between the aqueous phases on either side, by enzymes and substrate-specific carriers that may shuttle across or may be effectively plugged through the osmotic barrier layer.

We shall concentrate attention here on certain osmotic barrier–osmotic link functions of metabolically active lipid membranes in prokaryotic and eukaryotic cells, and will not be much concerned with chemical composition and mechanical and other functions of membranes, discussed in recent reviews (Salton, 1967; Lang, 1968; Bartnicki-

Garcia, 1968; Whittaker, 1968; Northcote, 1968; Korn, 1969) or dealt with by other contributors to this Symposium (Echlin, 1970; Ellar, 1970; Hughes, Lloyd & Brightwell, 1970).

The osmotic barrier/osmotic link function of a lipid membrane does, of course, depend upon the intrinsic physical stability of the membrane system between the aqueous phases on either side of it (Danielli, 1966; Bangham & Haydon, 1968; Wallach & Gordon, 1968), and on the appropriate location of enzymes and catalytic carriers in relation to the osmotic barrier or M phase (Mitchell, 1961 a, b). I do not propose to discuss this aspect of membrane structure and function in any detail here, but it may be helpful to remark that the factors contributing to the stability of structure and composition of membranes in living cells may be presumed to be fundamentally similar to those factors of close packing, electric charge neutralization and segregation of hydrophobic and hydrophilic groups that have been shown to be dominant in deter-mining the tertiary folding and sub-unit assembly of smaller scale bio-chemical complexes such as proteins and nucleic acids. For this reason, the functional inclusion of a given component in a lipid membrane may be less dependent upon its atomic composition—for example, whether it is classed as a protein (polypeptide) or lipid—than upon its configura-tion and its polar or non-polar secondary bonding characteristics. Inci-dentally, the recent elegant work of Schatz & Saltzgaber (1969) on the heterogeneous origin of mitochondrial 'structural protein' shows that a specialized structural protein species is not a major component of mitochondrial cristae membranes, and a revision of current notions of the origin and role of 'structural protein' is required.

The biochemical interpretation of biological transport phenomena

Progress in the understanding of transport phenomena and progress in the acquisition of knowledge of the catalytic structure and function of membranes in biochemical systems has been rather slow compared with progress in certain other biochemical fields. In addition to the special technical difficulties associated with biochemical work on fragile mem-brane-containing systems, the main stumbling block has been the problem of making the intellectual transition from the classically scalar idiom of metabolic enzymology, in which cells or subcellular vesicles were traditionally regarded as bags of enzymes in 'homogeneous' solution, to the idiom of vectorial metabolism, in which the directional aspect of the diffusion of substrates through the catalytic carrier systems in the lipid membranes between the aqueous compartments must be taken into account, and the directional aspect of group transfer (i.e.

group translocation) in the anisotropic enzymic environment of the lipid membranes must also be described (see Mitchell, 1967a, b, 1968; Robertson, 1968; Greville, 1969).

The general conception of vectorial metabolism in lipid membrane systems which I have endeavoured to foster had its origins in the work of Lundegardh (1945), Ussing (1947), Robertson & Wilkins (1948), Davies & Ogston (1950), Conway (1951), Danielli (1954) and others. These early studies, reviewed by Robertson (1968), were initially influenced by work of Langmuir (1938, 1939) on the mobility (or turning over) of molecules in thin lipid films, which enabled biochemists to appreciate how substrate-specific permeation through natural membranes might occur by specific translational or rotational movements of catalytic carriers in the hydrophobic fabric of the lipid phase of the membranes. The rather abstract theoretical discussion of chemical cells by Guggenheim (1933) also played a part (see Mitchell, 1968, 1969a) through the elegant work of Rosenberg (1948) and Ussing (1949). The theory of absolute reaction rates (Glasstone, Laidler & Eyring, 1941)— applicable to any reaction involving a rearrangement of matter—and the concept of group potential in relation to chemical group transfer reactions, discussed by Lipmann (1941), led to an obvious comparison between the diffusion of solute molecules through the lipid phase of a membrane, requiring the breaking and making of secondary valencies, and the diffusion of chemical groups from donors to acceptors in the environment of the active centre region of an enzyme, requiring the breaking and making of primary valencies, which I described in the context of Rosenberg's elegant work as follows (Mitchell, 1954a):

'The view, expressed nicely by Rosenberg (1948), that by virtue of specific permeability properties, the natural membranes act as connecting links between particular components of the phases which they separate has its counterpart in the view of the enzymes as couplers of reactions which can proceed only on or in the enzyme molecules. Rosenberg's treatment shows, in fact, that the energetics of the reactions in two phases connected by a membrane can be described in the same terms as "homogeneous" enzyme-linked reactions; the important implication being that the efficiency (or reversibility) of transport reactions is determined by the specificity of membrane permeability, exactly as the efficiency of coupled enzyme reactions is determined by the enzyme-substrate and enzyme-carrier specificities. This merging of the terms of description has appropriately coincided with the realization that the permeability properties of membranes to the substrates which they transport may be dominated by enzymic specificities. In complex biochemical systems, such as those carrying out oxidative phosphoryla-tion (e.g. Slater & Cleland, 1953), the osmotic and enzymic specificities appear to be equally important and may be practically synonymous.'

The comparison (and distinction) implied here between the catalysis of solute translocation on the one hand and of group translocation on the other was subsequently more explicitly defined (Mitchell, 1957, 1961*a*; Mitchell & Moyle, 1958), and I have suggested that the specific catalysts of solute translocation, which are not involved in primary bond exchange (Ussing, 1947, 1949; Danielli, 1954), may conveniently be called translocaters or porters (Mitchell, 1963*a*, 1967*a*) whereas the enzymes or catalytic carriers involved in group translocation may conveniently be described according to the orthodox biochemical nomenclature. The following equations represent porter-catalysed solute (S) translocation and enzyme-catalysed group (G) translocation, respectively, where D and A stand for donor and acceptor groups, and X represents the trans-location centre or active centre of the porter or enzyme, respectively.

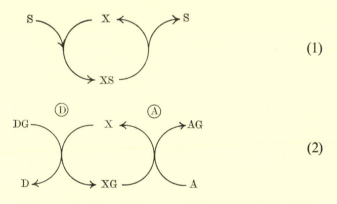

$$\text{(1)}$$

$$\text{(2)}$$

The central region of these equations represents the lipid (or M) phase of the membrane through which translocation occurs between the left (L) and right (R) aqueous phases according to the direction of the natural diffusion tendency of the transitional complex involving S or G. The difference between the solute translocation of equation (1) and the group translocation of equation (2) is that the former need involve only the opening and closing of secondary and ionic or non-covalent bonds such as those participating in the ionization or hydration of S (which are not indicated in the equation), whereas the latter must involve the opening and closing of the covalent bonds linking G with D or A. In the solute translocation of equation (1) the field of force on S across the membrane generally includes only pressure, concentration and electric components, whereas in the group translocation of equation (2) the field of force on G includes also a component corresponding to the chemical group potential difference of G in the compounds DG and AG. Thus, the equilibration of S across the membrane via the porter system of equation (1) does not depend on a chemical reaction as usually understood—i.e. it does not depend on a covalent bond exchange. The catalytic function of the porter in facilitating the opening and closing of the secondary and ionic bonds required to enable S to pass through the lipid phase is isotropic and causes an equalization of the total chemical potential of S across the membrane (Mitchell, 1961a, 1963a). On the other hand the equilibration of G across the membrane via the enzyme system of equation (2) depends on the chemical reaction

$$DG + A \rightarrow D + AG \tag{3}$$

and when the catalytic property of the enzyme is anisotropic so that the transfer of the group G is specific only for the group D on the left and only for the group A on the right [as indicated by Ⓓ and Ⓐ in equation (2)], the distribution of G across the membrane is not a function of the distribution of either DG or AG but depends on the difference between the group potential of D on the left and the group potential of A on the right, which is equal to the group potential difference of G across the system. Thus the intrinsic anisotropy or sidedness of the donor-acceptor specificity of the group translocation system of equation (2) is the primary requirement for coupling between metabolism and transport and for the generation, through secondary mechanisms, of the asymmetries of distribution of various components that are characteristic of living cells and are the basis of morphogenesis (Mitchell, 1961a, 1962, 1963a).

Many of the phenomena of biological transport can be explained in terms of appropriate combinations of the secondary and primary types of translocation reaction represented by equations (1) and (2), respectively. The usefulness of this rationale springs from the fact that, by the application of well-established physical and chemical principles, it provides the means of describing the vectorial aspect of solute diffusion and group transfer and gives spatial dimensions to the formerly scalar biochemical conception of metabolism.

Some comment on the notion of 'permease' appears to be required here in the context of the biochemical interpretation of membrane transport, more especially because this notion gave rise to a blurring of communication between microbiologists and others interested in biological transport which has not yet been completely resolved. The word permease was introduced by Cohen & Monod (1957) to describe 'stereospecific permeation systems, functionally specialized and distinct from metabolic enzymes' and alternatively to describe 'the specific inducible protein component of the system, while the expression "permease system" implies all the components'. They added, 'It is not excluded, of course, that the system may comprise a sequence of two (or more) inducible proteins catalysing successive steps in the *entry* reaction.' And further, 'When there is work performed in the process, it is virtually certain that covalent bonds are broken or transferred and, to that extent, the permease acts like an enzyme in activating their breakage or transfer.' Davis (1961) and Kennedy (1966) have drawn attention to the ambivalence of the definition of permease given by Rickenberg, Cohen, Buttin & Monod (1956) and by Cohen & Monod (1957) and, in amplification of my own views (Mitchell, 1959*a*, 1961*a*, 1962), I cannot do better than quote the following paragraph from the well-reasoned criticism by Kennedy (1966). 'It is clear that the word permease as originally defined is the designation of a *system* and in this sense is analogous to designations such as *galactozymase*, formerly applied to the entire complex of enzymes involved in the fermentation of a sugar. However, the term permease has been widely used in the literature to designate, not only the entire system, but also the specific, as yet hypothetical enzyme postulated by Monod and his collaborators to play an essential role in lactose transport. This substitution of the part for the whole has been the source of very real confusion, since it has carried the implication that the transport system consists of a single protein, coded by the *y* gene in the *lac* operon. Such usage is clearly at variance with the original definition of the term and may have hindered progress in this field by masking the possible degree of biochemical complexity of the system.' This and other related conceptual inconsistencies discussed by Kennedy (1966) have not been resolved by subsequent work of the permease school (see, for example, Kepes, 1960, 1961, 1964; Kepes & Cohen, 1962; Koch, 1964, 1967; Egan & Morse, 1966). It is particularly relevant to the present discussion that the permease classification lumps together protein catalysts of solute translocation and enzymes catalysing group translocation but excludes translocation catalysts that are not proteins. Amongst other critics of the use of the word permease, Christensen (1962) pointed out that the introduction of the permease nomenclature to describe essential protein components of substrate-specific membrane-transport systems was not justified by any corresponding rationalization in the interpretation of membrane-transport phenomena, because it had long been generally accepted that enzymes were involved in metabolically coupled transport processes—as shown, for example, in reviews by Lundegardh (1954), Conway

(1954), Wilbrandt (1954), Rothstein (1954) and Danielli (1954). Christensen (1962) also drew attention to the recommendation by the Commission on Enzymes of the International Union of Biochemistry, see Enzyme Nomenclature (1965) that the termination 'ase' should be reserved for the names of enzymes catalysing given primary bond exchanges. These criticisms should not be allowed to obscure the very considerable practical value of the discovery by the Paris school (Cohen & Monod, 1957; Kepes & Cohen, 1962), and independently by Green & Davis (see Davis, 1956), that certain enzymic or other protein components essential to the activity of some substrate-specific transport systems in micro-organisms are inducible. On the other hand, I think it must be admitted that the permease nomenclature introduced by the Paris school has tended to confuse rather than to clarify the biochemical interpretation of transport phenomena.

Paired group translocation reactions: proton-translocating oxidoreduction and hydrodehydration loops

Group translocation reactions involving the elements of water are especially important in that they can be arranged in pairs or loops (Mitchell, 1966a, b, 1968, 1969a) so that the net result is the translocation of protons. The arrangement of oxidoreduction reactions (Lundegardh, 1945; Davies & Ogston, 1950; Davies, 1957; Conway, 1951; Robertson, 1960; Mitchell, 1961c) and hydrodehydration reactions (Mitchell, 1961c, 1966a, b) in this way is illustrated by the following equations:

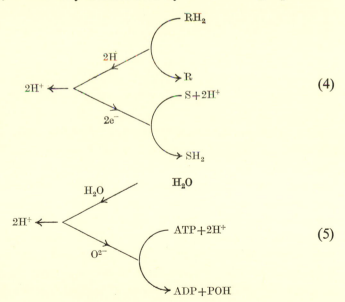

$$(4)$$

$$(5)$$

The equilibration of the particles undergoing translocation in the electrochemical field across the membrane has been discussed elsewhere (Mitchell, 1968, 1969 a).

It has been suggested that the respiratory chain systems, located in the cristae membrane of mitochondria and in the plasma membrane of bacteria, and the photo-oxidoreduction chain systems of the grana membrane of chloroplasts and the chromatophore membrane (or plasma membrane) of photosynthetic bacteria are made up of several proton-translocating oxidoreduction loops, like that of equation (4), arranged in series (Mitchell, 1961 c, 1962, 1963 a, 1966 a, b). Similarly it has been suggested that the reversible ATPases, located in the same membranes, correspond to the proton-translocating hydrodehydration loop of equation (5). It is beyond the scope of this review to discuss the extensive experimental evidence bearing on this conception; see reviews by: Slater (1967), Jagendorf & Uribe (1966), Robertson (1968), Witt, Rumberg & Junge (1968), Azzone, Rossi & Scarpa (1968), Crofts (1968), Greville (1969), Mitchell (1966 a, b, 1967 c, 1968, 1969 a, b). It is appropriate to mention, however, that recent observations on the cytochrome oxidase region of the respiratory chain of rat liver mitochondria (Mitchell & Moyle, 1967 c, 1969 c) leave little doubt that the arrangement in the membrane corresponds to Loop 3 of the chemiosmotic hypothesis (Mitchell, 1966 a, b) which can be represented as follows:

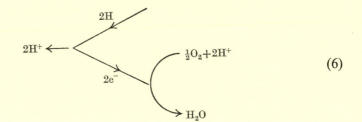

$$(6)$$

where the lower (electron-carrying) arm of the loop consists of cytochromes c_1, c, a and a_3, and the upper (hydrogen-carrying) arm of the loop is probably coenzyme Q or an enzyme-coenzyme Q complex. Thus, respiration and proton translocation occur by one and the same process in the oxygen-terminal part of the respiratory chain system of rat liver mitochondria. The establishment of this fact eliminates the possibility, suggested by Slater (1967), that respiration-driven proton translocation generally occurs by means of a proton pump actuated by a chemical intermediate synthesized during respiration; the oxidoreduction loop mechanism therefore appears to afford the only chemically-defined explanation of the observed proton translocation stoichiometries and other ion translocation data at present available (Mitchell & Moyle, 1967b, c, 1968, 1969a; Izawa & Hind, 1967; Crofts, 1968; Jackson, Crofts & von Stedingk, 1968; Dilley & Shavit, 1968; Schwartz, 1968; Witt et al. 1968; and see Greville, 1969). The respiratory chain systems of mitochondria from eukaryotic cells of different types and the respiratory chain systems of the plasma membrane of prokaryotic cells of different types exhibit differences of detailed enzymic and carrier composition, but probably conform to a fundamentally similar pattern of organization and function.

It has recently been shown that the ATPase of the plasma membrane of *Streptococcus faecalis* exhibits very similar properties to that of the mitochondrial cristae membrane (Harold & Baarda, 1969; Harold, Baarda, Baron & Abrams, 1969), and there appears to be little doubt that although oxidative phosphorylation does not occur in this organism the ATPase couples proton translocation across the membrane to the hydrolysis of ATP.

Secondary translocation of Na^+ and K^+ by the Na^+/K^+ antiporter ATPase

The intensive study of the Na^+/K^+ antiporter ATPase of plasma membranes (Skou, 1965; Post, 1968; Glynn, 1968; Skou & Hilberg, 1969; Baker et al. 1969) has shown clearly that the translocation of passengers which do not participate directly in the covalent bond exchanges of a chemical reaction can be coupled reversibly (Garrahan & Glynn, 1967) to the process of group transfer. This type of coupling is more complicated than the primary type of coupling given by pairs of group translocation reactions (see previous section) because it requires a system of secondary and electrovalent bonding and packing relationships that are so articulated as to translocate the passenger species in a particular direction relative to the flow of the chemical reactants which participate in the covalent bond exchanges (Mitchell, 1963a, 1967a). There is evidence that a phosphorylated intermediate (or possibly two such intermediates) is involved in the Na^+/K^+ antiporter ATPase reaction (Bader, Post & Bond, 1968; Skou & Hilberg, 1969; Siegel, Koval & Albers, 1969) and that Na^+ ions promote the synthesis of the intermediate, while K^+ ions promote its hydrolysis (Skou, 1965; Garrahan, Pouchan & Rega, 1969); but the details of the mechanisms by which the translocations of Na^+ and K^+ are linked to the ATPase reaction are far from clear (Yoshida, Nagi, Ohashi & Nakagawa, 1969; Baker et al. 1969; Skou & Hilberg, 1969).

During a chemical reaction that is secondarily coupled to the translocation of passengers such as Na^+ and K^+, the movements of different parts of the catalytic system (including the reactants) will presumably have different directions in space, depending upon the way the components are packed together as they pass through the transition states. Consequently the direction of translocation of an ion or molecule involved secondarily in the transitional complex can as readily be at right angles to the direction of translocation of a given group undergoing primary bond exchange as parallel to it (Mitchell, 1963a). This circumstance adds to the difficulty of relating the translocation of Na^+ and K^+ to particular chemical events in the case of the Na^+/K^+ antiporter ATPase.

Following the original scheme of Shaw (quoted by Glynn, 1957), the chemical transformations involved in the ATPase reaction were generally assumed to be separate in space and time from the physical translocation of the Na^+ and K^+ ions across the membrane. I have advocated the alternative view that the Na^+ and K^+ ions may be involved in the transitional intermediates of the ATPase reaction and that the translocation of Na^+ and K^+ may therefore occur in concert with the group transfer process catalysed by the ATPase system (Mitchell, 1961e, 1963a). Baker & Connelly (1966) and Garrahan & Glynn (1967) described experimental observations in favour of this view; the alternative interpretations have been discussed further by Mitchell (1967a), Stone (1968) and Glynn (1968). The following equations represent alternative basal reaction mechanisms that appear to be in accord with the experimental facts at present available (Glynn, 1968; Post, 1968; Fahn, Koval & Albers, 1968; Lindenmayer, Laughter & Schwartz, 1968; Priestland & Whittam, 1968; Yoshida $et\ al.$ 1969; Baker $et\ al.$ 1969; Skou & Hilberg, 1969):

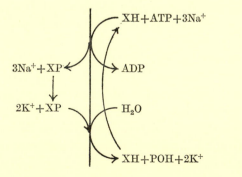

$$(7)$$

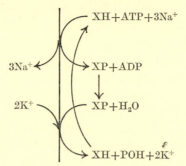

$$(8)$$

where XH stands for a group that can be phosphorylated and P stands for phosphoryl. In each equation the upper reaction represents the Na^+-activated XH phosphokinase activity and the lower reaction represents the K^+-activated XP phosphatase activity of the ATPase system. Equation (7) shows the translocation of phosphoryl across the membrane outwards in the phosphokinase reaction and inwards in the phosphatase reaction as in a previous scheme (Mitchell, 1967a) which corresponds to the basis of that elaborated by Stone (1968). It has the distinction (shared by the scheme of equation (8)) of accounting for the observed ATP-requiring Na^+ exchange across the membrane in absence of external K^+ and at low internal Na^+ concentration when the XP/XH couple would be expected to be fairly centrally poised in equilibrium with the Na^+-translocating phosphokinase reaction (see Mitchell, 1967a; Stone, 1968). Equation (8) illustrates that there need be no net macroscopic component of phosphoryl group transfer from one side of the membrane to the other (although in each enzyme complex there must nevertheless be a polarity of phosphoryl translocation orientated in some direction relative to the translocation of Na^+ and K^+ across the membrane). Equation (8) also emphasizes that the translocater centre (of the phosphokinase) involved in Na^+ translocation may be distinct from that (of the phosphatase) involved in K^+ translocation in the sense that the one does not have to be the chemical or physical derivative of the other.

Intrinsic anisotropy of vectorial enzyme systems catalysing chemiosmotic reactions

The enzyme-catalysed translocation reactions described in the previous two sections depend upon intrinsic anisotropic properties of the enzyme and carrier systems with respect to the M phase of the membrane in which they reside. The proton-translocating respiratory chain, photooxidoreduction and ATPase systems are macroscopically anisotropic in that they exhibit an intrinsic sidedness with respect to the direction of proton translocation and to the accessibility of other reactants (Löw & Vallin, 1963; Mitchell, 1963 b, 1966 a, b, 1968; Mitchell & Moyle, 1965 a, b, 1969 c; Lee & Ernster, 1966; Malviya, Parsa, Yodaiken & Elliott, 1968; Jackson et al. 1968; Scholes, Mitchell & Moyle, 1969; Harold et al. 1969; and see Greville, 1969).

In the case of the secondary coupling of solutes (such as Na^+ and K^+) to a group transfer reaction, as indicated in the previous section, the chemical field associated with group transfer need not be orientated across the membrane but must be orientated relative to the direction of the secondarily coupled solute translocation in *each individual* enzyme complex in the membrane. It follows that there need be no macroscopic orientation of the chemical field associated with the group transfer *in an assembly* of such enzyme complexes—although, there must, of course, be a macroscopic polarity of the secondary field associated with the solute (Na^+, K^+) translocation across the membrane. The properties of sidedness exhibited by the Na^+/K^+ antiporter ATPase have been reviewed by Skou (1965), Mitchell (1967 a) and Glynn (1968).

Catalysts of secondary translocation: uniport, antiport and symport

The kind of translocation catalyst that catalyses the noncoupled equilibration of a given solute across the M phase has been described as a uniporter (Mitchell, 1967 a). The best examples of uniporters in physiologically normal membranes are the systems specific for D-glucose and L-leucine translocation in the mature red blood cells of mammals (see Mitchell, 1967 a). Further work on these and other related systems was recently reviewed by Rothstein (1968). In the last few years, certain physiologically abnormal uniporters have been identified, and these promise to give new impetus to the understanding of translocation catalysis in lipid membranes because their mechanism of action can be explained quite simply (see Pressman, 1968; Mitchell, 1968).

Proton conductors

The well-known uncoupling agents 2,4-dinitrophenol (DNP), dicoumarol and azide were the first potent unphysiological ion conductors to be identified (Mitchell, 1961 d). These agents consist of a mixture of a weak acid (AH) and its corresponding anionic base form (A^-), and it is now generally agreed that their proton-conducting activity is attributable to a circulating carrier type of mechanism illustrated by the following equation:

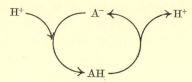

$$(9)$$

The π-bonded structure of the anion, which is characteristic of the un-coupling agents (Mitchell, 1961d), has the effect of delocalizing the electronic charge and permitting the anion to be relatively soluble in the non-polar medium of the M phase of the membrane without a charge-neutralizing partner (Mitchell, 1966a, b, 1968). This interpretation has been amply confirmed by studies of the proton-conducting effects of DNP and other classical uncoupling agents on mitochondrial cristae membranes (Mitchell & Moyle, 1967a–c; Carafoli & Rossi, 1967), on chloroplast grana membranes (Jagendorf & Neumann, 1965; Rumberg, Reinwald, Schröder & Siggel, 1968), on chromatophore membranes (Jackson $et\ al.$ 1968; Scholes $et\ al.$ 1969), on bacterial plasma membranes (Harold & Baarda, 1968b; Pavlasova & Harold, 1969), on red cell membranes (Chappell & Crofts, 1966; Harris & Pressman, 1967), on artificial lipid membranes (Chappell & Haarhoff, 1967; Hopfer, Leh-ninger & Thompson, 1968; Skulachev, Sharaf & Liberman, 1967; Liberman, Mochova, Skulachev & Topaly, 1968; Liberman & Topaly, 1968). Hopfer, Lehninger & Thompson (1968) and Liberman & Topaly (1968) have further shown that the agents of the DNP class have a high specificity for H^+ translocation, and the latter workers showed that the proton conductance is optimal when AH is about half dissociated to A^-, as expected on the basis of the circulating carrier model of equation (9). We should note that the conduction of H^+ ions across the M phase by the DNP class of agent involves the formation and breakdown of the covalent intermediate AH, but that the unusually high velocity of protonation-deprotonation reactions in aqueous media (Bell, 1959) causes this reaction to behave kinetically more like the reactions of secondary and ionic bonds than of covalent bonds of normal stability.

Cation conductors

The discovery by Pressman and co-workers that valinomycin and the gramicidins profoundly affect ion translocation in mitochondria (Press-man, 1963; Moore & Pressman, 1964) led to the observation by Chappell & Crofts (1965, 1966) that these antibiotics affected mito-chondrial swelling as though they catalysed specific alkali cation uniport reactions. This interpretation was confirmed by observations on the conductance of artificial lipid membranes to alkali metal ions (Chappell & Haarhoff, 1967; Lev & Buzhinsky, 1967; Mueller & Rudin, 1967; Andreoli, Tieffenberg & Tosteson, 1967). It is now generally agreed that the specific conduction of K^+ and other monovalent cations across lipid membranes by the macrocyclic antibiotics and chemically simpler polyethers occurs as a result of the expected abilities of these substances to solubilize cations of appropriate size in low dielectric media and thus act as carriers for such ions through the non-polar medium of the M phase (Eisenman, Ciani & Szabo, 1968; Liberman & Topaly, 1968; Mitchell, 1968; Tosteson, 1968; Henderson, McGivan & Chappell, 1969; Pinkerton, Steinrauf & Dawkins, 1969; Wipf & Simon, 1969; Ivanov *et al.* 1969). These reagents are themselves uncharged but their cation complexes are ions of large radius, and there is a fairly close analogy between their mechanism of action and the mechanism by which the proton conductors of the DNP class catalyse proton conduction, as illustrated by the following equation representing K^+ ion conduction by the valinomycin (V) class of reagent.

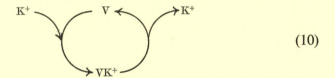

$$(10)$$

The mechanisms of equations (9) and (10) differ in the sign of the charged species crossing the M phase. There are, however, other distinguishing features. The specificity of bonding of H^+ by the proton-conducting reagent depends largely on the chemical property required for the formation of the covalent acid—a property that is lacking in the valinomycin type of reagent. The specificity of bonding of the cation with the valinomycin type of reagent depends on ion-dipole interactions responsible for hydration of the ion and carrier on the one hand, and for association between ion and carrier on the other. Other more complex factors of conformation, discussed by Pressman (1968), also affect the specificity. It is noteworthy that certain of the cation-conducting reagents of low specificity, notably the gramicidins, bind (and conduct) H_3O^+ ions as well as alkali metal ions (Henderson *et al.* 1969).

Antiporters and symporters

The concept of exchange diffusion or counter-transport (Ussing, 1947, 1949; Widdas, 1952) and the complementary concept of co-diffusion (see Crane, Miller & Bihler, 1961) are particular cases of the general principle that the flows of pairs of particles are sym-coupled if the particles reside together in a mobile complex or compound in a membrane, and are anti-coupled if they compete with one another (Mitchell, 1963a). This general principle of coupling does not necessarily require any structural or chemical similarity between the ions or molecules undergoing antiport or symport, as can readily be explained in terms of the linked function concept of Wyman (1948). The following equations illustrate antiport and symport of n moles of A and m moles of B, respectively, via the translocater X.

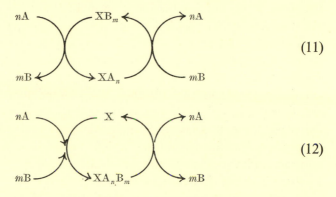

$$(11)$$

$$(12)$$

These equations are the osmotically-coupled analogues of the non-coupled uniport reaction of equation (1). When the component A is equilibrated through phase M between phases L and R by a uniporter, the total chemical potential difference of A across the M phase ($\Delta\bar{\mu}_A$) is zero, or

$$\Delta\bar{\mu}_A = 0. \tag{13}$$

But when components A and B equilibrate by tightly coupled antiporters or symporters as described by equations (11) and (12), respectively, it follows that for nA/mB antiport

$$n\Delta\bar{\mu}_A = m\Delta\bar{\mu}_B \tag{14}$$

and for nA/mB symport $\quad n\Delta\bar{\mu}_A = -m\Delta\bar{\mu}_B. \tag{15}$

Thus, the driving force on the translocation of A is balanced against that on the translocation of B by means of the coupling porter—just as the driving force on one chemical reaction is balanced against that on another by a coupling enzyme.

A number of antiporters and symporters have been described in the literature (see Heinz, 1967; Albers, 1967; Lehninger, Carafoli & Rossi, 1967; Klingenberg & Pfaff, 1968; Chappell & Robinson, 1968; Rothstein, 1968). Our interest here stems from the desire to identify the functional significance of these translocation catalysts in the coupling between one translocation and another, and in the coupling between metabolism and transport.

There are two main groups of translocation reactions coupled by porters which are at present receiving intensive study. On the one hand there is the group of porter-catalysed reactions associated with sugar and amino-acid absorption through the plasma membranes of certain eukaryotic cells, which are thought to be Na^+-sugar and Na^+-amino acid symport reactions (Crane *et al.* 1961; Crane, 1965; Heinz, 1967; Mitchell, 1967*a*; Kohn, Smyth & Wright, 1968; Thier, 1968; Daniels, Dawson, Newey & Smyth, 1969; Munck, 1968; Koopman & Schultz, 1969; Goldner, Schultz & Curran, 1969; Hauser, 1969; Bihler, 1969; Kittams & Vidaver, 1969; Christensen & Handlogten, 1969; and see Rothstein, 1968), or possibly Na^+ symport/K^+ antiport push-pull type reactions (Kuchler, 1967; Eddy, 1968*a*, *b*; Eddy, Hogg & Reid, 1969). On the other hand there is the group of porter-catalysed reactions associated with uptake of substrates and with ion movements through the plasma membranes of prokaryotic cells and through the cristae membranes of mitochondria, which appear to be proton-coupled symport and antiport reactions (Mitchell, 1961*c*, 1963*a*, 1967*a*, 1968, 1969*a*; Chappell & Crofts, 1966; Lynn & Brown, 1966; Chappell & Haarhoff, 1967; Lehninger *et al.* 1967; Chappell, Henderson, McGiven & Robinson, 1968; Jackson *et al.* 1968; Harold & Baarda, 1968*b*; Pavlasova & Harold, 1969; Fonyo & Bessman, 1968; Tyler, 1969; Palmieri & Quagliariello, 1969; Brierley, 1969; Kraayenhof & van Dam, 1969; Mitchell & Moyle, 1969*b*).

Recent work on antibiotics of the nigericin class (Lardy, Graven & Estrado-O, 1967; and see Pressman, 1968) has provided some insight about the mechanism by which tightly coupled H^+/cation antiport may occur. The members of the nigericin class of antibiotics are distinguished from those of the valinomycin class in that they contain a dissociable carboxyl group, and whereas tight cation binding (especially of K^+) occurs when the carboxyl group is deprotonated, it does not occur when the carboxyl group is protonated (Pressman, Harris, Jagger & Johnson, 1967; Pressman, 1968; Steinrauf, Pinkerton & Chamberlin, 1968). Nigericin induces tightly-coupled H^+/K^+ antiport in: mitochondrial membranes (Pressman *et al.* 1967), chloroplast grana membranes (Packer, 1967; Shavit, Dilley & San Pietro, 1968), chromatophore membranes (Jackson *et al.* 1968), bacterial plasma membranes (Harold & Baarda, 1968*a*, *b*), erythrocyte membranes (Harris & Pressman, 1967). The electrical neutrality of the H^+/K^+ antiport reaction has been confirmed by the observation that nigericin does not increase the net ion conductance of artificial lipid membranes under conditions where H^+/K^+ antiport occurs (Mueller & Rudin, quoted by Pressman *et al.* 1967). Thus, it is evident that, as in the agents of the DNP and valinomycin classes, the ion-conducting activity of the nigericin class of agent is attributable to the effective solubilization of H^+ ions and K^+ ions in the lipid phase. The H^+ ion is taken up by the anionic form of nigericin to give the uncharged lipid-soluble acid form, and alternatively the K^+ ion is taken up by the nigericin anion to give the uncharged lipid-soluble cation complex in which the hydrophilic properties of the K^+ ion are lost because of the substitution of other ligands for the normal hydration shell.

Natural H^+/alkali ion antiporters

The cristae membrane of rat liver mitochondria contains one or more systems catalysing tightly coupled electrically neutral H^+/K^+ and H^+/Na^+ antiport (Mitchell & Moyle, 1967b, 1969b; and see Christie, Ahmed, McLean & Judah, 1965). Near pH 7 the H^+/Na^+ antiport is more active than the H^+/K^+ antiport, and in either case activity increases at lower pH values. The rapid equilibration of sodium acetate across the cristae membrane of beef heart or rat liver mitochondria (Brierley, Settlemire & Knight, 1968; Mitchell & Moyle, 1969b) is explained by the following mechanism

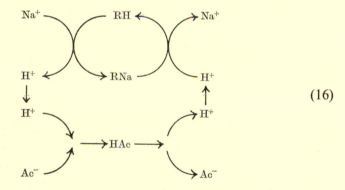

$$(16)$$

where R stands for the translocater centre of the H^+/Na^+ antiporter. The concentration of free acetic acid (HAc) in the membrane at neutral pH is sufficient to account for the uncatalysed rapid permeation of this species.

Circumstantial evidence suggests that H^+/K^+ and H^+/Na^+ antiporters occur in the plasma membranes of prokaryotes, as in the mitochondrial cristae membranes of eukaryotes. We may cite, for example, the characteristics of Na^+ and K^+ translocation, and especially the effects of proton-conducting reagents, in *Escherichia coli* (Schultz, Epstein &

Solomon, 1963; Damadian, 1968; Meury, 1969) and in *Streptococcus faecalis* (Zarlengo & Schultz, 1966; Harold & Baarda, 1968 *b*).

Natural H^+/substrate symporters and symport systems

The prototype of the proton-coupled symporters is that for 'phosphate' translocation through the cristae membrane of rat liver mitochondria which is equivalent to phosphoric acid uniport. By using light scattering to observe the osmotic swelling that accompanies salt entry through the cristae membrane of non-metabolizing mitochondria, Chappell & Crofts (1966) showed that ammonium phosphate entered rapidly; this can most simply be explained (Mitchell, 1967 *b*, 1968) by the following type of mechanism:

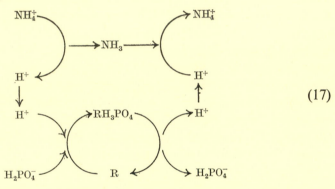

$$(17)$$

The net stoichiometry of this type of mechanism has been confirmed by other osmotic and ion-exchange measurements (Mitchell & Moyle, 1967 *b*, 1969 *b*). There are, however, variants of the mechanism of equation (17) which may account for the observed net stoichiometry. An OH^- exchange mechanism has been favoured by Chappell & Crofts (1966), but the reasons given for preferring such a mechanism are not valid (see Mitchell, 1968). A more likely mechanism that would be in accord with the experimental observations is shown in equation (17 *a*):

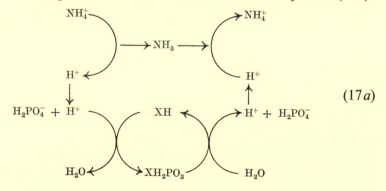

$$(17a)$$

In this case, translocation is assumed to involve the reversible formation of an intermediate XH_2PO_3 compound by the elimination of H_2O between H_3PO_4 and an XH group—which might, for example, be a thiol group—in the translocater centre of the porter. The mitochondrial phosphoric acid porter exhibits an alkaline pH optimum at high phosphate concentration (Mitchell & Moyle, 1969b), it exhibits a high temperature coefficient (Mitchell & Moyle, 1967b), it reacts with arsenate as well as with phosphate (Chappell & Crofts, 1966; Chappell & Haarhoff, 1967; Tyler, 1968, 1969) and it is specifically inhibited by mercurials and by certain other -SH reactors (Tyler, 1968, 1969; Bessman & Fonyo, 1968; Haugaard et al. 1969). In all these respects the mitochondrial phosphoric acid porter resembles the system catalysing phosphate exchange and uptake across the plasma membrane of *Staphylococcus aureus* (Mitchell, 1953, 1954a, b). The work on the PO_4 porter of *S. aureus* led to the conclusion that $H_2PO_4^-$ was translocated as a tightly-bonded complex or covalent intermediate (Mitchell, 1954a, b; Mitchell & Moyle, 1956b). It now seems likely that the covalent intermediate is XH_2PO_3 or H_3PO_4 and that the plasma membrane of *S. aureus* contains a phosphoric acid porter that is strictly analogous to that of the cristae membrane of mitochondria. There is circumstantial evidence for similar phosphoric acid porters in other bacteria, for example, in *Escherichia coli* (Weiden, Epstein & Schultz, 1967) and in *Bacillus cereus* (Rosenberg & La Nauze, 1968). The work on bacterial transport systems also prompted the suggestion that during the uptake of substrates phosphate may act as an exchanger, passing out in exchange for a reactant substrate on one translocater, and returning by another translocater (Mitchell, 1959a). This is illustrated for an anionic substrate A^- as follows:

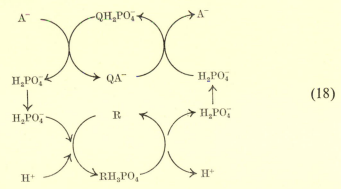

(18)

where Q stands for the translocater centre of the $H_2PO_4^-/A^-$ antiporter and R stands for that of the phosphoric acid porter.

Chappell & Haarhoff (1967) identified several proton-linked (or hydroxyl-linked) anion porter systems specific for dicarboxylic and tricarboxylic acids in mitochondria (see also Azzi & Azzone, 1967; Haslam & Griffiths, 1968; Azzone *et al.* 1968). The properties of these systems broadly resemble those of the phosphoric acid porter (Chappell, 1968), but net dicarboxylate (e.g. succinate or L-malate) translocation requires 'activation' by phosphate, and net tricarboxylate (e.g. citrate) translocation requires 'activation' by phosphate + L-malate. These and other related observations (Chappell & Haarhoff, 1967; Chappell & Robinson, 1968; Chappell *et al.* 1968; Haslam & Krebs, 1968; De Haan & Tager, 1968; Harris, 1968; Meijer & Tager, 1969) imply that dicarboxylate translocation may be attributable to an electrically neutral phosphate/dicarboxylate antiporter, while tricarboxylate translocation may be attributable to an electrically neutral L-malate/tricarboxylate antiporter. These antiporters correspond to the phosphate-linked anion antiporter of equation (18). It is important to note that, while there is some evidence of electrical neutrality of the proposed antiport reactions, it is not yet known what phosphate/anion stoichiometries these systems may have, or in what state of ionization or bonding the phosphate and anions may be translocated. The general principle of coupling by these systems is illustrated in the following abbreviated equation representing net citric acid translocation.

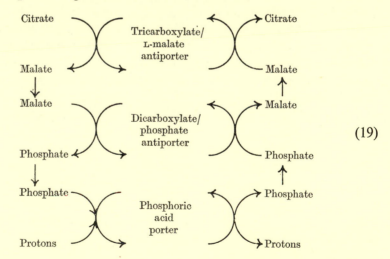

$$(19)$$

The possibility should be borne in mind that the tricarboxylate/dicarboxylate antiporter and the dicarboxylate/phosphate antiporter may represent acid uniporters which have relatively low binding constants for their acid substrates but broad substrate specificity; and that they may thus catalyse acid uniport at low substrate concentration but, under conditions of substrate saturation, may catalyse only rapid exchange. Circumstantial evidence has been obtained for porters catalysing translocation of glutamate and aspartate (Azzi, Chappell & Robinson, 1967), arginine (Keller, 1968), leucine (Buchanan, Popovitch & Tapley, 1969), α-glycerophosphate (Hansford & Chappell, 1967; and see Mitchell, 1968) and pyruvate (Simpson & Frenkel, 1969) through mitochondrial cristae membranes. It is noteworthy that the mitochondria of certain insects lack the porters for dicarboxylic and tricarboxylic acids (see Chappell, 1968).

The translocation of sulphate through the cristae membrane of rat liver mitochondria shows similar characteristics to that of phosphate and may be accounted for by the presence of a specific sulphuric acid porter system (Mitchell & Moyle, 1969b). These examples must suffice to illustrate the occurrence of proton-coupled porter systems which catalyse the translocation of several different types of substrate through mitochondrial cristae membranes. Analogous systems appear to be responsible for proton-coupled substrate uptake through the plasma membranes of certain prokaryotic cells. The direct uncoupling of such porter systems by proton-conducting reagents of the DNP class is a valuable diagnostic characteristic.

As I pointed out some years ago (Mitchell, 1963a), the equilibration of the concentration of galactose and of β-galactosides across the plasma membrane of *Escherichia coli* on addition of DNP (Horecker *et al.* 1961) indicates that the translocation of these sugars through the plasma membrane is catalysed by proton-sugar symporters. The kinetic observations of Winkler & Wilson (1966, 1967), Robbie & Wilson (1969) and West (1969) are consistent with my formulation; and Pavlasova & Harold (1969) have shown that the equilibration of β-galactoside distribution by DNP is not attributable to depletion of the ATP content of the cells. Kennedy (1966) and Scarborough, Rumley & Kennedy (1968) have isolated and purified a protein (the M protein) which appears to be implicated in β-galactoside translocation in *E. coli*, but which lacks chemical catalytic activity. This would also be consistent with my formulation of the mechanism of β-galactoside translocation if the M protein were the H^+-β-galactoside symporter or an essential part of it. The DNP-sensitivity of the accumulation of the neutral amino acids, leucine, isoleucine and valine, in *E. coli* (see Cohen & Monod, 1957; Schwartz, Maas & Simon, 1959) suggests that the neutral amino acid binding proteins isolated from the periplasm (Mitchell, 1961b) of osmotically shocked *E. coli*, by Piperno & Oxender (1966) and by

Anraku (1968), may be a functional part of an H^+-amino acid symport system. Likewise, the characteristics of the uptake of α-glycerophosphate in *E. coli* suggest that an H^+-α-glycerophosphate symport system is involved (Hayashi, Koch & Lin, 1964).

The sulphate translocation system of *Salmonella typhimurium* described by Dreyfuss (1964) has yielded a sulphate-binding protein (Pardee, 1966) which is situated in the surface of the plasma membrane of the cells (Pardee & Watanabe, 1968). Pardee (1966) found that the uptake of sulphate by the specific protein was not much affected over the range pH 5·3 to 8·3 in his assay system. Nevertheless, the sulphate binding protein may be an essential part of a system corresponding to the sulphuric acid porter system of rat liver mitochondria (Mitchell & Moyle, 1969b).

The uptake of citrate, isocitrate and *cis*-aconitate through the plasma membrane of *Aerobacter aerogenes* is mediated by a proton conductor-sensitive translocation reaction (Davis, 1956; Clarke & Meadow, 1959; Villarreal-Moguel & Ruiz-Herrera, 1969), and this system may also be analogous to the proton-coupled system of the cristae membrane of mitochondria from mammalian tissues. This last example illustrates that, by analogy with the two or three stage systems of equations (18) and (19), one or more circulating substrate components may possibly be involved in a given proton-coupled anion porter system. It will generally be necessary to explore such possibilities in each case of proton-coupled substrate translocation.

Na$^+$-coupled symporters

Of the considerable number of Na^+-coupled symport reactions for sugars, amino acids and other substrates referred to above (see p. 136) limitations of space permit the mention of only one example. Glycine translocation across the membrane of pigeon erythrocytes is mainly mediated by a highly specific porter, the activity of which is second order with respect to Na^+ concentration, and which may therefore be presumed to catalyse Na_2^+/glycine symport (Vidaver, 1964; Vidaver, Romain & Haurowitz, 1964; Eavenson & Christensen, 1967). This porter system is susceptible to certain inhibitors of translocation (Kittams & Vidaver, 1969), and there can be little doubt that it can be taken as representative of a whole class of porters which couple Na^+ translocation across the plasma membranes of eukaryotic cells to that of other substrates.

Physiological function of multienzyme multiporter systems

The rationale of the classification of the catalysts of chemical and translocation reactions in the earlier sections of this article was based on the premise that coupling phenomena in chemical and osmotic reactions depend on the catalysis of the equilibration of chemical groups and solutes between donor and acceptor groups and between donor and acceptor phases by appropriate assemblies of enzymes and porters in chemiosmotic systems. I propose, in this section, to make certain generalizations or speculations, the object of which is to help to prepare the ground for future experimental exploration. We will concentrate attention on two different types of coupling membrane system that are particularly relevant to the subject of this Symposium. (1) The plasma membrane system of certain eukaryotic cells, characterized by the presence of the Na^+/K^+ antiporter ATPase, generating a corresponding Na^+/K^+ anisotropy. (2) The plasma membrane system (and chromatophore membrane system) of prokaryotic cells and the cristae membrane system of mitochondria, characterized by the presence of the proton translocating ATPase and by proton translocating oxidoreduction systems, generating a corresponding H^+ anisotropy.

Energy transduction by alkali metal ion current in
eukaryotic cells of animals

The ouabain-sensitive Na^+/K^+ antiporter ATPase is present in the plasma membrane of a very wide range of eukaryotic cells from animal tissues (see Skou, 1965; Glynn, 1968; Rothstein, 1968), but this type of ATPase has not been found in higher plants, algae, moulds or in most of the bacteria that have been examined. Hafkenscheid & Bonting (1968) described an (Na^+-K^+) activated ATPase in *Escherichia coli*, but a large proportion of the ATPase activity was characteristic of a single cation-activated Mg^{2+}-ATPase (Hafkenscheid & Bonting, 1969) and it is doubtful whether the slight additional activity obtained in presence of $Na^+ + K^+$ should be attributed to a Na^+/K^+ antiporter ATPase of the type isolated from animal cell plasma membranes. The translocation of ions through the plasma membranes of a number of types of eukaryotes—for example, *Saccharomyces cerevisiae* (Dee & Conway, 1968; Borst-Pauwels & Jager, 1969; Peña, Cino, Gomez, Puyou & Tuena, 1969), *Neurospora* sp. (Slayman & Tatum, 1965; Slayman & Slayman, 1968), *Chlorella pyrenoidosa* (Barber, 1968), *Nitella* (Kitasato, 1968) and carrot root cells (Cram, 1969)—is insensitive to ouabain but is apparently directly uncoupled by proton conductors such as DNP. Thus in the plasma membranes of these cells the translocation mechanisms may possibly be analogous to the proton-coupled systems of the prokaryotic cells and mitochondria.

The well-known anisotropy with respect to Na^+ and K^+ distribution, and with respect to the membrane potential that is maintained by the electrogenic vectorial ATPase reaction (Adrian & Slayman, 1966; Harris & Ochs, 1966; Geduldig, 1968; Rang & Ritchie, 1968; Carpenter & Alving, 1968) in the plasma membranes of animal cells, is responsible for the 'excitability' and transmission of impulses in nerve and certain other types of cell (see Cole, 1968). The Na^+-motive force is also responsible for the translocation of various substrates, through the mediation of the symporters discussed above. Thus, in certain eukaryotic cells the duplex system of the Na^+/K^+ antiporter ATPase and the Na^+/substrate symporters coupled by the Na^+ ion current can account for excitability, and for the coupling of the uptake of solutes through the plasma membrane of the cells to fermentative or aerobic respiratory metabolism through the mediation of ATP.

Energy transduction by proton current in mitochondria and prokaryotic cells

The proton-translocating respiratory chain system present in the cristae membranes of the mitochondria of eukaryotic cells and in the plasma membranes of most aerobic prokaryotic cells catalyses an outwardly directed translocation of protons which is actually due to an outward flow of hydrogen atoms and an inward flow of electrons through the oxidoreduction loops (Mitchell, 1961c, 1962, 1963b; Mitchell & Moyle, 1965a, b, 1967b; Rossi & Azzone, 1965; Carafoli, Gamble, Rossi & Lehninger, 1967; Scholes et al. 1969; Edwards & Bovell, 1969; Cummins, Strand & Vaughan, 1969; and see Greville, 1969). Likewise, the proton-translocating ATPase systems of the cristae membranes of mitochondria and of the plasma membranes of certain prokaryotic cells catalyse an outwardly directed proton translocation during ATP hydrolysis (Mitchell & Moyle, 1965a, 1968; Rossi et al. 1967; Scholes et al. 1969). Oxidoreduction coupled through substrate-level phosphorylation may, of course, drive the outward proton translocation via the proton-translocating ATPase, as is probably the case, for example, in *Streptococcus faecalis*, where there is no cytochrome system (Harold & Baarda, 1968a, 1969; Harold et al. 1969) or in organisms such as *Escherichia coli* which possess both fermentative and cytochrome systems (Pavlasova & Harold, 1969). Thus, under normal physiological conditions, oxidoreductive metabolism—either directly, or through substrate-level phosphorylation and the proton-translocating ATPase—maintains a protonmotive force across the coupling membrane as illustrated in Fig. 1 a and b, respectively. I suggest, by analogy, that fermentative metabolism, involving oxidoreduction between pairs of organic substrates (see Stadtman, 1966), and chemolithotrophic or autotrophic metabolism, involving oxidoreduction of inorganic oxidants and/or reductants (see Peck, 1968; Kiesow, 1967), may be coupled through proton-translocating oxidoreduction loops, illustrated generally for a one electron equivalent oxidoreduction as follows:

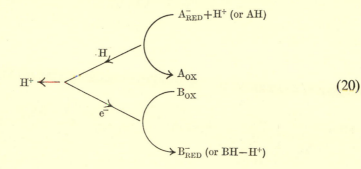

$$(20)$$

and shown in an abbreviated form for a two electron equivalent transfer in Fig. 1 c. The corresponding photosynthetic systems (Mitchell, 1966 a, b), incorporating a photoelectric reaction in the electron-carrying arm of the loop, are shown in Fig. 1 d, e.

The relatively low ion permeability of the coupling membranes enables the proton-translocating oxidoreduction reaction and the proton-translocating ATPase reaction to approach equilibrium with the proton-motive force across the membrane. When the proton-translocating oxidoreduction and ATPase reactions are catalysed in the same coupling membrane they approach equilibrium with each other (Mitchell & Moyle, 1969 a), and oxidoreductive metabolism may thus poise the $[ATP]/[ADP] \times [POH]$ level so that utilization of ATP in energy-requiring reactions results in its regeneration by the circulation of the proton current between the oxidoreduction loops and the ATPase system (Mitchell, 1966 a, b, 1968, 1969 a; and see Greville, 1969).

In a system at uniform hydrostatic pressure the total protonmotive force Δp (conveniently given in mV) is made up of the membrane potential $\Delta \psi$ (inner phase negative) and the chemical activity difference $-Z\Delta pH$ (ΔpH denoting outer pH minus inner pH) as follows

$$\Delta p = \Delta \psi - Z\Delta pH, \qquad (21)$$

where the factor Z is about 60. In mitochondria and certain bacteria $\Delta \psi$

is thought to be the major component of Δp under normal physiological conditions (Mitchell, 1961 c, 1966, 1968; Jackson *et al.* 1968; Mitchell & Moyle, 1969 a) but in prokaryotic organisms that grow in acid environ-

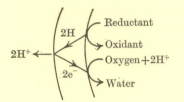

(a) Proton-translocating
aerobic respiration

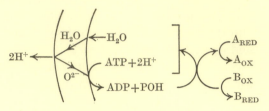

(b) Proton-translocating anaerobic
oxidoreduction (indirect)

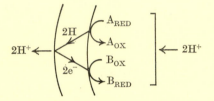

(c) Proton-translocating anaerobic
oxidoreduction (direct)

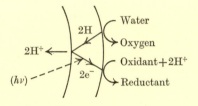

(d) Proton-translocating dehydrogenation
of water (chloroplast)

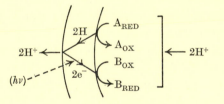

(e) Proton-translocating oxidoreduction
(cyclic or non-cyclic)

Fig. 1. Generation of the protonmotive force through oxidoreduction or hydro-dehydration loops in various ways: *a*, by aerobic respiration; *b*, by oxidoreduction coupled through substrate-level phosphorylation; *c*, by anaerobic oxidoreduction; *d*, by photolysis of water; *e*, by photo-oxidoreduction, not involving water as reductant. In all cases, two electron equivalent transfers are shown, but other transfer stoichiometries through the loops are possible. With the exception of *d*, the right side is represented as the inner phase or closed side of the membrane system. In *e*, the reaction corresponds to cyclic photo-oxidoreduction if B_{RED} is oxidized (directly or indirectly) by A_{OX}.

ments (pH 3 to 5), $-Z\Delta$ pH might well be the main component of Δp. In bacteria, such as staphylocci, where there is a considerable hydrostatic pressure difference across the plasma membrane (Mitchell &

Moyle, 1956 a), it may be necessary to include a pressure term in Δp; but the quantitative treatment of this matter has not yet been dealt with in the context of the chemiosmotic coupling theory (Mitchell, 1968, 1969 a). The uncoupling of oxidative or photosynthetic phosphorylation in bacteria by lytic or proton-conducting reagents (see e.g. Hamilton, 1968; Fields & Luria, 1969 a, b) is easily explained on the basis of the chemiosmotic hypothesis of the coupling mechanism.

The electrogenic outward translocation of protons by the vectorial metabolic systems of the coupling membrane of prokaryotic cells, or mitochondria in a salt medium, results in the leakage of cations inwards through the membrane and must cause a corresponding rise in the inner pH value unless these cations are continuously translocated out or are metabolized (Mitchell, 1961 c, 1968). In practice alkali metal ions are translocated out of mitochondria and may likewise be translocated out of prokaryotic micro-organisms (see Rothstein, 1968) by H^+/cation antiporters such as the H^+/Na^+ antiporter of rat liver mitochondria (Mitchell & Moyle, 1967 b, 1969 b) discussed above. As shown by equation (21) and discussed in detail elsewhere (Mitchell, 1968), this type of electrically neutral antiporter has the effect of collapsing the ΔpH component of the protonmotive force and increasing the $\Delta\psi$ component correspondingly in the steady state. The concentration of a given cation under steady-state conditions is dependent on the balance between the rate of entry (in the cationic form) down the electric gradient and the rate of exit by proton-linked antiport. For example, the relative abundance of K^+ in the inner phase of rat liver mitochondria (Gamble & Hess, 1966) in media containing Na^+ and K^+ can probably be accounted for by the fact that H^+/Na^+ antiport is more active than H^+/K^+ antiport (Mitchell & Moyle, 1967 b, 1969 b). As illustrated by the effects of valinomycin discussed above, entry of cations down the electric gradient may be specifically catalysed by appropriate uniporters.

Anions tend to leak from the inner phase of prokaryotic micro-organisms and mitochondria when a large membrane potential is maintained. For a membrane potential of 180 mV (negative inside) the concentration of a divalent anion (such as succinate) at equilibrium in the inner phase would be only about 10^{-6} times the concentration in the outer medium. However, the uptake of the required inorganic and organic anions is catalysed by electrically neutral proton/anion symport (or acid uniport) systems that rapidly equilibrate the effectively protonated form of the anions across the membrane (discussed in the previous section).

The basic thesis is that the protonmotive force created across the membrane by the vectorial oxidoreduction or ATPase reaction and by the exit of carbon dioxide and other acidic metabolic end-products may be utilized to cause an electrochemical potential difference of a variety of passengers across the membrane by appropriate proton-coupled porters (or porter systems) in the membrane. This duplex type of system is particularly adaptable and would seem to possess special evolutionary advantages in that although the vectorial enzyme system required to generate the protonmotive force between the inner and outer phases must have rather complex properties, including intrinsic structural

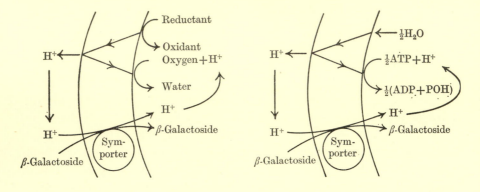

(a) β-Galactoside translocation coupled to respiration

(b) β-Galactoside translocation coupled to ATP hydrolysis

Fig. 2. Suggested H⁺-symport mechanism of β-galactoside uptake in *Escherichia coli*, after Mitchell (1933 *a*): *a*, when the proton current is due to respiration; *b*, when the proton current is due to ATP hydrolysis. These diagrams are intended to show the mutual dependence of the flows but not the stoichiometric relationships between them.

and functional anisotropy, the porters need not be intrinsically anisotropic and are simply required to couple the flow of H^+ ions to the flows of specific passengers by the secondary mechanisms discussed above.

In Fig. 2*a* the symporter mechanism previously suggested for β-galactoside uptake in respiring *Escherichia coli* is reproduced, and in Fig. 2*b* we illustrate the corresponding system operated by ATP hydrolysis. These linked duplex systems may be taken as examples of a general mechanism that is probably responsible for the proton-linked translocation of a wide range of organic and inorganic substrates, as illustrated in the diagram of Fig. 3, which is essentially a composite version of proton circuit diagrams published previously (Mitchell, 1963*a*, 1966*a*, *b*).

Since the environment generally contains facilities for maintaining oxidoreductive metabolism, either as a supply of light or of oxidants and

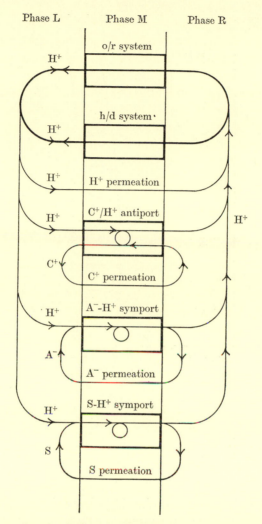

Fig. 3. Composite proton circuit diagram illustrating coupling between metabolism and transport as it is thought to occur in mitochondria and in certain prokaryotic cells, after Mitchell (1966 *a*, *b*). Translocation of protons through oxidoreduction (o/r system) is shown poised against proton translocation through the reversible ATPase (h/d system). For simplicity, the oxidoreduction reactants for the o/r system and the hydrodehydration reactants (i.e. ATP, ADP, POH and H_2O) for the h/d system have been omitted from the diagram. Dissipation of part of the proton current occurs by the translocation of H^+ through substrate-specific antiporter systems for certain cations C^+ (e.g. Na^+ or basic amino acids and symporter systems for certain anions A^- (e.g. phosphate or Krebs cycle acids) and neutral substrates S (e.g. sugars or neutral amino acids). Some of the proton-coupled porter systems may be complex, as explained in the text. The rate of dissipation of the proton current through the porter-coupled reactions in the steady state is dependent on anion, cation and neutral substrate permeation as indicated. The symbols A^- and C^+ do not denote the valency of the anions and cations and the stoichiometry of translocation is not indicated for A^-, C^+ or S.

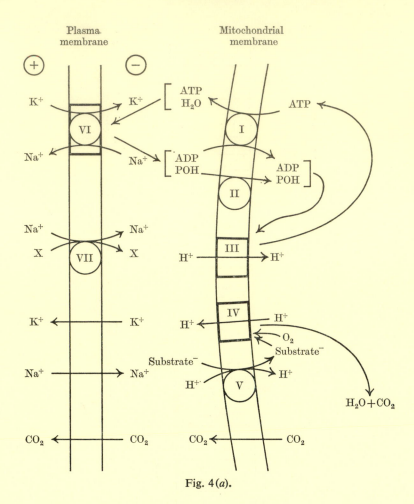

Fig. 4 (*a*).

reductants, chemiosmotic coupling between oxidoreduction loops and the reversible proton-translocating ATPase system provides a remarkably versatile method of coupling oxidoreductive metabolism to ATP synthesis. It is an evolutionarily attractive proposition (Mitchell, 1968) that the proton-translocating oxidoreduction loop system and the reversible proton-translocating ATPase may have arisen separately as alternatives for generating the pH difference and membrane potential required for nutrient uptake and ionic regulation via porter systems in primitive prokaryotic cells, and that the accidental occurrence of both systems in the same cell may then have provided the means of storing the free energy of oxidoreduction in ATP synthesized by the reversal of the ATPase, or in some other anhydride, such as pyrophosphate

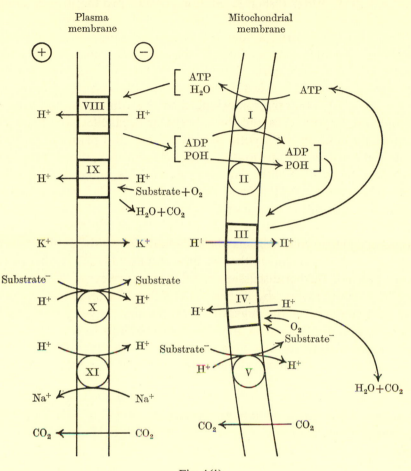

Fig. 4 (*b*).

Fig. 4. Qualitative flow diagram illustrating the general principles of coupling between metabolism and translocation by ion currents across the plasma membrane and mito-chondrial (cristae) membrane in eukaryotic cells: *a*, of animals; and *b*, of certain other organisms including moulds and plants. The following porters and vectorial enzyme systems are represented: I, the atractyloside-sensitive ATP/ADP antiporter; II, the phosphoric acid uniporter; III, the proton-translocating ATPase; IV, the proton-translocating oxidore-duction chain; V, a proton-coupled substrate (anion) symporter or porter system; VI, the Na$^+$/K$^+$ antiporter ATPase; VII, a sodium-coupled substrate (X) symporter (for which X might correspond to substrate$^-$ of V); VIII, a proton-translocating ATPase; IX, a proton-translocating oxidoreduction system; X, a proton-coupled substrate (anion) symporter or porter system; XI, a Na$^+$/H$^+$ antiporter. The un-numbered arrows indicate permeation (or uniport). The inclusion of systems VIII and IX in the plasma membrane in *b* is specula-tive. The polarity of the electric potential across the plasma membrane is indicated in the circles. The mitochondrial cristae membrane system is negative inside. The stoichiometries of translocation are not represented in this simplified diagram, the object of which is to indicate the general types of flow pattern that may occur in different types of eukaryote.

(Baltscheffsky, 1967; Fisher & Guillory, 1969), produced by a similar mechanism.

The remarkable correspondence between the proton-coupled systems of the plasma membranes of certain bacteria and those of the cristae membrane of mitochondria should lead to a very helpful diffusion of knowledge and ideas between the two fields of study: it also gives weight to the speculation that there may indeed be an evolutionary link between aerobic prokaryotic cells and the subcellular organelles now known as mitochondria in the eukaryotic cells of animals and plants.

As indicated in the previous section, certain types of eukaryotic cell including those of higher plants, algae, moulds and yeasts lack the Na^+/K^+ antiporter ATPase and differ from eukaryotic animal cells in that ion translocation appears to be proton coupled. The translocation of certain organic substrates through the plasma membranes of these non-animal eukaryotic cells may be accounted for by the duplex proton-coupled type of mechanism that we have discussed here. Limitations of space preclude further discussion of this matter; relevant references are available in the reviews by Robertson (1968) and Rothstein (1968).

Fig. 4 summarizes the suggested relationships between the primary translocation and secondary translocation reactions in eukaryotic cells containing mitochondria: a, in the case of animal cells; and b, in the case of the class of non-animal cells lacking the Na^+/K^+ antiporter ATPase.

Substrate uptake by group translocation

The diversity of substrate uptake mechanisms in living cells should not be obscured by the preoccupation of the present article with the generalizations concerning the versatility of the duplex systems coupled by currents of H^+ ions or Na^+ ions in prokaryotic and eukaryotic cells. It has long been recognized, for example, that the characteristics of uptake of certain sugars and sugar derivatives by bacteria are different from the characteristics of the proton-coupled bacterial systems discussed here, particularly with respect to uncoupling by proton-conducting reagents such as DNP. It has recently been found that the uptake of α-methyl-glucoside and certain other sugars and sugar derivatives through the plasma membrane of *Escherichia coli* and *Aerobacter aerogenes* is catalysed by a phosphorylating enzyme system coupled to metabolism through phosphoenol pyruvate (Kundig, Kundig, Anderson & Roseman, 1966; Simoni *et al.* 1967; Tanaka & Lin, 1967; Tanaka, Fraenkel & Lin, 1967; Simoni, Smith & Roseman, 1968). The work of Laue & Macdonald (1968) on β-galactoside uptake in *Staphylococcus aureus* led them to conclude that 'the suggestion of Kennedy & Scarborough (1967) that the phosphorylated sugar is the substrate of the staphylococcal "β-galactosidase", taken together with the experiments presented here, lead to the hypothesis that in Staphylococcus, galactoside permeation may depend upon the first reaction of galactoside metabolism'. A similar conclusion has been reached in the case of deoxyglucose uptake by yeast cells (Van Steveninck, 1968). Thus, it may turn out that the translocation of the sugar is integral with the process of phosphorylation, and that the mechanism of uptake is of the primary group-translocation type (see Mitchell, 1963a). However, the details of the phosphorylative uptake mechanism, which involves at least four catalytic components (Simoni et al. 1968) are not yet completely worked out, and it is possible that the substrate initially passes through the M phase of the membrane via a porter before being phosphorylated in the inner phase, as appears to be the case in glycerol uptake in E. coli (Hayashi & Lin, 1965; Sanno, Wilson & Lin, 1968).

CONCLUSION

In this review the relationship between metabolism and transport in certain membrane systems has been described by treating the pathways of group transfer (i.e. of chemical reaction) as extensions of the routes of escape of the chemical particles (reactants and resultants) that are common to metabolism and transport; and we have attributed a spatially directed (vectorial) group-translocation property to certain metabolic group-transfer enzymes or enzyme systems which may thereby act as prime movers of transport. The coupling of transport to metabolism depends on the anisotropy of the chemical reaction system as a whole, and does not necessarily require group translocation at the level of the enzyme or enzyme complex (see Mitchell, 1961a, 1967b; Kedem, 1961; Bright, 1969). Three basal types of catalytic unit appear to be involved in the physiological systems that organize the integral processes of metabolism and transport in prokaryotic and eukaryotic cells.

(1) Effectively isotropic enzymes or enzyme complexes which catalyse chemical change in certain regions or phases, providing a macroscopic source or sink for specific chemicals.

(2) Vectorial enzymes or enzyme complexes which catalyse the translocation equilibria of specific chemical groups across, or orientated relative to, an osmotic barrier phase.

(3) Intrinsically isotropic porters or porter systems which catalyse the secondary and ionic bond exchanges required for uniport, symport and antiport reactions across an osmotic barrier phase.

These basal catalytic units are the essential building modules which, when assembled in different permutations and combinations in physio-logical osmotic systems, can display a variety of physiological charac-teristics such as energy transduction, ionic regulation, active transport, excitability and morphogenesis. With regard to morphogenesis, the initial anisotropy or sidedness of the membrane systems which couple the chemical changes of metabolism to the generation of macroscopic differences of composition between the phases on either side, depends on the specific orientation or location of the enzymes in a semi-crystal-line membrane fabric. The question of the mechanism of segregation of the appropriate enzymes and other catalytic and structural components by appropriate locational valencies in the membranes and aqueous com-partments of cells (Mitchell, 1959 b, 1962) is beyond the scope of this article; but it is appropriate to conclude by remarking that the flame-like dynamic morphological properties of living organisms are produced by a kind of continuous filtration of environmental chemical components through the membranes and other spatially orientated catalytic struc-tures of the living cells. In non-photosynthetic cells, this complex filtra-tion of solutes and chemical groups is due entirely to the spontaneous process of diffusion; but in photosynthetic cells light energy may con-tribute not only indirectly to the maintenance of the diffusional flow, but may directly energize the translocation of electrons through the membranes containing the light-absorbing pigments, as in a solar cell (see Mitchell, 1967 c).

I wish to acknowledge my indebtedness to Drs Peter Hinkle, Jennifer Moyle and Peter Scholes for stimulating and helpful discussions on the subject of this paper. I thank Miss Stephanie Phillips for patient secretarial help, and Glynn Research Ltd. for financial support.

REFERENCES

ADRIAN, R. H. & SLAYMAN, C. L. (1966). Membrane potential and conductance during transport of sodium, potassium and rubidium in frog muscle. *J. Physiol.* **184**, 970.

ALBERS, R. W. (1967). Biochemical aspects of active transport. *A. Rev. Biochem.* **36**, 727.

ANDREOLI, T. E., TIEFFENBERG, M. & TOSTESON, D. C. (1967). The effect of valino-mycin on the ionic permeability of thin lipid membranes. *J. gen. Physiol.* **50**, 2527.

ANRAKU, Y. (1968). Transport of sugars and amino acids in bacteria. III. Studies on the restoration of active transport. *J. biol. Chem.* **243**, 3128.

AZZI, A. & AZZONE, G. F. (1967). Swelling and shrinkage phenomena in liver mitochondria. VI. Metabolism-independent swelling coupled to ion movement. *Biochim. biophys. Acta* **131**, 468.

AZZI, A. CHAPPELL, J. B. & ROBINSON, B. H. (1967). Penetration of the mitochondrial membrane by glutamate and aspartate. *Biochem. biophys. Res. Commun.* **29**, 148.

AZZONE, G. F., ROSSI, E. & SCARPA, A. (1968). Osmotic coupling in ion translocation. In *Regulatory Functions of Biological Membranes*, p. 236. Ed. J. Järnefelt. Amsterdam: Elsevier.

BADER, H., POST, R. L. & BOND, G. H. (1968). Comparison of sources of a phosphorylated intermediate in transport ATPase. *Biochim. biophys. Acta* **150**, 41.

BAKER, P. F., BLAUSTEIN, M. P., KEYNES, R. D., MANIL, J., SHAW, T. I. & STEINHARDT, R. A. (1969). The ouabain-sensitive fluxes of sodium and potassium in squid giant axons. *J. Physiol.* **200**, 459.

BAKER, P. F. & CONNELLY, C. M. (1966). Some properties of the external activation site of the sodium pump in crab nerve. *J. Physiol.* **185**, 270.

BALTSCHEFFSKY, M. (1967). Inorganic pyrophosphate and ATP as energy donors in chromatophores from *Rhodospirillum rubrum*. *Nature, Lond.* **216**, 241.

BANGHAM, A. D. & HAYDON, D. A. (1968). Ultrastructure of membranes: bimolecular organization. *Br. med. Bull.* **24**, 124.

BARBER, J. (1968). The influx of potassium into *Chlorella pyrenoidosa*. *Biochim. biophys. Acta* **163**, 141.

BARTNICKI-GARCIA, S. (1968). Cell wall chemistry, morphogenesis and taxonomy of fungi. *A. Rev. Microbiol.* **22**, 87.

BELL, R. P. (1959). *The Proton in Chemistry*, 1st ed. London: Methuen.

BIHLER, I. (1969). Intestinal sugar transport: ionic activation and chemical specificity. *Biochim. biophys. Acta* **183**, 169.

BORST-PAUWELS, G. W. F. H. & JAGER, S. (1969). Inhibition of phosphate and arsenate uptake in yeast by monoiodoacetate, fluoride, 2,4-dinitrophenol and acetate. *Biochim. biophys. Acta* **172**, 399.

BRIERLEY, G. P. (1969). Energy linked alterations of mitochondrial permeability to anions. *Biochem. biophys. Res. Commun.* **35**, 396.

BRIERLEY, G. P., SETTLEMIRE, C. T. & KNIGHT, V. A. (1968). Ion transport by heart mitochondria. *Archs Biochem. Biophys.* **126**, 276.

BRIGHT, P. B. (1969). Concerning 'Anisotropic' contributions of chemical reactions to flow equations and active transport. *J. theor. Biol.* **23**, 135.

BUCHANAN, J., POPOVITCH, J. R. & TAPLEY, D. F. (1969). Leucine transport by rat liver mitochondria *in vitro*. *Biochim. biophys. Acta* **173**, 532.

CARAFOLI, E., GAMBLE, R. L., ROSSI, C. S. & LEHNINGER, A. L. (1967). Super stoichiometric ratios between ion movements and electron transport in rat liver mitochondria. *J. biol. Chem.* **242**, 1199.

CARAFOLI, E. & ROSSI, C. S. (1967). The effect of dinitrophenol on the permeability of the mitochondrial membrane. *Biochem. biophys. Res. Commun.* **29**, 153.

CARPENTER, D. O. & ALVING, B. O. (1968). A contribution of an electrogenic Na^+ pump to membrane potential in *Aplysia* neurons. *J. gen. Physiol.* **52**, 1.

CHAPPELL, J. B. (1968). Systems used for the transport of substrates into mitochondria. *Br. med. Bull.* **24**, 150.

CHAPPELL, J. B. & CROFTS, A. R. (1965). Gramicidin and ion transport in isolated liver mitochondria. *Biochem. J.* **95**, 393.

CHAPPELL, J. B. & CROFTS, A. R. (1966). Ion transport and reversible volume changes of isolated mitochondria. In *Regulation of Metabolic Processes in Mitochondria*, p. 293. Eds. J. M. Tager, S. Papa, E. Quagliariello and E. C. Slater. Amsterdam: Elsevier.

CHAPPELL, J. B. & HAARHOFF, K. N. (1967). The penetration of the mitochondrial membrane by anions and cations. In *Biochemistry of Mitochondria*, p. 75. Eds. E. C. Slater, Z. Kaniuga and L. Wojtczak. London: Academic Press.

CHAPPELL, J. B., HENDERSON, P. J. F., McGIVAN, J. D. & ROBINSON, B. H. (1968). The effect of drugs on mitochondrial function. In *The Interaction of Drugs and Subcellular Components in Animal Cells*, p. 71. Ed. P. N. Campbell. London: Churchill.

CHAPPELL, J. B. & ROBINSON, B. H. (1968). Penetration of the mitochondrial membrane by tricarboxylic acid anions. *Biochem. Soc. Symp.* **27**, 123.

CHRISTENSEN, H. N. (1962). *Biological Transport*, 1st ed. New York: Benjamin.

CHRISTENSEN, H. N. & HANDLOGTEN, M. E. (1969). Reactions of neutral amino acids plus Na^+ with a cationic amino acid transport system. *Febs Letters* **3**, 14.

CHRISTIE, G. S., AHMED, K., McLEAN, A. E. M. & JUDAH, J. D. (1965). Active transport of potassium by mitochondria. I. Exchange of K^+ and H^+. *Biochim. biophys. Acta*, **94**, 432.

CLARKE, P. H. & MEADOW, P. M. (1959). Evidence for the occurrence of permeases for tricarboxylic acid cycle intermediates in *Pseudomonas aeruginosa*. *J. gen. Microbiol.* **20**, 144.

COHEN, G. N. & MONOD, J. (1957). Bacterial permeases. *Bact. Rev.* **21**, 169.

COLE, K. S. (1968). *Membranes, Ions and Impulses*, 1st ed. Berkeley: University of California Press.

CONWAY, E. J. (1951). The biological performance of osmotic work. A redox pump. *Science, N.Y.* **113**, 270.

CONWAY, E. J. (1954). Some aspects of ion transport through membranes. *Symp. Soc. exp. Biol.* **8**, 297.

CRAM, W. J. (1969). Respiration and energy-dependent movements of chloride at plasmalemma and tonoplast of carrot root cells. *Biochim. biophys. Acta* **173**, 213.

CRANE, R. K. (1965). Na^+-dependent transport in the intestine and other animal tissues. *Fedn Proc. Fedn Am. Socs exp. Biol.* **24**, 1000.

CRANE, R. K., MILLER, D. & BIHLER, I. (1961). The restrictions on possible mechanisms of intestinal active transport of sugars. In *Membrane Transport and Metabolism*, p. 439. Eds. A. Kleinzeller and A. Kotyk. New York: Academic Press.

CROFTS, A. R. (1968). Ammonium ion uptake by chloroplasts, and the high-energy state. In *Regulatory Functions of Biological Membranes*, p. 247. Ed. J. Järnefelt. Amsterdam: Elsevier.

CUMMINS, J. T., STRAND, J. A. & VAUGHAN, B. E. (1969). The movement of H^+ and other ions at the onset of photosynthesis in Ulva. *Biochim. biophys. Acta* **173**, 198.

DAMADIAN, R. (1968). Ion metabolism in a potassium accumulation mutant of *Escherichia coli* B. *J. Bact.* **95**, 113.

DANIELLI, J. F. (1954). Morphological and molecular aspects of active transport. *Symp. Soc. exp. Biol.* **8**, 502.

DANIELLI, J. F. (1966). On the thickness of lipid membranes. *J. theor. Biol.* **12**, 439.

DANIELS, V. G., DAWSON, A. G., NEWEY, H. & SMYTH, D. H. (1969). Effect of carbon chain length and amino group position on neutral amino acid transport systems in rat small intestine. *Biochim. biophys. Acta* **173**, 575.

DAVIES, R. E. (1957). Gastric hydrochloric acid production—the present position. In *Metabolic Aspects of Transport across Cell Membranes*, p. 244. Ed. Q. R. Murphy. Madison: University of Wisconsin Press.

DAVIES, R. E. & OGSTON, A. G. (1950). On the mechanism of secretion of ions by gastric mucosae and by other tissues. *Biochem. J.* **46**, 324.

DAVIS, B. D. (1956). Relationships between enzymes and permeability (membrane transport) in bacteria. In *Enzymes: Units of Biological Structure and Function*, p. 509. Ed. O. H. Gaebler. New York: Academic Press.

DAVIS, B. D. (1961). Specific membrane transport and its adaptation: Chairman's Introduction. In *Biological Structure and Function*, vol. 2, p. 571. Eds. T. W. Goodwin and O. Lindberg. New York: Academic Press.

DEE, E. & CONWAY, E. J. (1968). The relation between sodium ion content and efflux of labelled sodium ions from yeast. *Biochem. J.* **107**, 265.

DE HAAN, E. J. & TAGER, J. M. (1968). Evidence for a permeability barrier for α-oxoglutarate in rat-liver mitochondria. *Biochim. biophys. Acta* **153**, 98.

DILLEY, R. A. & SHAVIT, N. (1968). On the relationship of H^+ transport to photophosphorylation in spinach chloroplasts. *Biochim. biophys. Acta* **162**, 86.

DREYFUSS, J. (1964). Characterization of a sulfate- and thiosulfate-transporting system in *Salmonella typhimurium*. *J. biol Chem.* **239**, 2292.

EAVENSON, E. & CHRISTENSEN, H. N. (1967). Transport systems for neutral amino acids in the pigeon erythrocyte. *J. biol. Chem.* **242**, 5386.

ECHLIN, P. (1970). The photosynthetic apparatus in prokaryotes and eukaryotes. This Symposium, p. 221.

EDDY, A. A. (1968a). A net gain of sodium ions and a net loss of potassium ions accompanying the uptake of glycine by mouse ascites-tumour cells in the presence of sodium cyanide. *Biochem. J.* **108**, 195.

EDDY, A. A. (1968b). The effects of varying the cellular and extracellular concentrations of sodium and potassium ions on the uptake of glycine by mouse ascites-tumour cells in the presence and absence of sodium cyanide. *Biochem. J.* **108**, 489.

EDDY, A. A., HOGG, C. & REID, M. (1969). Ion gradients and the accumulation of amino acids by mouse ascites tumour cells depleted of adenosine triphosphate. *Biochem. J.* **112**, 11 P.

EDWARDS, G. & BOVELL, C. R. (1969). Characteristics of a light-dependent proton transport in cells of *Rhodospirillum rubrum*. *Biochim. biophys. Acta* **172**, 126.

EGAN, J. B. & MORSE, M. L. (1966). Carbohydrate transport in *Staphylococcus aurens*. III. Studies on the transport process. *Biochim. biophys. Acta* **112**, 63.

EISENMAN, G., CIANI, S. M. & SZABO, G. (1968). Some theoretically expected and experimentally observed properties of lipid bilayer membranes containing neutral molecular carriers of ions. *Fedn Proc. Fedn Am. Socs exp. Biol.* **27**, 1289.

ELLAR, D. (1965). *Enzyme Nomenclature*. Amsterdam: Elsevier.

ELLAR, D. (1970). Protective surface structures. This Symposium, p. 167.

FAHN, S., KOVAL, G. J. & ALBERS, W. (1968). Sodium-potassium-activated adenosine triphosphatase of *Electrophorus* electric organ. *J. biol. Chem.* **243**, 1993.

FIELDS, K. L. & LURIA, S. E. (1969a). Effects of colicins El and K on transport systems. *J. Bact.* **97**, 57.

FIELDS, K. L. & LURIA, S. E. (1969b). Effects of colicins El and K on cellular metabolism. *J. Bact.* **97**, 64.

FISHER, R. R. & GUILLORY, R. J. (1969). Partial resolution of energy-linked reactions in *Rhodospirillum rubrum* chromatophores. *Febs Letters* **3**, 27.

FONYO, A. & BESSMAN, S. P. (1968). Inhibition of inorganic phosphate penetration into liver mitochondria by p-mercuribenzoate. *Biochem. Med.* **2**, 145.

GAMBLE, J. L. & HESS, R. C. (1966). Mitochondrial electrolytes. *J. biol. Chem.* **210**, 765.

GARRAHAN, P. J. & GLYNN, I. M. (1967). The incorporation of inorganic phosphate into adenosine triphosphate by reversal of the sodium pump. *J. Physiol.* **192**, 237.

GARRAHAN, P. J., POUCHAN, M. I. & REGA, A. F. (1969). Potassium activated phosphatase from human red blood cells. The mechanism of potassium activation. *J. Physiol.* **202**, 305.

GEDULDIG, D. (1968). A ouabain-sensitive membrane conductance. *J. Physiol.* **194**, 521.

GLASSTONE, S., LAIDLER, K. J. & EYRING, H. (1941). *The Theory of Rate Processes*, 1st ed. New York: McGraw-Hill.

GLYNN, I. M. (1957). The ionic permeability of the red cell membrane. *Progr. Biophys. biophys. Chem.* **8**, 241.

GLYNN, I. M. (1968). Membrane adenosine triphosphatase and cation transport. *Br. med. Bull.* **24**, 165.

GOLDNER, A. M., SCHULTZ, S. G. & CURRAN, P. F. (1969). Sodium and sugar fluxes across the mucosal border of rabbit ileum. *J. gen. Physiol.* **53**, 362.

GREVILLE, G. D. (1969). A scrutiny of Mitchell's chemiosmotic hypothesis of respiratory chain and photosynthetic phosphorylation. In *Current Topics in Bioenergetics*, vol. 3, p. 1. Ed. D. R. Sanadi. New York: Academic Press.

GUGGENHEIM, E. A. (1933). *Modern Thermodynamics by the Methods of Willard Gibbs*, 1st ed. London: Methuen.

HAFKENSCHEID, J. C. M. & BONTING, S. L. (1968). Studies on (Na$^+$-K$^+$)-activated ATPase. XIX. Occurrence and properties of a (Na$^+$-K$^+$)-activated ATPase in *Escherichia coli*. *Biochim. biophys. Acta* **151**, 204.

HAFKENSCHEID, J. C. M. & BONTING, S. L. (1969). Studies on (Na$^+$-K$^+$)-activated ATPase. XXIII. A Mg^{2+}-ATPase in *Escherichia coli*, activated by monovalent cations. *Biochim. biophys. Acta* **178**, 128.

HAMILTON, W. A. (1968). The mechanism of the bacteriostatic action of tetrachlorosalicylanilide: a membrane-active antibacterial compound. *J. gen. Microbiol.* **50**, 441.

HANSFORD, R. G. & CHAPPELL, J. B. (1967). The effect of Ca^{2+} on the oxidation of glycerol phosphate by blowfly flight-muscle mitochondria. *Biochem. biophys. Res. Commun.* **27**, 686.

HAROLD, F. M. & BAARDA, J. R. (1968a). Effects of nigericin and monactin on cation permeability of *Streptococcus faecalis* and metabolic capacities of potassium-depleted cells. *J. Bact.* **95**, 816.

HAROLD, F. M. & BAARDA, J. R. (1968b). Inhibition of membrane transport in *Streptococcus faecalis* by uncouplers of oxidative phosphorylation and its relationship to proton conduction. *J. Bact.* **96**, 2025.

HAROLD, F. M. & BAARDA, J. R. (1969). Inhibition of membrane-bound adenosine triphosphatase and of cation transport in *Streptococcus faecalis* by N,N'-dicyclohexylcarbodiimide. *J. biol. Chem.* **244**, 2261.

HAROLD, F. M., BAARDA, J. R., BARON, C. & ABRAMS, A. (1969). Dio 9 and chlorhexidine: inhibitors of membrane-bound ATPase and of cation transport in *Streptococcus faecalis*. *Biochim. biophys. Acta* **183**, 129.

HARRIS, E. J. (1968). The dependence on dicarboxylic acids and energy of citrate accumulation in depleted rat liver mitochondria. *Biochem. J.* **109**, 247.

HARRIS, E. J. & OCHS, S. (1966). Effects of sodium extrusion and local anaesthetics on muscle membrane resistance and potential. *J. Physiol.* **187**, 5.

HARRIS, E. J. & PRESSMAN, B. C. (1967). Obligate cation exchanges in red cells. *Nature, Lond.* **216**, 918.

HASLAM, J. M. & GRIFFITHS, D. E. (1968). Factors affecting the translocation of oxaloacetate and L-malate into rat liver mitochondria. *Biochem. J.* **109**, 921.

HASLAM, J. M. & KREBS, H. A. (1968). The permeability of mitochondria to oxaloacetate and malate. *Biochem. J.* **107**, 659.

HAUGAARD, N., LEE, N. H., KOSTRZEWA, R., HORN, R. S. & HAUGAARD, E. S. (1969). The rate of sulfhydryl groups in oxidative phosphorylation and ion transport by rat liver mitochondria. *Biochim. biophys. Acta* **172**, 198.

HAUSER, G. (1969). Myo-inosital transport in slices of rat kidney cortex. II. Effect of the ionic composition of the medium. *Biochim. biophys. Acta* **173**, 267.

HAYASHI, S., KOCH, J. P. & LIN, E. C. C. (1964). Active transport of L-α-glycerophosphate in *Escherichia coli*. *J. biol. Chem.* **239**, 3098.

HAYASHI, S. & LIN, E. C. C. (1965). Capture of glycerol by cells of *Escherichia coli*. *Biochim. biophys. Acta* **94**, 479.

HEINZ, E. (1967). Transport through biological membranes. *A. Rev. Physiol.* **29**, 21.

HENDERSON, P. J. F., McGIVAN, J. D. & CHAPPELL, J. B. (1969). The action of certain antibiotics on mitochondrial, erythrocyte and artificial phospholipid membranes. *Biochem. J.* **111**, 521.

HOPFER, U., LEHNINGER, A. L. & THOMPSON, T. E. (1968). Protonic conductance across phospholipid bilayer membranes induced by uncoupling agents for oxidative phosphorylation. *Proc. natn. Acad. Sci. U.S.A.* **59**, 484.

HOREKER, B. L., OSBORN, M. J., McLELLAN, W. L., AVIGAD, G. & ASENSIO, C. (1961). The role of bacterial permeases in metabolism. In *Membrane Transport and Metabolism*, p. 378. Eds. A. Kleinzeller and A. Kotyk. New York: Academic Press. Prague: Czechoslovak Publishing House.

HUGHES, D. E., LLOYD, D. & BRIGHTWELL, R. (1970). Structure, function and distribution of organelles in prokaryotic and eukaryotic cells. This Symposium, p. 295.

IVANOV, V. T., LAINE, I. A., ABDULAEV, N. D., SENYAVINA, L. B., POPOV, E. M., OVCHINNIKOV, YU. A. & SHEMYAKIN, M. M. (1969). The physicochemical basis of the functioning of biological membranes: The conformation of valinomycin and its K^+ complex in solution. *Biochem. biophys. Res. Commun.* **34**, 803.

IZAWA, S. & HIND, G. (1967). The kinetics of the pH rise in illuminated chloroplast suspensions. *Biochim. biophys. Acta* **143**, 377.

JACKSON, J. B., CROFTS, A. R. & VON STEDINGK, L.-V. (1968). Ion transport induced by light and antibiotics in chromatophores from *Rhodospirillum rubrum*. *European J. Biochem.* **6**, 41.

JAGENDORF, A. T. & NEUMANN, J. (1965). Effect of uncouplers on the light-induced pH rise with spinach chloroplasts. *J. biol. Chem.* **240**, 3210.

JAGENDORF, A. T. & URIBE, E. (1966). Photophosphorylation and the chemiosmotic hypothesis. *Brookhaven Symp. Biol.* **19**, 215.

KEDEM, O. (1961). Criteria of active transport. In *Membrane Transport and Metabolism*, p. 87. Eds. A. Kleinzeller and A. Kotyk. New York: Academic Press.

KELLER, D. M. (1968). Accumulation of arginine by dog kidney cortex mitochondria. *Biochim. biophys. Acta* **153**, 113.

KENNEDY, E. P. (1966). Biochemical aspects of membrane function. In *Current Aspects of Biochemical Energetics*, p. 433. Eds. N. O. Kaplan and E. P. Kennedy. New York; Academic Press.

KENNEDY, E. P. & SCARBOROUGH, G. A. (1967). Mechanism of hydrolysis of O-nitrophenyl-β-galactoside in *Staphylococcus aureus* and its significance for theories of sugar transport. *Proc. natn. Acad. Sci. U.S.A.* **58**, 225.

KEPES, A. (1960). Études cinétiques sur la galactoside-perméase d'*Escherichia coli*. *Biochim. biophys. Acta* **40**, 70.

KEPES, A. (1961). Bacterial permeases. *Colloquium Ges. physiol. Chem.* **12**, 100.

KEPES, A. (1964). The place of permeases in cellular organization. In *The Cellular Functions of Membrane Transport*, p. 155. Ed. J. H. Hoffman. New Jersey: Prentice-Hall.

KEPES, A. & COHEN, G. N. (1962). Permeation. In *The Bacteria*, Vol. 4, p. 179. Eds. I. C. Gunsalus and R. Stanier. New York: Academic Press.

KIESOW, L. A. (1967). Energy-linked reactions in chemoautotrophic organisms. In *Current Topics in Bioenergetics*, vol. 2, p. 195. Ed. D. R. Sanadi. New York: Academic Press.

KITASATO, H. (1968). The influence of H^+ on the membrane potential and ion fluxes of *Nitella*. *J. gen. Physiol.* **52**, 60.

KITTAMS, D. W. & VIDAVER, G. A. (1969). Inhibition by β-phenylethylamine and similar compounds of glycine transport by pigeon red cells. *Biochim. biophys. Acta* **173**, 540.

KLINGENBERG, M. & PFAFF, E. (1968). Metabolic control in mitochondria by adenine nucleotide translocation. *Biochem. Soc. Symp.* **27**, 105.

KOCH, A. L. (1964). The role of permease in transport. *Biochim. biophys. Acta* **79**, 177.

KOCH, A. L. (1967). Kinetics of permease catalysed transport. *J. theor. Biol.* **14**, 103.

KOHN, P. G., SMYTH, D. H. & WRIGHT, E. M. (1968). Effects of amino acids, dipeptides and disaccharides on the electric potential across rat small intestine. *J. Physiol.* **196**, 723.

KOOPMAN, W. & SCHULTZ, S. G. (1969). The effect of sugars and amino acids on mucosal Na^+ and K^+ concentrations in rabbit ileum. *Biochem. biophys. Acta* **173**, 338.

KORN, E. (1969). Biological membranes. In *Theoretical and Experimental Biophysics*, p. 1. Ed. A. Cole. New York and London: Marcel Dekker.

KRAAYENHOF, R. & VAN DAM, K. (1969). Interaction between uncouplers and substrates in rat-liver mitochondria. *Biochim. biophys. Acta* **172**, 189.

KUCHLER, R. J. (1967). The role of sodium and potassium in regulating amino acid accumulation and protein synthesis in LM-strain mouse fibroblasts. *Biochim. biophys. Acta* **136**, 473.

KUNDIG, W., KUNDIG, F. D., ANDERSON, B. & ROSEMAN, S. (1966). Restoration of active transport of glycosides in *Escherichia coli* by a component of a phosphotransferase system. *J. biol. Chem.* **241**, 3243.

LANG, N. J. (1968). The fine structure of blue green algae. *A. Rev. Microbiol.* **22**, 15.

LANGMUIR, I. (1938). Overturning and anchoring of monolayers. *Science, N.Y.* **87**, 493.

LANGMUIR, I. (1939). Molecular layers. *Proc. Roy. Soc. Lond.* A **170**, 1.

LARDY, H. A., GRAVEN, S. N. & ESTRADA-O, S. (1967). Specific induction and inhibition of cation and anion transport in mitochondria. *Fedn Proc. Fedn Am. Socs exp. Biol.* **26**, 1355.

LAUE, P. & MACDONALD, R. E. (1968). Studies on the relation of thiomethyl-β-D-galactoside accumulation to thiomethyl-β-D-galactoside phosphorylation in *Staphylococcus aureus*. *Biochim. biophys. Acta* **165**, 410.

LEE, C. P. & ERNSTER, L. (1966). The energy-linked nicotinamide nucleotide transhydrogenase reaction: its characteristics and its use as a tool for the study of oxidative phosphorylation. In *Regulation of Metabolic Processes in Mitochondria*, p. 218. Eds. J. M. Tager, S. Papa, E. Quagliariello and E. C. Slater. Amsterdam: Elsevier.

LEHNINGER, A. L., CARAFOLI, E. & ROSSI, C. S. (1967). Energy-linked ion movements in mitochondrial systems. *Adv. Enzymol.* **29**, 259.

LEV, A. A. & BUZHINSKY, E. P. (1967). Cation specificity of model bimolecular phospholipid membranes containing valinomycin. *Cytology, U.S.S.R.* **9**, 102.

LIBERMAN, E. A., MOCHOVA, E. N., SKULACHEV, V. P. & TOPALY, B. P. (1968). Action of uncouplers of oxidative phosphorylation on bimolecular phospholipid membranes. *Biofizika, U.S.S.R.* **13**, 188.

LIBERMAN, E. A. & TOPALY, V. P. (1968). Selective transport of ions through bimolecular phospholipid membranes. *Biochim. biophys. Acta* **163**, 125.

LINDENMAYER, G. E., LAUGHTER, A. H. & SCHWARTZ, A. (1968). Incorporation of inorganic phosphate-32 into a Na^+, K^+-ATPase preparation: stimulation by ouabain. *Archs Biochem. Biophys.* **127**, 187.

LIPMANN, F. (1941). Metabolic generation and utilization of phosphate bond energy. *Adv. Enzymol.* **1**, 99.

Löw, H. & Vallin, I. (1963). Succinate-linked diphosphopyridine nucleotide reduction in submitochondrial particles. *Biochim. biophys. Acta* **69**, 361.

Lundegardh, H. (1945). Absorption, transport and exudation of inorganic ions by the roots. *Ark. Bot.* **32** A, 1.

Lundegardh, H. (1954). Anion respiration. *Symp. Soc. exp. Biol.* **8**, 262.

Lynn, W. S. & Brown, R. H. (1966). Accumulation of divalent organic anions by mitochondria. *Archs Biochem. Biophys.* **114**, 260.

Malviya, A. N., Parsa, B., Yodaiken, R. E. & Elliott, W. B. (1968). Ultrastructure of sonic and digitonin fragments from beef heart mitochondria. *Biochim. biophys. Acta* **162**, 195.

Meijer, A. J. & Tager, J. M. (1969). Effect of butyl malonate and inersalyl on anion–exchange reactions in rat-liver mitochondria. *Biochim. biophys. Acta* **189**, 136.

Meury, J. (1969). Action de deux inhibiteurs (2,4-dinitrophénol et p-chloromercuribenzoate) sur la recharge du potassium après jeûne glucidique chez la bactérie *Escherichia coli* B 163. *Biochim. biophys. Acta* **177**, 184.

Mitchell, P. (1953). Transport of phosphate across the surface of *Micrococcus pyogenes*: nature of the cell 'inorganic phosphate'. *J. gen. Microbiol.* **9**, 237.

Mitchell, P. (1954*a*). Transport of phosphate through an osmotic barrier. *Symp. Soc. exp. Biol.* **8**, 254.

Mitchell, P. (1954*b*). Transport of phosphate across the osmotic barrier of *Micrococcus pyogenes*: specificity and kinetics. *J. gen. Microbiol.* **11**, 73.

Mitchell, P. (1957). A general theory of membrane transport from studies of bacteria. *Nature, Lond.* **180**, 134.

Mitchell, P. (1959*a*). Structure and function in microorganisms. *Biochem. Soc. Symp.* **16**, 73.

Mitchell, P. (1959*b*). Biochemical cytology of microorganisms. *A. Rev. Microbiol.* **13**, 407.

Mitchell, P. (1961*a*). Biological transport phenomena and the spatially anisotropic characteristics of metabolism. In *Membrane Transport and Metabolism*, p. 22. Eds. A. Kleinzeller and A. Kotyk. New York: Academic Press.

Mitchell, P. (1961*b*). Approaches to the analysis of specific membrane transport. In *Biological Structure and Function*, vol. 2, p. 581. Eds. T. W. Goodwin and O. Lindberg. New York: Academic Press.

Mitchell, P. (1961*c*). Coupling of phosphorylation to electron and hydrogen transfer by a chemiosmotic type of mechanism. *Nature, Lond.* **191**, 144.

Mitchell, P. (1961*d*). Conduction of protons through the membranes of mitochondria and bacteria by uncouplers of oxidative phosphorylation. *Biochem. J.* **81**, 24 P.

Mitchell, P. (1961*e*). Contribution to discussion. In *Membrane Transport and Metabolism*, p. 318. Eds. A. Kleinzeller and A. Kotyk. New York: Academic Press.

Mitchell, P. (1962). Metabolism, transport and morphogenesis: Which drives which? *J. gen. Microbiol.* **29**, 25.

Mitchell, P. (1963*a*). Molecule, group and electron translocation through natural membranes. *Biochem. Soc. Symp.* **22**, 142,

Mitchell, P. (1963*b*). The chemical asymmetry of membrane transport processes. In *Cell Interface Reactions*, p. 33. Ed. H. D. Brown. New York: Scholar's Library.

Mitchell, P. (1966*a*). *Chemiosmotic Coupling in Oxidative and Photosynthetic Phosphorylation*, 1st ed. Bodmin: Glynn Research.

Mitchell, P. (1966*b*). Chemiosmotic coupling in oxidative and photosynthetic phosphorylation. *Biol. Rev.* **41**, 445.

Mitchell, P. (1967*a*). Translocations through natural membranes. *Adv. Enzymol.* **29**, 33.

MITCHELL, P. (1967*b*). Active transport and ion accumulation. In *Comprehensive Biochemistry*, vol. 22, p. 167. Eds. M. Florkin and E. H. Stotz. Amsterdam: Elsevier.

MITCHELL, P. (1967*c*). Proton translocation phosphorylation in mitochondria, chloroplasts and bacteria: Natural fuel cells and solar cells. *Fedn Proc. Fedn Am. Socs exp. Biol.* **26**, 1370.

MITCHELL, P. (1968). *Chemiosmotic Coupling and Energy Transduction*, 1st ed. Bodmin: Glynn Research.

MITCHELL, P. (1969*a*). Chemiosmotic coupling and energy transduction. In *Theoretical and Experimental Biophysics*, vol. 2, p. 159. Ed. A. Cole. New York and London: Marcel Dekker Inc.

MITCHELL, P. (1969*b*). The chemical and electrical components of the electrochemical potential of H^+ ions across the mitochondrial cristae membrane. *Febs Symposia* **17**, 219.

MITCHELL, P. & MOYLE, J. (1956*a*). Osmotic function and structure in bacteria. *Symp. Soc. gen. Microbiol.* **6**, 150.

MITCHELL, P. & MOYLE, J. (1956*b*). Permeation mechanisms in bacterial membranes. *Discuss. Faraday Soc.* **21**, 258.

MITCHELL, P. & MOYLE, J. (1958). Group translocation: a consequence of enzyme-catalysed group-transfer. *Nature, Lond.* **182**, 372.

MITCHELL, P. & MOYLE, J. (1965*a*). Stoichiometry of proton translocation through the respiratory chain and adenosine triphosphatase systems of rat liver mitochondria. *Nature, Lond.* **208**, 147.

MITCHELL, P. & MOYLE, J. (1965*b*). Evidence discriminating between the chemical and chemiosmotic mechanisms of electron transport phosphorylation. *Nature, Lond.* **208**, 1205.

MITCHELL, P. & MOYLE, J. (1967*a*). Acid-base titration across the membrane system of rat-liver mitochondria. Catalysis by uncouplers. *Biochem. J.* **104**, 588.

MITCHELL, P. & MOYLE, J. (1967*b*). Respiration-driven proton translocation in rat-liver mitochondria. *Biochem. J.* **105**, 1147.

MITCHELL, P. & MOYLE, J. (1967*c*). Proton-transport phosphorylation: some experimental tests. In *Biochemistry of Mitochondria*, p. 53. Eds. E. C. Slater, Z. Kaniuga and L. Wojtczak. London: Academic Press.

MITCHELL, P. & MOYLE, J. (1968). Proton translocation coupled to ATP hydrolysis in rat liver mitochondria. *European J. Biochem.* **4**, 530.

MITCHELL, P. & MOYLE, J. (1969*a*). Estimation of membrane potential and pH difference across the cristae membrane of rat liver mitochondria. *European J. Biochem.* **7**, 471.

MITCHELL, P. & MOYLE, J. (1969*b*). Translocation of some anions, cations and acids in rat liver mitochondria. *European J. Biochem.* **9**, 149.

MITCHELL, P. MOYLE, J. (1969*c*). The intrinsic anisotropy of the cytochrome oxidase region of the mitochondrial respiratory chain and the consequent vectorial property of respiration. *4th Round Table Discussion on Electron Transport and Energy Conservation, Bari, Italy*. (In the Press.)

MOORE, C. & PRESSMAN, B. C. (1964). Mechanism of action of valinomycin on mitochondria. *Biochem. biophys. Res. Commun.* **15**, 562.

MUELLER, P. & RUDIN, D. O. (1967). Development of K^+-Na^+ discrimination in experimental bimolecular lipid membranes by macrocyclic antibiotics. *Biochem. biophys. Res. Commun.* **26**, 398.

MUNCK, B. G. (1968). Amino acid transport by the small intestine in the rat. Evidence against interactions between sugars and amino acids at the carrier level. *Biochim. biophys. Acta* **156**, 192.

NORTHCOTE, D. H. (1968). Structure and function of plant-cell membranes. *Br. Med. Bull.* **24**, 107.

PACKER, L. (1967). Effect of nigericin upon light-dependent monovalent cation transport in chloroplasts. *Biochem. biophys. Res. Commun.* **28**, 1022.

PALMIERI, F. & QUAGLIARIELLO, E. (1969). Correlation between anion uptake and the movement of K^+ and H^+ across the mitochondrial membrane. *European J. Biochem.* **8**, 473.

PARDEE, A. B. (1966). Purification and properties of a sulfate-binding protein from *Salmonella typhimurium*. *J. biol. Chem.* **241**, 5886.

PARDEE, A. B. & WATANABE, K. (1968). Location of sulfate-binding protein in *Salmonella typhimurium*. *J. Bact.* **96**, 1049.

PAVLASOVA, E. & HAROLD, F. M. (1969). Energy coupling in the transport of β-galactosides by *Escherichia coli*: Effect of proton conductors. *J. Bact.* **98**, 198.

PECK, H. D. (1968). Energy coupling mechanisms in chemolithotrophic bacteria. *A. Rev. Microbiol.* **22**, 489.

PEÑA, A., CINO, G., GÓMEZ PUYOU, A. & TUENA, M. (1969). Studies on the mechanism of the stimulation of glycolysis and respiration by K^+ in *Saccharomyces cerevisiae*. *Biochim. biophys. Acta* **180**, 1.

PINKERTON, M., STEINRAUF, L. K. & DAWKINS, P. (1969). The molecular structure and some transport properties of valinomycin. *Biochem. biophys. Res. Commun.* **35**, 512.

PIPERNO, J. R. & OXENDER, D. L. (1966). Amino-acid binding protein released from *Escherichia coli* by osmotic shock. *J. biol. Chem.* **241**, 5732.

POST, R. L. (1968). The salt pump of animal cell membranes. In *Regulatory Functions of Biological Membranes*, p. 163. Ed. J. Järnefelt. Amsterdam: Elsevier.

PRESSMAN, B. C. (1963). Specific inhibitors of energy transfer. In *Energy-Linked Functions of Mitochondria*, p. 181. Ed. B. Chance. New York: Academic Press.

PRESSMAN, B. C. (1968). Ionophorous antibiotics as models for biological transport. *Fedn Proc. Fedn Am. Socs exp. Biol.* **27**, 1283.

PRESSMAN, B. C., HARRIS, E. J., JAGGER, W. S. & JOHNSON, J. H. (1967). Antibiotic-mediated transport of alkali ions across lipid barriers. *Proc. natn. Acad. Sci. U.S.A.* **58**, 1949.

PRIESTLAND, R. N. & WHITTAM, R. (1968). The influence of external sodium ions on the sodium pump in erythrocytes. *Biochem. J.* **109**, 369.

RANG, H. P. & RITCHIE, J. M. (1968). On the electrogenic sodium pump in mammalian non-myelinated nerve fibres and its activation by various external cations. *J. Physiol.* **196**, 183.

RICKENBERG, H. W., COHEN, G. N., BUTTIN, G. & MONOD, J. (1956). La galactoside-perméase d'*Escherichia coli*. *Annls Inst. Pasteur, Paris* **91**, 829.

ROBBIE, J. P. & WILSON, T. H. (1969). Transmembrane effects of β-galactosides on thiomethyl-β-galactoside transport in *Escherichia coli*. *Biochim. biophys. Acta* **173**, 234.

ROBERTSON, R. N. (1960). Ion transport and respiration. *Biol. Rev.* **35**, 231.

ROBERTSON, R. N. (1968). *Protons, Electrons, Phosphorylation and Active Transport*, 1st ed. Cambridge University Press.

ROBERTSON, R. N. & WILKINS, M. (1948). Quantitative relation between salt accumulation and salt respiration in plant cells. *Nature, Lond.* **161**, 101.

ROSENBERG, T. (1948). On accumulation and active transport in biological systems. I. Thermodynamic considerations. *Acta chem. scand.* **2**, 14.

ROSENBERG, H. & LA NAUZE, J. M. (1968). The isolation of a mutant of *Bacillus cereus* deficient in phosphate uptake. *Biochim. biophys. Acta* **156**, 381.

ROSSI, C. & AZZONE, G. F. (1965). H^+/O ratio during Ca^{2+} uptake in rat-liver mitochondria. *Biochim. biophys. Acta*, **110**, 434.

ROSSI, C. S., SILIPRANDI, N., CARAFOLI, E., BIELAWSKI, J. & LEHNINGER, A. L. (1967). Proton movements across the mitochondrial membrane supported by hydrolysis of adenosine triphosphate. *European J. Biochem.* **2**, 332.

ROTHSTEIN, A. (1954). Enzyme systems of the cell surface involved in the uptake of sugars by yeast. *Symp. Soc. exp. Biol.* **8**, 165.

ROTHSTEIN, A. (1968). Membrane phenomena. *A. Rev. Physiol.* **30**, 15.

RUMBERG, B., REINWALD, E., SCHRÖDER, H. & SIGGEL, U. (1968). Correlation between electron flow, proton translocation and phosphorylation in chloroplasts. *Naturwissenschaften* **55**, 77.

SALTON, M. R. J. (1967). Structure and function of bacterial cell membranes. *A. Rev. Microbiol.* **21**, 417.

SANNO, Y., WILSON, T. H. & LIN, E. C. C. (1968). Control of permeation to glycerol in cells of *Escherichia coli*. *Biochem. biophys. Res. Commun.* **32**, 344.

SCARBOROUGH, G. A., RUMLEY, M. K. & KENNEDY, E. P. (1968). The function of adenosine 5'-triphosphate in the lactose transport system of *Escherichia coli*. *Proc. natn. Acad. Sci. U.S.A.* **60**, 951.

SCHATZ, G. & SALTZGABER, J. (1969). Identification of denatured mitochondrial ATPase in 'structural protein' from beef heart mitochondria. *Biochim. biophys. Acta* **180**, 186.

SCHOLES, P., MITCHELL, P. & MOYLE, J. (1969). The polarity of proton translocation in some photosynthetic microorganisms. *European J. Biochem.* **8**, 450.

SCHULTZ, S. G., EPSTEIN, W. & SOLOMON, A. K. (1963). Cation transport in *Escherichia coli*. IV. Kinetics of net K uptake. *J. gen. Physiol.* **47**, 329.

SCHWARTZ, J. H., MAAS, W. K. & SIMON, E. (1959). An impaired concentrating mechanism for amino acids in mutants of *Escherichia coli* resistant to L-canavanine and D-serine. *Biochim. biophys. Acta* **32**, 582.

SCHWARTZ, M. (1968). Light induced proton gradient links electron transport and phosphorylation. *Nature, Lond.* **219**, 915.

SHAVIT, N., DILLEY, R. A. & SAN PIETRO, A. (1968). Ion translocation in isolated chloroplasts. Uncoupling of photophosphorylation and translocation of K$^+$ and H$^+$ ions induced by nigericin. *Biochemistry* **7**, 2356.

SIEGEL, G. J., KOVAL, G. J. & ALBERS, R. W. (1969). Sodium-potassium activated adenosine triphosphatase. VI. Characteristics of the phosphoprotein formed from orthophosphate in the presence of ouabain. *J. biol. Chem.* **244**, 3269.

SIMONI, R. D., LEVINTHAL, M., KUNDIG, F. D., KUNDIG, W., ANDERSON, B., HARTMAN, P. E. & ROSEMAN, S. (1967). Genetic evidence for the role of a bacterial phosphotransferase system in sugar transport. *Proc. natn. Acad. Sci. U.S.A.* **58**, 1963.

SIMONI, R. D., SMITH, M. F. & ROSEMAN, S. (1968). Resolution of a staphylococcal phosphotransferase system into four protein components and its relation to sugar transport. *Biochem. biophys. Res. Commun.* **31**, 804.

SIMPSON, E. R. & FRENKEL, R. (1969). Substrate-induced efflux of anions from bovine adrenal cortex mitochondria and its relationship to steroidogenesis. *Biochem. biophys. Res. Commun.* **35**, 765.

SKOU, J. C. (1965). Enzymatic basis for active transport of Na$^+$ and K$^+$ across a cell membrane. *Physiol. Rev.* **45**, 596.

SKOU, J. C. & HILBERG, C. (1969). The effect of cations, g-strophanthin and digomycin on the labelling from [^{32}P]-ATP of the (Na$^+$+K$^+$)-activated enzyme system and the effect of cations and g-strophanthin on the labelling from [^{32}P] ITP and ^{32}Pi. *Biochim. biophys. Acta* **185**, 198.

SKULACHEV, V. P., SHARAF, A. A. & LIBERMAN, E. A. (1967). Proton conductors in the respiratory chain and artificial membranes. *Nature, Lond.* **216**, 718.

SLATER, E. C. (1967). An evaluation of the Mitchell hypothesis of chemiosmotic coupling in oxidative and photosynthetic phosphorylation. *European J. Biochem.* **1**, 317.

SLATER, E. C. & CLELAND, K. W. (1953). The effect of the tonicity of the medium on the respiratory and phosphorylative activity of heart-muscle sarcosomes. *Biochem. J.* **53**, 557.

SLAYMAN, C. L. & SLAYMAN, C. W. (1968). Net uptake of potassium in *Neurospora*. Exchange for sodium and hydrogen ions. *J. gen. Physiol.* **52**, 424.

SLAYMAN, C. W. & TATUM, E. L. (1965). Potassium transport in *Neurospora*. II. Measurement of steady-state potassium fluxes. *Biochim. biophys. Acta* **102**, 149.

STADTMAN, E. R. (1966). Some considerations of the energy metabolism of anaerobic bacteria. In *Current Aspects of Biochemical Energetics*, p. 39. Eds. N. O. Kaplan and E. P. Kennedy. New York: Academic Press.

STEINRAUF, L. K., PINKERTON, M. & CHAMBERLIN, J. W. (1968). The structure of nigericin. *Biochem. biophys. Res. Commun.* **33**, 29.

STONE, A. J. (1968). A proposed model for the Na^+ pump. *Biochim. biophys. Acta* **150**, 578.

TANAKA, S., FRAENKEL, D. G. & LIN, E. C. C. (1967). The enzymatic lesion of strain MM-6, a pleiotropic carbohydrate-negative mutant of *Escherichia coli*. *Biochem. biophys. Res. Commun.* **27**, 63.

TANAKA, S. & LIN, E. C. C. (1967). Two classes of pleiotropic mutants of *aerobacter aerogenes* lacking components of a phosphoenolpyruvate-dependent phosphotransferase system. *Proc. natn. Acad. Sci. U.S.A.* **57**, 913.

THIER, S. O. (1968). Amino acid accumulation in the toad bladder: relationship to transepithelial sodium transport. *Biochim. biophys. Acta* **150**, 253.

TOSTESON, D. C. (1968). Effect of macrocyclic compounds on the ionic permeability of artificial and natural membranes. *Fedn Proc. Fedn Am. Socs exp. Biol.* **27**, 1269.

TYLER, D. D. (1968). The inhibition of phosphate entry into rat liver mitochondria by organic mercurials and by formaldehyde. *Biochem. J.* **107**, 121.

TYLER, D. D. (1969). Evidence of a phosphate-transporter system in the inner membrane of isolated mitochondria. *Biochem. J.* **111**, 665.

USSING, H. H. (1947). Interpretation of the exchange of radio-sodium in isolated muscle. *Nature, Lond.* **160**, 262.

USSING, H. H. (1949). Transport of ions across cellular membranes. *Physiol. Rev.* **29**, 217.

VAN STEVENINCK, J. (1968). Transport and transport-associated phosphorylation of 2-deoxy-D-glucose in yeast. *Biochim. biophys. Acta* **163**, 386.

VIDAVER, G. A. (1964). Transport of glycine by pigeon red cells. *Biochemistry*, **3**, 662.

VIDAVER, G. A., ROMAIN, L. F. & HAUROWITZ, F. (1964). Some studies on the specificity of amino acid entry routes in pigeon erythrocytes. *Archs Biochem. Biophys.* **107**, 82.

VILLARREAL-MOGUEL, E. I. & RUIZ-HERRERA, J. (1969). Induction and properties of the citrate transport system in *Aerobacter aerogenes*. *J. Bact.* **98**, 552.

WALLACH, D. F. H. & GORDON, A. S. (1968). Lipid-protein interactions in cellular membranes. In *Regulatory Functions of Biological Membranes*, p. 87. Ed. J. Järnefelt. Amsterdam: Elsevier.

WEIDEN, P. L., EPSTEIN, W. & SCHULTZ, S. G. (1967). Cation transport in *Escherichia coli*. VII. Potassium requirement from phosphate uptake. *J. gen. Physiol.* **50**, 1641.

WEST, I. C. (1969). The site of action of adenosine-5'-triphosphate on β-galactoside transport in *Escherichia coli*. *Febs Letters* **4**, 69.

WHITTAKER, V. P. (1968). Structure and function of animal cell membranes. *Br. Med. Bull.* **24**, 101.

WIDDAS, W. F. (1952). Inability of diffusion to account for placental glucose transfer in the sheep and consideration of the kinetics of a possible carrier transfer. *J. Physiol.* **118**, 23.

WILBRANDT, W. (1954). Secretion and transport of non-electrolytes. *Symp. Soc. exp. Biol.* **8**, 136.

WINKLER, H. H. & WILSON, T. H. (1966). The role of energy coupling in the transport of β-galactosides by *Escherichia coli*. *J. biol. Chem.* **241**, 2200.

WINKLER, H. H. & WILSON, T. H. (1967). Inhibition of β-galactoside transport by substrates of the glucose transport system in *Escherichia coli*. *Biochim. biophys. Acta* **135**, 1030.

WIPF, H.-K. & SIMON, W. (1969). Selective K^+ transport through synthetic membranes using antibiotics in a potential gradient. *Biochem. biophys. Res. Commun.* **34**, 707.

WITT, H. T., RUMBERG, B. & JUNGE, W. (1968). Electron transfer, field changes, proton translocation and phosphorylation in photosynthesis. *Colloquium Ges. biol. Chem.* **19**, 262.

WYMAN, J. (1948). Heme proteins. *Adv. Protein Chem.* **4**, 407.

YOSHIDA, H., NAGI, K., OHASHI, T. & NAKAGAWA, Y. (1969). K^+-dependent phosphatase activity observed in the presence of both adenosine triphosphate and Na^+. *Biochim. biophys. Acta* **171**, 178.

ZARLENGO, M. H. & SCHULTZ, S. G. (1966). Cation transport and metabolism in *Streptococcus faecalis*. *Biochim. biophys. Acta* **126**, 308.

THE BIOSYNTHESIS OF PROTECTIVE SURFACE STRUCTURES OF PROKARYOTIC AND EUKARYOTIC CELLS

D. J. ELLAR

Sub-Department of Chemical Microbiology,
Department of Biochemistry, University of Cambridge

Almost 30 years ago Joseph Needham (1942) observed that considerations of form were not the perquisite of the morphologist and that no sharp distinction could be drawn between morphology and biochemistry. Two decades later H. J. Rogers (1965) appealed for more attention to be paid to the problem of vectorial assembly. The differences which are apparent between prokaryotic and eukaryotic cells have been clearly stated (Stanier & van Niel, 1962). Nevertheless in the assembly of those components which occur on the outside of the plasma membrane, and which to a greater or lesser extent govern the form of the cell, both cell types are faced with the same problems of precursor synthesis and polymerization, membrane transport and polymer modification. A study of these processes and their control will require the combined efforts of biochemists, geneticists, biophysicists and other specialists. With such inter-disciplinary collaboration, the outlook is promising in that investigations during the next two decades should equip us to discuss morphology in terms of integrated chemical reactions which reflect the cell's genetic potential.

In considering those factors which control the shape of prokaryotic and eukaryotic cells it is clear that in many cases the form-determining components are insoluble structural polymers located outside the plasma membrane. These polymers give a rigidity to the cell and shield the protoplast from osmotic and physical damage. In different cell types these functions are achieved by various polymers which range from the peptidoglycan which fulfils the major role in determining the shape of bacteria, to the insoluble polymers of cellulose, xylan, chitin and silica which characterize the surface of many eukaryotic cells. If each cell surface possessed only one structural polymer, the problems associated with understanding its vectorial synthesis and precise role in form determination would be difficult enough. However, both prokaryotic and eukaryotic cells possess additional polymers; these range from teichoic acids in bacteria to glycoproteins in animal cells, whose role, if any, in protecting the synthetic apparatus of the cell and controlling its physical form, is not yet defined. Even in Gram-positive bacteria, where the role of peptidoglycan seems clear, the situation has been complicated by the results of Brinton, McNary & Carnahan (1969). These workers purified a protein which occurs on the outside of the peptidoglycan of *Bacillus brevis* in the form of a tetragonal polymer similar to the outer layer of *B. polymyxa* (Nermut & Murray, 1967) and of *B. cereus* (Ellar & Lundgren, 1967). The pure protein re-assembles *in vitro* to a cylindrical network with a diameter equal to that of the bacterium from which it was isolated. What role this protein plays in the determination of the bacillary form remains to be determined.

To what extent macromolecules classically located in the plasma membrane of prokaryotic and eukaryotic cells contribute to the maintenance of cell shape, especially in cells which lack complex insoluble polymers, is largely unresolved. Thus the unique biconcave shape of the erythrocyte has been attributed to various causes. One suggestion (Murphy, 1965) arises from the demonstration that cholesterol is concentrated around the periphery of the biconcave disc. Such an uneven sterol distribution might produce inequalities in surface tension which could affect cell shape. Also of interest is the finding of Op Den Kamp, van Iterson & van Deenen (1967) that after exposure of the Gram-positive bacterium *Bacillus megaterium* to an acidic environment, removal of the cell wall by lysozyme produced protoplasts which in hypotonic conditions retained the rod shape of the intact cell. The exposure to low pH values caused a marked decrease in the phosphatidylglycerol content of the plasma membrane. This decrease was almost exactly paralleled by the appearance of a new lipid, glucosaminylphosphatidylglycerol which accounted for 32% of the total membrane lipid.

The first problem in discussing the distribution and function of macromolecules at cell surfaces is one of definition. As pointed out by Rogers & Perkins (1968), no sharp boundaries can be drawn between cytoplasm, membrane and wall structures. Even the boundary between wall and environment is not easily discerned. This is especially true for animal and plant cells which may be closely apposed in tissues and specialized structures. Orthodox electron microscope procedures have often failed to show a surface coat in animal cells and only specific stains and refined techniques reveal these external layers. By the use of such techniques the surfaces of protozoa and many vertebrate cell types have been shown to possess characteristic layers of amorphous material external to the plasma membrane and which may be continuous or in patches (Revel & Ito, 1967). Attempts to identify the chemical composition of these surface coats have lead to the suggestion that layers which are characteristically rich in glycoprotein and polysaccharide residues are a common feature of vertebrate cells (Rambourg & Leblond, 1967). It is not yet clear whether these surface layers of animal cells are structures physically and chemically distinct from the underlying plasma membrane, which can be detached without changing the form or permeability properties of the cell. However, some of the compounds which can be located in these surface layers of animal cells, such as the glycoproteins, ought more appropriately to be considered as an integral part of the plasma membrane (Cook, 1968). As well as the difficulty in deciding what constitutes a 'surface structure', it is also necessary to define what is meant by a 'protective' surface. The protective properties of the insoluble form-determining polymers are obvious and have been cited; but the surface layers of prokaryotic and eukaryotic cells contain more components which in some cases are essential for cell function. It seems probable that most of these components play a role in protecting and maintaining the cell in its particular environment. In animal cells such compounds as the sialic acids have been implicated in the transport of K^+ ions across the plasma membrane (Glick & Githens, 1965). The function of the glycosyldiglycerides which are associated with the membrane of Gram-positive bacteria is not known. It has been suggested that association of the hydrophilic regions of these molecules might lead to the formation of 'pores' through which certain ions and water-soluble compounds could pass (Brundish, Shaw & Baddiley, 1967). Molecules and molecular aggregates which function in cellular transport, intercellular contact, phagocytosis and determination of antigenic character are in a sense protecting the cell. For the performance of some of these functions it is conceivable that these macromolecules may be more closely associated with the plasma membrane than was previously thought.

Perhaps in attempting to allocate rigid boundaries between wall chemistry and membrane chemistry we are begging the question, since we cannot yet describe the macromolecular structure of a cell membrane or the three-dimensional structure of many of the polymers which comprise the major wall component of many cells. An understanding of the ordered structure of the plasma membrane and the polymers arranged on its surface may come from studies of how, and to what extent, the membrane functions in the synthesis and ordering of these polymers. Besides the plasma membrane, many eukaryotic cells have a multiplicity of internal membranes whose role in the assembly and synthesis of surface structures is being intensively studied. It is pertinent to examine how the membranes of prokaryotic and eukaryotic cells are concerned with synthesis of surface components. Sharp differences in the chemical composition of these structures may separate the two kinds of cell, but at the synthetic level they have certain common features.

Biosynthesis of peptidoglycan in bacteria

Following the development of methods for the isolation of bacterial cell walls (Salton & Horne, 1951) rapid progress has been made in elucidating their chemical composition. The major form-determining component in bacterial cells is a rigid heteropolymer composed of β-1, 4 linked linear polysaccharide chains cross-linked through peptides. These peptidoglycan chains consist of alternating units of two N-acetyl-hexosamines, N-acetyl-D-glucosamine and N-acetylmuramic acid with peptide sub-units attached through the carboxyl group of the latter. Adjacent peptidoglycan strands may then be further linked to each other by peptide cross-bridges. The chemistry of this heteropolymer and the variations which occur in different bacterial species have been reviewed by Ghuysen, Strominger & Tipper (1968) and Ghuysen (1968). The emphasis in the present contribution is on studies of the enzymic synthesis of the peptidoglycan in cell-free systems. The overall synthetic process may conveniently divide into several stages: (1) precursor synthesis; (2) the association of precursors with the plasma membrane; (3) transport of precursors through the membrane; (4) polymerization into peptidoglycan; (5) polymer modification. As for many polysaccharide-containing polymers, the primary event in synthesis is the formation of nucleoside diphosphate glycosyl precursors from nucleotide triphosphate and a sugar phosphate. In peptidoglycan synthesis the immediate precursors are N-acetylglucosamine and N-acetylmuramyl-pentapeptide. UTP and N-acetylglucosamine-1-phosphate combine to form UDP-N-acetylglucosamine which is then believed to undergo condensation with phosphoenolpyruvate, followed by reduction to produce its 3-O-D-lactic ether, N-acetylmuramic acid. Amino acids are now added to the muramic acid in separate steps to form the muramylpentapeptide. Amino acids are added by separate enzymes, each of which is specific for the uridine nucleotide and the amino acid substrate. These enzymes require either Mg^{2+} or Mn^{2+} and ATP, and all appear to occur in the soluble cell fraction (Ghuysen, 1968). This finding that the membrane apparently has no role in the first stage of nucleotide precursor synthesis is common to the synthetic schemes for a variety of surface polymers. An exception to this is the synthesis of a teichoic acid polymer containing N-acetylglucosamine by *Staphylococcus lactis* I3 (Blumson, Douglas & Baddiley, 1966). These polymers will be discussed separately, but it is interesting that both the enzymes responsible for the synthesis of nucleotide precursors *and* those responsible for their subsequent polymerization were reported to occur in a particulate enzyme preparation composed of membrane fragments, It should be emphasized that the description of an enzyme as 'soluble' is often arbitrary; in some instances where the description is based entirely on speed of centrifugation, this appears to be unjustified. Sedimentation of small membrane fragments and mesosome preparations from bacteria requires centrifugation at much higher speeds for longer times (Ellar & Freer, 1969) than has been used in some cases to purify 'soluble' enzymes.

The first reports of the *in vitro* utilization of the nucleotide precursors UDP-*N*-acetylmuramyl-pentapeptide and UDP-*N*-acetylglucosamine were those of Chatterjee & Park (1964) and Meadow, Anderson & Strominger (1964). In these and subsequent reports it was shown that synthesis of uncross-linked peptidoglycan strands is catalysed by an enzyme complex in a particulate fraction which appears to consist entirely of membrane fragments (Matz & Strominger, 1968; D. J. Ellar & P. E. Reynolds, unpublished observations). Although there is no evidence for the involvement of ribosomes or primer substances in this synthesis, both ribosomes and residual fragments of previously synthesized peptidoglycan are undoubtedly present in many of these particulate preparations. In relatively few instances have the particles been analysed in detail, which makes comparison difficult. P. E. Reynolds (unpublished observations), however, has obtained substantial peptidoglycan synthesis (7·5 n-mole radioactive substrate incorporated/ hr/mg. protein) with a membrane preparation from *Bacillus megaterium* KM protoplasts, which was shown by isotope-labelling to contain less than 0·1 % of wall material. Both Chatterjee & Park (1964) and Meadow *et al.* (1964) used mechanical breakage to prepare their peptidoglycan-synthesizing system from *Staphylococcus aureus*. The former workers used glass beads to rupture the cocci. Unbroken cocci and wall fragments were removed by low speed centrifugation and a crude particulate preparation obtained by centrifugation at 105,000 *g* for 45 min. To eliminate ribosomes from this preparation, it was further fractionated by suspension in magnesium-free buffer and sedimentated at 38,000 *g* to obtain membrane fragments. Meadow *et al.* (1964) used ultrasonic treatment to break their cocci and to obtain a small-particle fraction which sedimented at 100,000 *g*. There were other significant differences of detail between the two preparative methods, notably in the degree of washing, the presence or absence of Mg^{2+} in the washing buffer and in the choice of centrifugation speeds for removal of intact cocci and wall fragments. These differences may in part explain the different properties which characterized the particulate preparations of these two groups. In both instances when UDP-acetylmuramyl-Ala-Glu-[^{14}C]-Lys-Ala-Ala and UDP-N-acetylglucosamine were incubated with the particulate preparation, radioactivity was incorporated into material with characteristics of peptidoglycan. The product obtained by Meadow *et al.* (1964) was hydrolysed by lysozyme and remained at the origin after chromatography to separate the original precursors. Moreover, it was only produced when the incubation was done on filter paper in a humid atmosphere, and not when the incubation mixture was in a test tube. In the assay of Chatterjee & Park (1964) incubation in a test tube produced an insoluble product which was not broken down by lysozyme. Glycine, which forms the polypeptide cross-bridges linking peptidoglycan strands in *S. aureus*, was also incorporated into this insoluble material. It was suggested this additional cross-linking might explain the insensitivity of the product to lysozyme, but subsequent work has provided a more plausible explanation (Anderson, Matsuhashi, Haskin & Strominger, 1965; Matsuhashi, Dietrich & Strominger, 1965). For their preparation Meadow *et al.* (1964) reported that the nucleotide products of the polymerization were UMP derived from UDP-N-acetylmuramyl-pentapeptide and UDP derived from UDP-N-acetylglucosamine. For the Chatterjee & Park (1964) system it was stated (Park, 1964) that UDP was produced from both nucleotides. Struve & Neuhaus (1965) repeated the assays made by both these groups and examined some of the differences between them. The experiments of Struve & Neuhaus showed that the enzyme system catalysed a UDP-N-acetylglucosamine-independent transfer of phospho-N-acetylmuramyl-pentapeptide to some acceptor in the enzyme preparation. This transfer was accompanied by a stoichiometric release of UMP. When UMP was added to particles which had previously incorporated radioactive phospho-N-acetylmuramyl-pentapeptide, a rapid de-labelling of the latter occurred. Moreover the addition of UMP at the start of the incubation produced a drastic inhibition of the transfer reaction. To account for these findings Struve & Neuhaus (1965) suggested the following reaction scheme in which the phospho-N-acetylmuramyl-pentapeptide is transferred to an acceptor associated with the membrane particles:

UDP-N-acetylmuramyl-pentapeptide + acceptor $\rightleftharpoons$

$$\text{acceptor} -\!\!\text{O}-\!\!\underset{\underset{\text{O}^-}{|}}{\overset{\overset{\text{O}}{\|}}{\text{P}}}-\!\!\text{O}-\ N\text{-acetylmuramyl-pentapeptide} + \text{UMP}.$$

The production of UMP and the reversibility of this reaction was supported by the fact that radioactive UMP exchanged with the UMP moiety of UDP-N-acetylmuramyl-pentapeptide. The next stage in peptidoglycan synthesis would then involve a second reaction in which UDP-N-acetylglucosamine was incorporated into the polymer, with the elimination of UDP and Pi.

Confirmation of an acceptor was contained in the reports of Anderson *et al.* (1965) and Matsuhashi *et al.* (1965). While the newly synthesized radioactive peptidoglycan remained at the origin in the chromatographic assay, a small amount of radioactive material migrated near the solvent front. This fast-moving component was rapidly synthesized and reached a maximum value before a significant amount of peptidoglycan had been formed. It also became labelled when [^{32}P]UDP-N-acetylmuramyl-pentapeptide was the substrate; from its solubility in various organic solvents, it was concluded that the material was a lipid. In the absence of UDP-N-acetylglucosamine the lipid intermediate was synthesized from an incubation mixture containing UDP-N-acetylmuramyl-penta-peptide. No peptidoglycan was formed under these conditions. When this mixture was re-incubated with UDP-N-acetylglucosamine, pepti-doglycan synthesis occurred; when UMP was added to the mixture at the outset, the intermediate was converted back to UDP-N-acetyl-muramyl-pentapeptide. Further labelling experiments (Anderson *et al.* 1965) confirmed that the lipid intermediate functioned catalytically as a carrier of phosphodisaccharide-pentapeptide units according to the following scheme:

$$\text{UDP-}N\text{-acetylmuramyl-pentapeptide} + \text{P-phospholipid} \rightleftharpoons \text{N-acetylmuramyl-pentapeptide-P-P-phospholipid} + \text{UMP} \quad (1)$$

$$\text{N-acetylmuramyl-pentapeptide-P-P-phospholipid} + \text{UDP-N-acetyl-glucosamine} \rightarrow \text{N-acetylglucosamine-N-acetylmuramyl-pentapeptide-P-P-phospholipid} + \text{UDP} \quad (2)$$

$$\text{N-acetylglucosamine-N-acetylmuramyl-pentapeptide-P-P-phospholipid} + \text{acceptor} \rightarrow \text{N-acetyl-glucosamine-N-acetyl-muramyl-pentapeptide-acceptor} + \text{P-phospholipid} + \text{Pi} \quad (3)$$

The chemical nature of the final acceptor was presumed to be an incomplete glycopeptide chain associated with the membrane particles.

The overall reaction required Mg^{2+} and was dependent on the physical nature of the particulate fraction. Particles prepared by grinding with alumina showed efficient peptidoglycan synthesis in test tube assays, whereas those prepared by ultrasonic treatment were relatively inactive unless the incubation was done on filter paper. The lysozyme insensitivity of the Chatterjee & Park (1964) product may be explained by the fact that a preparation, obtained by using the technique of these authors, by Matsuhashi *et al.* (1965) catalysed the formation of lipid intermediate, but was relatively inefficient in peptidoglycan synthesis.

In subsequent studies the disaccharide-pentapeptide-P-P-lipid was purified from *Micrococcus lysodeikticus* enzyme preparations and the lipid portion shown (Higashi, Strominger & Sweeley, 1967) to be a C_{55} isoprenoid alcohol with the following structure:

$$CH_3C{=}CHCH_2(CH_2\overset{\overset{\displaystyle CH_3}{|}}{C}{=}CHCH_2)_9CH_2\overset{\overset{\displaystyle CH_3}{|}}{C}{=}CHCH_2OH$$

(with CH_3 above the first carbon)

The lipid intermediate is clearly relevant to the problems of orientation and membrane transport. Furthermore, this compound is not unique to peptidoglycan synthesis or to prokaryotic cells. Identical or closely similar intermediates occur in the synthesis of bacterial lipopolysaccharide (Wright, Dankert, Fennessey & Robbins, 1967) and bacterial mannan (Scher, Lennarz & Sweeley, 1968) and have been tentatively identified in the synthesis of mannose-containing glycoproteins in normal animal cells and in tumour tissue (Caccam, Jackson & Eylar, 1969), in teichoic acid polymers (Brooks & Baddiley, 1969), bacterial cellulose (Kahn & Colvin, 1961), plant cellulose (Colvin, 1961; Villemez, McNab & Albersheim, 1968) and fungal chitin (Glaser & Brown, 1957). These will be discussed later. The possibility that in prokaryotic and in eukaryotic cells these lipid intermediates function in membrane transport is attractive. Intracellular precursors which are themselves unable to cross the membrane might be linked to the lipid moiety located at the inner surface of the membrane and be transported to the polymerization site on the outside of the membrane. Our inability to appreciate the precise molecular architecture of the membrane prevents us from describing this transport process in more detail. This problem is increased by the difficulty of detaching the relevant enzymes from their sites in the membrane, although progress in this direction has been reported (Heydanek & Neuhaus, 1969); the work to date shows that to obtain *in vitro* rates of peptidoglycan synthesis which compare with those which must occur *in vivo*, the particulate enzyme preparations

must retain some structural integrity. Considerable differences in preparative techniques used by different workers make difficult the understanding of this structural requirement. In addition to the differences
already mentioned as between particles prepared by ballistic disintegration, ultrasonic treatment and alumina grinding, Siewert & Strominger
(1968) reported that use of the lytic enzyme lysostaphin to prepare
enzyme particles from *Staphylococcus aureus* produced a preparation
which catalysed lipid-intermediate synthesis but functioned very
inefficiently in the synthesis of peptidoglycan.

The size of the membrane particles is probably not the crucial factor,
since Matz & Strominger (1968) observed that the alumina-ground
particles from *Staphylococcus aureus* which catalysed an efficient peptidoglycan synthesis were smaller than particles obtained by the use of
lytic enzymes. By homogenizing exponentially growing bacilli with
polystyrene beads Reynolds (1968) obtained a particle fraction with a
specific activity of peptidoglycan synthesis (30 n-mole/radioactive substrate incorporated/mg. protein/hr) appreciably greater than reported
for *S. aureus* and *Micrococcus lysodeikticus*. When a membrane preparation was obtained from protoplasts of *S. aureus* by using lysozyme, the
specific activity decreased to 7·5 n-mole/radioactive substrate incor-
porated/mg. protein/hr (P. E. Reynolds, unpublished observations). The
latter preparation contained less than 0·1 % of previously synthesized
peptidoglycan. The difficulty experienced in 'solubilizing' the enzymes
concerned in the synthesis of this and other surface polymers does not
allow us to conclude anything about their mode of attachment or insertion into the membrane complex. In *S. aureus* the disaccharide pentapeptide is modified while attached to the lipid intermediate, first by the
amidation of the α-carboxyl group of glutamic acid in a reaction requiring ATP and ammonium ion (Siewert & Strominger, 1968) and
then by the addition of five glycine residues to the ϵ-amino group of
lysine to form the pentaglycine bridging-peptide. Glycyl-*t*RNA is the
glycyl donor in this reaction which does not require the presence of
ribosomes (Bumsted, Dahl, Söll & Strominger, 1968; Niyomporn
Bunrueang, Dahl & Strominger, 1968). Matsuhashi *et al.* (1965) and
Siewert & Strominger (1968) reported that low concentrations of
deoxycholate did not affect the formation of the glycine-lipid intermediate but blocked the amidation reaction and prevented the utilization of the intermediate for peptidoglycan synthesis.

In *Micrococcus lysodeikticus* the only modification to the lipid-
disaccharide-pentapeptide before it was utilized for the synthesis of
peptidoglycan strands was the addition of one glycine residue to the

α-carboxyl group of glutamic acid (Katz, Matsuhashi, Dietrich & Stromonger, 1967); ATP appeared to be required in this reaction. With particles from *M. lysodeikticus*, peptidoglycan synthesis was completely inhibited by *n*-octanol, but the effect on lipid intermediate synthesis was slight. Interestingly, the efficiency of glycine addition to the lipid intermediate was increased under these conditions (Katz *et al.* 1967). As with *Staphylococcus aureus*, deoxycholate prevented peptidoglycan synthesis. In the synthesis of the lipid intermediate it had no effect on the incorporation of disaccharide pentapeptide, but glycine incorporation was blocked. Since these enzyme preparations were predominantly membrane fragments, it is not surprising that efficient peptidoglycan synthesis appeared to require a particular structural configuration of lipid and protein. Studies by Siewert & Strominger (1967) and Heydanek, Struve & Neuhaus (1969) confirm this. When the lipid was extracted from active membrane particles with organic solvents, reconstitution of enzymic activity by reincorporation of the lipid required the presence of low concentrations of a surface active compound such as deoxycholate. This condition of micelle formation or dispersion of the lipid throughout the lipid-free enzyme preparation is frequently required for the reconstitution or enhancement of both prokaryotic and eukaryotic membrane-associated polymerizing systems (Lennarz & Talamo, 1966; Pieringer, 1968). Particularly interesting in this respect is the demonstration of the probable *in vivo* role of the endogenous anionic surface active agents in *M. lysodeikticus* membranes (Lennarz & Talamo, 1966). An initial report that the enzyme responsible for transfer of the disaccharide pentapeptide to the lipid intermediate in *S. aureus* enzyme preparations could be removed with Triton X-100 (Struve, Sinha & Neuhaus, 1966), has been followed by two further reports on the 'solubilization' and properties of this enzyme (Heydanek *et al.* 1969; Heydanek & Neuhaus, 1969). Such diverse reagents as 10 M-urea, 0·08 M-KOH and sodium laurylsarcosinate (3 μmole/mg. protein) can be used to detach substantial amounts of the 'translocase' in an active form. The system requires further purification, but represents an encouraging start to a fractionation of the polymerizing system. It appears that the translocase can be 'solubilized', but demonstration of activity requires the re-aggregation of the 'soluble' enzyme into membrane-like fragments. This occurs during dialysis to remove the urea of sodium laurylsarcosinate.

The final step in the formation of a three-dimensional peptidoglycan polymer is the formation of the peptide cross-link between peptide chains. This is brought about by a transpeptidation in which the bond energy of the terminal D-alanyl-D-alanine bond is used to link the penultimate D-alanine in the muramyl-pentapeptide with the free amino group of a pentaglycine chain; the terminal D-alanine of the pentapeptide is released in this reaction. Since no external energy source is required, this reaction could occur away from the membrane at a site in the wall where ATP is not available. For reasons which are not known, this transpeptidation has not been observed in particle preparations from *Staphylococcus aureus* or *Micrococcus lysodeikticus*, but can be shown to occur in crude preparations of *Escherichia coli* envelopes, prepared by ultrasonic treatment in presence of EDTA (Araki *et al.* 1966; Izaki, Matsuhashi & Strominger, 1966). The structural and biochemical roles of the membrane in the synthesis of this peptidoglycan are being investigated with osmotically stable protoplasts and L-forms. Landman & Halle (1963) and Landman & Forman (1968) removed the wall peptidoglycan from *Bacillus subtilis* and obtained stable L-forms which were propagated on defined soft-agar media. When this medium was supplemented with 2·5 % agar or 25 % gelatin, peptidoglycan synthesis was initiated and the wall-less forms reverted to the bacillary form. The reason for this is not known. One possibility is that the agar or gelatin forms a barrier which prevents diffusion of peptidoglycan components away from the membrane before they can undergo crosslinking. Alternatively, as suggested by Robbins, Wright & Dankert (1966), these solidifying agents may change the conformation of the membrane to a state which is required for the vectorial synthesis of the peptidoglycan polymer. Clive & Landman (1968) obtained similar results when the agar or gelatin was replaced by membrane filters. Chatterjee, Ward & Perkins (1967) have clearly shown that although the growing L-forms of *Staphylococcus aureus* do not synthesize peptidoglycan, membranes isolated from ultrasonically treated organisms retain the capacity to synthesize peptidoglycan intermediates when supplied with the appropriate substrates.

The important contribution of the use of antibiotics to our understanding of peptidoglycan synthesis (Reynolds, 1966) will not be discussed here. With regard to the role of the membrane, the observation of Rogers & Perkins (1968) is relevant: under some circumstances antibiotics may have important effects on bacteria besides their effect on peptidoglycan synthesis. Bacitracin, for example, which inhibits dephosphorylation of the lipid intermediate, causes extensive damage to artificial lipid membranes at equivalent concentrations (D. J. Ellar, unpublished observations). Since our understanding of the attachment and organization of the peptidoglycan synthesizing system is still inadequate, we may also be underestimating the activity lost during preparation of the particles by mechanical disruption, chelating agents and cation deficiencies, etc. Thus while a divalent cation is clearly required for synthesis to occur, it cannot be assumed that cations are not required for the preservation and attachment of the polymerizing enzymes as they occur *in vivo*. This type of ionic binding has been observed for a number of enzymes associated with the bacterial membrane (Muñoz, Nachbar, Schor & Salton, 1968; Ellar, 1969). Siewert & Strominger (1968) reported that a tenfold increase in the concentration of Mg^{2+} in the buffer used to lyse *Staphylococcus aureus* treated with lysostaphin resulted in a particle preparation containing almost twice the amount of protein of particles prepared by using the lower Mg^{2+} concentration.

Biosynthesis of lipopolysaccharides in bacteria

While the peptidoglycan polymer is largely responsible for the rigidity of the outer surface of Gram-positive and Gram-negative bacteria, it constitutes a much smaller fraction of the total wall of the Gram-negatives (Salton, 1964). Gram-negative bacterial cell walls contain additional protein, lipid and polysaccharide components which the electron microscope reveals as projecting outward from the plasma membrane in an organized 'envelope' of layers and membrane-like profiles (Glauert & Thornley, 1969). Among these components which comprise the protective surface of Gram-negative bacteria there is one class of substances whose function and synthesis have been extensively examined. These are the lipopolysaccharides which determine such properties of Gram-negative bacteria as antigenicity, toxigenicity and sensitivity to phage infection. These properties, together with the detailed structural chemistry of these complex heteropolysaccharides have been thoroughly reviewed by Lüderitz, Jann & Wheat (1968). The emphasis here is placed on investigations which have shown that these protective surface polymers, like peptidoglycan, can be synthesized *in vitro* by preparations which contain plasma-membrane fragments.

Lipopolysaccharides have the following general structure:

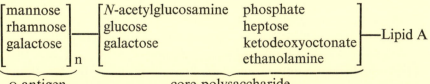

The 'core' region is a complex polysaccharide consisting of a backbone of phosphate, heptose, ethanolamine and 2-keto-3-deoxyoctonate covalently linked through the latter to lipid A (Osborn *et al.* 1964; Lüderitz, Staub & Westphal, 1966; Heath & Ghalambor, 1963). The remainder of the core region is composed of side-chains containing glucose, galactose and *N*-acetylglucosamine. The o-antigen chains are linked in an unknown way to the core structure and vary in composition between species. Lipid A appears to be a polyglucosamine phosphate, acetylated to some extent with β-hydroxy fatty acids (Lüderitz *et al.* 1968). Although sugar nucleotides are the precursors in both o-antigen and core polysaccharide synthesis, the actual pathways of synthesis are quite different.

The incorporation of specific glucose and galactose residues into the core polysaccharide has been studied (Rothfield, Osborn & Horecker, 1964) with mutants which cannot synthesize UDP-glucose or UDP-galactose (Nikaido, 1962; Osborn, Rosen, Rothfield & Horecker, 1962). The experiments show that the core is assembled by successive transfer of monosaccharide precursors to the incomplete core in a series of reactions catalysed by specific glycosyl transferases as follows:

$$\text{glucose-deficient lipopolysaccharide} + \text{UDP-glucose} \rightarrow \text{glycosyl-lipopolysaccharide} + \text{UDP} \qquad (1)$$

$$\text{glucosyl-lipopolysaccharide} + \text{UDP-galactose} \rightarrow \text{galactosyl-glucosyl-lipopolysaccharide} + \text{UDP} \qquad (2)$$

At each stage in synthesis, the new glycosyl unit is added directly to the non-reducing end of a specific incomplete lipopolysaccharide primer. No additional glycosyl intermediates are involved in this synthesis, but the physical and structural requirements for polymerization allow us to understand something of the way in which the cell surface can influence the progress of a reaction by virtue of its physical state.

When Gram-negative bacteria were treated ultrasonically, most of the glycosyl transferase activity was found in the cell envelope fraction sedimenting at 12,000 g (Rothfield *et al.* 1964b); this fraction also contained the plasma membrane of the cells. After removal of cell envelopes in this way, a significant portion of glucosyl and galactosyl transferase activity was also located in the 'soluble' supernatant fluid after centrifugation at 105,000 g (Rothfield *et al.* 1964). Initially this activity appeared to be dependent upon the addition of envelope fractions containing lipopolysaccharide deficient in the appropriate monosaccharide. These 'soluble' enzymes showed no activity with purified lipopolysaccharides, indicating the need for some additional component in the envelope fraction. Rothfield & Horecker (1964) and Rothfield *et al.* (1964) showed that this component was contained in the crude lipid extracts of the cell envelope. When this was added to isolated lipopolysaccharide, the latter functioned as acceptor in the transferase reactions. The special requirements for this recombination of lipid and lipopolysaccharide indicate that the acceptor activity of the latter is the result of a physical interaction in which the lipopolysaccharide–lipid complex assumes the structural form for binding the glycosyl transferases. This ordered structure must exist *in vivo* in the envelope, but *in vitro* it was achieved when the lipid and lipopolysaccharide were heated and slowly cooled together. The active constituent of the lipid extract was found to be phosphatidylethanolamine; subsequent studies showed that the structural requirements for activity resided in the fatty acid residues and in the component attached to the phosphate group. The precise molecular arrangement of this lipid-polysaccharide complex is not known, but it has been suggested that the annealing process leads to the formation of a common bimolecular leaflet (Rothfield, Takeshita, Pearlman & Horne, 1966). *In vivo*, this polymerizing system shows the familiar requirement for divalent cations. Experiments with the isolated lipid-lipopolysaccharide complex have shown that Mg^{2+} is required for the binding of the 'soluble' glycosyl transferases to this complex. Here again, therefore, it is important to note that Mg^{2+} was not present during the ultrasonic treatment and the preparation of the 'soluble' glycosyl transferases. In terms of the vectorial synthesis of this prokaryotic surface polymer, these results suggest that the transferase enzymes are specifically attached to growing lipopolysaccharide chains which in turn are structurally linked to phospholipid in the membrane.

The biosynthesis of the long chains of repeating oligosaccharide units which constitute the o-antigen proceeds in a different manner, and requires a polyisoprenoid lipid carrier of the same type as that involved in peptidoglycan synthesis (Nikaido & Nikaido, 1965; Zeleznick *et al.* 1965; Robbins, Wright & Bellows, 1964; Wright *et al.* 1965; Wright, Dankert, Fennessey & Robbins, 1967). The sequence of reactions involved is, again, dependent on the presence of the cell-envelope fraction. This sequence as it occurs in *Salmonella newington* begins with the synthesis of the mannosyl-rhamnosyl-galactosyl repeating unit from GDP-mannose, TDP-rhamnose and UDP-galactose. In a series of steps, the sugars are linked to the polyisoprenoid carrier lipid in the envelope *via* a pyrophosphate bond to form lipid-trisaccharide units.

These are then polymerized to form long lipid-linked o-antigen chains which are transferred to the core region to form the complete lipopolysaccharide. As with the particulate enzymes which synthesize peptidoglycan, the manner in which the envelope fraction is prepared and treated markedly affects its subsequent activity in polymerization. Thus preparations obtained by freezing and thawing in the presence of EDTA, or by breakage in a Hughes press followed by EDTA treatment are active in synthesising long chains of galactosyl-mannosyl-rhamnosyl repeating units (Robbins et al. 1964). Envelope preparations which have been treated ultrasonically are less active.

Biosynthesis of glycoproteins in animal cells

Glycoproteins are found at the surfaces of most animal cells. They are classically divided into two types: those which are firmly bound to the plasma membrane and those which are excreted and therefore considered to be extracellular (Eylar, 1965). Evidence which shows the membrane glycoproteins to be involved in such cellular properties as ion transport, cell antigenicity and intercellular contact clearly points to their protective role. These and many other aspects of membrane glycoproteins have been comprehensively reviewed by Cook (1968). This section will be devoted to a consideration of the biosynthesis of these eukaryotic surface components with emphasis on the role played by membranes.

Glycoproteins of animal cells bear a marked resemblance to bacterial peptidoglycan in that they consist of oligosaccharide chains covalently linked to polypeptides. As yet, however, the precise chemistry of the glycoproteins is not as well understood as that of the prokaryotic peptidoglycan. The major carbohydrates which are found in these polymers are D-galactose, D-mannose, D-galactosamine, N-acetyl-D-glucosamine and L-fucose, together with sialic acids. As with prokaryotic polymers, the biosynthesis of membrane glycoproteins involves the utilization of nucleotide sugar phosphates via specific glycosyl transferases. At some stage these oligosaccharides or monosaccharides must be linked to the polypeptide moiety. The nature of the carbohydrate-protein link appears to be similar in different groups of glycoproteins. One group is characterized by an asparagine-carbohydrate link; serine, threonine, glutamic acid and aspartic acid have been found linked to carbohydrate in other groups.

A major obstacle to locating the site of glycoprotein synthesis arises from the multiplicity of intracytoplasmic membranes which characterize the majority of eukaryotic cells. Consequently biosynthetic studies with cell-free systems have had to await the development of suitable pro-cedures for separating these membrane systems (Eylar & Cook, 1965; Bosmann, Hagopian & Eylar, 1968a). These procedures permit experi-ments which reveal the involvement of intracytoplasmic membranes in the synthesis of eukaryotic surface structures. Cook, Laico & Eylar (1965) separated the membranes of Ehrlich ascites carcinoma cells into smooth and rough endoplasmic reticulum. The smooth fraction was not homogeneous and probably contained the Golgi apparatus and plasma membranes, in addition to reticulum material. The incorporation of glucosamine and amino acids into these fractions as a measure of glycoprotein and protein synthesis was then studied. Analyses showed that the membrane glycoprotein was located in the membranes of the smooth reticulum fraction, but not in the membranes bearing ribo-somes (rough membranes). Isotopically labelled glucosamine was in-corporated into both types of membrane, but was not found in the polysomes or in the ribosomes after they were dissociated from the membrane. Incorporation of glucosamine and leucine into rough membranes increased steadily during the first hour of incubation, after which there was a marked increase in glucosamine incorporation. An initial lag was observed in glucosamine and leucine incorporation into smooth membranes. The termination of this lag coincided with the increase in glucosamine incorporation in the rough membranes, suggest-ing that glycoprotein synthesis in the smooth membranes was limited by carbohydrate synthesis. Experiments with inhibitors indicated that carbohydrate biosynthesis differed from protein synthesis in that it was not immediately inhibited by puromycin. It was concluded that mem-brane glycoprotein was probably synthesized by a non-synchronous process, in which a polypeptide receptor was first made on the poly-somes and later linked to oligosaccharides synthesized elsewhere. To determine where oligosaccharide synthesis occurred and where in the cell the linkage with protein was made, Eylar & Cook (1965) fractionated Ehrlich ascites cells and obtained a smooth membrane fraction that contained the specific glucosamine and galactose transferases.

As mentioned earlier, the 'smooth' membrane fractions cover a 'multitude of preparative sins' and their use is not likely to reveal the location of the eukaryotic polymerizing systems. Recently Bosmann et al. (1968a) have extensively purified HeLa cell membranes to separate plasma membrane from intracytoplasmic smooth membrane. The re-sults indicated that the two types of membrane differed in several respects. Of particular interest here are experiments which were sub-sequently made to assess the role of the purified membrane fractions in glycoprotein synthesis (Hagopian, Bosmann & Eylar, 1968; Hagopian & Eylar, 1968; Bosmann, Hagopian & Eylar, 1968b). The glycosyl transferases involved in the addition of carbohydrate units to the polypeptide receptor in the synthesis of the HeLa cell membrane glycoprotein are located in the fraction composed of smooth internal membranes. These enzymes form a multienzyme group which can be detached from the membrane with Triton X-100. An interesting contrast was provided by the enzyme collagen-glucosyl transferase which catalyses the addition of a terminal glucose residue to the carbohydrate moiety of collagen. The enzyme concerned with the synthesis of this *secreted* protein was exclusively located in the plasma membrane. Another interesting difference was that this enzyme displayed its maximum activity when bound to the membrane, whereas the membrane glycoprotein glycosyl-transferases only functioned maximally after 'solubilization' with Triton. Some caution is required in evaluating the action of detergents on membrane systems generally. It may be important to distinguish an increase in specific activity due to 'solubilization' of a particular enzyme from that produced by the removal of other non-catalytic proteins. Further characterization of this multienzyme group involved in HeLa cell membrane glycoprotein synthesis has shown that it includes a polypeptidyl-N-acetyl-galactosaminyl transferase, a glycoprotein-galactosyl transferase and two glycoprotein-fucosyl transferases. The polypeptidyl transferase attaches the first monosaccharide to the polypeptide and thereby converts it into a glycoprotein. In the case of the synthesis of bovine submaxillary protein, the transferase specifically catalyses the formation of glycosidic linkages between N-acetylgalactosamine and certain hydroxyamino acids in the polypeptide. The receptor activity of these amino acids appears to depend on the sequence or conformation in that particular region of the peptide. Following this initial step, the other monosaccharides are attached in sequence to the terminal N-acetylgalactosamine by the specific glycosyl transferases. All the enzymes show a requirement for divalent cation.

To account for these results Hagopian & Eylar (1968) proposed that after synthesis of the polypeptide receptor on the ribosomes of the rough endoplasmic reticulum, it 'migrates' to the smooth intracytoplasmic membranes where the carbohydrate chains are attached and extended by the specific glycosyl transferases. Other workers, however, have obtained results which suggest that the first carbohydrate may in part be linked to the polypeptide while still attached to the ribosomes (Molnar, Robinson & Winzler, 1965). In these experiments, however, membrane purification does not appear to have been sufficient to permit unequivocal conclusions. At this stage, in the eukaryotic cell, the membrane glycoprotein has still to be incorporated into the plasma membrane. How this occurs remains a problem, but it is believed that the Golgi apparatus may be involved. A considerable body of autoradiographic and other data suggests that the Golgi membranes and vesicles in eukaryotic cells may be sites of assembly for the plasma membrane and for products which the cell secretes into its environment. Although the Golgi apparatus forms a major part of the intracytoplasmic smooth fraction (Bosman *et al.* 1968*a*), its structure is destroyed during the purification which includes homogenization in the presence of EDTA. The assembly of plasma membrane or the insertion into it of membrane glycoproteins may involve a vectorial flow of nascent membrane from the Golgi apparatus to the cell surface. If this be the case, there may be only gradual changes in membrane composition and not sharp differences which would make identification easier.

With one exception, no glycosyl intermediates other than nucleotide sugars have been implicated in glycoprotein biosynthesis. In this respect the process resembles that for the synthesis of the core portion of lipopolysaccharide. It will be of interest to determine whether the process also requires phospholipid at some stage in monosaccharide transfer. The interesting exception to the rule is reported by Caccam *et al.* (1969), who studied the synthesis of mannose-containing glycoproteins which are *secreted* to the surface of liver, oviduct and myeloma cells. As with membrane glycoproteins, the glycosyl transferases are located in the smooth intracytoplasmic membranes. By contrast, however, incorporation studies suggest that GDP-mannose is first incorporated into a lipid intermediate, from which it is transferred to the growing glycoprotein. Preliminary studies indicate that this eukaryotic lipid is a polyisoprenoid, similar if not identical to the intermediates involved in lipopolysaccharide and peptidoglycan synthesis.

Biosynthesis of glycosaminoglycans in prokaryotes and eukaryotes

The term glycosaminoglycan is used to describe protein-polysaccharide polymers which, in contrast to glycoproteins, contain uronic acids and frequently also sulphate. In one instance, at least, they are synthesized in prokaryotes and in eukaryotes; this fact has provided the opportunity to examine the biosynthetic role of the different membranes which are found in cells of these two types. Such an examination is contained in a review by Stoolmiller & Dorfman (1969*a*), therefore only a few points of special interest will be made here. The evidence indicates considerable similarity in structure and biosynthesis between glycosaminoglycans and glycoproteins. Glycosaminoglycans synthesized by eukaryotic cells include chondroitin, heparin and hyaluronic acid. The latter compound is also synthesized by organisms of the prokaryote genus *Streptococcus*.

The immediate precursors of the polysaccharide portion of the molecule are the uronic acid, glucosamine and galactosamine nucleotide sugars. Polymerization of these monosaccharides has been studied in both prokaryotic and eukaryotic cell systems. From the results it appears that the mechanisms of polysaccharide chain extension are the same in both systems (Stoolmiller & Dorfman, 1969*b*). Hyaluronic acid synthesis has been achieved with a particulate system from a Group A *Streptococcus* sp. (Markovitz & Dorfman, 1962). Cell-free preparations were obtained from ultrasonically disrupted cells and from protoplasts freed from wall material by the action of a lytic enzyme. Both preparations catalysed the net synthesis of hyaluronic acid polymer from UDP-glucuronic acid and UDP-N-acetylglucosamine in the presence of Mg^{2+}. From the similar specific activities of the two preparations, it seems clear that the protoplast membrane is the site of synthesis of this polymer in the prokaryote. Analogous requirements for hyaluronic acid synthesis in eukaryotic cells were reported for a cell-free particulate enzyme preparation from embryonic rat skin (Schiller, 1964). No 'solubilization' of enzyme activity in either the prokaryote or eukaryote preparations resulted from ultrasonic treatment or treatment with detergents, trypsin or phospholipase. In some instances these treatments inactivated the enzymes. When papain was used, about 40 % of the hyaluronic acid synthetase activity appeared in the 105,000 g supernatant fluid; this effect may have been due to the disruption of previously unbroken cells however.

In the case of a number of glycosaminoglycans, the nature of the protein-carbohydrate linkage and the mechanism of its formation is fairly well understood (Stoolmiller & Dorfman, 1969*a*). The situation is reminiscent of glycoprotein synthesis, with the first monosaccharide being combined in an o-glycosidic linkage to an amino acid in a preformed polypeptide. In the case of chondroitin-4-sulphate, UDP-xylose is first linked to the hydroxyl of serine. Two galactose residues are then added sequentially to form the galactosyl-galactosyl-xylose-protein. The repeating portion of the chondroitin sulphate chain is then initiated by the transfer of glucuronic acid to the terminal galactosyl residue (Roden & Armand, 1966; Stoolmiller & Dorfman, 1969*a*). The structure of one such linkage region is as follows:

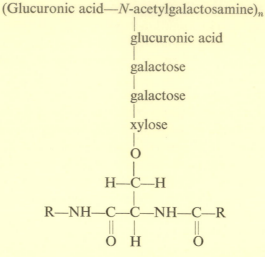

(Glucuronic acid—N-acetylgalactosamine)$_n$

glucuronic acid

galactose

galactose

xylose

O

H—C—H

R—NH—C—C—NH—C—R

O   H   O

Prokaryotic hyaluronic acid is also thought to be bound to protein (Hamerman, Rojkind & Sandson, 1966), but the situation is not as clear as with the eukaryotic system. Purified streptococcal hyaluronic acid does contain a limited number of amino acids and galactosamine (Stoolmiller & Dorfman, 1969b) but it shows a marked difference from that of eukaryotic cells in its response to antibiotics which inhibit protein synthesis. The synthesis of eukaryotic hyaluronic acid and chondroitin sulphate in tissue cultures is inhibited by puromycin and cycloheximide (Matalon & Dorfman, 1966), indicating that polysaccharide synthesis is to some extent dependent on protein synthesis. In strains of *Streptococcus*, however, puromycin and chloramphenicol have no effect on hyaluronic acid synthesis. This may mean that the prokaryotic polymerizing system does not require concomitant protein synthesis; but there are other possibilities to explain the insensitivity to antibiotics (Stoolmiller & Dorman, 1969b). It may also be relevant that the polymer is probably synthesized at different sites in prokaryotic and eukaryotic cells. Streptoccal hyaluronic acid is synthesized at the plasma membrane, whereas in eukaryotic cells it is the intracytoplasmic membranes that are active. The data indicate that both rough and smooth endoplasmic reticulum function in a similar manner to their role in membrane glycoprotein synthesis. It is thought that the initial attachment of carbohydrate to polypeptide receptor occurs on the rough membranes bearing ribosomes, and that extension of the carbohydrate chain via specific glycosyl transferases occurs in the smooth endoplasmic reticulum (Horwitz & Dosfman, 1968). Further fractionation of the smooth membranes as done by Bosmann *et al.* (1968a) is needed to distinguish the individual contributions of the plasma membrane and Golgi apparatus.

For the hyaluronic acid synthetase of *Streptococcus* spp. and chick fibroblasts, the evidence is strongly against the involvement of a polyisoprenoid lipid intermediate (Ishimoto, Temin & Strominger, 1966; Stoolmiller & Dorfman, 1969*b*). Lack of inhibition of polymer synthesis by bacitracin is by itself not indicative of the absence of such an intermediate. Scher & Lennarz (1969) reported the isolation of a polyisoprenoid lipid carrier, active in mannan biosynthesis, which contained only one phosphate group. As expected from the results of Siewart & Strominger (1967), bacitracin did not affect mannan biosynthesis in this system.

Biosynthesis of pneumococcal capsular polysaccharides

The polysaccharides which form the extensive capsule of the prokaryote *Diplococcus pneumoniae* (Pneumococcus) are responsible for determining the virulence and antigenic specificity of the different pneumococcal types. Reports describe almost 80 different polysaccharide types within the species (Mills & Smith, 1962). The available information on the structure of these polymers was reviewed by How, Brimacombe & Stacey (1964). The few studies on the cell-free synthesis of these surface polymers indicate a requirement for particulate fractions which are not homogeneous, but plasma membrane fragments appear to represent the major component. Smith & Mills (1962) and Smith, Mills, Bernheimer & Austrian (1960) studied the biosynthesis of Type I and Type III capsular polysaccharide, respectively. In both cases the polymer-synthesizing enzymes were located in fractions sedimenting between 30,000 g and 100,000 g. Type III polysaccharide, which is a polymer of cellobiuronic acid (β-1,4-glucuronosido-glucose) was synthesized when UDP-glucose and UDP-glucuronic acid were incubated with the particulate preparation in the presence of Mg^{2+}; without Mg^{2+}, no synthesis occurred. The precise structure of Type XIV polysaccharide is not known but it is a branched polymer composed of *N*-acetylglucosamine, glucose and galactose in the ratio 2:1:3 (Distler & Roseman, 1964). After disruption of pneumococci in a French pressure cell Distler & Roseman (1964) obtained a particle fraction sedimenting between 10,000 g and 105,000 g which in the presence of Mg^{2+} catalysed the incorporation of *N*-acetylglucosamine, glucose and galactose into antigenically active Type XIV polysaccharide. The results did not distinguish between *de novo* synthesis and synthesis by addition of sugars to endogenous primer polymer. This preparation also catalysed the synthesis of two glycolipids which were extracted into *n*-butanol (Kaufman, Kundig, Distler & Roseman, 1965). These lipids are similar in structure and mode of synthesis to the mannolipids in *Micrococcus lysodeikticus* (Lennarz & Talamo, 1966). Their role, if any, in capsular polysaccharide synthesis is not known.

Biosynthesis of bacterial mannan

Macfarlane (1964) reported the presence of the homopolysaccharide mannan in the plasma membrane of the prokaryote *Micrococcus lysodeikticus*. The reason for its location at the cell surface of this organism is not apparent, but it is considered here because it provides still another example of the involvement of a polyisoprenoid lipid carrier in the assembly of a surface polymer.

In their investigations of the biosynthesis of glycolipids in *Micrococcus lysodeikticus* Lennarz & Talamo (1966) showed that particulate preparations from this organism catalysed the incorporation of mannose from GDP-mannose into three mannose-containing lipids and mannan. Two of the lipids were examined and found to be α-D-mannosyl-(1 → 3′)-diglyceride and α-D-mannosyl-(1 → 3)-α-D-mannosyl-(1 → 3′)-diglyceride. These particle preparations were obtained after breakage of the cocci in a French pressure cell without added cation. An enzyme was found in the crude fraction which transferred mannose from GDP-mannose to the hydroxyl group of an exogenous 1,2-diglyceride. This enzyme was dependent for activity on Mg^{2+} and an anionic surface active agent. The unusual requirement for a surface-active agent evoked even more interest when it was found to be best satisfied by the use of 12- and 13-methyltetradecanoic acid salts. These two branched-chain fatty acids are the major fatty acids in the membrane of this organism. Perhaps this represents another example of a lipid or lipid component filling a structural role, as in the biosynthesis of the lipopolysaccharide core (see above). Two structural roles have been suggested (Lennarz & Talamo, 1966). One possibility is that the surface-active agent promotes the interaction of the hydrophobic diglyceride substrate with the enzyme, by modifying the physical state of the diglyceride; the other possibility is that the enzyme and surface-active agent combine to form a hydrophobic environment necessary for association with the substrate. A second enzyme was found in the crude cell extract. This catalysed the synthesis of the dimannosyl-diglyceride from GDP-mannose and mannosyl-diglyceride. Activity was Mg^{2+}-dependent but no requirement for a surface-active agent was found. Scher *et al.* (1968) investigated the chemistry of the third lipid. They found that this compound was a mannosyl-1-phosphoryl-polyisoprenol with a lipid moiety identical with that found in the intermediates involved in lipopolysaccharide and peptidoglycan synthesis (Wright *et al.* 1967; Higashi *et al.* 1967). Results reported recently confirm the expectation that mannosyl-1-phosphoryl-polyisoprenol is an obligatory intermediate in mannan biosynthesis (Scher *et al.* 1968; Scher & Lennarz, 1969). An enzyme was found in the particulate fraction of *M. lysodeikticus* which catalysed the reversible synthesis of mannosyl-1-phosphoryl-polyisoprenol from GDP-mannose and polyisoprenyl phosphate. This compound then donated mannosyl groups to the terminal non-reducing residues of endogenous mannan. This incorporation has so far been limited to the addition of single mannosyl units to the endogenous polymer. In this instance the lipid intermediate contains only one phosphate group as distinct from the similar intermediates cited above. This is reflected in the failure of bacitracin to inhibit mannan synthesis. An interesting and perhaps useful differential effect of detergent and chelating agent was noted in this overall process. Triton X-100 inhibited the transfer of mannose from the lipid intermediate to mannan, but did not affect the synthesis of the intermediate. In contrast, EDTA had no effect on mannose transfer, but markedly inhibited the synthesis of the lipid intermediate.

Biosynthesis of plant polysaccharides and bacterial cellulose

The biosynthesis of polysaccharides in plants and of cellulose in bacteria has been reviewed by Siegel (1968) and Hassid (1969). Studies of the cell-free synthesis of cellulose in particular afford another opportunity to compare the synthesis of a surface polymer in eukaryotic and prokaryotic cells.

The prokaryotic micro-organism *Acetobacter xylinum* synthesizes cellulose chains which appear to be identical with the cellulose synthesized by plant cells. Glaser (1958) was one of the first to achieve the cell-free synthesis of cellulose with *A. xylinum*. From his results it was clear that polymer synthesis was closely involved with an envelope fraction which contained the plasma membrane. This fraction was obtained after cell breakage by extensive ultrasonic treatment in the presence of EDTA and Mg^{2+}. After removal of cell debris by centrifugation at 15,000 *g*, the active preparation was sedimented at 140,000 *g* for 1 hr. Incorporation of UDP-glucose into material with the properties of cellulose was shown; no dependence on a divalent cation was observed. Synthesis was however dependent on the presence of a cellulosic primer. No significant incorporation occurred when the enzyme was first treated with cellulase. Synthesis was resumed when exogenous primer, in the form of soluble cellodextrins, was added. Attempts to 'solubilize' the synthetase with n-butanol or by preparing acetone powders were unsuccessful. Treatment with digitonin in the presence of EDTA, or with pancreatic lipase, produced an active preparation which was not sedimented at 140,000 g in 1 hr, but was extremely labile. While these findings indicate that the cellulose-synthesizing system was firmly associated with the envelope structures, subsequent studies by Colvin (1959) and Brown & Gascoigne (1960) suggested that an extra-cellular enzyme catalysed the final stages in cellulose assembly. This enzyme was prepared after ultrafiltration of an ethanol extract of whole bacteria to remove cell debris and cellulose microfibrils. The results showed that this culture filtrate contained an ethanol-soluble cellulose precursor which could be polymerized into insoluble cellulose by an enzyme also present in the culture fluid. Other experiments showed that a similar cell-free poly-merization could be achieved without the ethanol extraction. The culture fluid was filtered to remove bacteria and cellulose fibres. When this filtrate was incubated at 28° for 1 hr, cellulose formation was observed.

Of special interest are the experiments of Kahn & Colvin (1961), who fractionated ethanol extracts of *Acetobacter xylinum* which had pre-viously been incubated for 4 to 5 min. in the presence of [^{14}C]glucose. One of their column fractions was found to contain a radioactive lipid which appeared to be homogeneous. When this material was incubated with an aqueous fraction of ultrafiltered culture supernatant fluid, cellulose-like material was formed. They concluded that the lipid material was an immediate precursor of bacterial cellulose, which was polymerized by an *extracellular* enzyme present in the culture fluid. It was proposed that the lipid-glucose precursor functioned in transporting the glucose from the cytoplasm across the membrane to an extracellular location where it was incorporated into cellulose. Since this lipid did not accumulate in the medium, it was suggested that the process was cyclic. Colvin (1961) described a eukaryotic system with similar proper-ties. It was shown that the presumed polymerizing enzyme from *A. xylinum* would promote cellulose formation in the eukaryotic extract. The similarities between this system and the others already described, which involve lipid intermediates in sugar transport, suggest that further experiments might prove interesting. This is emphasized by the results of experiments with eukaryotic systems which synthesize cellu-lose (see later in this section) which also indicate the involvement of a glucolipid in cellulose synthesis. Before leaving the discussion of *A. xylinum* it should be added that the *in vivo* assembly of cellulose microfibrils in this prokaryote may be dependent on the structural organization of the various layers in the cell envelope. When this structural integrity was destroyed by treating bacteria with penicillin or lysozyme + EDTA, the resulting spheroplasts did not synthesize cellulose (Brown & Gascoigne, 1960).

The study of polysaccharide synthesis in cell-free systems from plants introduces complications which at first may seem bizarre to those who are more familiar with prokaryotic systems. For example, the season of the year at which the plants are harvested may determine the presence or absence of a particular enzyme (Hassid, 1969). However, from the results of continuous culture experiments with prokaryotes one is impressed more by the similarities than by any differences. Difficulty is encountered in evaluating cell-free cellulose synthesizing systems from plants because, in the absence of extensive fractionation, homogenates of plant tissues will contain a range of eukaryotic membranes. Only recently have such purifications been attempted (Villemez et al. 1968); to date much of the evidence for the involvement of various membranes in plant polysaccharide synthesis derives from electron microscopic and autoradiographic studies (Northcote, 1969).

Elbein, Barber & Hassid (1964) prepared a particle fraction sedimenting between 1,000 g and 20,000 g from mung beans (Phaseolus aureus) which catalysed the synthesis of cellulose from GDP-glucose, but not from UDP-, TDP- or CDP-glucose. The experiments did not distinguish between de novo cellulose synthesis and extension of endogenous primer. Brummond & Gibbons (1964, 1965), by using a cell-free preparation of Lupinus albus, obtained evidence that both UDP-glucose and GDP-glucose were incorporated into a range of polysaccharides including cellobiose. The use of UDP- and GDP-glucose in similar experiments was reported by Hassid (1969) to yield a polysaccharide containing only β-1,3-glucosyl linkages. Polysaccharide material of this type was also synthesized from UDP-glucose by a preparation from bean seedlings (Feingold, Neufeld & Hassid, 1958). Ordin & Hall (1967) homogenized sections of oat coleoptiles and collected a fraction sedimenting between 10,000 g and 140,000 g. Both UDP- and GDPK-glucose were incorporated into polymer, although the former was the more effective glucose donor. With GDP-glucose only cellobiose was formed, whereas with UDP-glucose additional β-1,3-linked material was detected. Similar results were obtained with bean seedling homogenates by Villemez, Franz & Hassid (1967).

In an attempt to characterize these various enzyme preparations more fully Villemez et al. (1968) made an extensive fractionation of bean shoot cells. After sucrose gradient separation, the resulting fractions were examined for polysaccharide formation from a variety of radio-active nucleotide sugars. Synthesis of polygalacturonic acid, galactan and cellulose was shown in particle fractions sedimenting over a wide range. The authors concluded that the polymerases originated from the plasma membrane which was broken into fragments of different sizes by the purification procedures. In support of this conclusion, they reported the absence of polymerizing activity from cell-wall prepara-tions, proplastids, and a preparation from onion stem cells which may represent Golgi bodies. Cell membranes of many types tend to alter their conformation when the cell is ruptured. Vesicular structures often form about each other and alter the *in vivo* relationship of such struc-tures as Golgi bodies and lysosomes to the plasma membrane and reticulum. In the absence of specific markers for the different membrane types, caution is necessary in assigning enzymic functions. Nevertheless, these results are encouraging and do furnish evidence for a different site of synthesis for two of the polymers synthesized from UDP-glucose. The bulk of the enzyme which synthesized alkali-soluble material (presumed to be callose) was found in the light particle fraction ($40,000\,g$ to $100,000\,g$), whereas the alkali insoluble polymer was synthesized by enzymes largely present in the heavier fractions ($1,000\,g$ to $40,000\,g$). Evidence was also obtained for the synthesis of a glucolipid from UDP-glucose. For two other eukaryotic cell-free systems there are reports of the possible involvement of an intermediate which is soluble in organic solvents. Feingold, Neufeld & Hassid (1959) and Bailey & Hassid (1966) described preparations which were active in xylan synthesis. They reported the incorporation of radioactivity from a mixture of UDP-xylose and UDP-arabinose into a product insoluble in water, but soluble in boiling ethanol. The suggestion was made that this compound may represent a hydrophobic acceptor in xylan synthesis. The second report is that of Glaser & Brown (1957), who showed the incorporation of radioactivity from UDP-N-acetylglucosamine into chitin by cell-free extracts of the fungus *Neurospora crassa*. Mycelium was homogenized in the presence of EDTA $+$ Mg^{2+}. The enzyme frac-tion sedimented in 1 hr at $140,000\,g$, but did not sediment after 10 min. at $10,000\,g$. Active material was partially solubilized by *n*-butanol treatment but not by using Triton X-100 or digitonin. The solubilized enzyme preparations showed an absolute requirement for endogenous chitodextrin primer. The reaction appeared to be reversible:

$$\text{UDP-}N\text{-acetylglucosamine} + (\beta\text{-1,4-acetylglucosamine})_n$$
$$\rightleftharpoons \text{UDP} + (\beta\text{-1,4-acetylglucosamine})_{n+1}$$

Rogers & Perkins (1968) drew attention to the similarity between these observations and those reported for systems such as the peptidoglycan synthetase discussed above.

Microscopic analysis of the synthesis of plant surface components has provided a great deal of information on the problems of vectorial synthesis. The results so far point to the involvement of Golgi vesicles, plasma membrane and rough endoplasmic reticulum in the synthesis of particular polymer species (Northcote, 1969). The indications are that biochemical analysis of these various structures, perhaps by using plant cell 'protoplasts', should be profitable.

Biosynthesis of teichoic acids in bacteria

The chemistry and biosynthesis of the bacterial wall polymers the teichoic acids has been dealt with in comprehensive reviews by Baddiley (1968) and Archibald, Baddiley & Blumson (1968). The role of the plasma membrane in their synthesis will be reviewed. Teichoic acids are polymers of either ribitol phosphate or glycerol phosphate linked through phosphodiester bonds, with sugar and D-alanine ester substituents attached to the hydroxyl groups of the polyols. Among the sugar substituents which have been found are glucose, N-acetylglucosamine, N-acetylgalactosamine and galactose. These polymers, which are thought to occur in all Gram-positive bacteria, differ to some extent in their cellular location. The evidence indicates that either ribitol or glycerol polymers can occur in the walls of these prokaryotes (Baddiley, 1968). Other studies with protoplasts (Hay, Wicken & Baddiley, 1963) have shown the existence of a second group of teichoic acids, invariably glycerol phosphate polymers, which are apparently located between wall and membrane. They may be attached in some way to the membrane and for this reason they are referred to as 'membrane teichoic acids'. Several workers have looked for an attachment between wall teichoic acids and the peptidoglycan component; this may occur through a phosphodiester link between a glycerol phosphate and a hydroxyl group of muramic acid (Button, Archibald & Baddiley, 1966). The relatively harsh conditions required to extract teichoic acids from wall preparations suggest a covalent bond of this type.

Burger & Glaser (1964) reported the cell-free synthesis of polyglycerol phosphate from CDP-glycerol by particulate enzyme preparations from *Bacillus subtilis* and *B. licheniformis*. Fractions were prepared by ultrasonic treatment and from protoplasts. Active fractions from ultrasonically broken bacteria were collected at 105,000 g, as compared to 40,000 g for membranes from lysed protoplasts. This suggests that the latter method produced larger membrane fragments. Since the assays showed that the protoplast-membrane fragments were the most active, this might be significant. The enzyme system appeared to have two activity optima, at pH 7·2 and pH 9·0 and displayed a high cation requirement for Mg^{2+} or Ca^{2+}. The enzyme was not solubilized by various lipases, lysozyme, n-butanol, Triton or digitonin.

Glaser (1964) reported the synthesis of polyribitol phosphate in extracts of *Lactobacillus plantarum* using CDP-ribitol. The enzyme showed the same high requirement for divalent metal cation, but only one activity optimum, at pH 8·3. Because protoplasts could not be prepared from this organism, the precise location of the synthetase is not clear. Earlier experiments with *Staphylococcus aureus* (Ishimoto & Strominger, 1963), who used ultrasonically-broken preparations, had located a similar enzyme in the fraction which sedimented between 30,000 *g* and 100,000 *g*. Subsequent work on the biosynthesis of other wall teichoic acids (Baddiley, Blumsom & Douglas, 1968) pointed to the membrane as the probable enzyme site. Young (1967) reported that one of the enzymes involved in glucosylation of the *Bacillus subtilis* wall teichoic acid formed aggregates under conditions which promoted the reassembly of disaggregated membranes.

In a series of studies reviewed by Baddiley (1968), it was shown that carbohydrate residues were transferred to the polymer from the appropriate nucleoside diphosphate sugars. A close relationship exists during polymerization between the synthesis of the polymer and its substitution with carbohydrate residues. The glycosyl transferases which catalyse this incorporation are also found in the same fraction, which consists predominantly of membranes. The mechanism whereby alanine is incorporated into teichoic acids is not definitely known, but it is thought to occur as a final step in biosynthesis (Baddiley, 1968). The direction of polymer growth in the polyglycerol phosphate teichoic acid of *Bacillus subtilis* has been determined by Kennedy & Shaw (1968). With a membrane preparation from lysozyme-treated bacilli, they showed that polymerization occurred by the additions of units to the glycerol terminus, rather than to the phosphate terminus. The membranes in this instance were not treated with EDTA.

Besides mixed ribitol-glycerol polymers, there are other teichoic acids in which the carbohydrate residues form part of the polymer chain (Baddiley, 1968). *Staphylococcus lactis* I3 contains a polymer of this type in which N-acetylglucosamine-1-phosphate is linked through its phosphate residue to glycerol phosphate, which is in turn linked to the 3-hydroxyl group of the next N-acetylglucosamine. The biosynthesis of this polymer has been achieved in a cell-free system (Baddiley, Blumsom & Douglas, 1968). The bacteria were disintegrated mechanically with glass beads and the synthesizing system located in a fraction which sedimented between 15,000 g and 105,000 g. Polymer synthesis was obtained from CDP-glycerol and UDP-N-acetylglucosamine in the presence of a high Mg^{2+} concentration. Although this polymer also contains alanine linked to the 6-hydroxyl of the amino sugar, the addition of either ATP or alanine had no effect on polymer synthesis. Since no incorporation of labelled alanine was obtained, it can be concluded (Baddiley, Blumsom & Douglas, 1968) that it is not obligatory for polymer synthesis. Attempts to fractionate the enzyme preparation by gradient centrifugation or detergent treatment were unsuccessful. It was shown by analysis to contain peptidoglycan material together with wall and membrane teichoic acids. Rather surprisingly, the enzyme did not utilize CDP-glycerol for synthesis of the poly-glycerol phosphate membrane teichoic acid. In this system, the enzymes responsible for the synthesis of the nucleotide precursors are also attached to the membrane preparation (Baddiley, 1968). Subsequent studies with this system (Baddiley, 1968) have provided the first suggestion that a lipid intermediate may be involved in the synthesis of these prokaryotic polymers. Radioactivity from UDP-N-[^{14}C]acetylglucosamine was incorporated into a lipid fraction in the enzyme particles and could be chased into teichoic acid by subsequent addition of unlabelled nucleotide and CDP-glycerol. [^{32}P]labelled nucleotide sugar was also used to show that phosphate and glycosyl residues were both transferred to the lipid material. CDP-glycerol was incorporated into lipid, but to a much decreased extent, suggesting either that the glycerol phosphate moiety was incorporated by a different mechanism, or alternatively, that its incorporation into the lipid followed that of the amino sugar and the unit then rapidly transferred to the growing polymer (Baddiley, 1968). A similar lipid has been implicated in the synthesis of a teichoic acid-like polymer from a strain of *Staphylococcus lactis* (Brooks & Baddiley, 1969). This is a poly-(N-acetylglucosamine-1-phosphate) polymer containing no alanine or polyol. It consists of amino sugar repeating units linked through the 1 and 6 positions by phosphodiester bonds. Cell-free synthesis was catalysed by a fraction which sedimented between 15,000 g and 100,000 g after mechanical disintegration of the cells. Very high amounts of enzyme protein and Mg^{2+} at 70 mM were required for polymer synthesis; UMP inhibited synthesis markedly. This was said to be consistent with the finding that a lipid phosphate intermediate (probably similar to the polyisoprenoid intermediates already cited in previous sections) was associated with the synthesis of this polymer.

The teichoic acids are intimately involved in such cell properties as antigenicity, protection against autolysis and susceptibility to phage infection. Their high content of charged groups, combined with their apparent location in a region extending from the exterior of the plasma membrane to the outer limits of the peptidoglycan, might be expected to influence the movement of ions in this region. Although an active regulatory role in ion transport has yet to be demonstrated, the results of continuous culture experiments have shown interesting changes in polymer production in response to alterations in the ionic content of the microbial environment (Ellwood & Tempest, 1969).

Conclusion

In this review of the role of prokaryotic and eukaryotic membranes in the assembly of surface polymers, many topics have not been discussed. Among these are the role of membranes in the synthesis of flagella and cilia, pinocytosis, the role of microtubules, and the evidence which indicates considerable modification of surface polymers at a distance from the membrane in plant-cell walls (see Northcote, 1969). Despite the evolutionary divergence of prokaryotes and eukaryotes their cells apparently share a need for the surfaces and micro-environments which membranes provide. Research into the mechanisms of enzyme action has revealed the molecular architecture which creates such micro-environments. With regard to the synthesis of many of the polymers which have been examined here, the next problem seems to be the macromolecular architecture of the polymerizing enzymes as they occur in association with membranes. The solution of this problem depends on deeper knowledge of membrane structure *per se*. As this review has shown, the synthesis of surface polymers of many kinds involves not only the protein but also the lipid components of membranes.

REFERENCES

ANDERSON, J. S., MATSUHASHI, M., HASKIN, M. A. & STROMINGER, J. L. (1965). Lipid-phosphoacetylmuramyl-pentapeptide and lipid-phosphodisaccharide-pentapeptide: presumed membrane transport intermediates in cell wall synthesis. *Proc. natn. Acad. Sci. U.S.A.* **53**, 881.

ARAKI, Y., SHIRAI, R., SHIMADA, A., ISHIMOTO, N. & ITO, E. (1966). Enzymatic synthesis of cell wall mucopeptide in a particulate preparation of *Escherichia coli. Biochem. biophys. Res. Commun.* **23**, 466.

ARCHIBALD, A. R., BADDILEY, J. & BLUMSOM, N. L. (1968). The teichoic acids. *Adv. Enzymol.* **30**, 223.

BADDILEY, J. (1968). Teichoic acids and the molecular structure of bacterial walls. *Proc. Roy. Soc. Lond.* B **170**, 331.

BADDILEY, J., BLUMSOM, N. L. & DOUGLAS, JULIA (1968). The biosynthesis of the wall teichoic acid in *Staphylococcus lactis* I3. *Biochem J.* **110**, 565.

BAILEY, R. W. & HASSID, W. Z. (1966). Xylan synthesis from uridine-diphosphate-D-xylose by particulate preparations from immature corncobs *Proc. natn. Acad. Sci. U.S.A.* **56**, 1586.

BOSMANN, H. B., HAGOPIAN, A. & EYLAR, E. H. (1968a). Cellular membranes: the isolation and characterization of the plasma and smooth membranes of HeLa cells. *Archs Biochem. Biophys.* **128**, 51.

BOSMANN, H. B., HAGOPIAN, A. & EYLAR, E. H. (1968b). Glycoprotein biosynthesis: the characterization of two glycoprotein: fucosyl transferases in HeLa cells. *Archs Biochem. Biophys.* **128**, 470.

BRINTON, C. C., MCNARY, J. C. & CARNAHAN, J. (1969). Purification and *in vitro* assembly of a curved network of identical protein sub-units from the outer surface of a *Bacillus*. *Bact. Proc.* p. 48.

BROOKS, B. & BADDILEY, J. (1969). The mechanism of biosynthesis and direction of chain extension of a poly-(N-Acetylglucosamine-1-phosphate) from the walls of *Staphylococcus lactis* N.C.T.C. 2102. *Biochem. J.* **113**, 635.

BROWN, A. M. & GASCOIGNE, J. A. (1960). Biosynthesis of cellulose by *Acetobacter acetigenum*. *Nature, Lond.* **187**, 1010.

BRUMMOND, D. A. & GIBBONS, P. A. (1964). The enzymatic synthesis of cellulose by higher plants. *Biochem. biophys. Res. Commun.* **17**, 156.

BRUMMOND, D. A. & GIBBONS, P. A. (1965). Enzymatic cellulose synthesis from UDP-D-Glucose-^{14}C by *Lupinus albus*. *Biochem. Z.* **342**, 308.

BRUNDISH, D. E., SHAW, N. & BADDILEY, J. (1967). The structure and possible function of the glycolipid from *Staphylococcus lactis* I3. *Biochem. J.* **105**, 885.

BUMSTED, R. M., DAHL, J. L., SÖLL, D. & STROMINGER, J. L. (1968). Biosynthesis of the peptidoglycan of bacterial cell walls. X. Further study of the glycol transfer ribonucleic acids active in peptidoglycan synthesis in *Staphylococcus aureus*. *J. biol. Chem.* **243**, 779.

BURGER, M. M. & GLASER, L. (1964). The synthesis of teichoic acids. I. Polyglycerophosphate *J. biol. Chem.* **239**, 3168.

BUTTON, D., ARCHIBALD, A. R. & BADDILEY, J. (1966). The linkage between teichoic acid and glycosaminopeptide in the walls of a strain of *Staphylococcus lactis*. *Biochem. J.* **99**, 11–14c.

CACCAM, J. F., JACKSON, J. J. & EYLAR, E. H. (1969). The biosynthesis of mannose-containing glycoproteins: a possible lipid intermediate. *Biochem. biophys. Res. Commun.* **35**, 505.

CHATTERJEE, A. N. & PARK, J. T. (1964). Biosynthesis of cell wall mucopeptide by a particulate fraction from *Staphylococcus aureus*. *Proc. natn. Acad. Sci. U.S.A.* **51**, 9.

CHATTERJEE, A. N., WARD, J. B. & PERKINS, H. R. (1967). Synthesis of mucopeptide by L-form membranes. *Nature, Lond.* **214**, 1311.

CLIVE, D. & LANDMAN, O. E. (1968). Growth of *Bacillus subtilis* L-forms on membrane filters and stimulation of their reversion to bacilli by the filters and by added cell wall. *Bact. Proc.* p. 58.

COLVIN, J. R. (1959). Synthesis of cellulose in ethanol extracts of *Acetobacter xylinum*. *Nature, Lond.* **183**, 1135.

COLVIN, J. R. (1961). Synthesis of cellulose from ethanol-soluble precursors in green plants. *Can. J. Biochem. Physiol.* **39**, 1921.

COOK, G. M. W. (1968). Glycoproteins in membranes. *Biol. Rev.* **43**, 363.

COOK, G. M. W., LAICO, M. T. & EYLAR, E. H. (1965). Biosynthesis of glycoproteins of the Ehrlich ascites carcinoma cell membranes. *Proc. natn. Acad. Sci. U.S.A.* **54**, 247.

DISTLER, J. & ROSEMAN, S. (1964). Polysaccharide and glycolipid synthesis by cell-free preparations from Type XIV *Pneumococcus. Proc. natn. Acad. Sci. U.S.A.* **51**, 897.

ELBEIN, A. D., BARBER, G. A. & HASSID, W. Z. (1964). The synthesis of cellulose by an enzyme system from a higher plant. *J. Am. chem. Soc.* **86**, 309.

ELLAR, D. J. & LUNGREN, D. G. (1967). Ordered substructure in the cell wall of *Bacillus cereus. J. Bact.* **94**, 1778.

ELLAR, D. J. (1969). Structure of the cytoplasmic membrane and mesosomes. *J. gen. Microbiol.* (In Press.)

ELLAR, D. J. & FREER, J. H. (1969). The isolation and characterisation of mesosome material from *Micrococcus lysodeikticus. J. gen. Microbiol.* (In Press.)

ELLWOOD, D. C. & TEMPEST, D. W. (1969). Influence of growth environment on the cell wall anionic polymers in some Gram-positive bacteria. *J. gen. Microbiol.* (In Press.)

EYLAR, E. H. (1965). On the biological role of glycoproteins. *J. theor. Biol.* **10**, 89.

EYLAR, E. H. & COOK, G. M. W. (1965). The cell-free biosynthesis of the glycoprotein of membranes from Ehrlich ascites carcinoma cells. *Proc. natn. Acad. Sci. U.S.A.* **54**, 1678.

FEINGOLD, D. S., NEUFELD, E. F. & HASSID, W. Z. (1958). Synthesis of a β-1, 3-linked glucan by extracts of *Phaseolus aureus* seedlings. *J. biol. Chem.* **233**, 783.

FEINGOLD, D. S., NEUFELD, E. F. & HASSID, W. Z. (1959). Xylosyl transfer catalyzed by an asparagus extract. *J. biol. Chem.* **234**, 488.

GHUYSEN, J. M. (1968). Use of bacteriolytic enzymes in determination of wall structure and their role in cell metabolism. *Bact. Rev.* **32**, 425.

GHUYSEN, J. M., STROMINGER, J. L. & TIPPER, D. J. (1968). Bacterial cell walls. In *Comprehensive Biochemistry.* Eds. M. Florkin and E. H. Stotz. vol. 26 A, p. 53. Amsterdam: Elsevier.

GLASER, L. (1958). The synthesis of cellulose in cell-free extracts of *Acetobacter xylinum. J. biol. Chem.* **232**, 627.

GLASER, L. (1964). The synthesis of teichoic acids. II. Polyribitol phosphate. *J. biol. Chem.* **239**, 3178.

GLASER, L. & BROWN, D. H. (1957). The synthesis of chitin in cell-free extracts of *Neurospora crassa. J. biol. Chem.* **228**, 729.

GLAUERT, A. M. & THORNLEY, M. J. (1969). Topography of the bacterial cell wall. In *A. Rev. Microbiol.* Palo Alto: Ann. Rev. Inc.

GLICK, J. L. & GITHENS, S. (1965). Role of sialic acid in potassium transport of L1210 leukaemia cells. *Nature, Lond.* **208**, 88.

HAGOPIAN, A. & EYLAR, E. H. (1968). Glycoprotein biosynthesis: studies on the receptor specificity of the polypeptidyl-N-acetylgalactosaminyl transferase from bovine sub-maxillary glands. *Archs Biochem. Biophys.* **128**, 422.

HAGOPIAN, A., BOSMANN, H. B. & EYLAR, E. H. (1968). Glyco-protein biosynthesis: the localization of polypeptidyl-N-acetylgalactosaminyl, collagen: glucosyl and glyco-protein: galactosyl transferases in HeLa cell membrane fractions. *Archs Biochem. Biophys.* **128**, 387.

HAMERMAN, D., ROJKIND, M. & SANDSON, J. (1966). Protein bound to hyaluronate: chemical and immunological studies. *Fedn Proc. Fedn Am. Socs exp. Biol.* **25**, 1040.

HASSID, W. Z. (1969). Biosynthesis of oligosaccharides and polysaccharides in plants. *Science, N.Y.* **165**, 137.

HAY, J. B., WICKEN, A. J. & BADDILEY, J. (1963). The location of intracellular teichoic acids. *Biochim. biophys. Acta* **71**, 188.

HEATH, E. C. & GHALAMBOR, M. A. (1963). 2-Keto-3-deoxyoctonate, a constituent of cell wall lipopolysaccharide preparations obtained from *E. coli. Biochem. biophys. Res. Commun.* **10**, 340.

HEYDANEK, G. & NEUHAUS, F. C. (1969). The initial stage in peptidoglycan synthesis. IV. Solubilization of phospho-N-acetylmuramyl-pentapeptide translocase. *Biochemistry, N.Y.* **8**, 1474.

HEYDANEK, M. G., STRUVE, W. G. & NEUHAUS, F. C. (1969). On the initial stage in peptidoglycan synthesis. III. Kinetics and uncoupling of phospho-N-acetyl-muramyl-pentapeptide translocase (uridine 5′-phosphate). *Biochemistry, N.Y.* **8**, 1214.

HIGASHI, Y., STROMINGER, J. L. & SWEELEY, C. C. (1967). Structure of a lipid intermediate in cell wall peptidoglycan synthesis: a derivative of a C_{55} isoprenoid alcohol. *Proc. natn. Acad. Sci. U.S.A.* **57**, 1878.

HORWITZ, A. L. & DORFMAN, A. (1968). Subcellular sites for synthesis of chondromucoprotein of cartilage. *J. Cell Biol.* **38**, 358.

HOW, M. J., BRIMACOMBE, J. S. & STACEY, M. (1964). The pneumococcal polysaccharides. *Adv. Carbohydrate Chem.* **19**, 303.

ISHIMOTO, N. & STROMINGER, J. L. (1963). Enzymic synthesis of polyribitol phosphate. *Fedn Proc. Fedn Am. Socs exp. Biol.* **22**, 465.

ISHIMOTO, N., TEMIN, H. M. & STROMINGER, J. L. (1966). Studies of carcinogenesis by avian sarcoma viruses. II. Virus-induced increase in hyaluronic acid synthetase in chicken fibroblasts. *J. biol. Chem.* **241**, 2052.

IZAKI, K., MATSUHASHI, M. & STROMINGER, J. L. (1966). Glycopeptide transpeptidase and D-alanine carboxypeptidase: penicillin-sensitive enzymatic reactions. *Proc. natn. Acad. Sci. U.S.A.* **55**, 656.

KAHN, A. W. & COLVIN, J. R. (1961). Synthesis of bacterial cellulose from labelled precursor. *Science, N.Y.* **133**, 2014.

KATZ, W., MATSUHASHI, M., DIETRICH, C. P. & STROMINGER, J. L. (1967). Biosynthesis of the peptidoglycan of bacterial cell walls. IV. Incorporation of glycine in *Micrococcus lysodeikticus. J. biol. Chem.* **242**, 3207.

KAUFMAN, B., KUNDIG, F. D., DISTLER, J. & ROSEMAN, S. (1965). Enzymatic synthesis and structure of two glycolipids from Type XIV *Pneumococcus. Biochem. biophys. Res. Commun.* **18**, 312.

KENNEDY, L. D. & SHAW, D. R. D. (1968). Direction of polyglycerolphosphate chain growth in *Bacillus subtilis. Biochem. biophys. Res. Commun.* **32**, 861.

LANDMAN, O. E. & FORMAN, A. (1968). Stages in the gelatin-induced reversion of *Bacillus subtilis* protoplasts to the bacillary form. *Bact. Proc.* p. 57.

LANDMAN, O. E. & HALLE, S. (1963). Enzymically and physically induced inheritance changes in *Bacillus subtilis. J. molec. Biol.* **7**, 721.

LENNARZ, W. J. & TALAMO, BARBARA (1966). The chemical characterization and enzymatic synthesis of mannolipids in *Micrococcus lysodeikticus. J. biol. Chem.* **241**, 2707.

LÜDERITZ, O., JANN, K. & WHEAT, R. (1968). Somatic and capsular antigens of Gram-negative bacteria. In *Comprehensive Biochemistry*. Eds. M. Florkin and E. H. Stotz, vol. 26 A, p. 105. Amsterdam: Elsevier.

LÜDERITZ, O., STAUB, A. M. & WESTPHAL, O. (1966). Immunochemistry of O and R antigens of *Salmonella* and related Enterobacteriaceae. *Bact. Rev.* **30**, 192.

MACFARLANE, M. G. (1964). In *Metabolism and Physiological Significance of Lipids*. Eds. R. M. C. Dawson and D. N. Rhodes. New York: John Wiley & Sons, Inc.

MARKOVITZ, A. & DORFMAN, A. (1962). Synthesis of capsular polysaccharide (hyaluronic acid) by protoplast membrane preparations of Group A *Streptococcus. J. biol. Chem.* **237**, 273.

MATALON, R. & DORFMAN, A. (1966). Hurler's syndrome: biosynthesis of acid mucopolysaccharides in tissue culture. *Proc. natn. Acad. Sci. U.S.A.* **56**, 1310.

MATSUHASHI, M., DIETRICH, C. P. & STROMINGER, J. L. (1965). Incorporation of glycine into the cell wall glycopeptide in *Staphylococcus aureus*: Role of sRNA and lipid intermediates. *Proc. natn. Acad. Sci. U.S.A.* **54**, 587.

MATZ, L. & STROMINGER, J. L. (1968). Identification of peptidoglycan synthesising particles as membrane fragments. *Bact. Proc.* p. 64.

MEADOW, P. M., ANDERSON, J. S. & STROMINGER, J. L. (1964). Enzymatic polymerization of UDP-N-acetylmuramyl-L-Ala-D-Glu-L-Lys-D-Ala-D-Ala and UDP-acetylglucosamine by a particulate enzyme from *Staph. aureus* and its inhibition by antibiotics. *Biochem. biophys. Res. Commun.* **14**, 382.

MILLS, G. T. & SMITH, EVELYN, E. B. (1962). Biosynthesis of pneumoccocal capsular polysaccharides. *Fedn Proc. Fedn Am. Socs exp. Biol.* **21**, 1089.

MOLNAR, J., ROBINSON, G. B. & WINZLER, R. J. (1965). Biosynthesis of glycoproteins. IV. The sub-cellular sites of incorporation of glucosamine-1-^{14}C into glycoprotein in rat liver. *J. biol. Chem.* **240**, 1882.

MUÑOZ, E., NACHBAR, M. S., SCHOR, M. T. & SALTON, M. R. J. (1968). Adenosinetriphosphatase of *Micrococcus lysodeikticus*: selective release and relationship to membrane structure. *Biochem. biophys. Res. Commun.* **32**, 539.

MURPHY, J. R. (1965). Erythrocyte metabolism. VI. Cell shape and the location of cholesterol in the erythrocyte. *J. Lab. clin. Med.* **65**, 756.

NEEDHAM, J. (1942). In *Biochemistry and Morphogenesis*. Cambridge University Press.

NERMUT, M. V. & MURRAY, R. G. E. (1967). Ultrastructure of the cell wall of *Bacillus polymyxa. J. Bact.* **93**, 1949.

NIKAIDO, M. (1962). Studies on the biosynthesis of cell wall polysaccharide in mutant strains of *Salmonella* I and II. *Proc. natn. Acad. Sci. U.S.A.* **48**, 1337; 1542.

NIKAIDO, H. & NIKAIDO, K. (1965). Biosynthesis of cell wall polysaccharide in mutant strains of *Salmonella*. IV. Synthesis of S-specific side-chain. *Biochem. biophys. Res. Commun.* **19**, 322.

NIYOMPORN BUNRUEANG, DAHL, J. L. & STROMINGER, J. L. (1968). Biosynthesis of the peptidoglycan of bacterial cell walls. IX. Purification and properties of the glycyl transfer ribonucleic acid synthetase from *Staphylococcus aureus. J. biol. Chem.* **243**, 773.

NORTHCOTE, D. H. (1969). Fine structure of cytoplasm in relation to synthesis and secretion in plant cells. *Proc. Roy. Soc. Lond.* B **173**, 21.

OP DEN KAMP, J. A. F., ITERSON, W. VAN & DEENEN, L. L. M. VAN (1967). Studies on the phospholipids and morphology of protoplasts of *Bacillus megaterium. Biochim. biophys. Acta* **135**, 862.

ORDIN, L. & HALL, M. A. (1967). Studies on cellulose synthesis by a cell-free oat coleoptile enzyme system: inactivation by airborne oxidants. *Pl. Physiol., Lancaster* **42**, 205.

OSBORN, M. J., ROSEN, S. M., ROTHFIELD, L. & HORECKER, B. L. (1962). Biosynthesis of bacterial lipopolysaccharide. I. Enzymatic incorporation of galactose in a mutant strain of *Salmonella. Proc. natn. Acad. Sci. U.S.A.* **48**, 1831.

OSBORN, M. J., ROSEN, S. M., ROTHFIELD, L., ZELEZNICK, L. D. & HORECKER, B. L. (1964). Lipopolysaccharide of the Gram-negative cell wall: biosynthesis of a complex heteropolysaccharide occurs by successive addition of specific sugar residues. *Science, N.Y.* **145**, 783.

PARK, J. T. (1964). In *Abstr. Sixth Int. Congr. Biochem.* New York, **32**, 463.

PIERINGER, R. A. (1968). The metabolism of glyceride glycolipids. I. Biosynthesis of monoglucosyl diglyceride and diglucosyl diglyceride by glucosyltransferase pathways in *Streptococcus faecalis*. *J. biol. Chem.* **243**, 4894.

RAMBOURG, A. & LEBLOND, C. P. (1967). Electron microscope observations on the carbohydrate-rich cell coat present at the surface of cells in the rat. *J. Cell Biol.* **32**, 27.

REVEL, J. P. & ITO, S. (1967). The surface components of cells. In *The Specificity of Cell Surfaces*. Eds. B. D. Davis and L. Warren. New Jersey: Prentice-Hall, Inc.

REYNOLDS, P. E. (1966). Antibiotics affecting cell-wall synthesis. In *Biochemical Studies of Antimicrobial Drugs*. Eds. B. A. Newton and P. E. Reynolds. Cambridge University Press.

REYNOLDS, P. E. (1968). Synthesis of cell-wall mucopeptide by particulate preparations from *Bacillus megaterium* and *Bacillus stearothermophilus*. *J. gen. Microbiol.* **53**, iv.

ROBBINS, P. W., WRIGHT, A. & BELLOWS, J. L. (1964). Enzymatic synthesis of the Salmonella O-antigen. *Proc. natn. Acad. Sci. U.S.A.* **52**, 13202.

ROBBINS, P. W., WRIGHT, A. & DANKERT, M. (1966). Polysaccharide biosynthesis. *J. gen. Physiol.* **49**, 331.

RODEN, L. & ARMAND, G. (1966). Structure of the chondroitin 4-sulphate-protein. *J. biol. Chem.* **241**, 65.

ROGERS, H. J. (1965). The outer layers of bacteria: the biosynthesis of structure. In *Function and Structure in Micro-Organisms*. Eds. M. R. Pollock and M. H. Richmond. Cambridge University Press.

ROGERS, H. J. & PERKINS, H. R. (1968). In *Cell Walls and Membranes*. London: E. and F. N. Spon Ltd.

ROTHFIELD, L. & HORECKER, B. L. (1964). The role of cell-wall lipid in the biosynthesis of bacterial lipopolysaccharide. *Proc. natn. Acad. Sci. U.S.A.* **52**, 939.

ROTHFIELD, L., OSBORN, M. J. & HORECKER, B. L. (1964). Biosynthesis of bacterial lipopolysaccharide. II. Incorporation of glucose and galactose catalysed by particulate and soluble enzymes in *Salmonella*. *J. biol. Chem.* **239**, 2788.

ROTHFIELD, L., TAKESHITA, M., PEARLMAN, M. & HORNE, R. W. (1966). Role of phospholipids in the enzymatic synthesis of the bacterial cell envelope. *Fedn Proc. Fedn Am. Socs exp. Biol.* **25**, 1495.

SALTON, M. R. J. & HORNE, R. W. (1951). Methods of preparation and some properties of cell walls. *Biochim. biophys. Acta* **7**, 177.

SALTON, M. R. J. (1964). In *The Bacterial Cell Wall*. Amsterdam: Elsevier.

SCHER, M. & LENNARZ, W. J. (1969). Studies on the biosynthesis of mannan in *Micrococcus lysodeikticus*. I. Characterization of mannan-^{14}C formed enzymatically from mannosyl-1-phosphoryl-undecaprenol. *J. biol. Chem.* **244**, 2777.

SCHER, M., LENNARZ, W. J. & SWEELEY, C. C. (1968). The biosynthesis of mannosyl-1-phosphoryl polyisoprenol in *Micrococcus lysodeikticus* and its role in mannan synthesis. *Proc. natn. Acad. Sci. U.S.A.* **59**, 1313.

SCHILLER, S. (1964). Synthesis of hyaluronic acid by a soluble enzyme system from mammalian tissue. *Biochem. biophys. Res. Commun.* **15**, 250.

SIEGEL, S. M. (1968). Biochemistry of the plant cell wall. In *Comprehensive Biochemistry*. Eds. M. Florkin and E. H. Stotz, vol. 26 A, p. 1. Amsterdam: Elsevier.

SIEWERT, G. & STROMINGER, J. L. (1967). Bacitracin: an inhibitor of the dephosphorylation of lipid pyrophosphate, an intermediate in biosynthesis of the peptidoglycan of bacterial cell walls. *Proc. natn. Acad. Sci. U.S.A.* **57**, 767.

SIEWERT, G. & STROMINGER, J. L. (1968). Biosynthesis of the peptidoglycan of bacterial cell walls. XI. Formation of the isoglutamine amide group in the cell walls of *Staphylococcus aureus*. *J. biol. Chem.* **243**, 783.

SMITH, EVELYN E. B. & MILLS, G. T. (1962). Experiments on the biosynthesis of Type I pneumococcal capsular polysaccharide. *Biochem. J.* 42 P.

SMITH, EVELYN E. B., MILLS, G. T., BERNHEIMER, HARRIET P. & AUSTRIAN, R. (1960). The synthesis of Type III pneumococcal capsular polysaccharide from uridine nucleotides by a cell-free extract of *Diplococcus pneumoniae* Type III. *J. biol. Chem.* **235**, 1876.

STANIER, R. Y. & VAN NIEL, C. B. (1962). The concept of a bacterium. *Arch. Mikrobiol.* **42**, 17.

STOOLMILLER, A. C. & DORFMAN, A. (1969a). The metabolism of glycosaminoglycans. In *Comprehensive Biochemistry*. Eds. M. Florkin and E. H. Stotz. **17**, 241. Amsterdam: Elsevier.

STOOLMILLER, A. C. & DORFMAN, A. (1969b). The biosynthesis of hyaluronic acid by *Streptococcus*. *J. biol. Chem.* **244**, 236.

STRUVE, W. G. & NEUHAUS, F. C. (1965). Evidence for an initial acceptor of UDP-N-acetyl-muramyl-pentapeptide in the synthesis of bacterial mucopeptide. *Biochem. biophys. Res. Commun.* **18**, 6.

STRUVE, W. G., SINHA, R. K. & NEUHAUS, F. C. (1966). On the initial stage in peptidoglycan synthesis. Phospho-N-acetylmuramyl-pentapeptide translocase (uridine monophosphate). *Biochemistry, N.Y.* **5**, 82.

VILLEMEZ, C. L., FRANZ, G. & HASSID, W. Z. (1967). Biosynthesis of alkali-insoluble polysaccharide from UDP-D-glucose with particulate enzyme preparations from *Phaseolus aureus*. *Pl. Physiol.* **42**, 1219.

VILLEMEZ, C. L., McNAB, J. M. & ALBERSHEIM, P. (1968). Formation of plant cell wall polysaccharides. *Nature, Lond.* **218**, 878.

WRIGHT, A., DANKERT, M., FENNESSEY, P. & ROBBINS, P. W. (1967). Characterization of a polyisoprenoid compound functional in o-antigen biosynthesis. *Proc. natn. Acad. Sci. U.S.A.* **57**, 1798.

WRIGHT, A., DANKERT, M. & ROBBINS, P. W. (1965). Evidence for an intermediate stage in the biosynthesis of the *Salmonella* o-Antigen. *Proc. natn. Acad. Sci. U.S.A.* **54**, 235.

YOUNG, F. E. (1967). Requirement of glucosylated teichoic acid for adsorption of phage in *Bacillus subtilis* 168. *Proc. natn. Acad. Sci. U.S.A.* **58**, 2377.

ZELEZNICK, L. D., ROSEN, S. M., SALTMARSH-ANDREW, M., OSBORN, M. J. & HORECKER, B. L. (1965). Biosynthesis of bacterial lipopolysaccharide. IV. Enzymatic incorporation of mannose, rhamnose and galactose in a mutant strain of *Salmonella typhimurium*. *Proc. natn. Acad. Sci. U.S.A.* **53**, 207.

PHOTOSYNTHETIC MECHANISMS IN PROKARYOTES AND EUKARYOTES

M. C. W. EVANS AND F. R. WHATLEY

Botany Department, King's College, University of London, 68 Half Moon Lane, London, S.E.24

The division of living organisms into prokaryotes and eukaryotes, on the basis of differences in those cell organelles not directly concerned with photosynthesis, is supported by a consideration of the structural organization of the photosynthetic apparatus when it is present (Echlin, this Symposium). In the prokaryotic blue-green algae and in the photosynthetic bacteria it takes the form of a relatively simple lamellar system (which can be isolated as chromatophores from the bacteria), whereas in the eukaryotic algae and higher plants the photosynthetic apparatus is confined to specialized organelles, the chloroplasts, in which a more complex lamellar system is developed in close association with the enzyme systems responsible for carbon-dioxide fixation.

The overall reaction of photosynthesis is basically the same in all groups, an oxidation-reduction sequence in which carbon dioxide is reduced to the level of carbohydrate at the expense of a variety of hydrogen donors 'activated' by light reactions. It is well represented by the equation proposed by van Niel (1941):

$$2H_2A + CO_2 \rightarrow (CH_2O) + H_2O + 2A.$$

The nature of the compound H_2A permits a division of photosynthetic organisms into two groups:

(*a*) Organisms which are obligate anaerobes when grown in the light with carbon dioxide, and in which H_2A must be supplied as a substrate more reducing than water, such as hydrogen sulphide, thiosulphate or some organic compounds.

(*b*) Organisms which can grow aerobically in the light and in which H_2A can be water, with the consequence that oxygen is discarded as a waste product.

Two photochemical reactions are involved in the photosynthetic electron transport system of oxygen-evolving organisms (see Fig. 2). Photosystem I is very similar to the photochemical system in bacteria and may be the primitive system. Photosystem II specifically involves oxygen evolution. Although the clear advantage to photosynthetic organisms of being able to use water as a hydrogen donor is fundamental to the spread of living organisms in the world, this advance, involving the addition of a second photochemical reaction to the one in primitive organisms, did not take place in step with the advances in structural organization on passing from the prokaryotic to the eukaryotic level. The blue-green algae (Cyanophyceae) are prokaryotic oxygen-evolving organisms, occupying the middle ground between the obligately anaerobic photosynthetic bacteria and the eukaryotic oxygen-evolving plants.

It is the purpose of this article to examine particularly the position of the blue-green algae from a biochemical point of view and to draw attention to the problems of maintaining the view that the distinction between prokaryotes and eukaryotes is biochemically significant. These problems encompass not only oxygen evolution but also pigment systems, carbon-fixation systems and nitrogen fixation.

PIGMENT SYSTEMS

The mechanism by which light energy is converted to chemical energy appears to be fundamentally the same in all photosynthetic organisms. Light is absorbed by a magnesium porphyrin pigment, a chlorophyll, with the resultant oxidation of a specialized 'reaction-centre' molecule and the production of a low-potential reductant. Although all photosynthetic organisms contain chlorophylls there are some differences in the particular chlorophyll molecules found in different groups of organisms (Fig. 1). Chlorophyll a is present in all algae and higher plants. In addition to chlorophyll a (the 'basic' photosynthetic pigment) green algae and higher plants contain chlorophyll b (the 'accessory' photosynthetic pigment). In blue-green and red algae the chlorophyll b is replaced by a related phycobilin, principally phycocyanin in blue-green algae and principally phycoerythrin in red algae. In brown algae chlorophyll b is replaced by fucoxanthin. The accessory pigment in all these organisms is closely associated with the oxygen-evolving reactions of Photosystem II. Light energy absorbed by other pigments may also be transferred to the reaction centres.

There is a very large amount of chlorophyll in relation to the catalytic molecules of the electron transport chain, e.g. about 400 moles chlorophyll per mole cytochrome f in higher plants (Boardman, 1968). It is thought that most of the chlorophylls function solely in a light-harvesting system, transferring energy to chlorophyll a in a specialized environment, the 'reaction centre'. The reaction centre has been identified in Photosystem I of higher plants with an electron-transport component

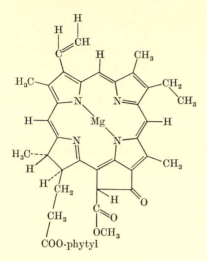

Chlorophyll a

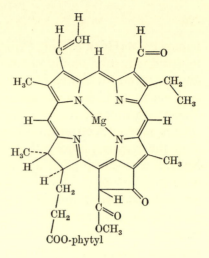

Chlorophyll b

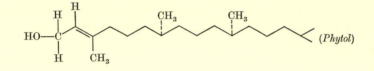

(Phytol)

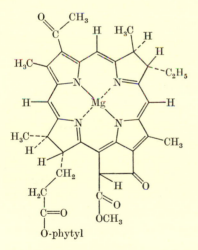

Bacteriochlorophyll a

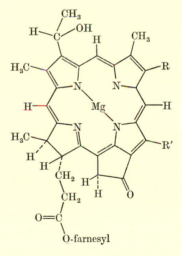

Chlorobium chlorophyll

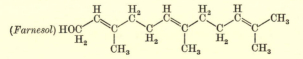

Fig. 1. Chlorophylls in prokaryotes and eukaryotes.

named P700 (Hoch & Kok, 1961). This component undergoes an oxidation and reduction on illumination that can be observed spectroscopically at 705 mμ. It can be identified kinetically with an electron transfer component producing an electron paramagnetic resonance signal in illuminated chloroplasts and appears to be the initial reactant in the photosynthetic electron transfer chain (Weaver, 1968).

In the purple photosynthetic bacteria the photosynthetic pigment equivalent to chlorophyll a is bacteriochlorophyll a (Jensen, Aasmundrud & Eimhjellen, 1964), which has absorption maxima between 800 and 1000 mμ in different species of bacteria *in vivo*, but is a single pigment when isolated. The reaction-centre chlorophyll analogous to P700 in Photosystem I of higher plants has been identified in *Rhodospirillum rubrum* as P890 (Clayton, 1965). The green bacteria contain small amounts of bacteriochlorophyll a (Hind & Olson, 1968), which may form the reaction centre, while the bulk light-harvesting chlorophyll is *Chlorobium* chlorophyll (bacteriochlorophylls c and d). The bacteria do not have accessory pigments associated with an oxygen-evolving apparatus as do algae and green plants, although light absorbed by other pigments, particularly carotenoids, may be used for photosynthesis. The variety of pigments which are used to capture light energy has made possible the growth of photosynthetic organisms at all wavelengths from about 400 to 1000 mμ, while restricting particular species to different ecological niches depending on light quality.

PHOTOSYNTHETIC ELECTRON TRANSPORT SYSTEMS

Green plants

Chloroplasts isolated from higher plants and green algae catalyse two reactions yielding ATP: (*a*) non-cyclic photophosphorylation, the synthesis of ATP from ADP and inorganic orthophosphate coupled to the light-driven flow of electrons from water to NADP; (*b*) cyclic photophosphorylation, the synthesis of ATP coupled to a cryptic electron flow driven by light (Whatley & Arnon, 1963). It is widely accepted that non-cyclic electron transport from water to NADP involves two light reactions operating in series (Fig. 2), electrons being transferred between the two photosystems by an intermediary electron transport chain containing cytochromes, quinones and the copper protein plastocyanin. The phosphorylation is coupled to oxidative electron transfer through the cytochrome chain in a manner analogous to that in oxidative phosphorylation. The P/2e ratio was for many years thought to be one; however, the use of chloroplasts prepared in different ways has given values greater than one, suggesting that there may be two phosphorylation sites in the non-cyclic electron transport chain (Izawa & Good, 1968). Three components of the inter-mediary electron transport chain have been isolated from chloroplasts and characterized, namely cytochrome *f*, plastoquinone and plasto-cyanin (all of which are found only in the photosynthetic tissues of green plants); the two '*b*-type' cytochromes which can also be identified

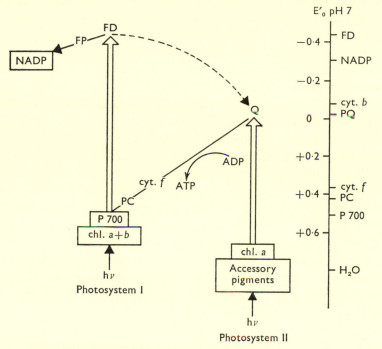

Fig. 2. The photosynthetic electron transport chain in green plants.
PC = Plastocyanin, PQ = Plastoquinone, FD = Ferredoxin.

spectroscopically have not yet been isolated (Bendall & Hill, 1968). The order in which these components are placed in Fig. 2 is based partly on the standard redox potential of the components and partly on spectro-scopic kinetic data. The only part of the electron transport chain for which there is complete biochemical evidence is in the terminal transfer of electrons from Photosystem I to ferredoxin and the subsequent reduction of NADP by ferredoxin-NADP reductase (Shin & Arnon, 1965). The ferredoxins isolated from higher plants and green algae are very similar; they have low redox potentials near that of the hydrogen electrode, contain two iron and two labile sulphur atoms per molecule, have molecular weights about 12,000 and are one-electron carriers.

Cyclic photophosphorylation is thought to be coupled to a cyclic electron flow catalysed by Photosystem I alone, and proceeding via ferredoxin and the cytochrome chain. The series formulation of the photosynthetic electron transport chain, involving two light reactions in plant photosynthesis, is supported by observation of the 'red drop' decrease in efficiency of photosynthesis at long wavelengths, and the Emerson enhancement effect (Emerson, Chalmers & Cedarstrand, 1957). It also provides an explanation for the ability of chloroplasts to catalyse partial reactions of photosynthesis, such as the reduction of electron acceptors with redox potentials near O mV and for the reduction of NADP by artificial donors with redox potentials near O mV (Losada, Whatley & Arnon, 1961).

The series formulation is not the only one which has been proposed. Arnon, Tsujimoto & McSwain (1965) suggested that non-cyclic electron flow involves a single light-reaction in which an electron is transferred directly from water to ferredoxin, and that the cytochromes and other intermediate carriers are involved only in a cyclic phosphorylation that is driven by a separate photoreaction. This scheme is not compatible with the evidence previously put forward for two light reactions. Arnold & Azzi (1968) subsequently proposed a mechanism with two photoreactions in which electrons are transferred from water to ferredoxin without an intermediate electron transport chain; this involves two chlorophyll molecules, one oxidizing water and the other reducing ferredoxin, the resulting oxidized and reduced chlorophyll molecules then interacting. The overall result is the transfer of an electron across a membrane from water to ferredoxin. The cytochrome chain is again thought to be involved only in cyclic phosphorylation.

Good rates of cyclic phosphorylation in isolated chloroplasts usually depend on the addition of artificial cofactors such as phenazonium methosulphate or of ferredoxin at ten times the concentration required to saturate non-cyclic electron flow; significant rates of endogenous photophosphorylation have, however, been obtained with rapidly-isolated chloroplasts (Nobel, 1967). The results of Kandler's experiments (1960) on glucose assimilation by *Chlorella*, as well as other pertinent evidence, indicate that cyclic phosphorylation does occur *in vivo* and it seems likely that a necessary component is usually lost during isolation of the chloroplasts.

As well as reduction of NADP, chloroplasts also catalyse the reduction of nitrate and nitrite to ammonia (Losada, Paneque, Ramirez & Del Campo, 1965). Nitrate reduction involves an NAD(P)-dependent flavoprotein, while nitrite reductase utilizes ferredoxin as electron donor.

The detailed mechanism by which ATP is synthesized by chloroplasts is not clear. Coupling factors and ATPases similar to those from mitochondria have been isolated and shown to be involved in the final stages of ATP synthesis (Avron & Neumann, 1968; Livne & Racker, 1968). The mechanism of the early stages of energy coupling is not understood. Chloroplast experiments have provided evidence to support the chemiosmotic hypothesis (Mitchell, this Symposium); for example, they establish large pH gradients on illumination and will catalyse the synthesis of ATP coupled to the breakdown of artificial pH gradients (Jagendorf, 1967). It is not however clear whether the pH gradient is a primary or secondary result of the energy trapping process.

The available evidence suggests that the energy trapping system and electron transport chain are very similar in all green plants and algae.

Blue-green algae

The overall photosynthetic reactions of the blue-green algae are essentially the same as those of higher plants. The preparation of photosynthetically active cell-free preparations of blue-green algae has proved relatively difficult. Various authors have, however, shown convincingly the occurrence of cyclic and non-cyclic photophosphorylation and the photoreduction of NADP with water as electron donor (Petrack & Lipmann, 1961; Biggins, 1967). The components of the photosynthetic electron transport chain which have been isolated from blue-green algae are very similar to those of green plants, apart from the accessory pigment phycocyanin. Three 'c-type' cytochromes have been isolated from *Anacystis nidulans* and one of these, cytochrome c_{553}, functions in place of the characteristically high molecular weight cytochrome f of higher plants (Holton & Myers, 1967). Plastocyanin is also present and functions in the non-cyclic electron transport chain (Lightbody & Krogmann, 1967). The ferredoxins in all blue-green algae investigated are of the same type as in higher plants; they have two iron and two labile sulphur atoms per molecule, are red in colour and are one-electron carriers (Evans, Hall, Bothe & Whatley, 1968). The presence of plastocyanin and plant-type ferredoxin thus seems to be characteristic of photosynthetic organisms which evolve oxygen.

When blue-green algae are grown in iron-deficient media the ferredoxin content decreases, and the synthesis of a flavoprotein (phytoflavin), which can replace ferredoxin in chloroplast reactions, is induced (Smillie, 1965). The physiological replacement of ferredoxin by another flavoprotein (flavodoxin) was first observed in *Clostridium pasteurianum* (Knight & Hardy, 1967); such a replacement has not been reported for green plants or algae.

Photosynthetic bacteria

Although bacterial photosynthesis can be represented by the overall van Niel equation for photosynthesis, it differs considerably from the system in organisms which evolve oxygen. The photosynthetic bacteria are strict anaerobes when growing photosynthetically; they use a wide variety of electron donors but not water. The inability to use water as electron donor results from the absence of Photosystem II (Fig. 2) and its related electron transport system.

The green bacteria (Chlorobacteriaceae), except for *Chloropseudomonas ethylicum*, are strict autotrophs, which use reduced sulphur compounds as electron donors. The purple sulphur bacteria (Thiorhodaceae) utilize either organic substrates, or grow autotrophically by using reduced sulphur compounds or molecular hydrogen as electron donor, but are unable to grow aerobically in the dark. The purple non-sulphur bacteria (Athiorhodaceae) cannot, in the main, use reduced sulphur compounds but grow either with organic substrates, or autotrophically with hydrogen as electron donor; they are facultatively photosynthetic and can grow aerobically in the dark, under which conditions the photosynthetic apparatus is repressed.

In cell-free preparations of purple bacteria the dominant photosynthetic reaction is cyclic photophosphorylation (Frenkel, 1958), which occurs with high rate in the absence of those electron carriers that must be added to plant and algal preparations to obtain high rates of ATP synthesis. However, some carriers, e.g. phenazonium methosulphate, can stimulate the rate of ATP synthesis by cell-free preparations of bacteria in some circumstances. NAD photoreduction has been shown in extracts of purple non-sulphur bacteria (Frenkel, 1958; Nozaki, Tagawa & Arnon, 1961) but this appears to be a reversed electron flow (of the type which can be shown in mitochondria), driven by high-energy intermediates generated during cyclic electron flow (Keister & Yike, 1967) and not a true non-cyclic electron flow.

Particles prepared from green photosynthetic bacteria show only low rates of cyclic photophosphorylation (Hughes, Conti & Fuller, 1963); in fact it is extremely difficult to obtain any phosphorylating activity at all. Particles from green bacteria are, however, able to catalyse the photoreduction of ferredoxin, although the rates of reduction are low. The reduced ferredoxin is utilized for carbon-dioxide fixation, either *indirectly via* a reduction of NAD catalysed by a soluble ferredoxin-NAD reductase (Evans, 1968), or *directly* in the fixation of carbon dioxide into keto acids, reactions catalysed by pyruvate synthase or α-ketoglutarate synthase (Evans, Buchanan & Arnon, 1966). It has not so far been possible to demonstrate phosphorylation coupled with the NAD reduction.

Photosynthetic bacteria do not have an accessory pigment system and do not show enhancement effects. It has therefore been generally considered that only one photoreaction occurs in photosynthetic bacteria. Spectroscopic studies of oxidation and reduction of cytochromes in green and in purple bacteria do suggest that two light reactions are involved in bacterial photosynthesis (Hind & Olson, 1968); they appear, however, not to operate in series as in green plants, but as separate

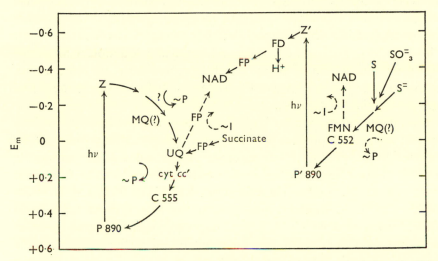

Fig. 3. Cyclic and non-cyclic electron transport systems in photosynthetic bacteria (Hind & Olson, 1968). UQ = Ubiquinone FP = Flavoprotein, FD = Ferredoxin.

systems for cyclic and non-cyclic electron flow (Fig. 3). There is evidence that ubiquinone may be the initial electron acceptor in the cyclic system (Vernon, 1968), whereas ferredoxin is presumably the acceptor for the non-cyclic system.

The components of the electron transport chain which have been isolated from photosynthetic bacteria are similar to those found in green plants, namely, cytochromes, ferredoxins and quinones. The purple bacteria contain a high-potential cytochrome c_2 (Bartsch, 1968) equivalent in potential to the cytochome f of higher plants. *Chromatium* has a second 'c type' cytochrome with three haem groups and one flavin mononucleotide (FMN) per molecule. The purple bacteria also contain cytochrome cc' with two haem groups per molecule. *Chlorobium thiosulfatophilum* contains three 'c type' cytochromes none of which has a high redox potential equivalent to cytochrome *f*. Cytochromes of the '*b* type' have been conclusively identified only in the non-sulphur purple bacteria and may perhaps be associated with the respiratory system rather than the photosynthetic electron transport chain. The ferredoxins of the photosynthetic bacteria differ from those of green plants and blue-green algae and are of the characteristic bacterial type; they are blackish-brown in colour with absorption maxima at 390 mμ, and contain 4 to 6 iron atoms per molecule (Bachoffen & Arnon, 1966). *Chromatium* ferredoxin is a two-electron carrier (Evans *et al.* 1968) and it is likely that ferredoxins from green bacteria also transfer two electrons per molecule. Ferredoxins from the green bacteria are very similar to those from *Clostridium* spp. with molecular weights of about 6000 and similar amino acid composition (Buchanan, Matsubara & Evans, 1969). On the other hand *Chromatium* ferredoxin has a molecular weight of 10,000, nearer that of plant ferredoxins, and contains more amino acids than the normal bacterial ferredoxins. It has a surprisingly low redox potential (-490 mV).

Chromatium also contains large amounts of a second non-haem iron protein with a high positive potential of $+350$ mV. This protein can be oxidized by the photosynthetic electron transport chain, but no specific function has yet been ascribed to it. Similar proteins have been isolated from non-sulphur purple bacteria (Dus, De Klerk, Sletten & Bartsch, 1967). The green bacteria also contain the iron protein rubredoxin (B. B. Buchanan & M. C. W. Evans, unpublished data) but no function is known for this protein. The quinones found in the photosynthetic systems of purple bacteria are ubiquinones, while the green bacteria contain menaquinone and *Chlorobium* quinone (Powls & Redfearn, 1969), the latter apparently being associated with sulphide oxidation. The only component isolated from the plant system for which there is no direct equivalent in the bacterial system is the copper protein plastocyanin.

DARK PHOTOSYNTHETIC REACTIONS

Carbon-dioxide fixation

The reductive pentose phosphate pathway for the fixation of carbon dioxide, in which a single primary carboxylation reaction, the carboxylation of ribulose diphosphate to give two molecules of phosphoglyceric acid, is involved, was originally proposed to describe the mechanism of carbon-dioxide fixation in the green alga *Chlorella* (Bassham & Calvin, 1962). Evidence for the operation of the cycle has been obtained by:

(1) kinetic analysis of $^{14}CO_2$ fixation experiments, with particular attention to the identification of 3-phosphoglyceric acid as the initial product of carbon-dioxide fixation; (2) demonstration that all the enzymes required for the cycle are present, particularly ribulose diphosphate carboxylase and ribulose-5-phosphokinase, which are the only enzymes not also involved in other metabolic pathways. Analysis of a wide variety of autotrophic prokaryotes and eukaryotes, both photosynthetic and chemosynthetic, indicates that the basic reductive pentose cycle is present in all autotrophs. No autotrophic organism has been described which does not contain ribulose diphosphate carboxylase. The importance of the cycle is also indicated by the observation that in facultatively autotrophic bacteria the synthesis of ribulose diphosphate carboxylase is induced under autotrophic conditions but completely repressed in the presence of organic substrates (Lascelles, 1961).

The range of primary products of carbon-dioxide fixation observed in short exposure experiments with photosynthetic bacteria may be extremely variable, especially as a result of variation in the ammonia content of the media used. In many experiments the primary products of carbon-dioxide fixation have been found to be amino acids, particularly alanine, aspartic acid and glutamate, but phosphoglyceric acid and glycollate have also been found. The three amino acids found are all closely related to keto acids involved in the oxidative tricarboxylic acid cycle. In the photosynthetic bacteria a reductive carboxylic acid cycle (Fig. 4) has been shown to operate (Evans et al. 1966). This is essentially a reversal of the oxidative tricarboxylic acid cycle in which the lipoic acid-dependent pyruvate oxidase and α-ketoglutarate oxidase systems are replaced by the ferredoxin-dependent enzymes pyruvate synthase (I) and α-ketoglutarate synthase (II), which catalyse the reductive carboxylation of acetyl- and succinyl-coenzyme A respectively.

I. Acetyl $CoA + CO_2 + FD_{red} \rightarrow$ Pyruvate $+ CoA + FD_{ox}$

II. Succinyl $CoA + CO_2 + FD_{red} \rightarrow$ α-ketoglutarate $+ CoA + FD_{ox}$

The overall cycle involves the fixation of four molecules of carbon dioxide. If oxalacetic acid is taken as the starting point of the cycle, one cycle results in the regeneration of the oxalacetate and the synthesis of a second molecule of oxalacetate. All the enzymes of the cycle have been shown in *Chlorobium thiosulfatophilum* (Evans et al. 1966) and in *Rhodospirillum rubrum* (Buchanan, Evans & Arnon, 1967); short-time exposure to carbon dioxide and degradation of the initial products of fixation also suggest that the cycle operates in these bacteria. The relative importance of the reductive carboxylic acid cycle and the pentose phosphate cycle during autotrophic growth is difficult to assess. The activities of the enzymes of both pathways have been reported for *Chlorobium thiosulfatophilum* (Smillie, Rigopoulos & Kelly, 1962; Evans *et al*. 1966). On this evidence neither cycle accounts for the rate of carbon-dioxide fixation by whole bacteria. However, the activity of both cycles together would be sufficient. Ribulose diphosphate carboxylase is present in purple bacteria only during autotrophic growth, while the carboxylation enzymes of the reductive carboxylic acid cycle are

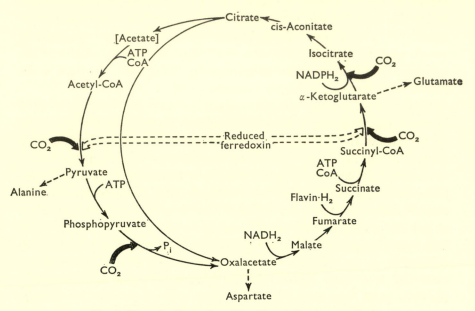

Fig. 4. The reductive carboxylic acid cycle in photosynthetic bacteria (Evans *et al*. 1966).

present under both autotrophic and heterotrophic conditions. Citrate lyase which is also required for the operation of the reductive cycle was, however, found only in extracts of autotrophically grown *Rhodospirillum rubrum*. No evidence has been obtained for the presence of α-ketoglutarate synthase in *Chromatium* and this organism also lacks malate dehydrogenase (Fuller, Smillie, Sisler & Kornberg, 1961). It may be significant that *Chromatium* is the only organism in which the enzymes of the pentose phosphate cycle are sufficiently active to account for the overall rate of carbon-dioxide fixation. It seems likely that in those organisms in which both cycles operate the relative activity of each cycle will depend on the physiological state of the organism. There is no evidence at present for the operation of the reductive carboxylic acid cycle in any oxygen-producing organism, whether prokaryote or eukaryote.

The primary product of carbon-dioxide fixation may vary with experimental conditions and with different organisms. For example, in green algae and higher plants at low carbon-dioxide concentrations, glycollic acid is the first detectable product of carbon-dioxide fixation; glycollate has also been reported from *Rhodospirillum rubrum*. However, no definitive mechanism for glycollate synthesis has been proposed, though the experiments of Vandor & Tolbert (1968) indicate that in isolated chloroplasts glycollate arises from fructose diphosphate. In certain tropical grasses and some species of the family Amaranthaceae the initial product of carbon-dioxide fixation is oxalacetate formed by the carboxylation of phosphoenolpyruvate (Hatch & Slack, 1966). In these plants this reaction appears to be the major pathway for carbon-dioxide fixation. It has been proposed that the C-4 of oxalacetate formed from phosphoenolpyruvate is transferred to an acceptor to form phosphoglyceric acid, the remaining 3-carbon skeleton being converted back to phosphoenolpyruvate.

The need for modification of the proposed mechanisms of the reductive pentose cycle is also suggested by the observation that in cell extracts the activities of one or more of the required enzymes are much too low to account for the overall carbon-dioxide fixing activity of the whole organism. In a survey of a wide range of organisms (Latzko & Gibbs, 1969) only a photosynthetic bacterium, *Chromatium* sp., contained all the necessary enzymes in amounts sufficient to account for the overall rate of carbon-dioxide fixation. The enzymes most frequently found to be seriously deficient are fructose-1,6-diphosphatase and ribulose-1,5-diphosphate carboxylase.

Nitrogen fixation and hydrogen evolution

The photosynthetic bacteria and the blue-green algae are both able to utilize nitrogen gas for growth. The fixation of nitrogen by blue-green algae is of considerable economic importance in tropical agriculture. Little is known, however, about the biochemistry of nitrogen fixation by these organisms. Cell-free nitrogen-fixing systems have been obtained from purple bacteria and shown to require for nitrogen fixation a low-potential reductant (which can be ferredoxin), together with ATP (Bulen, Burns & Leconte, 1965). These requirements are the same as those found for the better known systems from *Clostridium* and *Azotobacter*.

Fay, Stewart, Walsby & Fogg (1968) considered that nitrogen fixation by blue-green algae is associated with the formation of specialized cells, the heterocysts, which are apparently free from the accessory pigment phycocyanin which normally accompanies chlorophyll *a* in the non-specialized cells. It has been suggested that the photosynthetic apparatus of the heterocysts contains only Photosystem I and operates a bacterial type of photosynthesis, utilizing carbon compounds supplied by normal cells to donate electrons to nitrogen. The absence of an oxygen-evolving system is supposed to provide a suitable anaerobic environment for the oxygen-sensitive nitrogen-fixing enzymes. The evidence for these proposals is inconclusive at present.

Cell-free nitrogen-fixing systems have been prepared from blue-green algae (Cox & Fay, 1967). Unlike the bacterial system in which the enzymes are obtained in solution, the algal system appears to be associated with the photosynthetic lamellae. Nitrogen fixation was found to depend either on light or on the utilization of pyruvate as electron donor. The detailed mechanism of pyruvate oxidation is not known. The oxidation is presumably coupled to a low-potential electron carrier, perhaps ferredoxin, like the anaerobic bacterial system.

Under conditions of nitrogen deficiency the purple bacteria evolve hydrogen and carbon dioxide from organic substrates. The breakdown of the substrate proceeds through an anaerobic tricarboxylic acid cycle, (Ormerod, Ormerod & Gest, 1961). The function of this hydrogen evolution is not known. It may represent a malfunction of the nitrogen-fixing system, since it is prevented by the addition of molecular nitrogen.

HETEROTROPHY AND AUTOTROPHY

All photosynthetic organisms have the ability to grow as autotrophs with carbon dioxide as major carbon source. However, the great majority are not strict autotrophs but are able to assimilate organic substrates, either photosynthetically or by adapting to an aerobic oxidative metabolism when growth can occur in the dark, as in the non-sulphur purple bacteria and green algae.

Two groups of photosynthetic prokaryotes, the green sulphur bacteria of the genus *Chlorobium* and the blue-green algae, are however strict autotrophs, unable to grow without carbon dioxide as major carbon source. Both groups can assimilate simple organic molecules such as acetate into cell material, but most of the carbon must be supplied as carbon dioxide. The reason for this strict dependence on carbon dioxide as carbon source is not clear. It has been suggested that the blue-green algae are unable to grow heterotrophically because they lack an $NADH_2$ or $NADPH_2$ oxidase and have an incomplete tricarboxylic acid cycle (they apparently lack α-ketoglutarate oxidase; Smith, London & Stanier, 1967). Several workers have, however, demonstrated the oxidation of $NADH_2$ and $NADPH_2$ by extracts of blue-green algae, and such extracts also effect oxidative phosphorylation (Leach & Carr, 1969). The absence of α-ketoglutarate oxidase, although preventing dark aerobic growth, should not prevent photoheterotrophic growth; *Chromatium* for example lacks this enzyme. Growth on acetate as sole carbon source is known to occur only through the glyoxylate cycle, which is not present in blue-green algae. Pyruvate synthase is also absent from blue-green algae and they are therefore unable to carboxylate acetate to pyruvate. Acetate is in fact assimilated by blue-green algae through the citrate synthase reaction and is incorporated only into the amino acids of the glutamate family (Hoare, Hoare & Moore, 1967).

Species of the genus *Chlorobium* also lack the glyoxylate cycle but assimilate acetate through pyruvate synthase; they are however unable to utilize acetate as electron donor for the photoreduction of ferredoxin required for the pyruvate synthase reaction. The green bacteria appear to be unable to assimilate more complex organic molecules, presumably because these are not transported into the cell. The reasons for the strict autotrophy of the blue-green algae and of *Chlorobium* remain unclear; it seems likely that the main reasons are the absence of the enzyme systems necessary for growth on 2-carbon substrates and the impermeability of the cells to the majority of organic compounds.

CONCLUSION

The evidence for the photosynthetic activities which we have discussed confirms the proposition that the blue-green algae occupy a middle ground, resembling the photosynthetic bacteria in many ways and the green algae and higher plants in others. Although structurally the blue-green algae are closely allied with the bacteria they are clearly 'higher plants' in being able to use water as hydrogen donor for carbon-dioxide fixation. Morever, the components of their electron transport system are more closely allied to those of the green algae and higher plants than to the photosynthetic bacteria. On the other hand, the ability of many blue-green algae to fix atmospheric nitrogen relates them closely to the photosynthetic bacteria and differentiates them sharply from the eukaryotes, none of which have yet been shown to fix nitrogen. The ability to evolve or utilize hydrogen is found among blue-green algae and bacteria as well as in the green algae. It is absent from higher green plants, which appear to lack the enzyme hydrogenase. The particular pathways of carbon-dioxide fixation are variable throughout the groups. The reductive pentose cycle appears to be widely distributed (being present also in chemoautotrophs) and is the main pathway in *Chromatium* and higher plants, but it is accompanied by the reductive carboxylic acid cycle in some photosynthetic bacteria and by the phosphoenol-pyruvate carboxylase pathway in some higher plants (and perhaps also in *Chromatium*). It is reasonable to conclude that details of the pathways of carbon-dioxide fixation may prove to be of significance when the taxonomic classification of photosynthetic organisms is being considered.

REFERENCES

ARNOLD, W. & AZZI, J. R. (1968). Chlorophyll energy levels and electron flow in photosynthesis. *Proc. natn. Acad. Sci. U.S.A.* **61**, 29.

ARNON, D. I., TSUJIMOTO, H. & McSWAIN, B. D. (1965). Photosynthetic phosphorylation and electron transport. *Nature, Lond.* **207**, 1367.

AVRON, M. & NEUMANN, J. (1968). Photophosphorylation in Chloroplasts. *A. Rev. Pl. Physiol.* **19**, 137.

BACHOFFEN, R. & ARNON, D. I. (1966). Crystalline ferredoxin from the photosynthetic bacteria *Chromatium*. *Biochim. biophys. Acta* **120**, 259.

BASSHAM, J. A. & CALVIN, M. (1962). *Path of Carbon in Photosynthesis*. New York: Benjamin Press.

BENDALL, D. S. & HILL, R. (1968). Haem-proteins in photosynthesis. *A. Rev. Pl. Physiol.* **19**, 167.

BIGGINS, J. (1967). Photosynthetic reactions by lysed protoplasts and particle preparation from the blue green alga *Phormidium luridum*. *Pl. Physiol.* **42**, 1447.

BARTSCH, R. G. (1968). Bacterial Cytochromes. *A. Rev. Microbiol.* **22**, 181.

BOARDMAN, N. K. (1968). The photochemical systems of photosynthesis. *Adv. Enzymol.* **30**, 700.

BUCHANAN, B. B., EVANS, M. C. W. & ARNON, D. I. (1967). Ferredoxin-dependent carbon assimilation in *Rhodospirillum rubrum*. *Arch. Mikrobiol.* **59**, 32.

BUCHANAN, B. B., MATSUBARA, H. & EVANS, M. C. W. (1969). Ferredoxin from the photosynthetic bacterium *Chlorobium thiosulphatophilum*: a link to ferredoxin from non-photosynthetic bacteria. *Biochim. biophys. Acta* **189**, 46.

BULEN, W. A., BURNS, R. C. & LECONTE, R. J. (1965). Nitrogen fixation: hydrosulfite as electron donor with cell free preparation of *Azotobacter vinelandii* and *Rhodospirillum rubrum*. *Proc. natn. Acad. Sci. U.S.A.* **53**, 532.

CLAYTON, R. K. (1965). The biophysical problems of photosynthesis. *Science, N.Y.* **149**, 1346.

COX, R. M. & FAY, P. (1967). Nitrogen fixation and pyruvate metabolism in cell free preparation of *Anabaena cylindrica*. *Arch. Mikrobiol.* **58**, 357.

DUS, K., DE KLERK, M., SLETTEN, K. & BARTSCH, R. G. (1967). Chemical characterization of high potential iron proteins fror Chromatium and *Rhodopseudomonas gelatinosa*. *Biochim. biophys. Acta* **140**, 291.

EMERSON, R., CHALMERS, R. V. & CEDARSTRAND, C. (1957). Some factors influencing the long wave limit of photosynthesis. *Proc. natn. Acad. Sci. U.S.A.* **43**, 133.

EVANS, M. C. W. (1968). Ferredoxin–NAD reductase and the photoreduction of NAD by *Chlorobium thiosulfatophilum*. *Abstracts of International Congress of Photosynthesis Research*, p. 192.

EVANS, M. C. W., BUCHANAN, B. B. & ARNON, D. I. (1966). A new ferredoxin dependent carbon reduction cycle in a photosynthetic bacterium. *Proc. natn. Acad. Sci. U.S.A.* **55**, 928.

EVANS, M. C. W., HALL, D. O., BOTHE, H. & WHATLEY, F. R. (1968). The stoichiometry of electron transfer by bacterial and plant ferredoxins. *Biochem. J.* **110**, 485.

FAY, P., STEWART, W. D. P., WALSBY, A. E. & FOGG, G. E. (1968). Is the heterocyst the site of nitrogen fixation in blue green algae. *Nature, Lond.* **220**, 810.

FRENKEL, A. W. (1958). Light-induced reactions of chromatophores of *Rhodospirillum rubrum*. The photochemical apparatus, its structure and function. *Brookhaven Symp. Biol.* No. 11, p. 276.

FULLER, R. G., SMILLIE, R. M., SISLER, E. C. & KORNBERG, H. L. (1961). Carbon metabolism in *Chromatium*. *J. biol. Chem.* **236**, 2140.

HATCH, M. D. & SLACK, C. R. (1966). Photosynthesis by sugar cane leaves. A new carboxylation reaction and the pathway of sugar formation. *Biochem. J.* **101**, 103.

HIND, G. & OLSON, J. M. (1968). Electron transport pathways in photosynthesis. *A. Rev. Pl. Physiol.* **19**, 249.

HOARE, D. S., HOARE, S. L. & MOORE, R. B. (1967). The photoassimilation of organic compounds by autotrophic blue-green algae. *J. gen. Microbiol.* **49**, 351.

HOCH, G. & KOK, B. (1961). Photosynthesis. *A. Rev. Pl. Physiol.* **12**, 155.

HOLTON, R. W. & MYERS, J. (1967). Water soluble proteins from a blue-green alga. *Biochim. biophys. Acta* **131**, 362.

HUGHES, D. E., CONTI, S. F. & FULLER, R. C. (1963). Inorganic polyphosphate metabolism in *Chlorobium thiosulfatophilum*. *J. Bact.* **85**, 577.

IZAWA, S. & GOOD, N. E. (1968). The stoichiometric relation of phosphorylation to electron transport in isolated chloroplasts. *Biochim. biophys. Acta* **162**, 380.

JAGENDORF, A. T. (1967). Acid base transitions and phosphorylation by chloroplasts. *Fedn Proc. Fedn Am. Socs exp. Biol.* **26**, 1361.

JENSON, A., AASMUNDRUD, O. & EIMHJELLEN, K. E. (1964). Chlorophylls of photosynthetic bacteria. *Biochim. biophys. Acta* **88**, 466.

KANDLER, O. (1960). Energy transfer through phosphorylation mechanisms in photosynthesis. *A. Rev. Pl. Physiol.* **11**, 37.

KEISTER, D. I. & YIKE, N. J. (1967). Energy linked reactions in photosynthetic bacteria. *Archs Biochem. Biophys.* **121**, 415.

KNIGHT, E. & HARDY, R. W. F. (1967). Flavodoxin—chemical and biological properties. *J. biol. Chem.* **42**, 1370.

LASCELLES, J. (1961). The formation of ribulose 1:5-diphosphate carboxylase by growing cultures of Athiorhodaceae. *J. gen. Microbiol.* **23**, 499.

LATZKO, E. & GIBBS, M. (1969). Enzyme activities of the carbon reduction cycle in some photosynthetic organisms. *Pl. Physiol., Lancaster* **44**, 295.

LEACH, C. K. & CARR, N. G. (1969). Oxidative phosphorylation in an extract of *Anabaena variabilis*. *Biochem. J.* **121**, 125.

LIGHTBODY, J. J. & KROGMANN, D. W. (1967). Isolation and properties of plastocyanin from *Anabaena variabilis*. *Biochim. biophys. Acta* **131**, 508.

LIVNE, A. & RACKER, E. (1968). A new coupling factor for photophosphorylation. *Biochem. biophys. Res. Commun.* **32**, 1045.

LOSADA, M., PANEQUE, A., RAMIREZ, J. M., DEL CAMPO, F. F. (1965). Reduction of nitrate to ammonia in chloroplasts. In *Nonheme Iron Proteins*, p. 211. Ed. A. San Pietro. Yellow Springs: Antioch Press.

LOSADA, M., WHATLEY, F. R. & ARNON, D. I. (1961). Separation of two light reactions in noncyclic photophosphorylation of green plants. *Nature, Lond.* **190**, 606.

VAN NIEL, C. B. (1941). The bacterial photosyntheses and their importance for the general problem of photosynthesis. *Adv. Enzymol.* **1**, 263.

NOBEL, P. S. (1967). A rapid technique for isolating chloroplasts with high rates of endogenous photophosphorylation. *Pl. Physiol., Lancaster* **42**, 1389.

NOZAKI, M., TAGAWA, K. & ARNON, D. I. (1961). Non-cyclic photophosphorylation in photosynthetic bacteria. *Proc. natn. Acad. Sci. U.S.A.* **47**, 1334.

ORMEROD, J. G., ORMEROD, K. S. & GEST, H. (1961). Light dependent utilisation of organic compounds and photoproduction of molecular hydrogen by photosynthetic bacteria: relationship with nitrogen metabolism. *Archs Biochem. biophys.* **94**, 449.

PETRACK, B. & LIPMANN, F. (1961). Photophosphorylation and photohydrolysis in cell free preparations of blue-green algae. In *Light and Life*, p. 621. Eds. W. D. McElroy and B. Glass. Baltimore: John Hopkins Press.

POWLS, R. & REDFEARN, E. R. (1969). Quinones of the Chlorobacteriaceae, properties and possible functions. *Biochem. biophys. Acta* **172**, 429.

SHIN, M. & ARNON, D. I. (1965). Enzyme mechanisms of pyridine nucleotide reduction in chloroplasts. *J. biol. Chem.* **240**, 1405.

SMILLIE, R. M. (1965). Isolation of phytoflavin a flavoprotein with chloroplast ferredoxin activity. *Pl. Physiol., Lancaster* **40**, 1124.

SMILLIE, R. M., RIGOPOULOS, N. & KELLY, H. (1962). Enzymes of the reductive pentose phosphate cycle in the purple and in the green photosynthetic bacteria. *Biochim. biophys. Acta* **56**, 612.

SMITH, A. J., LONDON, J. & STANIER, R. Y. (1967). Biochemical basis of obligate autotrophy in blue-green algae and thiobacilli. *J. Bact.* **94**, 972.

VANDOR, S. L. & TOLBERT, N. E. (1968). *Pl. Physiol., Lancaster* **43**, 512.

VERNON, L. P. (1968). Photochemical and electron transport reactions of bacterial photosynthesis. *Bact. Rev.* **32**, 243.

WEAVER, E. C. (1968). EPR Studies of free radicals in photosynthetic systems. *A. Rev. Pl. Physiol.* **19**, 283.

WHATLEY, F. R. & ARNON, D. I. (1963). Photosynthetic phosphorylation in plants. In *Methods in Enzymology*, vol. 6, p. 308. Eds. S. P. Kolowick and N. O. Kaplan. London and New York: Academic Press.

THE PHOTOSYNTHETIC APPARATUS IN PROKARYOTES AND EUKARYOTES

P. ECHLIN

Botany Department, University of Cambridge

The vital parameters which distinguish the prokaryotes and eukaryotes have already been enumerated in this present Symposium. One of these characteristics is the form of the photosynthetic apparatus. This may be defined as the entire complex of membranous and particulate components which convert light energy into chemical energy. A wide variety of organisms carry out photosynthesis and the structure of the photosynthetic mechanism within such organisms is likewise diverse and may even show an evolutionary trend from morphological simplicity to complexity. In eukaryotic organisms this apparatus is located within a readily observable discrete membrane-bound structure, the chloroplast. In the prokaryotes this is less readily observable and would include the internal membranes associated with the photosynthetic pigments as well as other subcellular particulate components which are distributed throughout the contents of the cell. The description of the photosynthetic apparatus is usually limited, or confined to details of the membranous elements and their disposition and arrangement within the cell. It has now been firmly established that the photochemical reactions of photosynthesis, that is, the reactions whereby electromagnetic energy is converted to chemical energy utilizable by the cell, take place on lamellar systems. The other part of the photosynthetic process, in which the chemical energy manufactured by the light reaction is used to reduce inorganic or organic carbon to sugars, is not thought to be associated with membranous components, but is probably located in multienzyme complexes. Such complexes, although free in the non-membranous stroma of the photosynthetic apparatus, are nevertheless likely to be closely associated with the membranes.

The first part of this contribution is essentially a description of the form and patterning of the apparatus concerned with the light reaction of photosynthesis. Following the original description by Menke (1962), the basic membranous unit of all photosynthetic organisms is an enclosed sac or thylakoid. The form of the apparatus concerned with the dark reaction may be described only in terms of molecular morphology. In the prokaryotic organisms very little is known about such structures;

where details of the components of the non-membranous stroma in eukaryotic organisms are known, these will be mentioned as appropriate.

It is not intended to consider the configuration of the sub-cellular components concerned with the light reaction, although preliminary studies which have been confined to a small number of species would indicate that certain patterns and structural entities may be common in all photosynthetic systems. Details of the photochemical electron transfer reactions which take place in the membrane systems and the relation of these to the energy-transfer reactions and the structure of the membrane may be found in the article by Evans & Whatley (p. 203 in this Symposium) and in the following general studies and reviews: Vernon (1968), Menke (1966), Branton (1967, 1969), Park (1966), Bogorad (1967).

The second part of this paper will consider in broad terms the relationship between the photosynthetic apparatus of prokaryotes and eukaryotes, and how this may have evolved.

Much has been written on the structure of the photosynthetic apparatus and in a broad and comparative study such as this it is impossible to do more than high-light certain general features. Up-to-date references to all aspects of this topic, and indeed to photosynthesis as a whole, may be found in the reviews which appear periodically in *Photochemistry and Photobiology*, and in *Annual Reviews of Plant Physiology*, of *Microbiology* and of *Biochemistry*.

THE PHOTOSYNTHETIC APPARATUS
IN PROKARYOTIC CELLS

It is convenient to divide the photosynthetic prokaryotes into the photosynthetic bacteria and the blue-green algae. Although these two groups of organisms share many common characters, they are sufficiently diverse on morphological and on biochemical grounds to be considered separately; see Echlin & Morris (1965).

The photosynthetic bacteria

These bacteria are typically aquatic micro-organisms inhabiting anaerobic marine and freshwater environments. Unlike other photosynthetic organisms, these bacteria photosynthesize without oxygen evolution, and only under anaerobic conditions. Some may, however, live in the dark in the presence of oxygen, then obtaining their energy from respiration. They depend on the presence of external electron donors such as reduced sulphur or organic compounds and fall into three fairly distinct groups. (1) The Chlorobacteriaceae (green sulphur bacteria), which are predominantly obligate photoautotrophs. (2) The Thiorhodaceae (purple sulphur bacteria), which are photoautotrophs

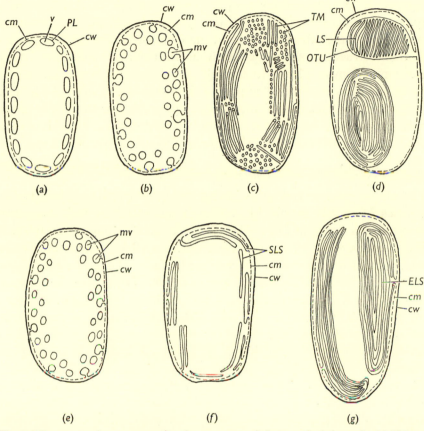

Fig. 1. Diagrams to show cross-sectional appearance of photosynthetic bacteria. (*a*) Chlorobacteriaceae. (*b*) Thiorhodaceae I, membranous vesicles (*mv*). (*c*) Thiorhodaceae II, tubular membranes (*TM*) (Eimhjellen, Steensland & Traetteberg, 1967). (*d*) Thiorhodaceae III, lamellar system (*LS*). (*e*) Athiorhodaceae I, membranous vesicles (*mv*). (*f*) Athiorhodaceae II, simple lamellar systems (*SLS*). (*g*) Athiorhodaceae III, elaborate lamellar system (*ELS*). (*cw* = cell wall; *cm* = cell membrane; *PL* = photosynthetic membrane; *v* = vesicle; *OTU* = outer thylakoid unit.)

capable of reducing carbon dioxide with the accompanying oxidation of H_2S to sulphide, though some may photo-assimilate organic compounds. (3) The Athiorhodaceae (purple non-sulphur bacteria), which photometabolize simple organic substances, but are inhibited by H_2S.

In the Thiorhodaceae and the Athiorhodaceae the photosynthetic apparatus appears to be a membranous extension of the cytoplasmic membrane. Pfennig (1967) considered the cytoplasmic membrane of bacteria to be far more complex than the equivalent structure in eukaryotic organisms, and that it can fulfil some of the functions in bacteria which are normally fulfilled by membranous organelles of eukaryotic organisms. Thus the various morphological forms of the photosynthetic apparatus in bacteria may be genetically and physiologically controlled extensions of the cell membrane. In the Chlorobacteriaceae, the photosynthetic structures are not continuous with the cell membrane; they are confined, by and large, to a number of elongate vesicles. The papers by Lascelles (1968) and Pfennig (1967) review some of the available data about the bacterial photosynthetic apparatus. In the present paper only some of the more recent findings will be discussed, together with a general statement about the morphology of the photosynthetic apparatus of each group.

Chlorobacteriaceae

The photosynthetic apparatus in this group of bacteria is confined to a series of membrane-bound vesicles. Cohen-Bazire (1963) examined three strains each of *Chlorobium limicola* and *C. thiosulphatophilum*, and found a series of discrete electron-transparent areas which lined the cortex of the cytoplasm between the surface cytoplasmic membrane system and the more central ribosomal region. These areas were found in all six strains examined and have been termed 'chlorobium vesicles'; they are 1000 to 1500 Å long and 300 to 400 Å wide. Cohen-Bazire isolated vesicular structures from the main pigment fraction of *C. thiosulphatophilum*; they were similar in form and dimension to the peripheral chlorobium vesicles found in the intact cell. Cohen-Bazire, Pfennig & Kunisawa (1964) extended these studies and found that the chlorobium vesicles, which constituted up to 25 % of the cell volume, were filled with fine fibrils only 15 to 20 Å wide. Holt, Conti & Fuller (1966) found similar structures in *Chloropseudomonas ethylicum*, and Pfennig & Cohen-Bazire (1967) found chlorobium vesicles in *Pelodictyon clathratiforme*, and also gas-vacuoles like those of blue-green algae. Earlier workers, including Vatter & Wolfe (1958) and Fuller, Conti & Mellin (1963), had considered that the photosynthetic apparatus in the green sulphur bacteria was in the form of discrete opaque particles 150 to 250 Å in diameter distributed throughout the cytoplasm. It is now clear that these holochrome particles are either a structural sub-unit of the chlorobium vesicle, or derived from it by comminution.

Thiorhodaceae

There is a considerable range of morphological variation in the appearance of the photosynthetic apparatus in Thiorhodaceae, though basically there are three types of vesiculate-lamellate structures, all of which appear to be outgrowths of the cell membrane.

(1) *Membranous vesicles.* Fuller & Conti (1963) and Fuller *et al.* (1963) examined the fine structure of *Chromatium* sp. strain D grown under controlled conditions of illumination and found chromatophore-like structures at the ends of invaginated cell membranes. In some cells these membranes were organized into simple lamellar systems. Cohen-Bazire (1963) examined the same organism, as well as *Chromatium okneii Thiospirillium jensenii* and a species of *Thiopedia*, and also found chroma-tophore-like membranous vesicles. A similar situation was found (Cohen-Bazire, 1963) in a species of *Thiocapsa*, but here the structure was complicated by the presence of a large area of lamellar pairs.

(2) *Tubular membranes.* These may appear either as a series of parallel straight tubes or as a three-dimensional network of irregularly branched tubes. Eimhjellen, Steensland & Traetteberg (1967) have shown that the photosynthetic pigments in a species of *Thiococcus* is associated with a unique internal membrane system made of branched tubes of uniform diameter (450 Å), continuous with the cell membrane. Osmotic lysis of the cell revealed a series of long ribbon-like tubules.

(3) *Lamellar system.* Lamellar systems, together with vesicles, have been demonstrated in *Thiocapsa*, but in some organisms of Thiorhodaceae the photosynthetic apparatus is composed entirely of a lamellate system. Raymond & Sistrom (1967) described an obligate halophilic photo-synthetic bacterium in which the photosynthetic apparatus was typically made of two discrete bundles of stacks of lamellae, each stack made of 8 to 12 closely adpressed membrane pairs. The most complicated arrange-ment of photosynthetic lamellae is seen in the photosynthetic bacterium recently isolated from mud flats in the Galapagos Islands and described by Holt, Trüper & Takacs (1968), Remsen, Watson, Waterbury & Trüper (1968) and Trüper (1968). These workers showed that the photo-synthetic apparatus was of the lamellar type and attached to the cell membrane. The lamellae were arranged in stacks which increased in number with increasing light intensity. The stacks were anchored to the outer cell membrane, and each stack in turn formed an integral unit within the cell. These authors considered these structures, which are analogous in form to the grana stacks in higher plants (see later), to be the most complex configurations yet seen in the photosynthetic bacteria.

Athiorhodaceae

More work has been done with organisms from this group than from the previous two groups, and it is thus perhaps not surprising that the Athiorhodaceae have shown the most diverse photosynthetic structures. Nevertheless, it is possible, as with the previous two groups, to put them in a number of broad classes. This was not, however, the view of Worden & Sistrom (1964), who considered that the structure of the photosynthetic apparatus in the Athiorhodaceae is the same in all forms, approximating to the lamellar system seen in *Rhodospirillum molischianum*.

(1) *Membranous vesicles*. Vatter & Wolfe (1958), Cohen-Bazire (1963) and Hickman & Frenkel (1959, 1965*a*) have shown that both *Rhodospirillum rubrum* and *Rhodopseudomonas spheroides* exhibit discrete membrane-bound vesicles of between 500 and 1000 Å diameter dispersed throughout the cytoplasm. Some of these vesicles are continuous with the plasma membrane and it was suggested that they arose by a simple invagination from this structure. In *Rhodopseudomonas gelatinosa*, described by Weckesser, Drews & Tauschel (1969), the photosynthetic apparatus is confined to a series of small individual invaginations of the cytoplasmic membrane, and it was suggested that the cytoplasmic membrane should be included as a structural entity of the photosynthetic apparatus. The earlier immunological evidence of Newton (1963), based largely on studies on *Rhodopseudomonas spheroides* and *Rhodospirillum rubrum*, would support this view. Membranous vesicles have also been found in *Rhodopseudomonas capsulata*, and it is now reasonably certain that the chromatophores of membranous vesicles in the Athiorhodaceae are formed by invagination of the plasma membrane, although it is not clear whether this physical connection remains between the fully-formed chromatophores and the engendering membrane.

(2) *Simple lamellar systems*. A lamellar system made of short membranous elements and arranged in discrete stacks in the periphery of the cell is considered to be characteristic of *Rhodospirillum molischianum* (Giesbrecht & Drews, 1962; Gibbs, Sistrom & Worden, 1965; Hickman & Frenkel, 1965*b*), *Rhodospirillum fulvum* and *Rhodospirillum photometricum* (Cohen-Bazire & Sistrom, 1966) and in some instances of *Rhodopseudomonas capsulata* (Drews, Lampe & Ladwig, 1969).

(3) *Elaborate lamellar system*. In *Rhodopseudomonas palustris* (Cohen-Bazire & Sistrom, 1966) and in *Rhodopseudomonas viridis* (Drews & Giesbrecht, 1965; Giesbrecht & Drews, 1966) the photosynthetic apparatus forms an extensive lamellar structure disposed as concentric

rings around the periphery of the cell wall. In *Rhodopseudomonas viridis* the thylakoids form in stacks which are considered to have as much unity as the grana of plastids. Up to eleven thylakoids may be found in a stack, and the stacks themselves may be either cylindrical or convex. There is a tendency for the thylakoid stacks to lie parallel to the long axis of the cell. In the budding photosynthetic bacterium *Rhodomicrobium vannielii* (Vatter, Douglas & Wolfe, 1959; Trentini & Starr, 1967) the photosynthetic apparatus at low light intensity is an extensive peripherally located system of lamellae, which at high light intensities becomes asymmetric and less extensive. It may be shown that a limited number of lamellae arise as unfoldings of the cytoplasmic membrane, which subsequently undergo proliferation by forking and branching back.

A comparative examination of the structures of the photochemical apparatus in photosynthetic bacteria has indicated a diversity of form ranging from a complex vesicular arrangement to simple chromatophores. It is apparent that much of the diversity and possibly some of the uncertainty of the structure of the photosynthetic apparatus in the photosynthetic bacteria has arisen from the different growth conditions used by various workers. Although a close relationship exists between the photosynthetic apparatus and the cell membrane, the size, shape and number of lamellae can, in many instances, be shown to vary with the photosynthetic pigment content of the organism. It is not at present possible to draw any conclusive phylogenetic significance from the arrangement of the photosynthetic apparatus in the photosynthetic bacteria.

The blue-green algae

The blue-green algae, Cyanophyceae, are one-celled to multicellular, coccoid to filamentous plants commonly found in plankton and most soils, although they occupy a range of somewhat specialized habitats. Although sharing many characteristics with the photosynthetic bacteria, they exhibit an important difference in that the photosynthetic mechanism is based on chlorophyll *a* and evolves oxygen. Two comprehensive review articles have appeared on the physiology and biochemistry (Holm-Hansen, 1968) and the fine structure (Lang, 1968) of the blue-green algae. Because of close morphological similarities and a common curious gliding movement, it is usual to include in the blue-green algae a large assemblage of non-photosynthetic organisms known as 'Farblose Algen', the colourless blue-green algae (Pringsheim, 1963). Depending on one's taxonomic persuasion, such colourless forms might include *Flexibacteria* and *Oscillospira*, together with some chemo-

synthetic forms such as *Beggiatoa* and *Thiothrix*. As these latter organisms are not photosynthetic, they will not be dealt with further here.

Studies of various representatives from the eight families comprising
the blue-green algae show that the basic membranous photosynthetic
element is the thylakoid. The thylakoids, which often have a very
striking arrangement within the cells, tend to lie parallel to one another,
and not in closely-packed stacks as in some of the photosynthetic
bacteria. Generally speaking, the lamellae in blue-green algae are less
formally organized than the lamellae in the photosynthetic bacteria, and
no evidence has yet been presented to show that there are stacks of

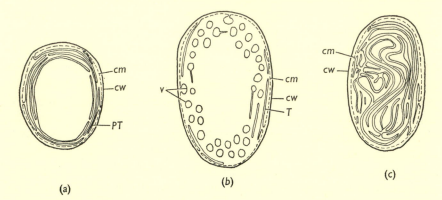

Fig. 2. Blue-green algae: diagrams of cross-sections. (*a*) peripherally arranged thylakoids
(*PT*) occasionally branching. (*b*) Merismopedia, vesicles (*v*), few thylakoids (*T*). (*c*) thylakoids branch and fill whole cell; internal space interconnected.

lamellae or groups of thylakoids. In the light microscope, the blue-green
algae look green throughout the cell, and fine-structure studies show
that the photosynthetic lamellae form a three-dimensional complex
throughout the entire cytoplasm. There are exceptions to this in some
of the unicellular blue-green algae, in which the photosynthetic lamellae
are located entirely at the periphery of the cell.

The lamellae tend to form large sheets and there are no records of
tubular thylakoids as shown to exist in a few photosynthetic bacteria.
In a few instances such as in *Merismopedia glauca* the photosynthetic
apparatus is replaced by a number of peripherally arranged membranous
vesicles; but as Echlin (1968*a*) has shown, these differences are probably
due to differences in age of the cultures.

The number, position and appearance of the thylakoids in any blue-
green alga depends on the species, its physiological condition and age,
and even on the way it has been prepared for electron microscopy

(Echlin, 1968 b). The sub-cellular structure of the photosynthetic apparatus is located in the membrane, and the indications are that it is the same as in higher plants. There is evidence that the photosynthetic lamellae in blue-green algae, like those in some of the photosynthetic bacteria, arise directly from the unicellular cell membrane. This has been shown in *Symploca* (Pankratz & Bowen, 1963), *Oscillatoria* (Jost, 1965) and in *Anacystis nidulans* (Allen, 1968). Menke (1961 a) showed that although thylakoid enlargement is probably by growth in area, the increase in number probably results from severance of the thylakoid membrane by cell walls and proliferation. Other mechanisms of thylakoid formation were discussed by Echlin & Morris (1965).

Unlike the case of the photosynthetic bacteria, information is now available about the non-membranous stroma or background cytoplasm in the immediate region of the photosynthetic lamellae of the blue-green algae. The two most important structural entities found in this region are the pro-lamellar bodies and the phycobilisomes. Lang & Rae (1967) have demonstrated the presence of a membranous configuration resembling the pro-lamellar body of greening eukaryotic chloroplasts, in old cells of blue-green algae. The morphological resemblance is quite striking although it is difficult to understand why these structures should only appear in old, and presumably fully-greened blue-green algal cells. Wildon & Mercer (1963) showed that in some blue-green algae the entire pigment complement could form in the dark. Pro-lamellar body development in eukaryotic chloroplasts is accompanied by chlorophyll synthesis and requires the presence of light. The blue-green algae contain lipid-soluble chlorophylls and carotenoids as photosynthetic pigments, together with water-soluble phycobilin pigments. Berns & Edwards (1965) showed that the phycobilins are in the form of hexamers approximately 150 Å in diameter. Gantt & Conti (1966, 1969) demonstrated the presence of larger 300 Å phycobilisomes attached at regular intervals along the inter-thylakoid space in a number of filamentous blue-green algae. Bourdu & Lefort (1967) and Echlin (1967) showed that the same structures may be seen in the inter-thylakoid space in the endo-cyanelles of *Glaucocystis* and *Cyanophora*. It is now clear that the earlier suggestions by Lefort (1959) and Echlin (1964) that phycobilin molecules were in the hydrophobic intra-thylakoid cavity were misinterpretations.

THE PHOTOSYNTHETIC APPARATUS IN
EUKARYOTIC CELLS

The characteristic feature of the photosynthetic apparatus in the eukaryotes is that it is contained within a discrete cell organelle, the chloroplast. Although most of the electron-microscope investigations on chloroplasts have been done with flowering plants, it appears that this organelle is the same for all vascular plants and the eukaryotic algae. The chloroplast consists basically of an ellipsoidal to circular body within a double membrane, between 3 and 10 μm along the long axis, situated in cells in those parts of the plant body exposed to light. Within the double membrane there is a slightly electron-dense granular matrix called the stroma which is traversed by a complex series of membranous elements. In most plants except the algae, the membranous elements are stacked into precisely orientated grana intercalated by less well organized stroma lamellae, and the whole lamellar system consti-tutes a single complicated three-dimensional membrane-enclosed cavity. The arrangement of the membranes within the chloroplast, and the envelopment of the chloroplast by membranes of the endoplasmic reticulum is characteristic for different groups of eukaryotic organisms. Details of these patterns will be discussed later.

Starch grains are another prominent feature of chloroplasts from plants which have been photosynthesizing for a few hours. The starch grains are approximately the same shape as the chloroplasts themselves, and are usually one to two μm long. There are several starch grains per chloroplast and they lie in the stroma close to the grana. A regular feature of the chloroplast includes a variable number of osmophilic globules 500 to 3000 Å in diameter, which are closely associated with the membranous elements. By using appropriate fixation and staining techniques it is possible to demonstrate the presence of ribosomes in the stroma, and in certain electron-transparent regions of the stroma, minute 25 Å fibrils of DNA. Other structures which have occasionally been found in different plant chloroplasts include the regularly packed fibrillar stromacentres described by Gunning (1965*a*) in *Avena*, and various inclusion bodies, some of which appear to have regular sub-structures.

The chloroplasts of higher plants grown entirely in the dark, although not of immediate importance to the present discussion, do nevertheless contain interesting structural components which do not appear in their light-grown counterparts. Such dark-grown chloroplasts, or strictly speaking 'plastids', are usually referred to as etioplasts, and

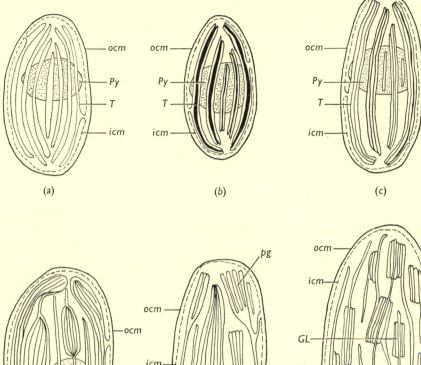

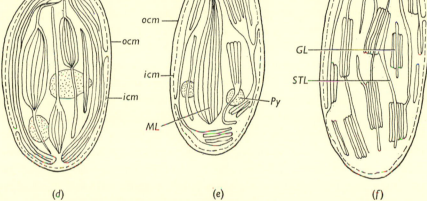

Fig. 3. Diagrams to show cross-sectional appearances of chloroplasts (endoplasmic reticulum omitted). (*a*) Red alga, chloroplasts with single thylakoids. (*b*) Cryptophyta, paired thylakoids (*T*). (*c*) Euglenas (some), Chrysophyta and some browns, triple thylakoids (*T*). (*d*) Green algae, multiple thylakoids; some differentiation into complex lamellae and single stroma lamellae. (*e*) Carteria, Chlamydobotrys; pseudograna (*pg*) and multiple lamellae (*ML*). (*f*) Higher plant chloroplasts; distinctive grana lamellae (*GL*) and stroma lamella (*STL*). (*ocm* = outer chloroplast membrane; *icm* = inner chloroplast membrane; *T* = parallel single thylakoids; *Py* = pyrenoid.)

when exposed to light develop into chloroplasts. Such etioplasts are filled with the typical homogeneous stroma, with ribosomes and DNA fibrils, and in place of the usual complex membranous configuration there are some paracrystalline pro-lamellar bodies. The pro-lamellar bodies have been found in the etioplasts from a number of different plants and have been adequately described by Gunning (1965b). They consist of a three-dimensional cubic lattice of interconnected tubules, which at their periphery are continuous with lamellae that extend into the stroma. It is fairly certain that the pro-lamellar body gives rise to the characteristic membranous components of the typical chloroplast.

Chloroplasts of the alga

Of all the groups of eukaryotic plants the algae show the greatest diversity in size, shape, form and number of chloroplasts. It is not possible to relate this diversity in form to any feature of the life history, ecology or metabolism of the algae, and it only rarely provides a cogent basis for seeking any significant phylogenetic relationships between the groups. Of the many tens of thousands of species of algae known to exist, the chloroplasts from only a few representatives from each class have been examined in any detail. There is also evidence that the form of the chloroplast is very dependent on the external environment, and there have been only a few studies in which different algae have been examined under a range of growth conditions. There may be a single chloroplast as in *Micromonas* to more than a hundred in *Euglena*. The chloroplasts may be ellipsoidal in *Euglena*, ribbon-like in *Spirogyra*, stellate in *Zygnema* and some desmids, and complexly irregular in the green alga *Oedogonium*. Algal chloroplasts are similar in most respects to higher plant chloroplasts, with two exceptions. Firstly, some algal chloroplasts either contain or are closely associated with pyrenoids, which are finely granular structures slightly denser than the stroma ground-substance and contain a dense central proteinaceous core, occasionally traversed by thylakoids. Pyrenoids are thought to be associated with carbohydrate metabolism. A more detailed consideration of their fine structure has been given by Gibbs (1962a, b), Bisalputra & Weier (1964), Manton (1966a, b) and Evans (1966). The second exception is in the arrangement of the photosynthetic thylakoids within the algal chloroplast. Some of these arrangements will now be described.

Single thylakoids

Single thylakoids are characteristic of the red algae and in the members of this group which have been investigated they lie parallel to each other and may run the entire length of the chloroplast. The chloroplast envelope is double-layered, and the pyrenoid where present is usually centrally located and traversed by a number of widely separated single thylakoids. By using special fixation procedures, it is possible to demonstrate the presence of electron-dense granules 320 Å in diameter on the outer surface of the thylakoid. The size, number and arrangement of these granules depends on age and growth conditions. These bodies, or phycobilisomes, which are variable in shape and size, contain the water-soluble phycobilin pigments, and probably mark the sites of photoreactive centres. Phycobilisomes have been reported in *Laurencia* (Bisalputra & Bisalputra, 1967) and *Porphyridium* (Gantt & Conti, 1965; Gantt, Edwards & Conti, 1968). Floridean starch, the carbohydrate reserve product typical of this group, is not formed inside the chloroplast.

Paired thylakoids

Paired thylakoids are characteristic of the Cryptophyta, and in those algae of this group which have been investigated (Greenwood, 1967) the thylakoids are arranged in closely appressed pairs. The thylakoids run along the whole length of the chloroplasts, but do not appear to traverse the pyrenoid. Although these algae contain phycobilin pigments, there are at present no reports of phycobilisomes similar to those shown in the red and blue-green algae. Gibbs (1962c) showed that the chloroplast was surrounded by an outer double-membrane envelope, outside the usual double-membrane envelope of the chloroplast. In the region where the chloroplast lies adjacent to the nucleus this outer envelope is continuous with the outer of the two nuclear membranes. As with the red algae, the polysaccharide reserve material lies outside the chloroplast.

Triple thylakoids

Triple thylakoids are by far the most common arrangement in the algae. The whole chloroplast is traversed by lamellae, which in cross-section appear as three closely-appressed thylakoids, although occasionally more lamellae are found. Like the previous two groups, the products of photosynthesis appear outside the chloroplast. In some instances there is a third membrane surrounding the chloroplast envelope which may or may not be associated with the nuclear membrane.

Euglenophyta. The biology of the euglenoid algae has recently been reviewed by Leedale (1967). Although the typical thylakoid number is three, it can be as great as twelve. The pyrenoids have a finely granular stroma traversed by two somewhat distended thylakoids.

Chrysophyta. The differences amongst members within this group appear to centre around the presence or absence of pyrenoids, and where this structure is present, how many thylakoids traverse it. In the diatoms, the pyrenoid is generally bounded by a single membrane and traversed by only one or two membranes (Drum & Pankratz, 1964) while in the Xanthophyceae a pyrenoid is only present in a few species. In papers by Falk (1967) and Falk & Kleinig (1968), the denser regions of the stroma devoid of lamellations and bulging from the chloroplast into the cytoplasm are interpreted as being pyrenoids. The situation is more complex in the Chrysophyceae, where Gibbs (1962c) and Manton (1966b) have shown that the pyrenoid is separated from the chloroplast by two superposed membranes, and that the outer of the two membranes is continuous both with the endoplasmic reticulum and the outer nuclear membrane. The dinoflagellates have the typical triple thylakoids, with the reserve polysaccharide deposited outside the chloroplast but close to the pyrenoid region. Leadbeater & Dodge (1966) have shown the presence of several pyrenoids in the marine dinoflagellate *Wolozsynskia micra*.

Phaeophyta. The triple thylakoids do not appear to be so tightly appressed and the rudimentary pyrenoids are only present in some of this group (Bouck, 1965; Evans, 1968). In some brown algae the reserve polysaccharide forms outside the chloroplast near the pyrenoid. A complex relationship appears to exist between the chloroplast membranes (outer and inner pairs), nuclear membranes, endoplasmic reticulum, and even the membranes of the dictyosome.

Multiple thylakoids and pseudograna. Multiple thylakoids and pseudograna are typical of the Chlorophyta; this algal group contains the most complicated internal structure, ranging from regularly arranged triple thylakoid bands in *Chlorella*, to eight bands in *Chlamydomonas*, and as many as 100 bands in *Nitella*. Unlike the other algae, there is usually little suggestion of regular bands, and the thylakoids coalesce and separate in an irregular fashion. The lamellar stacking approaches the complexity seen in higher plants, and in the desmid *Microasterias rotata* (Drawert & Mix, 1961) there can be several stacks of up to fifty thylakoids, with the thylakoid membranes extending through the stem from one stack to another. A similar situation is seen in *Acetabularia* and *Chara*, with complex lamellae separated by stroma lamellae, although not distinctly separable into the grana and stroma of higher plants (Crawley, 1964; Chambers & Mercer, 1964; Pickett-Heaps, 1967).

The term pseudograna has been applied to structures seen in *Carteria* (Joydon & Fott, 1964; Lembi & Lang, 1965) and *Chlamydobotrys* (Merrett, 1969; Wiessner & Ametunzen, 1969). In cross-section, these structures look like an individual thylakoid dilated and subsequently infolded to form a discrete stack of discs, similar to the grana stacks in higher plants.

Most members of the green algae have pyrenoids in the chloroplasts, which may be traversed or partially traversed by one or occasionally two thylakoids. Unlike *all* the other eukaryotic algae, in the green algae the reserve polysaccharide (starch) surrounds the pyrenoid and is contained within the chloroplast membrane. In this respect green algae are similar to higher plants.

Chloroplasts of the Bryophytes (mosses and liverworts)

Very little work has been done on the structure of the photosynthetic apparatus of the mosses and liverworts. The one liverwort which has been examined, *Anthoceros*, has a double-membrane bound chloroplast containing up to eight thylakoids in small stacks somewhat similar to the grana of higher plant chloroplasts (Manton, 1962; Menke, 1961b; Wilsenach, 1963). Unlike higher plants, the chloroplast of *Anthoceros* contains a pyrenoid bounded by a deeply invaginated single membrane which appears to partition the structure. It is thought that the chloroplast in *Anthoceros* may not be typical of all liverworts. Sun (1962) considered that the presence of partially differentiated grana lamellae and a pyrenoid places the *Anthoceros* chloroplast intermediate between those of the algae and those of higher plants.

The only work on the structure of moss chloroplasts is by Maier & Maier (1968), who found that in the sporophyte of *Polytrichum* there appear to be distinct grana as in higher plants.

Chloroplasts of lower vascular plants and gymnosperms

As with the Bryophytes, the Pteridophytes and Gymnosperms have not been examined in great detail. The studies which have been made on *Picea* (Von Wettstein, 1959), *Psilotum* (Sun, 1961), *Isoetes* (Paolillo, 1962), *Selaginella* (McHale, 1965), *Equisetum* (Sun, 1963) and *Matteuccia* the only fern (Gantt & Arnott, 1963), all indicate that the chloroplasts of this group of plants have the well differentiated grana and stroma lamellae seen in higher plants.

Chloroplasts of flowering plants

In this group of plants the organization of the internal membranes of the chloroplast is most highly differentiated and complex. The characteristic feature of these chloroplasts is the stacks of grana lamellae, which in cross-section look like piles of coins. The stroma lamellae, which are less regularly organized, extend from and interconnect with the grana stacks. There is a remarkable uniformity in the cross-sectional appearance of the chloroplasts from different plants, but it is clear from the few careful studies by investigators who have constructed three-dimensional models from serial sections that the chloroplasts of higher plants are anything but uniform. The models proposed by Ericksson, Kahn, Walles & Von Wettstein (1961) for *Spinacia* and *Phaseolus* showed the stroma lamellae as flat undissected sheets extending through the chloroplast and intersecting some grana stacks. Weier, Stocking, Thomson & Drever (1963) proposed that in *Nicotiana* the different grana stacks are connected by an open network of flattened tubules, while Heslop-Harrison (1963) with *Cannabis* and Wehrmeyer (1964) with *Spinacia* proposed that the individual grana within a stack are not isolated sacs, but are continuous with each other through the interconnecting stroma lamellae.

Paolillo, McKay & Griffiths (1969) have suggested that in all flowering plants the flat stroma membranes are spiralled around the cylindrical grana stacks like steps of a spiral staircase. Depending on the species, there may be up to 15 separate helices per grana stack. It is not possible here to discuss the validity of these various models; reference should be made to the original papers for details. Some of the grana within a stack do not extend into the stroma lamellae, while others may do so, on one or both sides, as seen in cross-section. Similarly, the stroma lamellae extending from the grana may end blindly or interconnect with grana from another stack. The result is the vastly complex interconnected internal membranous cavity, separate and distinct from the granular stroma, referred to earlier.

Not only may the appearance of the membranes within the chloroplast vary between species, but considerable variation may be observed within the same plant. A comparison of chloroplasts from mesophyll cells and the bundle-sheath cells of sugar cane shows that while the former are typical of higher plants, the latter lack grana and are like those of certain algae (Park, 1966). The same thing has been seen in maize (Hodge, McLean & Mercer, 1955). The guard cells of many grasses contain only partially differentiated chloroplasts lacking the fully-developed grana substructures which are present in the mesophyll cells (Brown & Johnson, 1962). Structural changes may occur as a result of development; Spurr & Harris (1968) have shown that all the grana and many of the stroma thylakoids disappear during maturation of *Capsicum annuum*. The nutritional status of plants markedly affects the arrangement of the membranes within the chloroplast. Thomson & Weier (1962) and Thomson, Weier & Drever (1964) in studies on mineral deficiencies in beans showed that the highly organized grana-stroma organization broke down to a lamellar system, and that in low phosphate the mesophyll plastids appeared similar to the alga-like plastids seen in normal bundle-sheath cells in monocotyledons.

Ashton, Gifford & Bisalputra (1963) showed that low concentrations of Atrazine, (2-chloro-4-ethylamino-6-isopropylamino-S-triazine), which blocks photosynthesis, caused disruption of the internal chloroplast membranes in *Phaseolus vulgaris*, although these structures remained unchanged when the plants were placed in the dark or in the absence of the drug. Schmid (1967) showed a change in the lamellar structure in various chlorophyll-deficient plants, and Shumway & Weier (1967) showed distinctive changes in chloroplast morphology in pigment-deficient mutants of maize. Finally, infection by pathogens can also induce structural changes within chloroplasts. Sun (1965) observed lamellar degradation in virus-infected *Abutilon*, and Harding, Williams & McNabola (1968) saw distinct changes in the chloroplast from *Brassica* plants infected with the fungus *Albugo*.

THE EVOLUTIONARY ORIGINS OF CHLOROPLASTS

There is some evidence that chloroplasts are able to enlarge and possibly divide, thus increasing in number, and various organelles within the plant body have been implicated in the formation of chloroplasts where these organelles are absent from the cell. An authoritative account of the various theories and proposals for the formation, growth and differentiation of chloroplasts may be found in the book edited by Kirk & Tilney-Bassett (1967).

There is now an increasing amount of circumstantial evidence which allows one to speculate on the evolutionary origin of chloroplasts. There are clear indications that the process of photosynthesis is an ancient one, and that the first photosynthetic organisms were on this planet some 3000 million years ago. Although ultrastructural details are lacking, it is thought that these organisms were similar in many respects to modern photosynthetic bacteria. Shortly after (within 300 million years!) blue-green algae were prominent photosynthetic organisms, and it was not until between 1500 million and 1000 million years ago that the first eukaryotic photosynthetic organisms appeared. Much has been speculated about the biology of the microflora of the Precambrian era and its significance to the origin of life and the evolution of cellular structures. The author has recently summed up some of the evidence (Echlin, 1969a), and more extended reports are in press (Echlin, 1969b, 1970). A major problem which has been intriguing biologists for some time is the origin of the eukaryotic cell; the recent paper by Sagan (1967) contains much valuable information. Part of the problem is the origin of the various organelles within the eukaryotic cells, and whether they arose endogenously or exogenously. It may be profitable to discuss these two aspects with reference to the chloroplast.

The endogenous origin of the chloroplast would involve an internal partitioning-off of the photosynthetic lamellae by membranes within the cell. There is at present no evidence about how this may have occurred; what follows is speculation. It is likely that this partitioning would have occurred in blue-green algae rather than in photosynthetic bacteria, since the latter have no eukaryotic counterparts. The cell membrane, which is in intimate contact with the photosynthetic lamellae in the blue-green algae, would proliferate to form photosynthetic lamellae and then pinch off, leaving a discrete membranous entity within the cell. The double membrane which surrounds *all* eukaryotic chloroplasts would then arise by an infolding proliferation and subsequent sealing-off of one of the outer photosynthetic lamellae. A more attractive theory would have the initial formation of the chloroplast as outlined above, but that other membranous components within the pro-eukaryotic cell would envelop the protochloroplast, thus forming the double envelope. It is of interest to note that lipid analysis of the inner and outer mitochondrial membranes shows that the outer membrane is chemically closer to the membranes of the endoplasmic reticulum than to the inner membrane, while the inner membrane shares certain characters with the bacterial cell membrane (Parsons & Yano, 1967). It would be of interest to know whether the same situation exists with regard to chloroplasts. These theories by no means satisfactorily explain the formation of the protochloroplast, for although they go some way in showing how compartmentalization of the photosynthetic membranes may have occurred, they do not explain the origin of other substructures in the chloroplast, such as DNA and ribosomes. It could be argued that the initial partitioning also included some blue-green algal nucleoplasm and cytoplasm, and it may be significant that eukaryotic chloroplast DNA and ribosomes share many characteristics of the DNA and ribosomes of prokaryotic cells.

Considerably more has been written about the exogenous origin of chloroplasts. The basis of this theory is that blue-green algae gave rise to chloroplasts by a process of endosymbiosis. This concept stems from genetic evidence for the existence of extrachromosomal genes and the discovery that chloroplasts contain DNA and ribosomes, and are capable of synthesizing protein *in vitro*. This idea was first considered by Mereschkowsky (1905) and Famintzin (1907) but the theory was not taken seriously until Ris & Plaut (1962), by using more modern techniques, showed that endosymbiosis might well be considered as an evolutionary step towards the formation of complex cell systems. There is no conclusive evidence about how this may have occurred, but present knowledge of symbiotic associations allows certain speculations to be made.

Going back to Precambrian times (about 1500×10^6 years ago), one may envisage a photosynthetic blue-green alga living in association with a non-photosynthetic organism, such as a fungal hypha or filamentous bacterium-like structure. The first stage in combining the two organisms is envisaged as a loose cellular association, such as that of alga and fungus in lichens. It is conceivable that the blue-green alga may initially have been an ectoparasite, and the presence of chitinases and/or cellulases would have permitted partial or complete penetration of the host cell wall. Pinocytosis would then lead to engulfment of the blue-green alga, thus effecting its entry into the host cytoplasm, though topologically it would be external to the cell. Alternatively, the blue-green alga might effect entry through a mechanical break in the cell wall, which was subsequently repaired. In any event, the actual entry of the blue-green alga into the host provides one of the most serious problems about the whole process. One may envisage subsequent stages of increasing inter-dependence, culminating in the engulfment of the blue-green alga by the filamentous organism or invasion of the filamentous organism by the blue-green alga, at which stage the 'host' alga would benefit fully from the products of photosynthesis. This process of entry seems likely to have been an all-or-nothing process; fossilized examples are likely to show the blue-green alga either *in* or *out*. The structures which now constituted the new organism then followed a separate evolution before combining to form a new cell.

There are living examples of such intracellular associations between blue-green algae and other organisms. Blue-green algae have for example been found growing within the diatom *Rhopalodia*, in the colourless flagellates *Peliaina* and *Cyanophora*, and in the amoeba *Pauleinella* (McLaughlin & Zahl, 1966; Geitler, 1959). These associations probably originated from a capacity to resist digestion by the host cell and it is usual to find a significant decrease in the thickness of the algal cell wall. Although the 'symbiotic' algae appear to have lost the power of survival outside the host cell, they are still able to divide more or less in synchrony with the host cell, thus ensuring replication into the daughter cells. Although much work has been done on the structure of these organisms, physiological data are insufficient to be able to state with certainty that the relationship is symbiotic, although it is tacitly assumed to be so. The selective advantage gained by the 'symbiotic' blue-green algae, other than protection, is not entirely clear, but there may well be nutritional interrelations which await elucidation. The presence of these blue-green algae may profoundly modify the host's metabolism, for in *Peliaina* carrying blue-greens plentiful amounts of reserve starch are produced, whereas individuals without blue-greens often lack starch completely.

Probably the most interesting of these symbiotic relationships is that seen in *Glaucocystis*. The morphology of this organism has been extensively studied by several authors (Lefort, 1965; Schnepf, Koch & Deichgraber, 1967; Echlin, 1967; Hall & Claus, 1967). Leaving aside its rather dubious taxonomic position, it would appear that *Glaucocystis* is a eukaryotic apoplastidic alga containing blue-green algae. It is able to grow autotrophically in the presence of light, and it would seem that the blue-green algae here act as chloroplasts. The blue-green algae which have a considerably diminished cell wall divide in synchrony with the host cell, but are apparently unable to divide outside the host.

There is a great similarity between the ultrastructure of chloroplasts and that of endocellular blue-green algae. The basic constituents of blue-green algal cells were shown by Ris & Plaut (1962) to be present in the chloroplast, the only significant difference being the arrangement of the photosynthetic lamellae. But, as already indicated, variations in the form of these structures is one of the characteristics of photosynthetic organisms.

A detailed examination of the biochemical similarities between blue-green algae and chloroplasts lends further support to the hypothesis of the exogenous origin of the chloroplasts. This has been an area of fruitful research; limitation of space will only allow a résumé of the evidence. It was shown by Ris & Plaut (1962), Kislev, Swift & Bogorad (1965) and Granick & Gibor (1967) that chloroplast and blue-green algal DNA were both of the prokaryotic type, appearing as a loose network of 25 Å diameter fibrils within the centre of the cell. Recent buoyant-density centrifugation studies (Edelman *et al.* 1967) on blue-green algal DNA and chloroplast DNA show that there is a greater correlation of DNA base-pair composition of blue-green algae and chloroplasts, than between chloroplast and nuclear DNA. Shipp, Kieras & Haselkorn (1965) were unable to detect sequence homologies between DNA from tobacco plant chloroplasts and DNA from tobacco cell nuclei, and the hybridization experiments of Craig, Leach & Carr (1969) indicate that a greater homology exists between the DNA of blue-green algae and *Euglena* chloroplasts than between that of blue-green algae and members of the Athiorhodaceae. Several pieces of evidence indicate that chloroplasts contain a DNA-dependent RNA polymerase and DNA polymerase, thus having the enzymes necessary for both DNA replication and for transcription and translation of genetic information. Scott, Shah & Smillie (1968) showed that DNA synthesis occurred in isolated chloroplasts and replication may occur independently of nuclear DNA replication.

The ribosomes of blue-green algae and chloroplasts are also very similar. Gibbs (1968) showed that *Ochromonas* chloroplast ribosomes increased ten-fold during chloroplast development, and that the chloroplast ribosomal RNA was made within the chloroplast, apparently at the site of the chloroplast DNA. Stutz & Noll (1967) showed that chloroplast ribosomes and bacterial ribosomes have identical sedimentation properties; whereas chloroplast ribosomes and bacterial ribosomes sedimented at 70s, the plant cytoplasmic ribosomes sedimented at 80s. They also showed that RNA extracted from chloroplast ribosomes sedimented at 23s and 16s and was indistinguishable from RNA of bacterial ribosomes. Ribosomal RNA from plant cytoplasm gives values of 25s and 16s. Odintsova & Yurina (1969) have shown that the protein composition of ribosomes is very different in the chloroplast and cytoplasm. There is more evidence along the same lines, demonstrating this close similarity in structure and biochemical properties between chloroplast and prokaryotic cell ribosomes. There are several papers which attest to the close similarity in ribosomal drug sensitivity of chloroplasts and prokaryotic cells. Thus Ellis (1969) found that chloroplast ribosomes from tobacco leaves showed the same stereospecificity of inhibition by chloramphenicol as did bacterial ribosomes, whereas cytoplasmic ribosomes were unaffected. Sensitivity of chloroplast, bacterial and cytoplasmic ribosomes to cycloheximide showed that only the cytoplasmic ribosomes were affected. A similar situation has been found (Ellis, 1969) in the selective sensitivity to actinomycin. In general, all these, and many other studies, have shown that there is a close similarity between chloroplasts and blue-green algae. Although many other experiments remain to be done, particularly about the process by which the chloroplast makes protein, it is unlikely that we shall ever know whether the origin of chloroplasts was exogenous or endogenous.

CONCLUSION

An examination of the structural features of the photosynthetic apparatus of prokaryotic and eukaryotic cells indicates that they are very diverse. Although different cells show variety in membrane arrangement, there is a common structure, the thylakoid, in all photosynthetic cells. As more data become available, it will not be surprising to find that the substructure of the thylakoids, including the orientation and disposition of the pigments and associated enzymes in them, will show an even greater unity. The little work which has been done appears to indicate that it is unlikely that the photosynthetic apparatus has more than one origin, and that the arrangement of the sub-units within the thylakoid is such as to give maximal efficiency of energy conversion. The significance of the variety in the arrangement of the thylakoids within the prokaryotic cells and chloroplast is less clear. This too may be reflected in increasing efficiency of energy conversion. It might therefore be possible to speak of the primitive photosynthetic apparatus in prokaryotic cells, as compared to the advanced configurations seen in the chloroplasts of higher plants. But these different thylakoid arrangements may represent different responses to changes in the environment surrounding the organism. As indicated earlier, the origin of the chloroplast remains a mystery, and although the theory of the endogenous origin is rather pedestrian, and the theory of the exogenous origin has a certain elegant and intriguing fascination, neither hypothesis has been finally proved.

REFERENCES

ALLEN, M. M. (1968). Photosynthetic membrane system in *Anacystis nidulans*. *J. Bact.* **96**, 836.

ASHTON, F. M., GIFFORD, E. M. & BISALPUTRA, T. (1963). Structural changes in *Phaseolus vulgaris* induced by atrazine. II. Effects on the fine structure of chloroplasts. *Bot. Gaz.* **124**, 336.

BACHMANN, M. D., ROBERTSON, D. S., BOWEN, C. C. & ANDERSON, I. C. (1967). Chloroplast development in pigment deficient mutants of maize. I. Strict anomalies in plastids of allelic mutants at W_3 locus. *J. Ultrastruct. Res.* **21**, 41.

BERNS, D. S. & EDWARDS, M. R. (1965). Electron micrographic investigations of C-phycocyanin. *Archs Biochem. Biophys.* **110**, 511.

BISALPUTRA, T. & WEIER, T. E. (1964). The pyrenoid of *Scenedesmus quadricola*. *Am. J. Bot.* **51**, 881.

BISALPUTRA, T. & BISALPUTRA, A. A. (1967). Occurrence of DNA fibrils in chloroplasts of *Laurencia spectabilis*. *J. Ultrastruct. Res.* **17**, 14.

BOGORAD, L. (1967). Chloroplast structure and development. In *Harvesting the Sun*. Eds. A. San Pietro, F. A. Greer and T. J. Army. London and New York: Academic Press.

BOUCK, C. B. (1965). Fine structure and organelle associations in brown algae. *J. Cell Biol.* **26**, 523.

BOURDU, R. & LEFORT, M. (1967). Structure fine, observée en cryodécapage, des lamelles photosynthetiques des Cyanophycées endosymbiotiques: *Glaucocystis nostochinearum* Itzigs. et *Cyanophora paradoxa. C. r. Sci. Nat.* **265** (Ser. D), 37.

BRANTON, D. (1967). Structural units of chloroplast membranes. In *Le Chloroplast.* Ed. C. Sironval. Paris: Masson et Cie.

BRANTON, D. (1968). Structure of the photosynthetic apparatus. In *Photophysiology III.* Ed. A. C. Giese. London & New York: Academic Press.

BRANTON, D. (1969). Membrane structure. *A. Rev. Pl. Physiol.* **20**. (In Press.)

BROWN, W. V. & JOHNSON, S. C. (1962). The fine structure of the grass guard cell. *Am. J. Bot.* **49**, 110.

CHAMBERS, T. C. & MERCER, F. V. (1964). Studies on the comparative physiology of *Chara australis.* II. The fine structure of the protoplast. *Aust. J. Biol. Sci.* **17**, 372.

COHEN-BAZIRE, G. (1963). Some observations on the organization of the photosynthetic apparatus in purple and green sulphur bacteria. In *Bacterial Photosynthesis.* Eds. H. Gest, A. San Pietro and L. P. Vernon. Ohio: Antioch Press.

COHEN-BAZIRE, G., PFENNIG, N. & KUNISAWA, R. (1964). The fine structure of the green bacteria. *J. Cell Biol.* **22**, 207.

COHEN-BAZIRE, G. & SISTROM, W. R. (1966). The procaryotic photosynthetic apparatus. In *The Chlorophylls.* Eds. L. P. Vernon and G. P. Seeley. London and New York: Academic Press.

CRAIG, I. W., LEACH, C. K. & CARR, N. G. (1969). Studies with DNA from blue-green algae. *Arch. Mikrobiol.* **65**, 218.

CRAWLEY, J. C. W. (1964). Cytoplasmic fine structure in *Acetabularia. Expl Cell Res.* **35**, 497.

DRAWERT, H. & MIX, M. (1961). Licht und Elektronenmikroskopische untersuchungen an desmidiacean. *Planta* **56**, 648.

DREWS, G. & GIESBRECHT, P. (1965). Die Thylakoidstrukturen von *Rhodopseudomonas* spec. *Arch. Mikrobiol.* **52**, 242.

DREWS, G., LAMPE, H. H. & LADWIG, R. (1969). Die Entwicklung des Photosynthesapparates in Dunkelkulturen von *Rhodopseudomonas capsulata. Arch. Mikrobiol.* **65**, 12.

DRUM, R. W. & PANKRATZ, H. S. (1964). Pyrenoids, raphes and other fine structures in diatoms. *Am. J. Bot.* **51**, 405.

ECHLIN, P. (1964). The fine structure of the photosynthetic apparatus in blue-green algae. *Proc. 10th Int. Bot. Congr.* Edinburgh.

ECHLIN, P. (1967). The biology of *Glaucocystis.* I. The morphology and fine structure. *Br. phycol. Bull.* **3**, 225.

ECHLIN, P. (1968*a*). The fine structure and taxonomy of uni-cellular blue-green algae. *Proceedings 4th European Regional Conference on Electron Microscopy* **2**, 389.

ECHLIN, P. (1968*b*). The culture of the blue-green algae and the use of the electron microscope in identification and classification. In *Identification Methods for Microbiologists.* Part B. Eds. B. M. Gibbs and D. A. Shapton. London and New York: Academic Press.

ECHLIN, P. (1969*a*). The origins of plants. *New Scientist* **42**, 286.

ECHLIN, P. (1969*b*). Primitive photosynthetic organisms. *Proceedings 2nd Int. Conf. of Geochemistry London.* London: Pergamon Press.

ECHLIN, P. (1970). The origins of plants. In *Phytochemical Phylogeny.* April 1969 Symposium Phytochemical Society. London and New York: Academic Press.

ECHLIN, P. & MORRIS, I. (1965). The relationship between blue-green algae and bacteria. *Biol. Rev.* **40**, 143.

EDELMAN, M., SWINTON, D., SCHIFF, J. A., EPSTEIN, H. T. & ZELDIN, B. (1967). DNA of the blue-green algae (Cyanophyta). *Bact. Rev.* **31**, 315.

EIMHJELLEN, K. E., STEENSLAND, H. & TRAETTEBERG, J. (1967). A *Thiococcus* sp. nov. gen., its pigments and internal membrane system. *Arch. Mikrobiol.* **59**, 82.

ELLIS, R. J. (1969). Chloroplast ribosomes: stereospecificity of inhibition by chloramphenicol. *Science, N.Y.* **163**, 477.

ERICKSSON, G., KAHN, A., WALLES, B. & WETTSTEIN, VON D. (1961). Zur makromolekularen Physiologie der Chloroplasten. *Ber. Dt. bot. Ges.* **74**, 221.

EVANS, L. V. (1966). Distribution of pyrenoids among some brown algae. *J. Cell Sci.* **1**, 449.

EVANS, L. V. (1968). Chloroplast morphology and fine structure in British fucoids. *New Phytol.* **67**, 173.

FALK, H. (1967). Zum feinbau von *Botrydium granulatum* Grev. (Xanthophyceae). *Arch. Mikrobiol.* **58**, 212.

FALK, H. & KLEINIG, H. (1968). Feinbau und Carotinoide von *Tribonema* (Xanthophyceae). *Arch. Mikrobiol.* **61**, 347.

FAMINTZIN, A. (1907). Die Symbiose als Mittel der Synthese von Organismen. *Biol. Zbl.* **27**, 353.

FULLER, R. C. & CONTI, S. F. (1963). The microbial photosynthetic apparatus. In *Microalgae and Photosynthetic Bacteria*. Tokyo. Japanese Society of Plant Physiologists.

FULLER, R. C., CONTI, S. F. & MELLIN, D. B. (1963). The structure of the photosynthetic apparatus in green and purple sulphur bacteria. In *Bacterial Photosynthesis*. Eds. H. Gest, A. San Pietro and L. P. Vernon. Ohio: Antioch Press.

GANTT, E. & ARNOTT, H. J. (1963). Chloroplast division in the gametophyte of the fern *Matteuccia struthiopteris* (L.) Todara. *J. Cell Biol.* **19**, 446.

GANTT, E. & CONTI, S. F. (1965). Ultrastructure of *Porphyridium cruentum*. *J. Cell Biol.* **26**, 365.

GANTT, E. & CONTI, S. F. (1966). Granules associated with the chloroplast lamellae of *Porphyridium cruentum*. *J. Cell Biol.* **29**, 423.

GANTT, E. & CONTI, S. F. (1969). Ultrastructure of blue-green algae. *J. Bact.* **97**, 1486.

GANTT, E., EDWARDS, M. R. & CONTI, S. F. (1968). Ultrastructure of *Porphyridium aeruginosum*: a blue-green coloured Rhodophyton. *J. Phycol.* **4**, 65.

GEITLER, L. (1959). Syncyanosen. In Ruland's *Handbuch der Pflanzenphysiologie*, vol. 2, p. 530. Berlin-Gottingen-Heidelberg: Springer-Verlag.

GIBBS, S. P. (1962a). The ultrastructure of the pyrenoids in algae exclusive of green algae. *J. Ultrastruct Res.* **7**, 247.

GIBBS, S. P. (1962b). The ultrastructure of the pyrenoids of green algae. *J. Ultrastruct. Res.* **7**, 262.

GIBBS, S. P. (1962c) Nuclear envelope-chloroplast relationship in algae. *J. Cell Biol.* **14**, 433.

GIBBS, S. P. (1968). Autoradiographic evidence for the in situ synthesis of chloroplast and mitochondrial RNA. *J. Cell Sci.* **3**, 327.

GIBBS, S. P., SISTROM, W. R. & WORDEN, P. B. (1965). The photosynthetic apparatus of *Rhodospirillum molischianum*. *J. Cell Biol.* **26**, 395.

GIESBRECHT, P. & DREWS, G. (1962). Elektronenmikroskopische Untersuchungen uber die Entwicklung der 'Chromatophoren' von *Rhodospirillum molischianum* Giesberger. *Arch. Mikrobiol.* **43**, 152.

GIESBRECHT, P. & DREWS, G. (1966). Uber die Organisation und die makromolekular Architektur der Thylakoide 'lebender' Bakterien. *Arch. Mikrobiol.* **54**, 297.

GREENWOOD, A. (1967). In *The Plastids*. Eds. J. T. O. Kirk and R. E. Tilney-Bassett. London: W. H. Freeman & Co.

GRANICK, S. & GIBOR, A. (1967). The DNA of chloroplasts, mitochondria and centrioles. *Progress in Nucleic Acid Research and Molecular Biology* **6**, 143.

GUNNING, B. E. S. (1965*a*). The fine structure of chloroplast stroma following aldehyde osmium tetroxide fixation. *J. Cell Biol.* **24**, 79.

GUNNING, B. E. S. (1965*b*). The greening process in plastids. I. The structure of the prolamellar body. *Protoplasma* **60**, 111.

HALL, W. T. & CLAUS, G. (1967). Ultrastructural studies on the cyanelles of *Glaucocystis nostochinearum* Itzigsohn. *J. Phycol.* **3**, 37.

HARDING, H., WILLIAMS, P. H. & McNABOLA, S. S. (1968). Chlorophyll changes, photosynthesis of chloroplasts in *Albugo candida* induced 'green islands' in detached *Brassica juncea* cotyledons. *Can. J. Bot.* **46**, 1229.

HESLOP-HARRISON, J. (1963). Structure and morphogenesis of lamellar systems in grana-containing chloroplasts. I. Membrane structure and lamellar architecture. *Planta* **60**, 243.

HICKMAN, D. D. & FRENKEL, A. W. (1959). The structure of *Rhodospirillum rubrum*. *J. Biochem. Biophys. Cytol.* **6**, 277.

HICKMAN, D. D. & FRENKEL, A. W. (1965*a*). Observations on the structure of *Rhodospirillum rubrum*. *J. Cell Biol.* **25**, 279.

HICKMAN, D. D. & FRENKEL, A. W. (1965*b*). Observations on the structure of *Rhodospirillum molischianum*. *J. Cell Biol.* **25**, 261.

HODGE, A. J., McLEAN, J. D. & MERCER, F. V. (1955). Ultrastructure of the lamellae and grana in the chloroplasts of *Zea mays* L. *J. Biophys. Biochem. Cytol.* **1**, 605.

HOLM-HANSEN, O. (1968). Ecology, physiology and biochemistry of blue-green algae. *A. Rev. Microbiol.* **22**, 47.

HOLT, S. C., CONTI, S. F. & FULLER, R. C. (1966). Photosynthetic apparatus in the green bacterium *Chloropseudomonas ethylicum*. *J. Bact.* **91**, 311.

HOLT, S. C., TRÜPER, H. G. & TAKACS, B. J. (1968). Photosynthesis of *Ectothiorhodospira mobilis* strain 8113 thylakoids: chemical fixation and freeze etching studies. *Arch. Mikrobiol.* **62**, 111.

JOST, M. (1965). Die Ultrastruktur von *Oscillatoria rubescens* D.C. *Arch. Mikrobiol.* **50**, 211.

JOYDON, L. & FOTT, B. (1964). Quelques particularites infrastructurales dué plaste des Carteria (Volvocales). *J. Microscopie* **3**, 159.

KIRK, J. T. O. & TILNEY-BASSETT, R. A. E. eds. (1967). *The Plastids*. London: W. H. Freeman & Co.

KISLEV, V., SWIFT, H. & BOGORAD, L. (1965). Nucleic acids of chloroplast and mitochondria in Swiss chard. *J. Cell Biol.* **25**, 327.

LANG, N. (1968). The fine structure of blue-green algae. *A. Rev. Microbiol.* **22**, 15.

LANG, N. & RAE, P. M. M. (1967). Structures in a blue-green alga resembling prolamellar bodies. *Protoplasma* **64**, 67.

LASCELLES, J. (1968). The bacterial photosynthetic apparatus. *Advances in Microbial Physiology* **2**, 1.

LEADBEATER, B. & DODGE, J. D. (1966). The fine structure of *Wolozsynskia micra* sp. nov.—a new marine dinoflagellate. *Br. Phycol. Bull.* **3**, 1.

LEEDALE, G. F. (1967). Euglenida—Euglenophyta. *A. Rev. Microbiol.* **21**, 31.

LEFORT, M. (1959). Aperçu sur la structure du chromatoplasma de diverses Cyanophycées. *Bull. Soc. fr. Physiol. vég.* **5**, 187.

LEFORT, M. (1965). Sur le chromatoplasma d'une Cyanophycée endosymbiotique: *Glaucocystis nostochinearum*. *C. r. Sci. Paris* **261**, 233.

LEMBI, C. A. & LANG, N. J. (1965). Electronmicroscopy of *Carteria* and *Chlamydomonas. Am. J. Bot.* **52**, 464.

MAIER, K. & MAIER, U. (1968). Zur Frage einer Neubildung von Mitochondrien aus Plastids. *Protoplasma* **65**, 239.

MANTON, I. (1962). Observations on plastid development in the meristem of *Anthoceros. J. exp. Bot.* **13**, 325.

MANTON, I. (1966a). Some possibly significant structural relations between chloroplasts and other cell components. In *Biochemistry of Chlorophylls*. Ed. T. W. Goodwin. London and New York: Academic Press.

MANTON, I. (1966b). Further observations on the fine structure of *Chrysochromulina chiton* with special reference to the pyrenoid. *J. Cell Sci.* **1**, 187.

MCHALE, J. T. (1965). An ultrastructural study of *Selaginella* chloroplasts. *Am. J. Bot.* **52**, 630.

MCLAUGHLIN, J. J. A. & ZAHL, P. A. (1966). *Endozoic Algae in Symbiosis*, vol. 1, p. 257. Ed. S. M. Henry. London and New York: Academic Press.

MENKE, W. (1961a). Uber das lamellarsystem des Chromatoplasmas von Cyanophyceen. *Z. Naturf.* **166**, 543.

MENKE, W. (1961b). Uber die Chloroplasten var. *Anthoceros punctatus. Z. Naturf.* **166**, 334.

MENKE, W. (1962). Structure and chemistry of plastids. *A. Rev. Pl. Physiol.* **13**, 27.

MENKE, W. (1966). The structure of the chloroplast. In *The Biochemistry of the Chlorophylls*, vol. 1. Ed. T. W. Goodwin. London and New York: Academic Press.

MERESCHKOWSKY, C. (1905). Uber Natur und Ursprung der Chromatophoren in Pflanzenreiche. *Biol. Zbl.* **25**, 593.

MERRETT, M. (1969). Observations on the fine structure of *Chlamydobotrys stellata* with particular reference to the unusual chloroplast structure. *Arch. Mikrobiol.* **65**, 1.

NEWTON, J. W. (1963). Composition of bacterial chromatophores. In *Bacterial Photosynthesis*. Ed. H. Gest, A. San Pietro and L. P. Vernon. Ohio: Antioch Press.

ODINTSOVA, M. S. & YURINA, N. P. (1969). Proteins of chloroplasts and cytoplasmic ribosomes. *J. molec. Biol.* **40**, 503.

PANKRATZ, H. S. & BOWEN, C. C. (1963). Cytology of blue-green algae. I. The cells of *Symploca muscorum. Am. J. Bot.* **50**, 387.

PAOLILLO, D. J. (1962). The plastids of *Isoetes howellii. Am. J. Bot.* **49**, 590.

PAOLILLO, D. J., MACKAY, N. C. & GRAFFIUS, J. R. (1969). The structure of grana in flowering plants. *Am. J. Bot.* **56**, 344.

PARK, R. B. (1966). Chloroplast structure. In *The Chlorophylls*. Ed. L. P. Vernon and G. R. Seeley. London and New York: Academic Press.

PARSONS, D. F. & YANO, Y. (1967). The cholesterol content of the inner and outer membranes of guinea-pig liver mitochondria. *Biochim. biophys. Acta* **135**, 362.

PFENNIG, N. (1967). Photosynthetic bacteria. *A. Rev. Microbiol.* **21**, 285.

PFENNIG, N. & COHEN-BAZIRE, G. (1967). Some properties of the green bacterium *Pelodictyon clathratiforme. Arch. Mikrobiol.* **59**, 226.

PICKETT-HEAPS, J. D. (1967). Ultrastructure and differentiation in *Chara*. I. Vegetative cells. *Aust. J. biol. Sci.* **20**, 539.

PRINGSHEIM, E. G. (1963). Farblose Algen. *Ein Beitrag zur Evolutionsforschung*. Stuttgart: Gustav Fischer.

RAYMOND, J. C. & SISTROM, W. R. (1967). Isolation and preliminary characterization of a halophilic photosynthetic bacterium. *Arch. Mikrobiol.* **59**, 255.

REMSEN, C. G., WATSON, S. W., WATERBURY, J. B. & TRÜPER, H. G. (1968). Fine structure of *Ectothiorhodospira mobilis*. Pelsh. *J. Bact.* **95**, 2374.

Ris, H. & Plaut, W. (1962). Ultrastructure of DNA-containing areas in the chloroplasts of *Chlamydomonas*. *J. Cell Biol.* **13**, 383.

Sagan, L. (1967). On the origin of mitosing cells. *J. theor. Biol.* **14**, 225.

Schmid, G. H. (1967). Photosynthetic capacity and lamellar structures in various chlorophyll deficient plants. *J. Microscopie* **6**, 485.

Schnepf, E., Koch, W. & Deichgraber, L. (1967). Zur Cytologie und Taxonomischen Einordung von *Glaucocystis*. *Arch. Mikrobiol.* **55**, 149.

Scott, N. S., Shah, V. C. & Smillie, R. M. (1968). Synthesis of chloroplast DNA in isolated chloroplasts. *J. Cell Biol.* **38**, 151.

Shipp, W. S., Kieras, F. J. & Haselkorn, R. (1965). DNA associated with tobacco chloroplasts. *Proc. natn. Acad. Sci. U.S.A.* **54**, 207.

Shumway, K. L. S. & Weier, T. E. (1967). The chloroplast structure of Iojap maize. *Am. J. Bot.* **54**, 773.

Spurr, A. R. & Harris, W. M. (1968). Ultrastructure of chromoplasts and chloroplasts in *Capsicum annuum*. I. Thylakoid membrane changes during fruit ripening. *Am. J. Bot.* **55**, 1210.

Stutz, E. & Noll, H. (1967). Characterization of cytoplasmic and chloroplast polysomes in plants: evidence for three classes of ribosomal RNA in nature. *Proc. natn. Acad. Sci. U.S.A.* **57**, 774.

Sun, C. N. (1961). Submicroscopic structure and development of the chloroplasts of *Psilotum triquetrum*. *Am. J. Bot.* **48**, 311.

Sun, C. N. (1962). Submicroscopic structure and development of chloroplast and pyrenoid in *Anthoceros laevis*. *Protoplasma* **55**, 89.

Sun, C. N. (1963). Submicroscopic structure and development of chloroplasts in *Equisetum hiemale*. *Protoplasma* **56**, 346.

Sun, C. N. (1965). Structural alterations of chloroplasts induced by virus in *Abutilon striatum* v. Thompson. *Protoplasma* **60**, 426.

Thomson, W. W. & Weier, T. E. (1962). Fine structure of chloroplasts from mineral deficient leaves of *Phaseolus vulgaris*. *Am. J. Bot.* **49**, 1047.

Thomson, W. W., Weier, T. E. & Drever, H. (1964). Electron microscope studies on chloroplasts from phosphate deficient plants. *Am. J. Bot.* **51**, 938.

Trentini, W. C. & Starr, M. P. (1967). Growth and ultrastructure of *Rhodomicrobium vannielii* as a function of light intensity. *J. Bact.* **93**, 1699.

Trüper, H. G. (1968). *Ectothiorhodospira mobilis* Pelsh: a photosynthetic sulphur bacterium depositing sulphur outside the cells. *J. Bact.* **95**, 1910.

Vatter, A. E. & Wolfe, R. S. (1958). The structure of photosynthetic bacteria. *J. Bact.* **75**, 480.

Vatter, A. E., Douglas, H. C. & Wolfe, R. S. (1959). Structure of *Rhodomicrobium vannielii*. *J. Bact.* **77**, 812.

Vernon, L. P. (1968). Photochemical and electron transport reactions of bacterial photosynthesis. *Bact. Rev.* **32**, 243.

Weckesser, J., Drews, G. & Tauschel, H. D. (1969). Zur Feinstuktur und Taxonomie von *Rhodopseudomonas gelatinosa*. *Arch. Mikrobiol.* **65**, 346.

Wehrmeyer, W. (1964). Uber Membranbildungsprozesse im Chloroplasten. *Planta* **63**, 13.

Weier, T. E., Stocking, C. R., Thomson, W. W. & Drever, H. (1963). The grana as structural units in chloroplasts of mesophyll of *Nicotiana rustica* and *Phaseolus vulgaris*. *J. ultrastruct. Res.* **8**, 122.

Wettstein, von D. (1959). Formation of plastid structures. *Brookhaven Symp. Biol.* **11**, 138.

Wiessner, W. & Ametunzen, F. (1969). Beziehungen Zwischen submikroskopischer Chloroplasten struktur und Aft der Kohlenstoffquelle unter phototrophen Ernahrungesbedingungen bei *Chlamydobotrys stellata*. *Arch. Mikrobiol.* **66**, 14.

WILDON, D. C. & MERCER, F. V. (1963). The ultrastructure of the vegetative cell of blue-green algae. *Aust. J. biol. Sci.* **16**, 585.

WILSENACH, R. (1963). Differentiation of the chloroplast of *Anthoceros. J. Cell Biol.* **18**, 419.

WORDEN, P. B. & SISTROM, W. R. (1964). Preparation and properties of bacterial chromatophore fractions. *J. Cell Biol.* **23**, 135.

ADDENDUM

Since this paper was originally written interesting new findings have been published about the evolutionary origins of chloroplasts. Allsopp (1969) considers that the red algae represent an intermediate group between the blue-green algae and higher eukaryotes, and believes that the more primitive red algae were originally derived from blue-green algae and in turn gave rise to all other eukaryotes. This hypothesis appears to be based on the type of evidence used in considering the endogenous or exogenous origins of chloroplasts. Allsopp discusses both these hypotheses and on balance considers that the cell organelles of eukaryotes were more likely to have arisen by progressive evolution of prokaryotes than by progressive endosymbiosis. Following the work of Taylor (1968) on symbiotic chloroplasts in opisthobranchs, Trench, Greene & Bystrom (1969) showed that chloroplasts remained functional for at least six weeks in several marine gastropod molluscs. Nass (1969) has shown that mouse fibroblasts in suspension culture will incorporate isolated spinach and African violet chloroplasts, these structures retaining their integrity for some time. Perhaps this phenomenon, together with the presence of endocyanelles in certain algae, should now best be considered as yet another of the multifarious properties of prokaryotes. It may simply be that these photosynthetic structures organisms have a curious ability which enables them to live free, as in the case of blue-green algae and some *Chlorella* species, or to become incorporated, sometimes irreversibly, into other seemingly alien host cells. Although in the short term this incorporation may appear to endow a great selective advantage, we may have been premature in thinking that this phenomenon has any true phylogenetic significance.

ALLSOPP, A. (1969). Phylogenetic relationships of the procaryota and the origin of the eucaryotic cell. *New Phytologist* **68**, 591.

NASS, M. (1969). Uptake of isolated chloroplasts by mammalian cells. *Science* **165**, 1128.

TAYLOR, D. L. (1968). Chloroplasts as symbiotic organelles in the digestive gland of *Elysia viridis* [Gastropoda: Opisthobranchia]. *J. Mar. Biol. Assoc. U.K.* **48**, 1.

TRENCH, R. K., GREENE, R. W. & BYSTROM, B. G. (1969). Chloroplasts as functional organelles in animal tissues. *J. Cell Biol.* **42**, 404.

PLASMIDS AND CHROMOSOMES IN PROKARYOTIC CELLS

M. H. RICHMOND

*Department of Bacteriology, University of Bristol,
University Walk, Bristol, BS8 1TD*

INTRODUCTION

For many years microbiologists, betraying their origins in botany and zoology, tried desperately to allocate all the bacteria they isolated to one of the various 'species' that had already been defined on the basis of a range of cultural characteristics. However, such was the variety encountered among bacteria that the number of 'distinct' species described had to be increased steadily, as the growth from edition to edition of such tomes as *Bergey's Manual of Determinative Bacteriology* shows.

Nowadays, many of the sources of bacterial variation are understood, to some extent at least, and the attitude of microbiologists has become somewhat different. Whereas before, attempts were made to show conformity, now much emphasis is placed on diversity, and it is possible these days to argue that certain species of bacteria might not be nearly so prevalent were it not for the activities of microbiologists who painstakingly re-isolate only the bacteria with characters that fit into the Bergey classification. Anyone who has attempted to 'identify' enteric bacteria isolated from faecal samples by reference to Bergey—or even to Cowan & Steel (1966)—knows the variation and overlap between the characters of such bacterial isolates, particularly when the temptation to throw out those that 'don't fit' is resisted. Nor is it possible nowadays to think of bacterial species as characterized by a tightly limited range of characters. If species means anything among bacteria these days, it relates more to the characters displayed by organisms occupying a given ecological niche than to anything else; and a limited alteration in the environment can be expected to lead to a culture of organisms with characteristics that could even allow its being classified as a new species.

Undoubtedly, one of the sources of variation that can play an important part in the bacterial cell is the extrachromosomal genetic elements known as plasmids. Whereas mutation can modify the nature of the cell's DNA with a consequent direct effect on phenotype, the change more often than not leads to the loss of a character, and this is usually deleterious, at least in the first instance. Acquisition of a plasmid, however, ensures that the cell receives a block of genetic material that, in some cases, may contain six to ten new complete genes. Such an addition to the genetic content of the cell may therefore be of the greatest advantage to that cell in its survival in an existing environment or in adjustment to a new one. In the sections that follow, therefore, the nature and transmission of bacterial plasmids will be discussed to try to assess the extent to which they play a part in the flexible adaptation by bacteria to a changing environment; and incidentally to try to undermine, to some extent, the 'old' idea of speciation among prokaryotes.

BACTERIAL PLASMIDS

Plasmids and chromosomes

Before describing the properties and behaviour of bacterial plasmids it is important to define the word *plasmid*, and to distinguish it, if possible, from the word *chromosome*. Both plasmids and the chromosome are capable of autonomous self-replication in a bacterial cell; that is, they are *replicons* according to the definition of Jacob & Brenner (1963). In practice two characters often distinguish plasmids from the chromosome and they have been used by some in an attempt to define the two types of replicon.

(1) The amount of genetic material carried by the plasmid is much less than that in the chromosome. In many plasmids, the amount of DNA may be as little as 0·1 to 0·2 % of that in the chromosome.

(2) Many plasmids can be lost from the cell without prejudicing the cell's ability to grow in a wide range of laboratory media. For example, in a strain of *Staphylococcus aureus* carrying a penicillinase plasmid, loss of the plasmid often leads to a slightly enhanced rate of growth in all media that will normally support the growth of staphylococci, so long as penicillin is not present in the medium (see, for example, Fig. 5 in Richmond, 1968).

Although these two plasmid characteristics make it attractive to define plasmids either as 'small replicons' or 'adventitious replicons', this is really unsatisfactory—the first since the distinction is arbitrary and the second since it is preferable to give a definition that holds for all growth conditions. It seems best, therefore, to define the chromosome as the replicon which carries the genetic information necessary for a particular essential cell character. Because the most typical cell character is ability to divide, the chromosome then can be defined as the replicon which carries the genetic information for cell division. In the light of this definition, then, plasmids are non-chromosomal or extrachromosomal replicons.

Another term much used in the past in discussion of bacterial plasmids is *episome*. As originally defined for use about bacteria, the term referred to genetic material that was capable of reversible transition between the chromosomal and the extrachromosomal states in such a manner that the genetic content of the cell with respect to the episomal markers was not altered (Jacob & Wollman, 1958). The last half of this definition was meant to exclude gene duplications and the like. While the term episome has been very useful, the genetic interactions of many extrachromosomal elements have now been found so complex that behaviour as a true episome is only one facet of their existence (see below). Because of this Hayes (1969a) and Wollman (1969) have suggested that the noun 'episome' should no longer be used, but that the adjective 'episomal' should be retained to describe a type of genetic interaction. The definition of episome has never included the character of *transmissibility* (see p. 269) although the classical episome, the F-factor, is, of course, normally transmissible.

Plasmid replication and distribution to daughter cells

A common feature of all the plasmids so far identified is that they are independent replicons and probably therefore consist of double-stranded DNA molecules (see p. 269). Most plasmids can be considered as consisting of two parts, one part concerned with replication of the replicon and its distribution to daughter cells, and the other part carrying genetic determinants that confer properties on the host cell but which are not themselves concerned directly with plasmid survival.

As far as the part involved in plasmid replication and distribution is concerned, no clear description of the molecular elements involved is yet available. Jacob, Brenner & Cuzin (1963) suggested that there must be a direct point of contact between the plasmid and the cell structure if accurate distribution is to be achieved and proposed that this occurs at 'membrane attachment sites'. In molecular terms, this implies an intermolecular bond between plasmid-DNA and some part of the cell structure. Consequently there must be two genetic regions involved in the process, one on the plasmid and presumably one on the chromosome. Two classes of mutants in which plasmid stability is impaired have been isolated—one class is plasmid-linked and the other is apparently chromosomal—from *Escherichia coli* (Jacob *et al.* 1963; Cuzin & Jacob, 1965a, b) and *Staphylococcus aureus* (Novick, 1967a). They may represent the two classes of genes involved in the interacting components of the attachment site.

In addition to this intermolecular 'attachment', plasmid survival probably also involves several enzymes. One such enzyme is certain to be a DNA polymerase molecule needed to replicate the plasmid-DNA and another might be an enzyme or nucleic acid factor necessary to initiate a new round of plasmid replication. Cuzin & Jacob (1965 *b*) reported at least three separate cistrons involved in the survival of F'-*lac* in *Escherichia coli* which may represent the relevant genes and their phenotypic effects.

Even less is known about the molecular mechanisms involved in the distribution of replicated plasmids. If the attachment site hypothesis is correct, distribution must presumably await the synthesis of a new attachment site or mechanism, and this in turn is likely to require protein synthesis. It might be at this stage that some of the compounds known to impair plasmid survival (such as acriflavine; Hirota, 1960) exert their effect, since many of them are known to be inhibitors of RNA polymerase (see the section on 'curing' below). Basically the problems of plasmid distribution have much in common with those concerning distribution of the chromosome (Lark, 1966; Pritchard, Barth & Collins, 1969).

In summary, therefore, plasmid replication and distribution is already known to involve several genes, some plasmid-borne and some chromosomal. Some of the genes may be involved in the interaction of the plasmid-DNA with the cell structure, but others may synthesize diffusible products such as polymerase enzymes; this suggests the possibility of complementation between two plasmids co-existing in the same cell. Such complementation has already been reported for pairs of mutant F-factors in *Escherichia coli* (Jacob *et al.* 1963) but the uncertainty about the precise genetic situation that exists when an F-carrying cell is superinfected with a further F plasmid (Dubnau & Maas, 1968) makes the interpretation of these experiments very uncertain.

'Curing'

Because plasmids are not essential to bacteria (at least under most conditions of growth) any agent that interferes with plasmid replication or distribution, without having an equivalent effect on the host, will lead to impaired survival of the plasmid in dividing bacteria. Several compounds have been claimed to 'cure', 'purge' or 'free' bacteria from plasmids in this way; these are listed in Table 1.

The first of these compounds to be described (and it is important to distinguish its effect from other 'curing' agents) was acriflavine (Hirota, 1960). This compound cures *Escherichia coli* of the F-factor, but is much less effective against other plasmids and other cells. In the case of the F-factor the effect is dramatic. Overnight growth of cells in the presence of acriflavine at pH 7·6 (the pH value is important) gives a culture from which the F-factor has completely disappeared. All the other curing agents tested so far exert a much less sharp effect (see Table 1). The next most effective compound would seem to be rifampicin (a rifamycin), which at 0·01 μg./ml. may yield a staphylococcal culture, 25 % of which is plasmid-free after overnight growth from an inoculum that contained at most 1 % of plasmid-free cells when the agent was added (J. H. Johnston and M. H. Richmond, unpublished experiments).

Table 1. *Some compounds reported to 'cure' prokaryotic cells of extrachromosomal elements*

Compound	Plasmid	References
Acriflavine and other acridines	F(G−)	Hirota (1960)
	Penicillinase (G+)	Hashimoto, Kono & Mitsuhashi (1964)
	R(G−)	Watanabe & Fukasawa (1961)
	Chloramphenicol resistance (G+)	Chabbert, Baudens & Gerbaud (1964)
Cobalt ions	F(G−)	Hirota (1960)
Ethidium bromide	Penicillinase (G+)	Bouanchaud, Scavizzi & Chabbert (1968)
	R(G−) and F(G−)	Bouanchaud *et al.* (1968)

G− = Gram-negative: G+ = Gram-positive.

At present the precise molecular action of curing agents is unknown. However, most of the active compounds either intercalate with DNA or interfere directly with DNA replication (Waring, 1966). This might be the origin of the curing effect, particularly if plasmid replication is more sensitive than chromosomal replication. Alternatively, the curing agents may interact specifically with RNA polymerase (rifampicin is known to inhibit bacterial RNA polymerase specifically; Wehrli, Knüsel, Schmid & Staehelin, 1968; Lancini, Pallanza & Silvestri, 1969), or indirectly through the inability of intercalated DNA to act as an adequate template for RNA polymerase (see Newton, 1966).

Types of plasmid

If the picture outlined above is true, all plasmids share a group of genes responsible for ensuring their replication and distribution to daughter cells. Certainly detailed differences may exist in this group of genes from plasmid to plasmid, particularly perhaps as between plasmids from Gram-positive and Gram-negative bacteria, but there may be no difference in principle. Against this common background, the identity of the different types of plasmid depends upon the nature and number of the other genes carried. On these grounds it is possible to group the plasmids of prokaryotic cells into several classes as shown in Table 2. However this classification is made, it is unsatisfactory, since, as we shall see later, recombination between different types of plasmid may occur freely, leading to transient hybrid plasmids containing genes derived from two parental structures (see p. 265). Most usually plasmids are classified in two main categories, transmissible or non-transmissible, according to whether or not they carry the genes necessary for conjugation. Because of this special property (see p. 269), transmissible plasmids are often thought of as a case apart, but this is unjustified so far as their *extrachromosomal* survival is concerned. Nor is it correct to think of non-transmissible plasmids as incapable of being transferred to other bacterial cells (see p. 269).

Table 2. *A classification of different types of bacterial plasmids and the organisms in which they are found*

(A) '*Transmissible*'	(B) '*Non-transmissible*'
Some R-factors (enteric bacteria)	Some R-factors (enteric bacteria)
Some *col*-factors (enteric bacteria)	Some *col*-factors (enteric bacteria)
F (fertility factor) (enteric bacteria)	Staphylococcal plasmids (*S. aureus*)
Haemolysin factor (enteric bacteria)	
FP-factor (Pseudomonas)	
? P-factor (*Vibrio cholerae*)*	

* For *V. cholerae* P-factor see Bhaskaran (1964).

Another method of classifying plasmids is into de-repressible and non-de-repressible types. Examples of the first class are prophage P1 (Lennox, 1955; Adams & Luria, 1958) and the colicinogenic factor responsible for the synthesis of colicin E1 (DeWitt & Helinski, 1965). Normally both these plasmids replicate in step with the bacterial chromosome and their replication is therefore matched in the multiplication of the cell, precisely as found with other plasmids. Under certain circumstances, however, uncontrolled plasmid replication occurs to produce a vegetative burst of DNA synthesis, and this kills the cell (Herschman & Helinski, 1967). In this respect prophage P1 is distinct from many other lysogenizing phages in which the phage genome exists in the lysogenic state as a part of the chromosome and not as a free replicon. Phage P1 has no chromosomal attachment site in the strict sense and can survive in a manner exactly analogous to other independent replicons (Boice & Luria, 1961).

Apart from these two methods of classification, which are based on special additional attributes of the plasmids, most plasmids are grouped by the nature of the genetic markers they carry. The majority of plasmids detected and studied in Gram-positive and in Gram-negative bacteria carry linked antibiotic resistance-determinants; these are called resistance factors, or R-factors, particularly when they occur in Gram-negative bacteria. A very wide range of this type of plasmid is known. Apart from factors which carry resistance-determinants, a few are also known that carry other genes; for example, some carry genes whose products have a biosynthetic or metabolic role. Thus the *lac* operon (Adelberg & Burns, 1959, 1960) and certain groups of arginine biosynthesis genes (Maas, Maas, Wiame & Glansdorff, 1964; Scaife, 1967) have been found as part of the F-plasmid in some strains of *Escherichia coli*; such plasmids have, of course, been widely used in microbial genetics. Other genetic determinants carried, at least under some circumstances, in the plasmid state, are those which specify the K88 antigen in some strains of *Escherichia coli* (Ørskov & Ørskov, 1966) and those involved in haemolysin production by some strains of *E. coli* pathogenic for pigs (Williams-Smith & Halls, 1967; Williams-Smith, 1969).

Physical properties of plasmids

Although plasmids exist in Gram-positive and in Gram-negative bacteria, it is only in Gram-negative bacteria that their physical properties have been examined in any detail. Genetic experiments have suggested clearly that plasmids in these bacteria are either circular or circularly-permuted linear molecules (see Fig. 1) for at least some stage of their existence (see Roth & Helinski, 1967). Examination of the plasmid-DNA from these cells by the electron microscope has shown this to be so. It appears to be characteristic of all plasmids in cells isolated from natural surroundings that a simple differential centrifugation designed to separate the plasmids from the chromosomal-DNA on the basis of buoyant density does not show the presence of any plasmid-DNA (Marmur *et al.* 1961). To demonstrate the plasmid as a separate entity it is first necessary to transfer it to a species whose chromosomal-DNA has a distinctly different base ratio (and consequently buoyant density) from the chromosomal-DNA of the original host. In practice transfer from *Escherichia coli* to a *Proteus* sp. or to *Serratia marcescens* has proved effective. When this is done, density-gradient centrifugation of DNA from the recipient shows the plasmid to be a circular DNA molecule with about 1 to 2 % of the length of the bacterial chromosome

(Falkow, Citarella, Wohlheiter & Watanabe, 1966; Rownd, Nakaya & Nakamura, 1966).

Recently Nisioka, Mitani & Clowes (1969) have studied R-factor DNA with the electron microscope a stage further by showing that cultures of cells carrying an R-factor may contain more than one type of circular plasmid molecule. Three contour lengths were found among

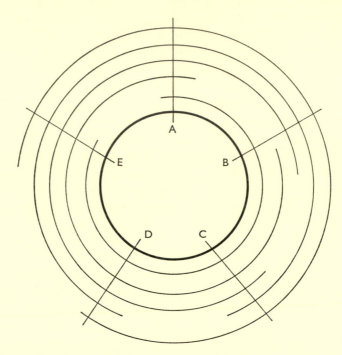

Fig. 1. A diagrammatic representation to show how linkage studies on a series of linearly permuted molecules with markers ABCDE, BCDEA, CDEAB, DEABC and EABCD can give the overall appearance of the presence of a circular linkage group.

the DNA molecules examined: 36 μm, 29 μm and 7 μm (Nisioka *et al.* 1969). After rejecting decisively the interpretation that these fragments were artifacts, the authors concluded that this particular R-factor exists in two forms in the strains of *Proteus mirabilis* used, and that the circular DNA molecules are related as follows:

$$36 \ \mu\text{m} \rightleftharpoons 29 \ \mu\text{m} + 7 \ \mu\text{m}.$$

A similar picture has been obtained for the *col* E1 plasmid (DeWitt & Helinski, 1965; Bazarel & Helinski, 1968*a, b*; Goebel & Helinski, 1968).

Although experiments on isolated plasmid-DNA have not yet been made with *Staphylococcus aureus*, all the genetic data suggest that the plasmids in these strains are either circular or circularly permuted (Richmond, 1967, 1968); indeed circularity may be an essential requirement for replicon survival in all prokaryotic cells.

Fragmentation of plasmids

The observations of Nisioka *et al.* (1969) suggest that certain plasmids may break down and re-form in the host cell. This conclusion is supported by observations on plasmids of enteric bacteria and of *Staphylococcus aureus*. In *Salmonella typhimurium* type 1*a* plasmid-carrying genes which confer resistance to ampicillin, streptomycin, sulphonamide and tetracycline, occasionally seem to break down to form a plasmid conferring resistance to ampicillin, sulphonamide and tetracycline, and another plasmid-conferring streptomycin resistance, since streptomycin-sensitive segregants can be isolated from the parent strain at low frequency (Anderson, 1968). Similarly, staphylococci carrying a penicillinase plasmid with the markers $pen^r.ero^r$ segregate penicillin-sensitive cells, albeit at a very low frequency (M. H. Richmond, unpublished obervations). This phenomenon is further discussed below in the section on plasmid/plasmid interactions (see p. 262).

Integration

Although plasmids are capable of autonomous replication and survival in the host cell, this does not imply that they always exist in this state. In many cases part or all of the plasmid can integrate into the chromosome so that the plasmid-DNA becomes strictly part of the chromosomal replicon, and the replication and distribution of the plasmid genes becomes a direct consequence of chromosomal replication and distribution (see Adelberg & Pittard, 1965; Scaife, 1967; Hayes, 1969*b* for detailed discussions of this process). In some cases, plasmid integration can be shown to occur and to be reversible; in other cases the interaction of the plasmid and the chromosome can merely be inferred from the 'passage' of certain genes from the chromosomal to the plasmid state.

Plasmid integration can be shown most clearly with the F-factor (Adelberg & Pittard, 1965). The mechanism of integration into the *Escherichia coli* chromosome appears to occur as shown diagrammatically in Fig. 2 (Scaife & Gross, 1963; Scaife & Pekhov, 1964). In this case there is experimental verification of a model originally proposed by Campbell (1962) to explain the integration of phage λ. There are two requirements for integration: (1) the existence of homology

9

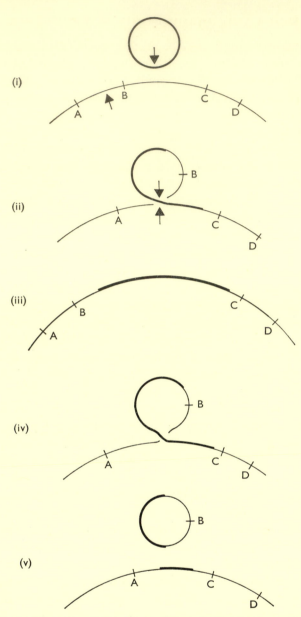

Fig. 2. A diagrammatic representation of a possible mechanism for the reversible integration of a plasmid (thick line) into a chromosome (thin line) with the concomitant excision of a chromosomal marker (B). (i). Plasmid and chromosome maintained independently in the same cell. (ii). Apposition of a chromosomal region with a homologous region on the plasmid (indicated by arrows) and the formation of a single cross-over at the point of homology. (iii). The plasmid now integrated as part of the chromosome. (iv). Apposition of the regions of homology once more. (v). Excision of part of the plasmid and part of the chromosome to form a 'new' replicon, now containing the marker 'B'. A, B, C, D: hypothetical genetic markers. This model was first described by Campbell (1962) to explain the integration of phage λ into the *Escherichia coli* chromosome and was subsequently adapted by Broda, Beckwith & Scaife (1964) to account for the mobilization of chromosomal markers by an F plasmid.

between regions on the plasmid and part of the chromosome; (2) the presence of an enzyme capable of catalysing the insertion of the plasmid-DNA, because this involves the breakage and rejoining of DNA strands.

It is difficult to see why any homology should exist between plasmid and chromosome unless the plasmid were originally derived from the chromosome, yet it is a fact that various F factors are capable of integrating at different chromosomal sites (Wollman, Jacob & Hayes, 1956) and there is undoubtedly some homology between an F plasmid and some part of the chromosome (Falkow & Citarella, 1965). Furthermore, an $F'.lac$ plasmid integrates less well into a cell with a lac deletion than into one with a point mutation in the chromosomal lac region (Broda, Beckwith & Scaife, 1964). With certain prophages, such as λ, integration involves the action of a prophage-specified 'integrase' enzyme (Thomas, 1968). Plasmid integration, on the other hand, seems to depend on the use of the cell's normal recombination enzymes since the integration of the F-factor in *Escherichia coli* is much impaired in recombination-deficient (rec^-) mutants (J. Scaife, personal communication).

Irreversible integration of plasmid fragments has been studied by Novick (1967b) with penicillinase plasmids in *Staphylococcus aureus*. In this case, ultraviolet irradiation of phage capable of transducing plasmids leads to the transfer of plasmid fragments rather than the whole plasmid. Some of the transduced fragments survive in the recipient in the extrachromosomal state, but the majority are found to have integrated into the chromosome. Once in this state the fragments seem unable to regain the plasmid state, and indeed transduction of these genes to a second, plasmid-less strain always yields transductants in which the transferred markers are chromosomal in the recipient (M. H. Richmond, unpublished observations). Together, these experiments suggest that such irreversibly integrated fragments may lack the plasmid component of the 'attachment site' mechanism, and consequently can only survive by using the chromosome for replication and distribution.

The reversible interaction of a plasmid with a chromosome is often inferred from experiments on gene 'mobilization'. For example, the F-factor can mobilize any chromosomal marker with a probability of about $1:10^4$ to $1:10^5$ of the F^+ population, but this value is calculated from the frequency of subsequent transfer of the genes concerned to an F^- recipient (Meynell, Meynell & Datta, 1968). Thus these experiments give only an indirect measure of the frequency of extraction of markers from the chromosome by the F-factor. Originally it was thought that R-factors in enteric bacteria mobilized chromosomal genes much less effectively than the F factor (Sugino & Hirota, 1962) but it is now clear that this low value is due rather to the poor efficiency of subsequent transfer to the R⁻ cell, rather than to any lower ability of the R-factor to extract genes from the chromosome (Meynell *et al*. 1968). Staphylococcal plasmids can also mobilize chromosomal markers. Thus an integrated erythromycin-resistance gene can be excised from the chromosome by a penicillinase plasmid lacking an *ero^r* marker (Richmond & Johnston, 1969*a*).

In summary, therefore, many plasmids indulge in transient interaction with the bacterial chromosome and this can lead to the extraction of chromosomal genes and their survival in the extrachromosomal state. Thus there is no clear distinction between 'plasmid' and 'chromosomal' genes; the situation is one of continuous flux. Apparently the gene remains fixed in the chromosome only in the absence of an extrachromosomal replicon with a base sequence homologous with the gene in question; but since extrachromosomal replicons are of frequent occurrence, one must assume that there is a continuous flow of genetic information from the chromosome to other replicons, and *vice versa* in most bacterial cells.

Plasmid compatibility

The presence of one type of plasmid in a cell often influences the acceptance by that cell of a second plasmid. For example, pairs of penicillinase plasmids may be assigned to one of two main classes, depending on whether or not they are compatible. Thus two plasmids which can co-exist in the same cell are said to be compatible and to belong to distinct compatibility groups (Novick & Richmond, 1965). Other pairs of plasmids which cannot co-exist, at least for any length of time, are said to be 'incompatible' and to belong to the same group.

Although with the staphylococcal plasmids the longevity of the bi-plasmid state was formerly discussed in terms of compatibility, the phenomenon may now be best understood in terms of the attachment site or mechanism used to maintain the plasmid in the cell. In these terms, plasmids of the same group cannot co-exist for long since they compete for the same maintenance site or system, whereas this competition does not arise between compatible elements because they may use separate systems (Richmond & Johnston, 1969*b*). This implies that there must be several distinct maintenance sites in *Staphylococcus aureus*; experimental evidence supports this view (Novick & Richmond, 1965; Novick, 1967*a*; Richmond & Johnston, 1969*b*). In many staphylococcal cells there appear to be at least four sites apart from that needed for chromosomal replication and distribution: one each for two types of penicillinase plasmids, one for tetracycline-resistant plasmids and one for a chloramphenicol-resistant plasmid.

There is no evidence, so far, that staphylococcal plasmids can be transferred from cell to cell by any means other than transduction. Apart from the maintenance considerations discussed above and the host range of the phage used, there appears to be no restriction on the uptake of staphylococcal plasmids into the cell. With the extrachromosomal elements of the enteric bacteria, however, the situation is much more complex and depends to a great extent on whether the plasmid is transferred by conjugation or by transduction. When conjugation is involved, transfer occurs infrequently to a cell already carrying an identical or closely related plasmid; but in many cases the presence of a different plasmid in the cell has little inhibitory effect. Where inhibition of this kind occurs, it may be of two types: either an exclusion due to a surface property of the female cell, or to the inability of the infecting plasmid-DNA to establish itself in the recipient. As an example of the first type of restriction, the F-plasmid can be transferred to DNA-less cells derived from an F^- parent, but not to cells where the parent was originally F^+, despite the absence of sex-pili from the surface of the latter (Cohen, Fisher, Curtis & Adler, 1969). Clearly therefore the presence of an F-plasmid in cells leads to some type of surface blockage to transfer, even in the absence of F sex-pili. The fact that this blockage disappears when F^+ cells are grown under starvation conditions (to produce an F^- phenocopy; Lederberg, Cavalli & Lederberg, 1952) and that the same effect can be produced by treating intact cells with periodate (Sneath & Lederberg, 1961) suggests that F^+ cells normally have a surface structure that prevents conjugation. As far as this blockage to transfer of the F-plasmid is concerned, the effective surface element seems to be determined by the F-plasmid in the recipient, but by no other type of plasmid. Thus the formation of mating pairs between an F^+ and any R^+ cell is unimpaired, even though the presence of the R-factor in the recipient inhibits the ultimate survival of the incoming F-element (see below; Watanabe et al. 1964). In most cases where a resident plasmid inhibits the transfer of a plasmid from a donor cell, the blockage is due to the inability of the incoming DNA to replicate in the recipient cell. This has been shown to occur with the transfer of F plasmids to an Hfr recipient (Scaife & Gross, 1962; Echols, 1963), for pairs of R factors (Takano, Watanabe & Fukusawa, 1966; Watanabe et al. 1966) and for pairs of col plasmids (Nagel de Zwaig, 1966). When the transfer of the plasmid occurs by transduction and not conjugation (Watanabe, 1963) then the inhibitory effect of the cell surface is bypassed and the survival of the incoming plasmid appears to be determined solely by its ability to replicate in the recipient cell.

In summary, the effect of a cell's resident plasmid on the acceptance and survival of an additional plasmid is extremely complex. In practically all cases the pattern is due either to the complete exclusion of the plasmid so that the infecting DNA never gets into the cell at all, or to a competition between the incoming element and the resident for the plasmid maintenance site. Further details of this highly complex aspect of plasmid behaviour have been reviewed by Meynell *et al.* (1968) and by Novick (1969).

Effects of plasmids on the phage pattern of the host cell

Many plasmids restrict the uptake of foreign non-plasmid DNA, such as phage particles, as well as other plasmids. Cells containing an F-plasmid (Zinder, 1960; Schell *et al.* 1963; Mäkelä, Mäkelä & Soikkelli, 1964; Hakura, Otsuji & Hirota, 1964), *col* plasmids (Watanabe & Okada, 1964; Strobel & Nomura, 1966) and R plasmids (Yoshikawa & Akiba, 1962; Watanabe *et al.* 1966) can all restrict the multiplication of superinfecting phage. This has considerable practical importance in the phage-typing of enteric bacteria (Felix, 1955; Anderson, 1965; Guinée, Scholtens & Willems, 1967). The presence of a plasmid can also affect phage-typing by providing a phage-receptor not present in plasmid-less cells. In the enteric bacteria many plasmids (transmissible type) specify the synthesis of a six-pilus which is a specific receptor for some phages (Dettori, Maccacaro & Piccinin, 1961; Dettori, Maccacaro & Turri, 1963; Dettori & Neri, 1965).

Plasmid/plasmid interactions

The integration of bacterial plasmids into the chromosome has been described (see p. 257). In essence this process is one of formation of a single replicon where two existed previously, and all the evidence suggests that the compound structure relies for its survival in the cell on the system normally used to replicate and distribute the chromosome. As well as integrating to become part of the chromosomal replicon, however, plasmids may merge, either transiently or permanently, with other plasmids when suitable ones are available. The molecular processes involved in this step are probably analogous to those of chromosomal integration (see Fig. 2), the only difference being in the size of the recipient replicon. The modification of the integration procedure needed to account for the merging of two plasmids to form a single structure is described diagrammatically in Fig. 3.

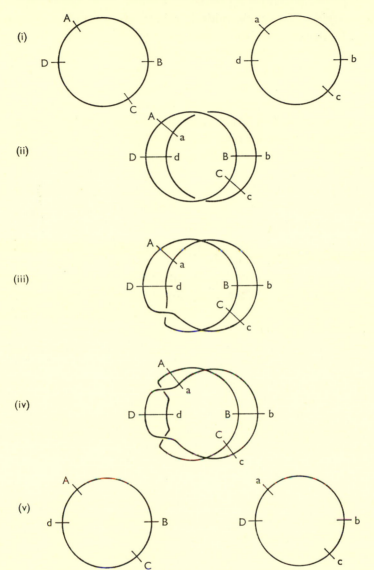

Fig. 3. A modification of the original Campbell (1962) hypothesis for the integration of phage λ to account for the 'association' and 'dissociation' of a pair of circular plasmids. (i) A pair of homologous plasmids maintained independently in a single cell. (ii) Apposition of the two plasmids so that regions of homology come to lie adjacent to one another. (iii) Formation of an associated single structure by one cross-over. (iv) Dissociation of the single plasmid into two separate plasmids once more by a second cross-over in the opposite sense from the first. (v) Two plasmids once more maintained independently in the cell.

Note that the outcome of this process is a complementary pair of recombinant plasmids. The precise distribution of the markers between the two plasmids is determined by the exact positions of the two cross-overs. The process will, however, never give two versions of the same marker on one plasmid: complementary pairs are always formed. A, B, C, D: hypothetical markers. Based on Richmond (1967).

The merging of two plasmids has been shown to occur in several systems. Under appropriate conditions, two penicillinase plasmids may merge (or 'associate') to form a double plasmid; but once formed this is free to re-form a pair of single plasmids (Richmond, 1967; Richmond & Johnston, 1969*b*). Among staphylococcal elements, plasmids of the same compatibility group seem to enter into this association more easily than two plasmids from distinct groups; but equally the associated 'double' structure reverts to two 'single' plasmids more easily when two plasmids from the same group are involved (Richmond & Johnston, 1969*b*).

Staphylococcal plasmids can also, as a relatively rare event, integrate with a defective piece of a phage genome (Novick, 1967*c*). Under these circumstances the phage strain carrying such a hybrid plasmid may act as a source of 'high-frequency transducing' phage when an appropriate recipient is available. Similarly, R-factors can integrate with various phage genomes to produce high frequency transducing lysates. So far this has been shown to occur with phages $\epsilon15$ (Kameda, Harada, Suzuki & Mitsuhashi, 1965), phage P22 (Dubnau & Stocker, 1964) and phage P1 (Kondo & Mitsuhashi, 1964, 1966).

Association and dissociation of plasmids has also been reported for some Gram-negative bacteria. The fragmentation of a single plasmid has been described above (see p. 257), but here we are more concerned with the free association and dissociation of two or more plasmids to form a single structure when further plasmids are added to a cell already carrying one plasmid. Complex associated plasmids have been obtained from *col* + F plasmids, from *col* + F'.*lac* plasmids, from two different types of *col* plasmid, from *col* + R plasmids, and from pairs of R plasmids of various kinds (for a review of the association of plasmids in Gram-negative bacteria, see Fredericq, 1969). Perhaps the most complex 'hybrid' plasmid yet described carries the markers *fer* (fertility), *colB,tonB* (resistance to phages T1 and ϕ80 and to *colB*,I and V).*trp*. *cysB*.*colV*.*R(S)*.*R(C)*.*R(T)*. This 'hybrid' plasmid was built by sequential transfer of separate plasmids, the first with the markers *R(SCT)*, the next with *colV* and the third with the remaining genes. The last of these three parental plasmids was itself constructed by using the *colB* plasmid to mobilize *tonB*.*trp*.*cysB* from the *Escherichia coli* chromosome (Fredericq, 1969). Once formed, these hybrid plasmids are far from permanent structures and apparently fragment easily. Thus Fredericq (1969) has shown that phage P1 can transduce fragments from the hybrid plasmid *R(SCT)colB*.*trp*.*colV* (see Fig. 4) and the plasmid *fer*.*colB*.*tonB*.*trp*.*cysB*.*colV*.*R(S)*.*R(C)*.*R(T)* mentioned above segregates fragments as shown in Fig. 5. It is not yet clear, however, whether the fragments of these hybrid plasmids can always survive together in a single cell or whether the source of the transduced or segregant fragments is always different cells. Since most of the hybrid plasmids are built from two or more replicons, each of which was originally capable of independent survival, there must be the possibility that each hybrid element will contain the necessary information to allow several fragments to survive as true replicons when the parental structure breaks up; but whether this actually happens is not yet clear.

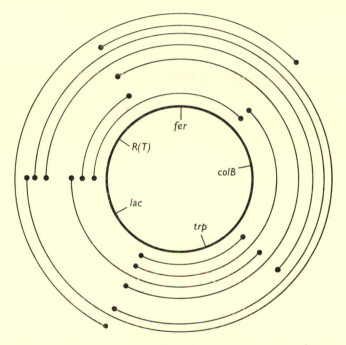

Fig. 4. Overlapping segments transferred by phage P1 grown on a donor carrying at $R(T).colB.trp.lac.$ hybrid episome. Fredericq (1969).

The association and dissociation of two or more plasmids as described ensures that there is a continual reassortment of genetic markers between plasmids in prokaryotic cells. In general, the progeny replicons which arise at the end of the association/dissociation step are both capable of independent survival and between them comprise all the markers originally present on the parents. Thus, with the two parental plasmids ABC and abc the associated single plasmid will be $ABCabc$ and the first generation of progeny plasmids may be $ABC+abc$, $ABc+abC$ or $Abc+aBC$ (for details see Richmond, 1968, 1969). When a second round of association and dissociation is allowed before segregation to the haploid state occurs, then a further range of genotypes may arise from the two parental sets of markers. In all these cases, however, the progeny replicons are complementary pairs, the markers on one plasmid always being matched by the markers on the other (Richmond, 1969).

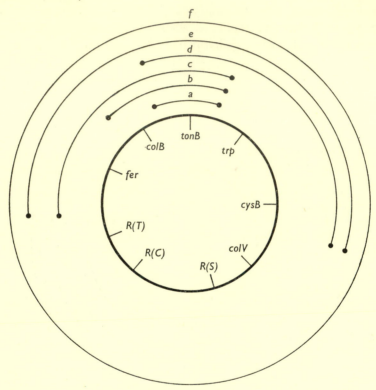

Fig. 5. Segregants of an $R(SCT).colB.trp.colV.cysB$.hybrid episome. The incidence of the various deletions is: a, 4·5 %; b, 1·0 %; c, 0·5 %; d. 0·5 %; e, 0·5 % and f (whole plasmid), 93 %. Fredericq (1969).

Occasionally, however, complementary pairs are not formed. For example, in a transductional cross between two penicillinase plasmids from the same compatibility group in *Staphylococcus aureus*, such as a cross between a plasmid with the markers ($i^-.p^+.Cd^s.ero^r$) and one with the markers ($i^+.p^-.Cd^r.ero_{del}$), most of the transductant clones contain a mixture of cells that is consistent with the suggestion that the two parental plasmids formed an intermediate associated double structure and then segregated to give complementary pairs of either parental or recombinant plasmids. In about 2 % of transductant clones, however, the colonies contain a mixture of cells, one type with the genotype $(i^+p^-.Cd^s.ero^r)$ and the other with $(i^+p^-.Cd^r.ero^r)$. Thus, in this case, two *unselected* markers (i and p) are homo-allelic in all cells of the clone (i^+p^- in both cases) while another unselected marker (Cd) is hetero-allelic (Cd^r/Cd^r). This is reminiscent of findings reported by Morse, Lederberg & Lederberg (1956) with transductional heterogenotes of phage λ. In that case a similar 'homogenotization' of markers occurred by a process then called 'automixis'. In practice, the effect of this process is the 'copying' of markers from one replicon to another *without their concomitant loss from the donor replicon*; in other words there is a gene duplication.

A further example of this process of gene 'copying' between replicons has recently been reported by Dr Elizabeth H. Asheshov (personal communication). She has studied *Staphylococcus aureus* strain PS 80 where, in her original strain, the penicillinase genes (i^+ and p_A^+) were chromosomal and a plasmid with the markers ($Cd^r.Hg^r$) was also present. Recently certain isolates of PS 80 have spontaneously changed their genotype so that they still carry ($i^+p_A^+$) in the chromosome, but in addition now have a plasmid with the markers ($i^+p_A^+.Cd^r.Hg^r$). Thus there has been a duplication of the penicillinase genes; the inference is that the chromosomal markers have been 'copied' on to the plasmid without concomitant loss from the donor replicon.

At present the molecular processes which underlie 'copying' are obscure, even though 'homogenotization' has been used for many years by those interested in studying the regulation of β-galactosidase synthesis to 'transfer' genetic markers from a chromosomal *lac* region to a superinfecting F'.*lac* plasmid in *Escherichia coli* (J. Scaife, personal communication). Originally it was thought that 'homogenotization' could only arise by normal recombination between the genetic components of heterogenotes where plasmid multiplication had occurred and two copies of each element were available for recombination (Morse *et al.* 1956). Now it seems possible that such progeny can arise by breakage and rejoining of *single* strands of the DNA double helices which compose two interacting replicons. In fact such single-strand breaks may be a normal occurrence in the recombination process (Meselson, 1967; Hayes, 1969*b*) and Scaife (1967) gave a model to show how inter-plasmid 'copying' might take place (see Fig. 6).

Replicons therefore seem to merge and separate with a continual redistribution of markers. The freedom with which markers flow varies widely from one pair of interacting plasmids to another; this presumably reflects the degree of genetic homology between the plasmids.

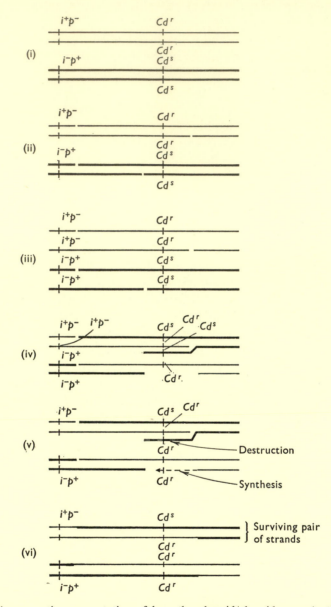

Fig. 6. Diagrammatic representation of how the plasmid/plasmid cross $(\alpha . i^- p^+ . Cd^s) \times (\alpha . i^+ p^- . Cd^r)$ can lead to a mixed clone of two cell types, one $(i^+ p^- . Cd^r)$ and the other $(i^+ p^- . Cd^s)$. Diagram based on Fig. 4 in Scaife (1967). (i) Apposition of two homologous regions on the plasmids. (ii) Various breaks occur in single strands. (iii) The strands separate. (iv) Duplexes re-form with some mismatching. (v) Overlapping regions are digested and gaps repaired; this leads to the 'loss' of one copy of Cd^s and a 'gain' of one copy of Cd^r. (vi) Repair of breaks in single strand.

However, positional effects also seem to be important since pairs of penicillinase plasmids (which must have much genetic homology) that are held at the same attachment site form associated double plasmids much more freely than when they are held at separate sites (Richmond, 1967; Richmond & Johnston, 1969b and unpublished observations).

PLASMID TRANSMISSIBILITY

Most of the emphasis in this article so far has been on the interactions between replicons within a single cell; there has been relatively little discussion of how this situation may arise. Often these interactions occur between structures already present in the cell as distinct entities, such as between a plasmid and the chromosome, or between two plasmids recently derived *de novo* from a single pre-existing structure (see p. 257). But in many cases the interactions follow the transfer of an element from one cell to become an additional replicon in a second. To facilitate this process, some plasmids carry the genetic information to cause the cell to form the molecular apparatus needed for plasmid transfer by cell-to-cell contact. In practice, this feature is confined to a class of plasmids, often known as 'transmissible plasmids' or 'sex-factors', from enteric bacteria and perhaps also from pseudomonads (Holloway, 1955, 1956) and has never been found among plasmids from Gram-positive organisms. However, to call the 'sex-factors' of the enteric bacteria 'transmissible' is somewhat misleading since plasmid transmissibility among Gram-positive bacteria does occur, but always depends upon transduction. Moreover, the efficiency of this process may not be much different from that of conjugation. For example, in *Escherichia coli*, a normal frequency of transfer of an R-factor by cell-to-cell contact may well be $1:10^3$ or $1:10^4$ R$^+$ cells, while in *Staphylococcus aureus* transfer by transduction may occur with a frequency of about $1:10^5$ of the recipient population. It might, therefore, be better to distinguish the behaviour of sex-factors from other plasmids by calling them 'self-transmissible' plasmids. In special circumstances, for example in cells carrying a de-repressed R-factor (Meynell & Datta, 1967), the frequency of transfer may be much higher than is normally found with an R$^+$ culture; but the same can also be said for the transfer of staphylococcal plasmids which have recombined to become part of a phage genome (Novick, 1967c). The transmissibility of plasmids between bacterial strains is therefore an important activity complementary to the genetic interactions of the elements within the individual cells.

In all cases the range of organisms to which transfer can take place seems to be restricted by the nature and characteristics of the transfer process itself. In Gram-positive and in Gram-negative bacteria the range of organisms that can act as recipients in transduction experiments is set by the specificity of the phage preparations involved, rather than by the ability of the recipient to accept foreign DNA, and the range of Gram-negative bacteria that can accept plasmids by conjugation is wider than when the transfer is by transduction (Watanabe, 1963; Datta, 1965; Meynell *et al.* 1968). In general, plasmid transfer between 'species' of Gram-negative bacteria occurs more freely than between Gram-positive 'species'; this may, however, just reflect the methods used to differentiate the 'species' involved (see p. 271).

THE POSSIBLE ROLE OF REPLICONS IN THE SURVIVAL AND EVOLUTION OF PROKARYOTIC CELLS

The ability of prokaryotic cells to exchange genetic material, whether by transduction or conjugation (or by any other means), is unquestionably of the greatest value to them in their adjustments to changes in their environment. For example, in the human alimentary tract the ability of invading pathogenic salmonellas to pick up antibiotic resistance-determinants from *Escherichia coli* already present certainly helps the invaders to colonize a gut containing a high concentration of antibiotic, whereas without the gene transfer the infection would be arrested.

If there is an advantage to organisms in being able to exchange genes or groups of genes freely, it is not so clear why it is advantageous to organize the material as a series of replicons. Possible reasons are as follows.

(1) The availability of independently replicating groups of genes allows the accumulation on one replicon of a series of genetic determinants required for the colonization of a given ecological niche. The accumulation of a collection of antibiotic resistance-determinants on a single replicon (an R-factor) in the enteric bacteria may occur in this way.

(2) The fact that genes of similar ecological significance (though of very different biochemical consequences) can accumulate on a single replicon allows the whole replicon to be 'lost' from the cell when the genes concerned are not required. This is almost certainly an advantage to the cell in terms of energy; at a minimum the cell would not have to synthesize the genes themselves, let alone their products. But this step would only remain an advantage if the replicon were recovered by at

least some cells of the population, if the environmental conditions changed and the genes became essential for survival.

(3) Since replicons are capable of surviving as independent genetic entities in prokaryotic cells and carry at least some of the information needed for their survival in the cell, it seems likely that an intact replicon is more able to establish itself in a recipient cell than is a fragment of a replicon, such as a gene or group of genes derived from the chromosome. The relatively high efficiency of transfer of a given gene as part of a plasmid, when compared with its transfer from a chromosomal source, supports this view (Arber, 1960).

In summary, therefore, it may be that the organization of given genes into a single replicon, coupled with the increased efficiency of transfer of this type of structure, could be part of a mechanism that has evolved to give maximum flexibility to a bacterial population in its response to changes in its environment.

If this role of replicons in bacterial cells is correct, an overall genetic picture of a prokaryotic cell emerges. The genes that are necessary for the cell to survive whatever the conditions of growth (such as those needed for cell division, ribosome synthesis, energy metabolism, etc.) accumulate on one replicon. This replicon will consequently be large and is what we call the chromosome. Outside the essential class of genes there will be others which tend to reflect the detailed environment in which the cell is found at that time. Some of these genes—those most continuously needed—may tend to join the essential genes on the most persistent replicon, that is, the chromosome, while the others may organize into a series of extrachromosomal replicons namely, plasmids. The distribution of genes between the chromosome and various plasmids probably varies in response to changes in the environment. If this genetic flexibility exists in all prokaryotic cells—and it is well established in the enteric bacteria and in *Staphylococcus aureus*—the extent to which organisms can be said to fall into 'species' inevitably depends greatly on the freedom of genetic exchange that can occur between them. The distinction between *Escherichia coli* and many salmonellas and shigellas is not great: it is already certain that they have many genes in common on their chromosomal replicons (Sanderson, 1967; Taylor & Trotter, 1967) and may often carry similar plasmids (Watanabe, 1963; Datta, 1965; Anderson, 1968). Similarly *Bacillus subtilis* is very like *Bacillus licheniformis* (Yoshikawa & Sueoka, 1963; Tyeryar, Lawton & MacQuillan, 1968). In these cases, therefore, attempts to separate groups of strains into 'species' may be relatively meaningless and the very existence of 'species' may not be unconnected with the activities of workers intent on 'speciation'. On the other hand, *Staphylococcus aureus* and *Streptococcus pyogenes* are much more distinct and this either reflects very different ecological niches that the organisms occupy and a consequent distinct content of genes, or a strict limit to the exchange of genetic information between the two organisms. Of these two possibilities, the second seems more likely in the case of *S. aureus* and *S. pyogenes*. It may well be that speciation in bacteria reflects the restrictions on replicon transmissibility more than any other single factor.

CONCLUSION

The genetic structure of all prokaryotic cells seems to involve a number of interpenetrating genetic pools. In certain extreme cases (e.g. infection with virulent phage) there is no genetic interaction at all between the phage and any replicon in the cell, and the uncontrolled replication of the phage genome kills the cell. In the majority of cases, however, the various gene pools in the cells are organized as replicons; that is, they survive in the cell as a series of autonomously replicating structures whose replication is adjusted to the rate of division of the cell. In practice this probably means that one of the replicons involved (the chromosome) sets the pace for the replication of the others. Another characteristic of the chromosome is that it carries genes which approximate to the minimum required for a given organism to occupy its 'normal' ecological position.

In some cases genetic recombination between replicons is rare, but in the majority of cases it occurs freely and may result in the association and dissociation of replicons, that is, in the merging or separation of gene pools. Furthermore, since selection pressure seems to ensure that genes of related ecological significance accumulate on single replicons, certain of these structures will not be needed under some growth conditions and may be 'lost' from the cell. Similarly, under other conditions, new replicons may be taken up by the cell by one of the means of gene transfer discussed above, and establish themselves as an addition to the 'basic' genetic content of the cell.

A prokaryotic cell is therefore in a state of continuous genetic flux, not caused so much by mutation (though this has a part) but by the continuous addition, subtraction and reassortment of genes among the various replicons. One gets the impression of a group of genes corresponding roughly to a minimum genetic requirement of the particular organism being carried on the chromosome and being relatively inflexible in this position, but with a continuous addition and deletion of material going in parallel, and occasionally impinging by 'integration' or 'excision'.

Interpenetration of gene pools is not confined to prokaryotic cells although it is here that the interactions are at their most fluid. Many eukaryotic cells carry organelles with their own gene pools but these seem to interact in a much more restricted way than those found in prokaryotic cells. This aspect of the subject is discussed by Dr D. Wilkie in this Symposium (see p. 381).

REFERENCES

ADAMS, J. N., Jr. & LURIA, S. E. (1958). Transmission by phage P1: abnormal phage function of the transduced particles. *Proc. natn. Acad. Sci. U.S.A.* **44**, 590.

ADELBERG, E. A. & BURNS, S. N. (1959). A variant sex-factor in *Escherichia coli. Genetics, Princeton* **44**, 497.

ADELBERG, E. A. & BURNS, S. N. (1960). Genetic variation in the sex-factor of *Escherichia coli. J. Bact.* **79**, 321.

ADELBERG, E. A. & PITTARD, J. (1965). Chromosome transfer in bacterial conjugation. *Bact. Rev.* **29**, 161.

ANDERSON, E. S. (1965). Origin of transmissible drug-resistance factors in the Enterobacteriaceae. *Br. med. J.* ii, 1289.

ANDERSON, E. S. (1968). The ecology of transferable drug resistance in the Enterobacteria. *A. rev. Microbiol.* **22**, 131.

ARBER, W. (1960). Transduction of chromosomal genes and episomes in *Escherichia coli. Virology* **11**, 273.

BAZAREL, M. & HELINSKI, D. R. (1968a). Characterization of multiple circular DNA forms of a colicinogenic factor E_1 from *Proteus mirabilis. Biochemistry* **7**, 3513.

BAZAREL, M. & HELINSKI, D. R. (1968b). Circular DNA forms of colicinogenic factors E_1, E_2 and E_3. *J. molec. Biol.* **36**, 185.

BHASKARAN, K. (1964). Segregation of genetic factors during recombination in *Vibrio cholerae*, Strain 162. *Bull. Wld Hlth Org.* **30**, 845.

BOICE, L. & LURIA, S. E. (1961). Transfer of transducing prophage Pl*dl* upon bacterial mating. *Bact. Proc.* p. 197.

BOUANCHAUD, D. H., SCAVIZZI, M. R. & CHABBERT, Y. A. (1968). Elimination by ethidium bromide of antibiotic resistance in enterobacteria and staphylococci. *J. gen. Microbiol.* **54**, 417.

BRODA, P. M. A., BECKWITH, J. R. & SCAIFE, J. (1964). The characterization of a new type of F-prime factor in *Escherichia coli* K12. *Genet. Res.* **5**, 144.

CAMPBELL, A. M. (1962). Episomes. *Adv. Genet.* **11**, 101.

CHABBERT, Y. A., BAUDENS, J. G. & GERBAUD, G. R. (1964). Variations sous l'influence de l'acriflavine, et transduction de la résistance à la kanamycine et au chloramphenicol chez les staphylocoques. *Annls Inst. Pasteur, Paris* **107**, 678.

COHEN, A., FISHER, W. D., CURTIS III, R. & ADLER, H. (1969). *Cold Spring Harb. Symp. quant. Biol.* (In Press.)

COWAN, S. T. & STEEL, K. J. (1966). *Manual for the Identification of Medical Bacteria.* Cambridge University Press.

CUZIN, F. & JACOB, F. (1965a). Existence chez *Escherichia coli* d'une unité génétique formée de différents réplicons. *Compt. Rend.* **260**, 5411.

Cuzin, F. & Jacob, F. (1965b). Analyse génétique fonctionelle de l'épisome sexuel d'*Escherichia coli* K12. *Compt. Rend.* **260**, 2087.

Datta, N. (1965). Infectious drug resistance. *Br. med. Bull.* **21**, 254.

Dettori, R., Maccacaro, G. A. & Piccinin, G. L. (1961). Sex-specific bacteriophages of *Escherichia coli* K12. *G. Microbiol.* **9**, 141.

Dettori, R., Maccacaro, G. A. & Turri, M. (1963). Sex-specific phages of *Escherichia coli* K12. IV. Host-specificity, pattern of lysis and lethality of phage μ2. *G. Microbiol.* **11**, 15.

Dettori, R. & Neri, M. G. (1965). Batteriofage filimentoso specifico per cellule *Hfr* ed F+ di *E. coli* K12. *G. Microbiol.* **13**, 111.

DeWitt, W. & Helinski, D. R. (1965). Characterization of colicinogenic factor E1 from a non-induced and a mitomycin C-induced *Proteus* strain. *J. molec. Biol.* **13**, 692.

Dubnau, E. & Maas, W. (1968). Inhibition of replication of an F'. *lac* episome in *Hfr* cells of *Escherichia coli*. *J. Bact.* **95**, 531.

Dubnau, E. & Stocker, B. A. D. (1964). Genetics of plasmids in *Salmonella typhimurium*. *Nature, Lond.* **204**, 1112.

Echols, H. (1963). Properties of F' strains of *Escherichia coli* superinfected with F lactose and F galactose episomes. *J. Bact.* **85**, 262.

Falkow, S. & Citarella, R. V. (1965). Molecular homology of F-merogenote DNA. *J. molec. Biol.* **12**, 138.

Falkow, S., Citarella, R. V., Wohlheiter, J. A. & Watanabe, T. (1966). The molecular nature of R-factors. *J. molec. Biol.* **17**, 102.

Felix, A. (1955). World survey of typhoid and paratyphoid B phage types. *Bull. Wld Hlth Org.* **13**, 109.

Fredericq, P. (1969). The recombination of colicinogenic factors with other episomes and plasmids. In *Bacterial Episomes and Plasmids*, p. 163. Eds. G. E. W. Wolstenholme and M. O'Connor. London: J. & A. Churchill.

Goebel, W. & Helinski, D. R. (1968). Generation of higher multiple circular DNA forms in bacteria. *Proc. natn. Acad. Sci. U.S.A.* **61**, 1406.

Guinée, P. A. M., Scholtens, R. T. & Willems, H. M. C. C. (1967). Influences of resistance factors on the phage types of *Salmonella panama*. *Antonie van Leeuwenhoek* **33**, 30.

Hakura, A., Otsuji, N. & Hirota, Y. (1964). A temperate phage specific for female strains of *Escherichia coli* K12. *J. gen. Microbiol.* **35**, 69.

Hashimoto, H., Kono, M. & Mitsuhashi, S. (1964). Elimination of penicillin resistance of *Staphylococcus aureus* by treatment with acriflavine. *J. Bact.* **88**, 261.

Hayes, W. (1969a). Introduction: What are episomes and plasmids? In CIBA Foundation Symposium *Bacterial Episomes and Plasmids*, p. 4. Eds. G. E. W. Wolstenholme and M. O'Connor. London: J. & A. Churchill.

Hayes, W. (1969b). In *The Genetics of Bacteria and their Viruses*. 2nd edn., p. 747. Oxford and Edinburgh: Blackwell Scientific Publications.

Herschman, H. R. & Helinski, D. R. (1967). Comparative study of the events associated with colicin induction. *J. Bact.* **94**, 691.

Hirota, Y. (1960). The effect of acridine dyes on mating-type factors in *Escherichia coli*. *Proc. natn. Acad. Sci. U.S.A.* **46**, 57.

Holloway, B. W. (1955). Genetic recombination in *Pseudomonas aeruginosa*. *J. gen. Microbiol.* **13**, 572.

Holloway, B. W. (1956). Self-fertility in *Pseudomonas aeruginosa*. *J. gen. Microbiol.* **15**, 221.

Jacob, F. & Brenner, S. (1963). Sur la régulation de la synthèse du DNA chez les bactéries: l'hypothèse du réplicon. *Compt. Rend.* **256**, 298.

JACOB, F., BRENNER, S. & CUZIN, F. (1963). On the regulation of DNA synthesis in bacteria. *Cold Spring Harb. symp. quant. Biol.* **28**, 329.

JACOB, F. & WOLLMAN, E. L. (1958). Les épisomes, éléments génétiques ajoutés. *Compt. Rend.* **247**, 154.

KAMEDA, M., HARADA, K., SUZUKI, M. & MITSUHASHI, S. (1965). Drug resistance of enteric bacteria. V. High frequency of transduction of R-factors with bacteriophage epsilon. *J. Bact.* **90**, 1174.

KONDO, E. & MITSUHASHI, S. (1964). Drug resistance of enteric bacteria. IV. Active transducing bacteriophage P1*CM* produced by the combination of R-factor with bacteriophage P1. *J. Bact.* **88**, 1266.

KONDO, E. & MITSUHASHI, S. (1966). Drug resistance of enteric bacteria. VI. Introduction of bacteriophage P1*CM* into *Salmonella typhi* and formation of P1d*CM* and F-*CM* elements. *J. Bact.* **91**, 1787.

LANCINI, G., PALLANZA, R. & SILVESTRI, L. G. (1969). Relationships between bacteriocidal effects and inhibition of ribonucleic acid nucleotidyltransferase by rifampicin in *Escherichia coli* K.12. *J. Bact.* **97**, 761.

LARK, K. G. (1966). Regulation of chromosome replication and segregation in bacteria. *Bact. Rev.* **30**, 1.

LEDERBERG, J., CAVALLI, L. L. & LEDERBERG, E. M. (1952). Sex compatibility in *Escherichia coli. Genetics, Princeton* **37**, 720.

LENNOX, E. S. (1955). Transduction of linked genetic characters of the host by bacteriophage P1. *Virology* **1**, 190.

MAAS, W. K., MAAS, R., WIAME, J. & GLANSDORFF, J. (1964). Studies on the mechanism of repression of arginine biosynthesis in *Escherichia coli*. I. Dominance of repressibility in zygotes. *J. molec. Biol.* **8**, 359.

MÄKELA, O., MÄKELA, P. H. & SOIKKELLI, P. (1964). Sex-specificity of the bacteriophage T7. *Annls Med. exp. Biol. Fenn.* **42**, 188.

MARMUR, J., ROWND, R., FALKOW, S., BARON, L. S., SCHILDKRAUT, C. & DOTY, P. (1961). The nature of intergeneric episomal infection. *Proc. natn. Acad. Sci. U.S.A.* **47**, 972.

MESELSON, M. (1967). The molecular basis of genetic recombination. In *Heritage of Mendel*, p. 81. Ed. R. A. Brink, Madison: University of Wisconsin Press.

MEYNELL, E. M. & DATTA, N. (1967). Mutant drug resistance factors of high transmissibility. *Nature, Lond.* **214**, 885.

MEYNELL, E., MEYNELL, G. G. & DATTA, N. (1968). Phylogenetic relationships of drug-resistance factors and other transmissible bacterial plasmids. *Bact. Rev.* **32**, 55.

MORSE, M. L., LEDERBERG, E. M. & LEDERBERG, J. (1956). Transductional heterogenotes in *Escherichia coli. Genetics, Princeton* **41**, 759.

NAGEL DE ZWAIG, R. (1966). Association between colicinogenic and fertility factors. *Genetics, Princeton* **54**, 381.

NEWTON, B. A. (1966). Effects of antrycide on nucleic acid synthesis and function. *Symp. Soc. gen. Microbiol.* **16**, 213.

NISIOKA, T., MITANI, M. & CLOWES, R. C. (1969). Composite circular forms of R-factor deoxyribonucleic acid molecules. *J. Bact.* **97**, 376.

NOVICK, R. P. (1967a). Mutations affecting replication and maintenance of penicillinase plasmids in *Staphylococcus aureus. Proc. VI Int. Congr. Chemotherapy, Vienna*, p. 269.

NOVICK, R. P. (1967b). Penicillinase plasmids of *Staphylococcus aureus. Fedn Proc. Fedn Am. Socs exp. Biol.* **26**, 29.

NOVICK, R. P. (1967c). Properties of a cryptic high-frequency transducing phage in *Staphylococcus aureus. Virology* **33**, 155.

NOVICK, R. P. (1969). Extrachromosomal inheritance in bacteria. *Bact. Rev.* **33**, 210.

NOVICK, R. P. & RICHMOND, M. H. (1965). Nature and interactions of the genetic elements governing penicillinase synthesis in *Staphylococcus aureus. J. Bact.* **90**, 467.

ØRSKOV, I. & ØRSKOV, F. (1966). Episome carried surface antigen of *Escherichia coli.* I. Transmission of the determinant of the K 88 antigen and the influence of the transfer on chromosomal markers. *J. Bact.* **91**, 69.

PRITCHARD, R. H., BARTH, P. T. & COLLINS, J. (1969). Control of DNA synthesis in bacteria. *Symp. Soc. gen. Microbiol.* **19**, 263.

RICHMOND, M. H. (1967). Associated diploids involving penicillinase plasmids in *Staphylococcus aureus. J. gen. Microbiol.* **46**, 85.

RICHMOND, H. M. (1968). The plasmids of *Staphylococcus aureus* and their relation to other extrachromosomal elements in bacteria. In *Advances in Microbial Physiology*, vol. 2, p. 43. Eds. A. H. Rose and J. P. Wilkinson. London and New York: Academic Press.

RICHMOND, M. H. (1969). Extrachromosomal elements and the spread of antibiotic resistance in bacteria. The Fourth Colworth Medal Lecture. *Biochem. J.* **113**, 225.

RICHMOND, M. H. & JOHNSTON, J. H. (1969*a*). The reversible transition of certain genes in *Staphylococcus aureus* between the integrated and the extrachromosomal state. *Genet. Res.* **13**, 267.

RICHMOND, M. H. & JOHNSTON, J. H. (1969*b*). The genetic interactions of penicillinase plasmids in *Staphylococcus aureus.* In the CIBA Foundation Symposium *Bacterial Episomes and Plasmids,* p. 179. Eds. G. E. W. Wolstenholme and M. O'Connor. London: J. & A. Churchill.

ROTH, T. F. & HELINSKI, D. R. (1967). Evidence for the circular DNA forms of a bacterial plasmid. *Proc. natn. Acad. Sci. U.S.A.* **58**, 650.

ROWND, R., NAKAYA, R. & NAKAMURA, A. (1966). Molecular nature of the drug resistance factors of the Enterobacteriaceae. *J. molec. Biol.* **17**, 376.

SANDERSON, K. E. (1967). Revised linkage map of *Salmonella typhimurium. Bact. Rev.* **31**, 354.

SCAIFE, J. (1967). Episomes. *A. Rev. Microbiol.* **21**, 601.

SCAIFE, J. & GROSS, J. D. (1962). Inhibition of multiplication of an F-*lac* factor in *Hfr* cells of *Escherichia coli* K12. *Biochem. biophys. Res. Commun.* **7**, 307.

SCAIFE, J. & GROSS, J. D. (1963). The mechanism of chromosome mobilization by an F-prime factor in *Escherichia coli* K12. *Genet. Res.* **4**, 325.

SCAIFE, J. & PEKHOV, A. P. (1964). Deletion of chromosomal markers in association with F-prime factor formation in *Escherichia coli* K12. *Genet. Res.* **5**, 495.

SCHELL, J., GLOVER, S. W., STACEY, K., BRODA, P. M. A. & SYMONDS, N. (1963). The restriction of the phage T3 by certain strains of *Escherichia coli. Genet. Res.* **4**, 483.

SNEATH, P. H. A. & LEDERBERG, J. (1961). Inhibition by periodate of mating in *Escherichia coli. Proc. natn. Acad. Sci. U.S.A.* **47**, 86.

STROBEL, M. & NOMURA, M. (1966). Restriction of the growth of the bacteriophage BF 23 by a colicine (colI-P9) factor. *Virology* **28**, 763.

SUGINO, Y. & HIROTA, Y. (1962). Conjugal fertility associated with a resistance factor in *Escherichia coli. J. Bact.* **84**, 902.

TAKANO, T., WATANABE, T. & FUKUSAWA, T. (1966). Specific inactivation of infectious λ DNA by sonicates of restrictive bacteria with R-factors. *Biochem. biophys. Res. Commun.* **25**, 192.

TAYLOR, A. L. & TROTTER, D. C. (1967). Revised linkage map of *Escherichia coli. Bact. Rev.* **31**, 332.

THOMAS, R. (1968). Lysogeny. *Symp. Soc. gen. Microbiol.* **18**, 315.

TYERYAR, F. J., Jr., LAWTON, W. D. & MACQUILLAN, A. M. (1968). Sequential replication of the chromosome of *Bacillus licheniformis. J. Bact.* **95**, 2062.

WARING, M. J. (1966). Cross-linking and intercalation in nucleic acids. *Symp. Soc. gen. Microbiol.* **16**, 235.

WATANABE, T. (1963). Infective heredity of multiple drug resistance in bacteria. *Bact. Rev.* **27**, 87.

WATANABE, T. & FUKASAWA, T. (1961). Elimination of resistance factors by acridine dyes. *J. Bact.* **81**, 679.

WATANABE, T., NISHIDA, H., OGATA, C., ARAI, T. & SATO, S. (1964). Episome-mediated transfer of drug resistance in Enterobacteriaceae. VII. Two types of naturally occurring R-factors. *J. Bact.* **88**, 716.

WATANABE, T. & OKADA, T. (1964). New types of sex-specific bacteriophage of *Escherichia coli. J. Bact.* **87**, 727.

WATANABE, T., TAKANO, T., ARAI, T., NISHIDA, H. & SATO, S. (1966). Episome mediated transfer of drug resistance in Enterobacteriaceae. X. Restriction and modification of phage by *fi⁻* R-factors. *J. Bact.* **92**, 477.

WEHRLI, W., KNÜSEL, F., SCHMID, K. & STAEHELIN, H. (1968). Interaction of rifamycin with bacterial DNA polymerase. *Proc. natn. Acad. Sci. U.S.A.* **61**, 667.

WILLIAMS-SMITH, H. (1969). Veterinary implications of transfer activity. In *Bacterial Episomes and Plasmids*, p. 213. Eds. G. E. W. Wolstenholme and M. O'Connor. London: J. & A. Churchill.

WILLIAMS-SMITH, H. & HALLS, S. (1967). The transmissible nature of the genetic factor in *Escherichia coli* that controls haemolysin production. *J. gen. Microbiol.* **47**, 153.

WOLLMAN, E. L. (1969). In discussion to Hayes (1969*a*).

WOLLMAN, E. L., JACOB, F. & HAYES, W. (1956). Conjugation and genetic recombination in *Escherichia coli. Cold Spring Harb. Symp. quant. Biol.* **21**, 141.

YOSHIKAWA, M. & AKIBA, T. (1962). Studies on transferable drug resistance in bacteria. 4. Suppression of plaque formation of phages by a resistance factor. *Jap. J. Microbiol.* **6**, 121.

YOSHIKAWA, H. & SUEOKA, N. (1963). Sequential replication of *Bacillus subtilis* chromosome. I. Comparison of marker frequencies in exponential and stationary phases. *Proc. natn. Acad. Sci. U.S.A.* **49**, 559.

ZINDER, N. (1960). Sexuality and mating in *Salmonella. Science, N.Y.* **131**, 924.

THE EVOLUTIONARY SIGNIFICANCE
OF RECOMBINATION IN PROKARYOTES*

WALTER F. BODMER

Department of Genetics, Stanford School of Medicine
Stanford University, Stanford, California 94305, U.S.A.

The major evolutionary feature of genetic recombination is that it facilitates the accumulation in a single individual of advantageous mutations which arose separately in different individuals. As pointed out by R. A. Fisher and H. J. Muller in the 1930s this should in most cases greatly accelerate the rate of evolution. This evolutionary advantage must be expressed at the level of the population and not that of the individual. If it is the main basis for the evolution of recombination mechanisms and sexual differentiation, then their evolution must have depended on inter-population rather than intra-population selection. Almost all evolutionary changes within a population depend on differential selection among individuals. It therefore seems likely that recombination may first have occurred as a by-product of some advantage conferred at the individual level and then, later, have been exploited by inter-population selection. Recombination and sexual reproduction are much more prevalent among eukaryotes than prokaryotes, suggesting that the increase in evolutionary rates afforded by recombination is less significant for prokaryotes.

In this paper, the major features of recombination in prokaryotes will first be reviewed and contrasted with recombination in eukaryotes. The evidence for an evolutionary role for recombination in prokaryotes will be considered next. A simple two-locus model illustrating the evolutionary advantage of recombination and its dependence on population size will then be analysed. Finally, the effect of recombination on the organization of the genetic material, in particular very close linkage between genes with related functions, will be discussed.

The major features of recombination in prokaryotes

Recombination in eukaryotes is almost always reciprocal, involving the complete genomes of the two partners. In prokaryotes, on the other hand, recombination is in almost all cases asymmetrical, involving a major contribution from one partner and a minor one from the other.

* Dedicated to Prof. Th. Dobzhansky on the occasion of his 70th birthday.

There exists in both classes of organisms a variety of mechanisms for mediating genetic recombination between different individuals. These are, in prokaryotes, associated with differences in the relative contributions of the two partners to recombinant offspring.

The three major mechanisms of exchange in bacteria are, in increasing order of contribution from the minor partner: (a) DNA mediated transformation; (b) phage-mediated transduction, either specialized (in the case of lysogenic phages) or generalized; (c) bacterial conjugation. Transformation involves the insertion of quite small lengths of donor DNA (1500 to 20,000 nucleotide pairs) into a single strand of recipient DNA. The size of the DNA inserted during transduction is, generally, somewhat larger than this. In the case of specialized transduction, only regions of the bacterial genome adjacent to the integration site of the lysogenic phage are transferred. Generalized transduction, on the other hand, involves the transfer of randomly selected sections of the bacterial genome, which are comparable in size to the transducing phage's genome. Conjugation may involve the exchange of large fractions of the bacterial genome, though the details of the process are less well understood. Transmission of plasmids, such as the R or drug-resistance factors, can be considered as a special case of conjugation.

Recombination between viruses follows mixed infection of a cell by genetically different virus particles. It is, in general, a symmetrical process which is in some respects more analogous to recombination in eukaryotes than are the bacterial recombination mechanisms. Moreover, while there is no evidence that recombination is in any way an important part of the bacterial life cycle, it is important for viral life cycles, especially for lysogenic phages and also, presumably, transformation-inducing animal viruses.

Molecular evidence suggests similar mechanisms for the association and exchange of genetic material at the DNA level in all the above processes. These mechanisms involve some combination of breakage of DNA strands (by nucleases), homologous base pairing, repair of single strand gaps (by polymerases) and formation of covalent bonds between recombinant sections (by ligases; review: Bodmer & Darlington, 1969). Similar mechanisms at the DNA level presumably exist in eukaryotes, though they are obscured by the complexities of the chromosomal organization needed to manipulate, during mitosis, meiosis and cell division, the much larger amounts of DNA found in eukaryotes.

Evidence for the evolutionary role of recombination in prokaryotes

All the forms of recombination in prokaryotes are either known to occur, or are presumed to occur, in Nature. The original discovery of transformation by Griffith, who used mixtures of live and heat-killed pneumococci inoculated into mice, demonstrated that transformation could occur following mixed infection by genetically different bacteria. Similarly, recombination *in vivo*, presumably by conjugation, has been observed following mixed infection of mice by enteric bacteria, even of different species (*Escherichia coli* and *Salmonella typhimurium*; Schneider, Formal & Baron, 1961). In these cases, mixed infection of the mice with genetically different bacteria is analogous to mixed infection of a bacterium by genetically different phages. The only known limitation to the natural occurrence of recombination between bacteria by transduction or between viruses following mixed infection is the frequency with which genetically different organisms will be associated.

Though recombination among prokaryotes is presumed to occur in Nature it is not easy to identify evolutionary changes which *must* have been the result of recombination. The most striking example is undoubtedly that of bacteria whose simultaneous resistance to a variety of unrelated drugs is due to the presence of a transmissible plasmid, the R factor. This plasmid can be transmitted infectively from one bacterium to another, leading to a rapid spread of multiple drug resistance. The presence in a single patient of fully-sensitive and multiply-resistant shigellas belonging to the same serological type is direct evidence for the natural occurrence of infective transfer. This transfer, is not, of course, recombination in the classical sense. However, it seems most likely that the R factors, which can be thought of as a form of defective phage, accumulate their various drug-resistance genes by recombination among themselves (review: Meynell, Meynell & Datta, 1968).

Lederberg & Edwards (1953) showed that Salmonella serotypes could be recombined by transduction. They were actually able to produce recombinants with serotypes which corresponded to previously identified strains. They suggested that many of the very large number of serotypes known to occur in Nature may have been produced by recombination. New serotypes are, presumably, almost always at an advantage because they are less likely to be overcome by a previously established immune response to Salmonella infection. Though it is hard to exclude mutation as a source of this variation, the fact that many of the serotypes are permutations or combinations of other serotypes, together with the fact that no known serotype has ever knowingly been produced in the laboratory by mutation, strongly supports a recombinational origin for this variability (J. Lederberg, personal communication).

Very indirect evidence for the importance of recombination by transformation in *Bacillus subtilis* comes from the connection between sporulation and the development of competence with respect to transformation (e.g. Young, 1967). Thus, it has been suggested that competence is analogous to a pre-sporulation state; it is known also that certain classes of mutants which are unable to produce spores cannot be made competent. Since sporulation is generally considered to be a response to adverse (minimal) conditions, the association of transformation competence with sporulation could be interpreted as a way to increase the probability of producing recombinants under adverse conditions, when they might be most needed.

Recombination has often been suggested to be a mechanism producing new variants of pathogenic viruses, though no conclusive evidence for this exists. Yamamoto (1969) described the experimental formation of a new recombinant bacteriophage species following simultaneous infection in Salmonella by two *unrelated* phages. There is no reason why such processes should not occur naturally; Yamamoto suggested they may play an important role in bacteriophage evolution. All of the recombination enzymes, such as nucleases, ligases and polymerases, suggested by the usual molecular models for recombination (e.g. Bodmer & Darlington, 1969) are involved in normal cellular processes other than recombination, namely DNA replication (e.g. Ganesan, 1968), and DNA repair following damage by ultraviolet radiation and other agents (Boyce & Howard-Flanders, 1964; Setlow & Carrier, 1964). Thus, in principle, recombination could be a side product of these other essential cellular processes. If, as emphasized above, recombination and sexual differentiation have been moulded by inter-population rather than intra-population selection, then recombination must have arisen initially as a side product of other evolutionary changes. It, in any case, requires the participation of several gene products, which cannot all have evolved simultaneously for the purpose of achieving recombination.

Transduction and viral recombination (in DNA viruses) have been found in essentially all the prokaryotes in which they have been looked for, and might be such a 'by-product' phenomenon. Conjugation and bacterial DNA mediated transformation have only been found in a limited number of organisms. Transformation depends on the existence of a competent state which, at least in the case of *Bacillus subtilis* as suggested above, might be a by-product of the sporulation process. Conjugation seems more likely to have evolved specifically for recombination purposes, perhaps especially in relation to the very rapid evolution that can occur by contact-mediated transmission of plasmids, such as the multiple drug-resistance R factors. A particular case in which recombination specific mechanisms are known to exist is the phage λ integration system (review: Signer 1968). Campbell (1962) originally suggested that integration of phage λ into the bacterial chromosome during lysogenization might occur by a recombination process. Much evidence has now been accumulated which supports Campbell's model for λ and other similar phages. Recent genetic evidence implies the existence of gene products whose sole function is to mediate the integration recombination process. This is, however, necessary for an essential part of the life cycle of the phage, and so has presumably evolved through individual, not inter-populational, selection processes. It is possible that mechanisms similar to the λ integration system exist in higher organisms, for the transposition of genetic material from one region of the chromosome to another for control purposes at various stages of development (A. D. Kaiser, personal communication; see also below). This would again provide a basis for an individual rather than a populational advantage of recombination.

The rate of evolution in sexual and asexual populations

There seems little doubt that, in general, recombination occurs much less frequently in prokaryotes than in eukaryotes. A simple model will now be described to suggest why recombination may be less important to prokaryotes than to eukaryotes from an evolutionary point of view. Two different advantageous mutations in an asexual population can only be incorporated into a single individual, and so into the population, if one of the mutations occurs in a descendant of an individual in which the other mutation occurred. On the other hand, in a sexual population, namely one with a recombination mechanism, the two mutations can be brought together in the same individual by recombination. The relative rate of formation of individuals carrying both advantageous mutants by double mutation, as opposed to recombination between single mutants, depends on the frequencies of the single mutants in the population. The rate of formation by recombination is, however, nearly always faster by a factor of at least two, as we shall show below. When multiplied over many loci, the rate of formation of new combinations by recombination must be some orders of magnitude greater than the rate achievable by sequential mutation. H. J. Muller has aptly characterized this difference as a contrast between evolution in series and evolution in parallel.

A simple two-gene haploid model which illustrates the different rates of evolution in sexual and asexual population will now be analysed, following the approach initially suggested by Muller (1964) and later developed by Crow & Kimura (1965). Let a and b represent the prevailing allelic states at two loci and A, B the corresponding new advantageous alleles. For simplicity, assume complete symmetry with respect to A and B, so that genotypes Ab and aB initially both occur with equal frequencies x_0 and have selective advantages $1 + s$, relative to 1 for ab. Assume, further, that the mutation rates $a \to A$ and $b \to B$ are both μ. Consider now the situation in which the double mutant has not yet arisen and genotypes Ab and aB both have frequency x_0. We shall calculate the approximate average number of generations needed to produce *one* individual of type AB assuming, in turn, sexual and asexual reproduction. Since x_0 is, presumably, very small and the frequency of ab nearly 1, almost all matings involving genotypes Ab and aB are $Ab \times ab$ and $aB \times ab$. The approximate frequencies of Ab and aB after n generations will therefore be $(1 + s)^n x_0$. If the population size is N and the recombination frequency between the loci r, then the average number of AB individuals produced by recombination in generation i, assuming random mating, will be

$$2x_0^2 (1+s)^{2i} \qquad\qquad \times \tfrac{1}{2}r \qquad\qquad \times N$$

$\begin{array}{c}\text{Frequency of mating}\\ Ab \times aB\end{array}$	$\begin{array}{c}\text{Probability of}\\ AB \text{ offspring}\end{array}$	$\begin{array}{c}\text{Population}\\ \text{size}\end{array}$

The number produced in n generations will be the sum of the above expression over all values of i from o to n. We require the value of n such that this sum is 1, which is the solution of the equation

$$Nrx_0^2 \sum_{i=0}^{n} (1+s)^{2i} = 1. \tag{1}$$

Summing the geometric series $\sum_{i=0}^{n} (1+s)^{2i}$

gives
$$(1+s)^{2n+1} = 1 + \frac{s(2+s)}{Nrx_0^2}. \tag{2}$$

Taking logarithms of both sides of equation (2) gives the following equation for n, the average number of generations required to produce one AB individual by recombination:

$$n = (\tfrac{1}{2}) \left[\frac{\log \left(1 + s \dfrac{(2+s)}{Nrx_0^2} \right)}{\log (1+s)} - 1 \right]. \tag{3}$$

The recombination fraction r here refers to the actual rate of production of recombinants between individuals in a population. In prokaryotes this must reflect the relative frequency of sexual versus asexual reproduction. Most simply, the recombination fraction observed following mating must, approximately, be multiplied by the relative probability that a mating takes place. This will, in general, substantially decrease the effective value of r to be used in equation (3).

The average number of AB individuals produced by mutation in the i th generation is $2x_0 (1+s)^i \mu N$, since mutation can occur either in Ab or aB.

Summing as above over i, the number of generations $\bar{n}$, required on the average to produce one AB individual by mutation is given by

$$2x_0\mu N \sum_{i=0}^{\bar{n}} (1+s)^i = 1. \tag{4}$$

Summing the geometric series gives

$$(1+s)^{\bar{n}+1} = 1 + \frac{s}{2x_0\mu N}. \tag{5}$$

Taking logarithms of both sides of equation (5) we have

$$\bar{n} = \frac{\log (1+s/[2x_0\mu N])}{\log (1+s)} - 1, \tag{6}$$

which is essentially the same as a formula originally given by Muller (1964). The assumption of simple geometric increase in the frequencies of Ab and aB used to derive equations (3) and (6) is, as pointed out by Muller (1964), only approximate, and works so long as $(1+s)^i$ is not too large. The approximation is, however, adequate for a qualitative comparison of evolutionary rates in sexual and asexual populations. Crow & Kimura (1965) gave a somewhat more precise formula for $\bar{n}$. The above treatment also ignores the effects of random drift, in particular with respect to the probability of survival of new genotypes.

From equations (3) and (6) it can be seen that

$$\bar{n} \gtrless 2n$$

according to whether $\quad s/(2x_0\mu N) \gtrless \dfrac{s(2+s)}{Nrx_0^2} \tag{7}$

or, approximately, $\quad\quad x_0 \gtrless \dfrac{4\mu}{r}, \tag{8}$

assuming s is small. Thus, in particular when linkage is loose and r is nearly $\frac{1}{2}$, and so long as $x_0 > 8\mu$, then $\bar{n} > 2n$ and so the rate of formation of AB individuals will be at least twice as fast in a sexual population as compared with an asexual population. For given x_0 both n and $\bar{n}$, as expected, decrease as s and N increase. The value of $\bar{n}$ also decreases as μ increases. The value of n, on the other hand, *increases* as r, the recombination fraction between the loci, decreases. Thus, as might be expected, tight linkage between the loci decreases markedly the rate of evolution by recombination.

If there is initially just one mutation of each type Ab and aB present in the population, then $x_0 = 1/N$. In this case, from equation (6), $\bar{n}$ is independent of N and x_0 while n, from equation (3), is given by

$$\frac{1}{2}\left(\frac{\log\,(1+N\,[s(2+s)/r])}{\log\,(1+s)} - 1\right)$$

and increases with increasing population size N. From (7) when $r = \frac{1}{2}$, then $\bar{n} > 2n$, if, approximately,

$$N < \frac{1}{8\mu}. \tag{9}$$

Recombination has a more favourable effect the *smaller* the population size. This is because the frequency of the mating $Ab \times aB$ depends on the square of the genotype frequencies, which increase as N decreases, given that $x_0 = 1/N$. If, for example, $x_0 = 1/N$, $s = 0\cdot01$ and $\mu = 10^{-6}$, then $\bar{n}$ is approximately 520 generations. When $N = 10^5$, then n is 365, while when N is 10^6, n is 480, and still less than $\bar{n}$, though not by much. Actually, μ here refers to mutation at a specific site within the gene, and so is likely to be of the order of 10^{-7} to 10^{-8}. The population size, on the other hand, refers to that group within which mating can effectively occur at random. Even for bacteria, this may not be much larger than 10^7 and is probably much smaller for most other organisms. Thus, in the majority of cases, $x_0 > 8\mu$ and so the rate of formation of the double mutant in sexual populations will be at least twice the corresponding rate in asexual populations. This result was predicted by Fisher (1930; p. 123), though characteristically without any accompanying analysis. He commented that the only populations in which sexual reproduction would not be advantageous would be those containing just one gene! The greater rate of evolution by recombination will increase geometrically with increasing numbers of genes, resulting in enormous advantages to organisms with many genes.

It can be shown that, once the double mutant has formed, and provided there are no selective interactions between the loci, the double mutant will evolve essentially independently of the single mutants in a sexual population. In this case, the rate of increase of the double mutants is the same in asexual and sexual populations. The advantage of sex is thus only with respect to the initial rate of production of the type AB, and not with respect to its rate of increase in the population once it has arisen (Maynard Smith, 1968; Crow & Kimura, 1969). If the original selective disadvantages of the genotypes Ab and aB were t, and a selection-mutation balance equilibrium had been reached, the expected frequency of the double mutant before the change would be approximately μ^2/t^2 in both sexual and asexual populations. However, with $\mu = 10^{-7}$ to 10^{-8} and t as low as $0 \cdot 01$, μ^2/t^2 is at most 10^{-10}, making it very unlikely that the double mutant even exists in any population, let alone has reached an equilibrium state. The analysis given above for the rate of formation of the double mutant is, therefore, still quite relevant to this situation. Where mechanisms exist that can maintain Ab and aB at frequencies less than they would be in a polymorphic population, recombination is even more favourable as a mechanism for producing the double mutant, since n decreases in proportion to x_0^2, while $\bar{n}$ decreases only in proportion to x_0 (see equations (3) and (6)). The same also applies to hybridization between populations in one of which Ab is rare and aB common, and vice versa in the other.

Muller (1932, 1964) and Crow & Kimura (1965) in their analysis of sexual and asexual rates of evolution ignored the length of time needed for recombining two mutants into the same individual in a sexual population. They therefore assumed that all the favourable mutants which are produced within the time needed for one new mutation to be incorporated into an asexual population (essentially equation (6) with $x_0 = 1/N$) could successfully be incorporated into a sexual population. They then used this number as the relative advantage of sexual over asexual reproduction. It would seem from the analysis given here, that their treatment is an oversimplification which must, in general, over-estimate the advantages of sex, especially for organisms with large numbers of genes.

One of the main qualitative conclusions from the simple model considered above is that, other things being equal, the evolutionary advantage of sexual reproduction is much greater in smaller populations. Population sizes of different organisms on the whole decrease with increasing complexity of the organism. The simplicity of prokaryotes, among other factors, results in very large population sizes, and presumably in a much diminished evolutionary advantage of recombination. This alone might account for the fact that there is, on the whole, little evidence for the specific evolution of highly developed recombination systems in prokaryotes. Nevertheless the existence of recombination in prokaryotes might well have laid the basis for the further evolution of highly differentiated obligatory sexual systems in higher organisms.

The effect of the genetic system on the spatial organization of the genetic material

It is well known that a number of biochemical pathways in *Escherichia coli*, *Salmonella* spp., *Bacillus subtilis* and other bacteria are controlled by adjacent sets of cistrons. The homologous genes have, however, been shown not to be closely linked in fungi and are presumed not to be so in other higher organisms (review: Bodmer & Parsons, 1962). The striking phenomenon of close linkage between metabolically related genes does not, however, extend to all bacteria. It has been shown, for example, that not all the genes controlling tryptophan metabolism which are adjacent in *E. coli*, *Salmonella* spp. and *B. subtilis* are closely linked in *Pseudomonas aeruginosa* (Holloway, Hodgins & Fargie, 1963; Gunsalus *et al.* 1968). On the other hand, not all the biosynthetic pathways in *E. coli* are controlled by adjacent genes. The one outstanding exception is the arginine pathway (Maas, 1961; Gorini, Gundersun & Burger 1961; Vogel, Bacon & Baich, 1963). Close linkage between the genes controlling flagella in *E. coli* (Iino & Lederberg, 1964) provides another striking example of close linkage between genes with related functions. Similar examples not involving biochemical pathways are well documented in higher organisms, notably the haemoglobin β, γ, δ and ϵ chains in man, the IgG H gamma-globulin chain in mouse and man, the H2 and HLA histocompatibility polymorphisms (and presumably other blood group polymorphisms) and the T-alleles in mice. These examples show that close linkage between genes with related functions most probably occurs in all organisms. An intriguing question, however, is why only in some bacteria are the genes controlling sequential steps in metabolic pathways adjacent, while probably in all other organisms they are not all closely linked.

A discussion of the way in which such 'clustered' metabolic pathways might have evolved was given by Horowitz (1965). Only the end product is generally needed by the organism. The intermediate stages of the pathway can, therefore, have no intrinsic selective value. How then can they have evolved? Horowitz suggested, as have many others, that primitive organisms started in a relatively rich environment, in which the required end-product was available. On this premise, the metabolic pathway must have evolved 'backwards', gradually making more substances available from which the end-product could be synthesized.

This requires an increase in genome size to provide the new genes needed to code for the new intermediate enzymes of the metabolic pathway. There are two contrasting possibilities for the origin of such a clustered group of genes. They may represent duplications of genes from different parts of the genome which have been brought together by translocation (or by inversion), presumably because of some basic advantage associated with their occurrence in a cluster. On the other hand, the genes may represent a series of tandem duplications from one primordial gene, followed by subsequent differentiation of the duplications to match the enzymic functions required for the metabolic pathway. It is well documented in *Drosophila* that tandem duplication favours the occurrence of unequal crossing-over, because of the existence of genetically homologous regions adjacent to each other, and so the further extension of duplications along the chromosome. The hypothesis of tandem duplication is the more likely one, since the probability will be very small of getting the right translocations or inversions to bring together duplicated genes from many different locations in the genome. An important corollary of such an explanation for the origin of the enzymes of a metabolic pathway is that they should show evidence of a common evolutionary origin although they may have relatively unrelated functions. The study of genetic homology between function-ally different proteins within the same species is of great interest, as it is the only way in which evidence for a common evolutionary origin of functionally unrelated proteins can be obtained.

According to Horowitz's (1965) model, the clustering of genes in a biochemical pathway follows from their evolution. The operon type of control discovered by Jacob and Monod for the *lac* region of *Escherichia coli* may also be a consequence rather than a cause of this evolutionary origin of a cluster. The lack of clustering for the same genes in some bacteria and in higher organisms implies that there must have existed selective forces favouring reorganization of the genes on the chromo-some. At the same time, alternatives to the operon method of control of these related functions must have evolved. To explain these changes, a model is needed for the way in which selection can favour reorganization of genes for reasons other than coordinate control within an operon.

A simple model for the way in which selection could favour closer linkage between interacting genes was proposed by Fisher (1930). He pointed out that if *a* and *b* are prevailing alleles at two different loci and *A*, *B* are new mutants at these loci, then under certain circumstances if the double heterozygotes *AB/ab* and *Ab/aB* are at an advantage as compared with the prevailing homozygote *ab/ab*, while the single heterozygotes Ab/ab and aB/ab are not at an advantage, then natural selection may favour the combination AB only if the two genes are sufficiently closely linked. Translocations placing these two genes nearer together may then be selected for. A considerable amount of work on the genetical theory of two linked loci with such types of interacting selective values has been published in recent years (see e.g. Bodmer and Felsenstein, 1967, for further references). In general, the combination AB will be favoured only if the recombination fraction between the two loci, r, is less than the difference in fitness between the double heterozygotes AB/ab and Ab/aB and the prevailing homozygote ab/ab. Whenever such interactive selective values occur, they will favour close linkage between the relevant mutations, either by its production, when the genes were not originally closely linked, or its maintenance if they were. This mechanism is rather different from that of a purely functional advantage of proximity between genes as required by the operon theory. The contrast between these two selective forces for gene rearrangement was pointed out by Bodmer & Parsons (1962) and re-emphasized by Stahl & Murray (1966). Bodmer & Parsons (1962) also pointed out that close linkage may be important in promoting the increase in frequency of the advantageous double-mutant combination AB, when the single mutants are at a disadvantage.

Recombination disrupts associations between mutually advantageous genes while at the same time increasing the rate of formation of new advantageous combinations. This duality of the recombination mechanism was pointed out long ago by Fisher, Darlington, Mather, Dobzhansky and others. On the basis of this duality, H. J. Muller emphasized that the advantage of sexual reproduction would be lost if non-additive selective interactions between genes were common. Genes which are very closely linked (see equation (3); when $r \to 0$, $n \to \infty$) behave essentially as if they were a section of an asexual organism, as Fisher (1930) pointed out. The advantages of both sexual and asexual systems can therefore be maintained, to some extent, according to the chromosomal arrangement of the genes of a sexual organism. The evolution of pairing between homologues and subsequent recombination within the chromosome must have been an essential part of the evolution of the sexual mechanism, since this is required in order to take full advantage of the increased opportunities for recombining mutants of different origin.

The duality of the effects of recombination can be illustrated by the following model situation, taken from Bodmer & Parsons (1962). Suppose in a haploid organism, assuming as before a two-locus two-allele model, Ab and aB are disadvantageous with respect to ab, and are each maintained at equilibrium frequencies μ/t by mutation selection balance. Suppose further, however, that alleles A and B interact in such a way that AB has a fitness $1+\alpha$ in comparison with ab. The increase in the frequency of AB is counteracted by recombination, which leads to its breakdown into the disadvantageous types Ab and aB. When AB is rare, almost all matings will be $AB \times ab$ and so provide the opportunity for loss of AB by recombination at a rate r. The genotype AB will therefore, in general, only increase in frequency if its selective advantage, $1+\alpha$, is enough to counteract recombination breakdown or, approximately, if

$$(1+\alpha)(1-r) > 1$$

or

$$r < \frac{\alpha}{1+\alpha}. \tag{10}$$

This result was first given by Bodmer & Parsons (1962) and later proved rigorously by Bodmer & Felsenstein (1967). It provides the simplest model example of an interaction between linkage and selection. Thus, given selective interaction such that AB is fitter than ab, but Ab and aB are not, AB will only increase in frequency if the two genes are sufficiently closely linked. In particular, from equation (10) the recombination fraction must, approximately, be less than the selective advantage of AB over ab. Now suppose that genotype AB has not yet been formed and consider the question: under what conditions will its rate of formation be greater by recombination that by mutation? The rate of formation of AB by mutation will be

$$2\mu/t . \mu = 2\mu^2/t.$$

The rate of formation by recombination, on the other hand, is

$$2\mu^2/t^2 \times 1/2\,r = r\,\mu^2/t^2.$$

Thus the rate of formation by recombination is higher than that by mutation if

$$r/t^2 > 2/t$$

or

$$r > 2t \tag{11}$$

Conditions (10) and (11) restrain r, approximately, to the range

$$\alpha > r > 2t. \tag{12}$$

Only if inequality (12) holds will sexual reproduction favour the production of the type AB and also allow it to increase in frequency.

A mixture of sexual and asexual reproduction, as is found in some prokaryotes and many plants but only in a few animal species, might seem to offer the best compromise. A new advantageous interacting genotype could then reproduce asexually without recombinational breakdown into less favourable combinations. The sexual mechanism would be held in reserve to produce new genetic changes which may be needed to combat new environmental circumstances. It seems, however, from the few animal species which have dual systems, that the temporary advantages of asexuality are not enough to be worth the modification of an exclusively sexual system, especially if this has evolved to an extent where it infringes on individual selective values. Perhaps also the balance between asexuality and sexuality would be too unstable. The former, due to its temporary advantage, might oust the latter, only to destroy the potential of the species for future adaptation.

The disruption of linked interacting gene clusters by recombination in sexual organisms may have been a contributing factor toward the evolution of alternatives to the operon polycistronic-mRNA control system found in at least some prokaryotes. A most important factor in this evolutionary trend must, surely, however, have been the greater complexity, and hence flexibility, of control mechanisms needed for the development of more complex organisms. It is now generally accepted that a major feature of cell differentiation is the activation of selected sub-sets of genes required for the characteristic activity of particular differentiated cells. Genes in differentiated tissues must be activated in co-ordinate sets. The operon theory of Jacob and Monod provides one possible answer to this control problem for the bacterial cell but is not, in its original form, generally applicable to higher organisms. Perhaps, in higher organisms for purposes of control, it would be advantageous to have several copies of the gene for a particular enzyme in different parts of the genome according to the stages of development and the type of differentiated cell in which the protein is needed. (This may be an important factor in accounting for the larger DNA content of higher organisms.) Given that differential and coordinate control of gene activity may depend on certain arrangements of genes on the chromosomes, though not necessarily in related clusters, an enzyme activity required in different differentiated cells with different combinations of other enzymes may best be controlled by duplicating the genetic information for this enzyme and placing it in different regions of the genome according to its associated activities. Once duplication has occurred, some subsequent differentiation of the duplicated genes is expected, at least within the restraint of maintaining the required enzyme activity. This provides an explanation for the advantage of duplicating the genetic information for particular enzyme activities in higher organisms. These factors might be the basis for the selective forces which have favoured chromosomal rearrangements. These selective forces favour a reorganization of the genes on the chromosome following their evolution by tandem duplication, according to different criteria than those imposed by the operon theory for the control of genetic activity in bacteria.

The greater internal organization of membrane structures in the cells of higher organisms may allow the transport of metabolites, and even immediate gene products, to different parts of the cell in such a way as to overcome the need for close proximity of genes whose activities are to be co-ordinately controlled. The chromosome may, for example, be folded in such a way that genes required to act together are placed physically close to each other on a membrane, allowing co-ordinate messenger RNA synthesis and subsequent association of the protein products into structures that may be required for carrying out a sequence of metabolic steps. An alternative possibility suggested by A. D. Kaiser (personal communication) and already mentioned above is that the genes themselves may be transposed in the somatic cells of higher organisms by a mechanism analogous to the insertion and excision of lysogenic phages.

While the arguments for selection favouring rearrangements of genes may at the present time seem vague, there can be little doubt that the organization of the genes on the chromosome of prokaryotes and eukaryotes is the product of natural selection throughout evolution. Genes do not occur on chromosomes at random, either with respect to each other or with respect to their time and place of action during development and differentiation.

This investigation was supported in part by U.S. Public Health Service Career Development Program Award GM 35002 and by a research grant GM 10452.

REFERENCES

BODMER, W. F. & DARLINGTON, A. J. (1969). Linkage and recombination at the molecular level. In *Genetic Organization*, vol. 1. Eds. E. Caspari and A. W. Ravin. New York and London: Academic Press (in Press).

BODMER, W. F. & FELSENSTEIN, J. (1967). Linkage and selection: theoretical analysis of the deterministic two locus random mating model. *Genetics, Princeton* 57, 237.

BODMER, W. F. & PARSONS, P. A. (1962). Linkage and recombination in evolution. *Adv. Genet.* 11, 1.

BOYCE, R. P. & HOWARD-FLANDERS, P. H. (1964). Release of ultraviolet light-induced thymine dimers from DNA in *E. coli* K-12. *Proc. natn. Acad. Sci. U.S.A.* 51, 293.

CAMPBELL, A. N. (1962). Episomes. *Adv. Genet.* 11, 101.

CROW, J. F. & KIMURA, M. (1965). Evolution in sexual and asexual populations. *Am. Nat.* **99**, 439.

CROW, J. F. & KIMURA, M. (1969). Evolution in sexual and asexual populations: a reply. *Am. Nat.* **103**, 89.

FISHER, R. A. (1930). *The Genetical Theory of Natural Selection*. Oxford University Press (also reprinted, 1958, by Dover publications).

GANESAN, A. T. (1968). Studies on *in vitro* replication of *Bacillus subtilis* DNA. *Cold Spring Harb. Symp. quant. Biol.* **33**, 45.

GORINI, L., GUNDERSUN, W. & BURGER, M. (1961). Genetics of regulation of enzyme synthesis in the arginine biosynthetic pathway of *Escherichia coli*. *Cold Spring Harb. Symp. quant. Biol.* **26**, 173.

GUNSALUS, I. C., GUNSALUS, C. F., CHAKRABARTY, A. M., SIKES, S. & CRAWFORD, I. P. (1968). Fine structure mapping of the tryptophan genes in *Pseudomonas putida*. *Genetics, Princeton* **60**, 419.

HOLLOWAY, B. W., HODGINS, L. & FARGIE, B. (1963). Unlinked loci affecting unrelated biosynthetic steps in *Pseudomonas aeruginosa*. *Nature, Lond.* **199**, 926.

HOROWITZ, N. (1965). The evolution of biochemical synthesis—retrospect and prospect. In *Evolving Genes and Proteins*. New York: Academic Press, Inc.

IINO, T. & LEDERBERG, J. (1964). Genetics of Salmonella. In *The World Problem of Salmonellosis*. Ed. E. van Oye. The Hague: Dr W. Junk Publishers.

LEDERBERG, J. & EDWARDS, P. R. (1953). Serotype recombination in Salmonella. *J. Immun.* **71**, 232.

MAAS, W. K. (1961). Studies on repression of arginine biosynthesis in *Escherichia coli*. *Cold Spring Harb. Symp. quant. Biol.* **26**, 183.

MAYNARD SMITH, J. (1968). Evolution in sexual and asexual populations. *Am. Nat.* **102**, 469.

MEYNELL, E., MEYNELL, G. & DATTA, N. (1968). Phylogenetic relationships of drug resistance factors and other transmissible bacterial plasmids. *Bact. Rev.* **32**, 55.

MULLER, H. J. (1932). Some genetic aspects of sex. *Am. Nat.* **8**, 118.

MULLER, H. J. (1964). The relation of recombination to mutational advance. *Mutat. Res.* **1**, 2.

SCHNEIDER, H., FORMAL, S. B. & BARON, L. S. (1961). Experimental recombination *in vivo* between *Escherichia coli* and *Salmonella typhimurium*. *J. exp. Med.* **114**, 141.

SETLOW, R. B. & CARRIER, W. L. (1964). The disappearance of thymine dimers from DNA: an error-correcting mechanism. *Proc. natn. Acad. Sci. U.S.A.* **51**, 226.

SIGNER, E. R. (1968). Lysogeny: the integration problem. *A. Rev. Microbiol.* **22**, 451.

STAHL, F. W. & MURRAY, N. (1966). The evolution of gene clusters and genetic circularity in micro-organisms. *Genetics, Princeton* **53**, 569.

VOGEL, H. J., BACON, D. F. & BAICH, A. (1963). Induction of acetylornithine Δ-transaminase during pathway-wide repression. In *Informational Macromolecules*. Eds. H. J. Vogel and V. Bryson. New York and London: Academic Press.

YAMAMOTO, N. (1969). Genetic evolution of bacteriophage. 1. Hybrids between unrelated bacteriophages P 22 and Fels-2. *Proc. natn. Acad. Sci. U.S.A.* **62**, 63.

YOUNG, F. E. (1967). Competence in *Bacillus subtilis* transformation system. *Nature, Lond.* **213**, 773.

STRUCTURE, FUNCTION AND DISTRIBUTION OF ORGANELLES IN PROKARYOTIC AND EUKARYOTIC MICROBES

D. E. HUGHES, D. LLOYD AND R. BRIGHTWELL

*Medical Research Council Group for Microbial Structure and
Function, Department of Microbiology,
University College, Cathays Park, Cardiff*

This article describes those organelles of eukaryotic microbes which are enclosed in membranes (mitochondria, kinetosomes, lysosomes, etc.) and the mesosomes and the plasma membrane which may perform analogous functions in prokaryotes. Organelles of locomotion and nuclei are not discussed. Overlap with other contributions has been minimized by omitting discussions of membrane structure and function at the molecular level. The interpretation of cell structure and function may be profoundly affected by the methods used, especially when information is gained by the fractionation of cell homogenates. For this reason a brief discussion of practical methods is included.

PRACTICAL METHODS

Histochemistry

Histochemical methods for locating chemical constituents or enzymes within cells by electron microscopy depend on heavy deposition of electron-dense material (up to 25% increase of dry weight) to produce contrast (Kay, 1965; Sjöstrand, 1967). For locating enzymes the reagents are usually too dilute to be located directly and contrast is enhanced by coupling; a simple example is the widely used method for phosphatase (Gomori, 1952), where the enzyme site is indicated by the deposition of electron-dense granules of metal sulphide. Both the size and location of the granules are determined by the rates of the various reactions, including the initial enzymic reaction, the concentration of reactants and the rates of diffusion of products towards and away from the enzyme site. The resolution under the most favourable conditions (1st order reactions, no permeability barriers, correct fixation, etc.) is approx. 0·2 μm (Holt & O'Sullivan, 1958). These limitations may also apply to the use of ferritin or fluorescent labelled antibodies (Beachey

& Cole, 1966), and to autoradiography (Reith, Schüler, Vogel & Klingenberg, 1967).

The main advantage of histochemical methods is that the enzyme site may be made visible at relatively low resolution (approx. 0·1 to 0·2 μm) against the background of the other cell structures which may be resolved down to about 30 to 40 Å (see Marr, 1960). Negative-staining for electron microscopy has greatly improved the resolution of protein and associated structures but this is relatively unspecific (Horne & Muscatello, 1968). Freeze-etching is beginning to yield interesting three-dimensional information especially about membrane and wall structures at medium resolution (Steeve, 1957; Moor & Mühlelthaler, 1963) but as yet cannot be coupled with methods for locating enzymes.

With the development of more rapid and successful fractionation methods there is a pressing need to accelerate the fixing, embedding and sectioning of large numbers of samples for electron microscopy, as for instance when following changes during growth and differentiation. Negative-staining does not replace the need for sectioning (Horne & Muscatello, 1968; Hughes, 1970).

Cell fractionation

The choice of the initial step for cell disintegration is still largely empirical, and is mainly between enzymic preparation and lysis of spheroplasts, or disintegration by solid or hydrodynamic shear (see Hughes, Wimpenny & Lloyd, 1970; Lloyd & Hughes, 1970). The method for preparing protoplasts is difficult to apply widely because of the lack of specific enzymes to dissolve cell walls; even when these are available this method cannot easily be applied to a large mass of cells. Further, important cell structures may change during enzymic treatment. Hydrodynamic gradients up to 10^8 sec.$^{-1}$ are necessary to break most bacteria, yeast and fungi because of the high tensile strength of their cell walls, whereas most cell organelles are damaged by gradients as low as 10^3 to 10^4 sec.$^{-1}$ (T. Coakley, unpublished results in this laboratory). Damage may be avoided by using continuous flow or other devices to remove quickly the disintegrated material from the centres of high shear, or by using more carefully controlled shearing devices. The choice of suitable eukaryotic cells without strong cell walls, e.g. the amoeba *Hartmanella castellanii*, has in part governed our own choice of experimental material.

Until recently, gradient centrifugation in tubes, chromatography and electrophoresis have been the main methods for separating organelles from homogenates. Usually only relatively small amounts of homogenate can be conveniently treated by these methods. This major dis-

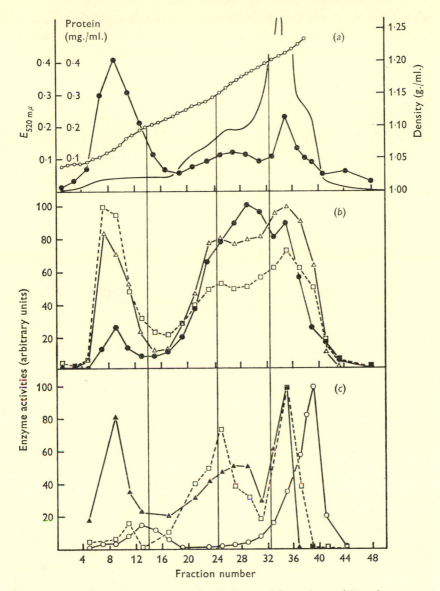

Fig. 1. Fractionation of homogenate of cells from a 6-day culture of *Tetrahymena pyriformis* in the BXIV zonal centrifuge rotor on a tris-buffered sucrose density gradient. The culture was harvested at 440,000 cells/ml., the cells were gently homogenized in a hand homogenizer; the homogenate contained 77 mg. protein. Overlay (9 % sucrose) volume was 60 ml., cushion (60 % sucrose) 150 ml. Centrifugation at 35,000 rev./min. for 2·75 hr, equivalent to 100,000 g for 60 min. in sample zone. (*a*) Sucrose gradient (—○—), protein (—●—) and light-scattering (continuous line). (*b*) Acid phosphatase (—●—), DNAase (—△—) and N-acetyl-β-D-glucosaminidase (----□----). (*c*) Catalase (—○—), NADPH- (----□----) and NADPH- (—▲—) cytochrome *c* oxidoreductases. Recoveries were: protein 120 %, acid phosphatase 97 %, DNAase 150 %, acetylglucosaminidase 95 %, catalase, 110 %, NADH- 83 % and NADPH- cytochrome *c* oxidoreductase 47 %. (Experiment with G. Turner and Susan E. Venables.)

advantage has been largely overcome by the introduction of zonal centrifugation (Anderson, 1966). The power of this method is illustrated by the fractionation of an homogenate of *Tetrahymena pyriformis* (Pl. 2, fig. 1 to 4).

Criteria for judging the integrity of separated intracellular organelles have been discussed fully elsewhere (Hughes *et al.* 1970; Chappell & Hansford, 1969; Roodyn, 1967*b*). In general, morphological examination by electron microscopy is combined with estimation of 'marker' enzymes and/or of specific chemical constituents. The identification of suitable markers is a major stage in such studies. The present discussions and doubts about the role of mesosomes in prokaryotes largely result from the lack of such specific markers (page 313).

SUBCELLULAR ORGANELLES

Mitochondria

The structure of the mitochondrion appears to be similar in all eukaryotes; the most variable feature is the complexity of organization of the invaginations of the inner membrane (cristae) which appears to be closely correlated with the energy needs of the cell. The separation of inner and outer membranes of isolated mitochondria (Sottocasa, Kuylenstierna, Ernster & Anders, 1967; Parsons *et al.* 1967) has given a detailed picture of the distribution of enzyme activities between these two membranes and the matrix space (Fig. 2) and has led to the conclusion that it is the inner membrane (and matrix) which is concerned with many of the principal cellular oxidation mechanisms (fatty acid oxidation, tricarboxylic acid cycle, respiratory chain electron transport) and coupled energy conservation (ATP synthesis by oxidative phosphorylation). It is also the inner membrane which functions as a permeability barrier to inorganic ions, adenine and pyridine nucleotides, carboxylic acids and amino acids (Chappell, 1968; Mitchell, this Symposium).

While most work on mitochondrial biochemistry has been done with mitochondria from mammals, there is an increasing interest in the mitochondria of eukaryotic micro-organisms, since these have some distinctive properties (Table 1). Aspects best studied in eukaryotic micro-organisms include the biogenesis of mitochondrial membranes, mitochondrial nucleic acid, lipid and protein synthesis, mitochondrial genetics, and variations of mitochondrial composition in relation to environmental conditions (Lloyd, 1969; Roodyn & Wilkie, 1968; Wilkie, p. 381 this Symposium).

Because of the absence of discrete respiratory organelles from pro-karyotes any comparison between mitochondria and prokaryotic structures is necessarily a comparison of the inner mitochondrial

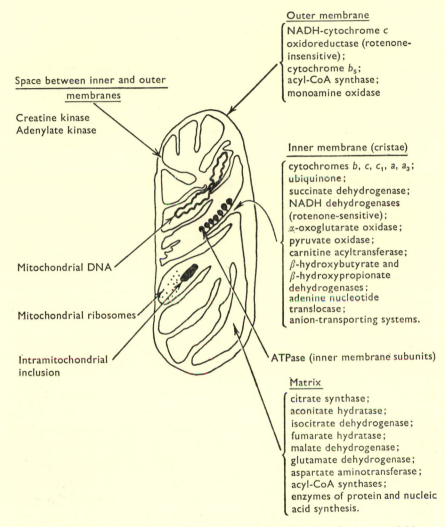

Outer membrane
NADH-cytochrome c oxidoreductase (rotenone-insensitive);
cytochrome b_5;
acyl-CoA synthase;
monoamine oxidase

Space between inner and outer membranes

Creatine kinase
Adenylate kinase

Inner membrane (cristae)
cytochromes b, c, c_1, a, a_3;
ubiquinone;
succinate dehydrogenase;
NADH dehydrogenases (rotenone-sensitive);
α-oxoglutarate oxidase;
pyruvate oxidase;
carnitine acyltransferase;
β-hydroxybutyrate and β-hydroxypropionate dehydrogenases;
adenine nucleotide translocase;
anion-transporting systems.

Mitochondrial DNA

Mitochondrial ribosomes

Intramitochondrial inclusion

ATPase (inner membrane subunits)

Matrix
citrate synthase;
aconitate hydratase;
isocitrate dehydrogenase;
fumarate hydratase;
malate dehydrogenase;
glutamate dehydrogenase;
aspartate aminotransferase;
acyl-CoA synthases;
enzymes of protein and nucleic acid synthesis.

Fig. 2. Diagram of the mitochondrial structure and localization of enzyme activities.

membrane with the bacterial cytoplasmic membrane (Table 2). While the overall process of ATP production is similar in principle in both types of situation, there are some interesting differences in the detailed mechanisms by which this process is accomplished (Gelman, Lukoyanova & Ostrovskii, 1967).

Table 1. *A comparison of some of the properties of mitochondria isolated from eukaryotic micro-organisms, higher plants and animals*

Organism	Occurrence of rotenone-sensitive site	Occurrence of Site-I phos-phorylation	Oxidation of exogenous NADH	Reference
Yeasts				
Saccharomyces cerevisiae	Absent	Absent	+	⎫ Ohnishi, Sottocasa
Saccharomyces carlsbergensis	Absent	Absent	+	⎭ & Ernster (1966)
Endomyces magnusii	Present	Present	.	Kotelnikova & Zvjagilskaja (1966)
Torulopsis utilis	Present	Present	+	⎫ Light, Ragan, Clegg
Torulopsis utilis (Fe-limited)	Absent	Absent	.	⎭ & Garland (1968)
Fungi				
Aspergillus niger	Present	Present	+	Watson & Smith (1967)
Algae				
Prototheca zopfii	Present	Present	+	Lloyd (1965)
Protozoa				
Polytomella caeca (flagellate)	Present	Present	+	Lloyd, Evans & Venables (1968)
Hartmanella castellanii (amoeba)	Absent	Absent	+	Lloyd & Griffiths (1968)
Tetrahymena pyriformis (ciliate)	Present	?	−	Turner & Lloyd (1968)
Higher plant mitochondria	Present	Present	+	—
Higher animal mitochondria	Present	Present	−	—

Kinetoplast

Trypanosomid protozoa possess an unusual organelle situated close to the base of the flagellum. This structure, the kinetoplast, may be regarded as a differentiated part of the large single mitochondrion which is found in these flagellates. The kinetoplast contains DNA which differs from nuclear DNA in several respects: it has a lower buoyant density, a lower melting temperature and it readily re-natures (Dubuy, Mattern & Riley, 1966). Variation in the structure and function of the mitochondrion during the complex life-cycles of these parasitic protozoa have been studied (see review by Vickerman, 1965). The stages of the life cycle which occur in the blood of vertebrate hosts have lost the classical pattern of terminal respiration components (tricarboxylic acid cycle, cytochrome system) which is found in the stages borne in the insect host or those which have been grown in axenic culture (Hill & White, 1968). The development of the protozoal mitochondria is dependent to some extent on the presence of kinetoplast-DNA, and 'dyskinetoplastic forms' induced by treatment with acridine dyes (Steinert & van Assel, 1967) or arising spontaneously can survive only in the blood stream (Mühlpfordt, 1963).

Table 2. *Similarities and differences between the energy-producing mitochondria of eukaryotes and the energy-producing membrane of prokaryotes*

Mitochondria	Membranes
Discrete organelles, some autonomy	Energy production chief feature of a polyfunctional membrane
Respiratory chain and oxidative phosphorylation mechanism an integral part of 55 Å inner membrane	Respiratory chain and oxidative phosphorylation mechanism integral part of a lipoprotein membrane (70 to 80 Å thick). Membrane-bound systems in anaerobes, e.g. hydrogenase
90 Å sub-units on inner side of inner membrane	90 Å sub-units on inner side of membrane
Permeases and ATPases in inner membrane	Permeases and ATPases in membrane or associated with membrane
ATP-generating system isolated from cytosol by permeability barrier (inner membrane); compartmentation important control mechanism	ATP-generating system exposed to cytosol; no compartmentation observed
Amount of energy-generating machinery in cell proportional to (1) number of mitochondria, (2) complexity of invagination of cristae	Total amount of energy-generating machinery proportional to complexity of invagination-forming mesosomes
Fairly constant composition with regard to content of cytochromes, ratio of cytochromes, lipid composition, but vary with growth conditions. Phospholipids involved in functional integrity of respiratory chain. Steroids absent from inner membrane	Very variable composition with regard to respiratory enzymes and lipid composition, depending on growth conditions. Phospholipids involved in integrity of respiratory chain; steroids absent (except in Mycoplasma)
Rate of ATP generation closely linked to rate of ATP utilization (respiratory control). Structure and activity of mitochondrial energy-generation under hormonal control. High efficiency of oxidative phosphorylation	Respiratory control only rarely demonstrated in bacteria; coarse control possible by modification of respiratory assemblies. Efficiency of oxidative phosphorylation uncertain; low efficiency in bacterial extracts
Ubiquinone, cytochromes b, c, c_1, a, a_3	Naphthoquinone Gram+, naphthoquinone and ubiquinone Gram−, many different cytochromes, often several oxidases
Classical electron transport inhibitors nearly always effective, e.g. rotenone, antimycin A, cyanide, CO, azide	Classical inhibitors of mitochondrial electron-transport often ineffective
Classical uncouplers (DNP) and inhibitors (oligomycin) effective	Dinitrophenol and oligomycin often ineffective

Peroxisomes (glyoxysomes)

Elegant fractionation studies of rat liver have shown that urate oxidase and D-amino acid oxidase are associated with subcellular particles which have sedimentation properties different from mitochondria and lysosomes and which correspond to the microbodies described by electron microscopists (De Duve & Baudhuin, 1966). These organelles, which have been named peroxisomes, have a diameter of about 0·5 μm, a sedimentation value of $4\cdot4\times10^3$ s and a median buoyant density

Table 3. *Properties of peroxisomes (glyoxysomes) from various sources*

Modified from Müller, Hogg & De Duve (1968)

Source	Median equilibrium density in sucrose gradient g./ml.	Catalase	Urate oxidase	D-amino acid oxidase	L-amino-acid oxidase	Glycollate oxidase
Rat liver[a]	1·23	+	+	+	+	+
Rat kidney[b]	1·24	+	−	+	+	+
Germinating castor-bean seeds[c]	1·25	+	.	.	.	+
Spinach, wheat and tobacco leaves[d]	1·24	+	.	.	.	+
Tetrahymena pyriformis[e,g]	1·24	+	−	+	−	+
Acanthamoeba (Hartmanella) castelanii[f,g,h]	1·20, 1·22	+	+	−	.	.
Polytomella caeca[i]	1·22, 1·24	+	−	−	.	.
Saccharomyces carlsbergensis[j]	1·20	+	.	.	.	.

Source	L-α-hydroxy acid oxidase	Glyoxylate oxidase	Citrate synthase	Isocitrate lyase	Malate synthase	Malate dehydrogenase (NAD linked)	Isocitrate dehydrogenase (NADP linked)	Glyoxylate reductase
Rat liver[a]	+	.	−	−	−	−	+	.
Rat kidney[b]	+	.	.	.	.	.	.	.
Germinating castor-bean seeds[c]	.	.	+	+	+	+	−	.
Spinach, wheat and tobacco leaves[d]	.	.	.	.	.	+	−	+
Tetrahymena pyriformis[e,g]	+	+	−	+	+	−	+	−
Acanthamoeba (Hartmanella) castellanii[f,g,h]	−	.	.	.	.	.	.	.
Polytomella caeca[i]	−	.	.	−	.	.	.	.
Saccharomyces carlsbergensis[j]	.	.	.	.	.	.	.	.

a, De Duve & Baudhuin (1966); b, Baudhuin, Muller, Poole & De Duve (1965); c, Breidenbach, Kahn & Beevers (1968); d, Tolbert *et al.* (1968); Kisaki & Tolbert (1969); e, Müller, Hogg & De Duve (1968); f, Müller & Møller (1967); g, Brightwell, Lloyd, Turner & Venables (1968); h, A. J. Griffiths, S. M. Bowen & D. Lloyd, unpublished; i, Cooper, Jones, Venables & Lloyd (1969); j, Cartledge, Burnett, Brightwell & Lloyd (1969).

of 1·23 g./ml. They are thus smaller and denser than mitochondria, and they may be separated from the latter either by rate of sedimentation or isopycnic centrifugation. In rat liver cells there are about four mitochondria to each microbody. The peroxisomes comprise 2·5% of the total liver protein i.e. 6·5 mg. protein/g. wet wt liver (Leighton *et al.* 1968).

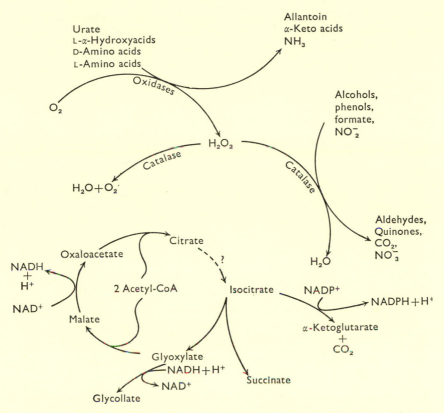

Fig. 3. Enzyme reactions associated with some peroxisomes. Distribution of these enzymes in peroxisomes from various sources shown in Table 3.

The presence of peroxisomes has now been demonstrated in a variety of plant tissues and also in several eukaryotic micro-organisms (Table 3).

Although present knowledge of these organelles only hints at their physiological functions, the comparative study of their enzyme complement throws light on their possible evolution. Some of the reactions catalysed by peroxisomal enzymes are shown in Fig. 3. The possible metabolic role of these enzymes has been considered under three headings by De Duve & Baudhuin (1966) as follows.

(*a*) Disposal of hydrogen peroxide. The compartmentation of hydrogen peroxide producing oxidases together with catalase immediately suggests that the latter enzyme has a protective function. However several considerations make it likely that the peroxisomal catalase is more important as a peroxidase, i.e. reduction of hydrogen peroxide to water at the expense of electron donors such as alcohols, phenols, formate or nitrite.

(b) Role in energy metabolism and oxidative metabolism. There is some evidence that peroxisomes contribute significantly to the respiration of liver slices, but there is no mechanism for ATP synthesis by oxidative phosphorylation in these organelles. Peroxisomes may also participate in the oxidation of reduced nicotinamide nucleotide generated in the cytosol, a function generally attributed to mitochondrial α-glycerophosphate or malate shuttles.

(c) Participation in specific pathways. It has been suggested that the enzymes of peroxisomes are involved in mammalian glyconeogenesis, providing pathways for the production of α-keto acids. The demonstration that the entire glyoxylate cycle is located in the peroxisomes of germinating seedlings (Breidenbach, Kahn & Beevers, 1968), where fat is being converted to carbohydrate, provides the most clear-cut metabolic role for peroxisomes. In *Tetrahymena pyriformis* the key enzymes of the glyoxylate cycle, isocitrate lyase and malate synthase, are also present in peroxisomes; but here the complete operation of the cycle during glyconeogenesis requires metabolic interplay with the mitochondria (Müller, Hogg & De Duve, 1968). Peroxisomes from photosynthetic tissues of higher plants contain an NADH-linked glyoxylate reductase in addition to catalase and glycollate oxidase and are thus able to oxidize internally generated NADH. A glutamate-glyoxylate transaminase is also localized in these peroxisomes (Tolbert *et al.* 1968). It has been suggested (De Duve & Baudhuin, 1966), that peroxisomes may be 'fossil organelles', since their distribution is restricted to only a few mammalian organs (kidney, liver, possibly bone) and catalase is a dispensable enzyme in humans. Urate oxidase is an enzyme that was lost early in primate evolution, and the metabolic role of D-amino acid oxidase is unknown. Peroxisomal respiration may represent an early type of oxidative metabolism which may have arisen during the evolutionary transition from ancestral anaerobic life. Such a system, although providing no ATP, would confer advantages to organisms exposed to oxygen produced by photosynthesis or flavoprotein enzymes, both by decomposing potentially toxic hydrogen peroxide and by the re-oxidation of reduced nicotinamide nucleotides. Roodyn & Wilkie (1968) suggested that the peroxisomal glyoxylate cycle and its associated side paths may

have been the evolutionary antecedent of the mitochondrial tricarboxylic acid cycle, the latter originating perhaps by symbiotic association of an invasive aerobic prokaryote with a large anaerobic hetero-
troph (Sagan, 1967). The importance of the peroxisome as a respiratory
organelle may then have decreased progressively as its functions were
taken over by the thermodynamically more efficient system derived
from a symbiotic aerobic prokaryote. Present day peroxisomes may
confer selective advantages to cells which possess them; even so, many
of their enzymes may be relics of their evolutionary past.

No compartmentation of enzymes such as catalase, flavoprotein
oxidases or the glyoxylate cycle enzymes has yet been demonstrated in
prokaryotes. Most anaerobes are catalase-negative, but catalase has
been demonstrated in most aerobic and facultatively aerobic bacteria
(Dolin, 1961).

Changes in peroxisomal enzymes during cell differentiation have not
yet been studied. Several systems might provide evidence about the
variability of peroxisomal enzyme composition. For instance, the discovery of sedimentable catalase in aerobically grown yeast (Avers &
Federman, 1968; Cartledge, Burnett, Brightwell & Lloyd, 1969)
suggests that changes occur in the structure and function of peroxisomes
during respiratory adaptation. The catalase of yeast is induced during
respiratory adaptation (Chantrenne & Courtois, 1954) and is repressed
under anaerobic conditions and at high glucose concentrations (Lorenc,
Dukwicz & Kwiek, 1968; Sulebele & Rege, 1968; T. G. Cartledge,
J. K. Burnett, R. Brightwell & D. Lloyd, unpublished observation).
The inducibility of glyoxylate cycle enzymes, for example during protozoal encystment (Bowen & Griffiths 1969), is another system which
merits study in this connection.

Endoplasmic reticulum

Virtually all eukaryotes have the cytoplasmic structure known as the
endoplasmic reticulum. Ribosomes or polysomes are bound to regions
of this double membrane which is then called rough endoplasmic
reticulum (Plate 3). On isolation both endoplasmic reticulum and free
polysomes will synthesize protein (Table 4).

Fragments of rough and of smooth reticulum, cytoplasmic membrane,
Golgi vesicles and small lysosomes sediment in sucrose homogenates of
mammalian tissues to yield the microsomal fraction. The differences
between rough and smooth reticulum are not clear-cut (Moure, 1968),
but rough and smooth can be separated by further centrifugation.
Glucose-6-phosphatase is the usual marker enzyme for endoplasmic

reticulum; other microsomal enzymes not shown in Table 4 were discussed by De Duve, Wattiaux & Baudhuin (1962).

Proteins of class (1) in Table 4 are transferred to the lumen of the smooth endoplasmic reticulum (Redman & Sabatini, 1966) and then released in vesicles formed from the membrane. These move to the cytoplasmic membrane (De Duve & Wattiaux, 1966) or to the Golgi apparatus (Jamieson & Palade, 1968). Proteins of class (3) could be transported similarly.

Table 4. *Functions and components of rough and smooth endoplasmic reticulum of rat liver cells*

Function	Components involved	Reference
Protein synthesis	Ribosomes on membrane (i.e. rough endoplasmic reticulum), mRNA cofactors	Campbell & Lawford (1968)
(1) Proteins for export, e.g. serum albumin, hydrolytic enzymes		Ragnotti, Lawford & Campbell (1969)
(2) Endoplasmic reticulum protein, e.g. NADPH-cytochrome c reductase		Campbell & Lawford (1968)
(3) Organelle protein, e.g. catalase for peroxisomes, cytochrome c for mitochondria, histone for nucleus (HeLa cells)		Gallwitz & Mueller (1969)
(4) Soluble cytoplasmic proteins		Campbell & Lawford (1968)
Hydroxylation of steroids and drugs, oxidation of hydrocarbons, β-oxidation of fatty acids	Probably flavoprotein NADPH-cytochrome c reductase and cyto-chrome P450 in smooth membranes	Orrenius et al. (1968); Miyake, Gaylor & Mason (1968)
Glycogen storage, lipid metabolism	Smooth membrane components	Moure (1968)

Endoplasmic reticulum in plant cells (reviewed by Northcote, 1968 a, b) surrounds plastids and is on either side of sieve plates in developing phloem. The surface of the double membrane closest to the plate has no ribosomes while the surface farthest from the plate has ribosomes. A microsomal fraction from pea epicotyl tissue synthesizes α-glucanase (Davies & Maclachlan, 1968).

Many fungi (Bracker, 1967; and Pl. 3) and yeasts (Marchant & Smith, 1968) have an endoplasmic reticulum. The membrane decreases during ascospore formation in *Saccharomyces cerevisiae* and proliferates during adaptation to aerobiosis. Cytochrome c peroxidase and tellurite reduction have been detected on the membrane or in the lumen in several yeasts. Vesicles apparently derived from the endoplasmic reticulum occur near sites of bud formation in yeast and of wall formation in fungi (Hawker & Gooday, 1967; Brenner & Carroll, 1968; McClure, Park & Robinson, 1968): these vesicles might be involved in cell-wall synthesis or, in yeast, cell-wall breakdown.

Rough and smooth membrane occurs in different classes of protozoa (Grimstone, 1966; Levy & Elliott, 1968; Bowers & Korn, 1968). NADH- and NADPH-cytochrome *c* reductases sediment with non-mitochondrial membranes in *Tetrahymena pyriformis* (Brightwell, Lloyd, Turner & Venables, 1968).

Endoplasmic reticulum is probably of major importance in the synthesis and transport of protein in differentiation and adaptation of eukaryotic micro-organisms; the purification and characterization of it should be pursued.

In bacteria, proteins are synthesized by ribosomes attached to the protoplast membranes, which probably include some mesosome membrane, e.g. from *Escherichia coli* (Tani & Hendler, 1964) and *Bacillus megaterium* (Schlessinger, 1963). Cytoplasmic bacterial ribosomes, which may be attached to a fine reticulum of non-unit membrane structure (Salton, 1968), synthesize polypeptides at higher and lower rates than do membrane-bound ribosomes, depending on the system (Hendler, 1965). Bacterial exoenzymes might be synthesized solely on membranes, as is mammalian serum albumin.

Golgi apparatus

Beams & Kessel (1968), in reviewing the cytoplasmic structure known as the Golgi apparatus, emphasized that the cisternae of lamellae formed from several parallel membranes are interconnected, and appear as a system of branched tubules. These tubules may pinch-off to form numerous vesicles which surround the Golgi body (Plate 3). There are 1 to 20 or more Golgi bodies in a cell. The 'forming face' of the stack of lamellae is that closer to the nucleus and endoplasmic reticulum in some organisms.

Appropriately fixed Golgi bodies from plants and Golgi vesicles from pancreas have been isolated by centrifugation. Thiamine pyrophosphatase is a possible marker enzyme for the Golgi apparatus.

The Golgi apparatus synthesizes, stores or concentrates products for secretion or incorporation into other parts of the cell, often outside the cytoplasmic membrane (Table 5). An example of synthesis by the Golgi apparatus is the synthesis of complex carbohydrates for cell walls in the alga *Pleurochrysis* (Table 5). Cell wall fragments are visible in the cisternae and vesicles closest to the cell membrane but not in those close to the nucleus. It has been suggested that the Golgi apparatus moves, laying down the cell wall progressively at different portions of the cell periphery (Brown, 1969).

Proteins stored and concentrated in the Golgi apparatus, such as those of the zymogen granules of mouse pancreas, are synthesized at the

Table 5. *Functions of the Golgi apparatus*

Function	Involvement of cisternae (c) or vesicles (v)	Organism	Reference
Concentration of pancreatic zymogen intended for secretion and synthesized at endoplasmic reticulum	v c, v	Guinea pig Rat, mouse	Jamieson & Palade (1968), Beams & Kessel (1968)
Concentration of proteins of azurophil granules (lysosomes) and specific granules in polymorphonuclear leucocytes. Each type of granule released at opposite pole of the same Golgi stack at different stages of cell maturation	c, v	Rabbit	Bainton & Farquhar (1966)
Storage or concentration of lysosomal enzymes synthesized at the endoplasmic reticulum	c, v v	Various mammalian cell types Euglena	De Duve & Wattiaux (1966) Beams & Kessel (1968)
Pigment formation:			
(i) melanin	v	Human melanoma	Beams & Kessel (1968)
(ii) oocytic pigment; secretion granules formed by fusion of Golgi apparatus and endoplasmic reticulum vesicles containing different products	v	Salamander	
Synthesis or modification of membranes. The membrane on release from the Golgi apparatus moves through the cell as the limiting membrane of a vesicle	c, v c, v	Rat (epithelial cell membrane) *Pythium ultimum*	Hicks (1968) Grove, Bracker & Morre (1968)
Synthesis or modification of complex carbohydrates, including glycoproteins	c, v	Rat	Rambourg et al. (1969); Neutra & Leblond (1969); Beams & Kessel (1968)
(i) for export from goblet cells and other secretory cells as mucus, cell coats and cartilage matrix			
(ii) for cell walls of outer root cap cells	v	Wheat	Northcote (1968a)
(iii) for theca	v	Green and brown algae	
(iv) for cell walls	c, v	*Pleurochrysis scherfelii.* (chrysophycean alga)	Brown (1969)

rough endoplasmic reticulum and move to the lumen of the smooth region, whose membrane 'pinches off' to form vesicles containing zymogen. These migrate to the forming face of the Golgi apparatus and fuse with or form lamellae. Vesicles containing zymogen are later reformed from the distal pole of the Golgi stack by 'pinching off' or change in shape of the distal lamella. Such vesicles usually fuse to form secretion granules; primary lysosomes may form by a similar process (Fig. 4). Other similar processes are shown in Table 5. If chemical modification of metabolic products takes place in the Golgi apparatus, the necessary enzymes could likewise be transferred to lamellae from the endoplasmic reticulum. We do not know the mechanism of control which must act when two or more products, such as lysosomal enzymes and cell wall, are synthesized concurrently, but not all secretory products pass through the Golgi apparatus, for example salamander oocyte pigment (Table 5). Because they have common enzymes some functions attributed to the Golgi apparatus could also be performed by endoplasmic reticulum.

Most protozoa have the Golgi apparatus (Beams & Kessel, 1968; Mollenhauer, Evans & Kogut, 1968; Bowers & Korn, 1968), but it has not been found in some ciliates; for example, *Tetrahymena pyriformis*. The number and size of Golgi bodies in the giant amoeba *Pelomyxa carolinensis* are decreased by starvation (Daniels, 1964). This author, but no one since, proposed that Golgi lamellae in this organism are formed from pinocytic vesicles, because vesicles near or fusing with the cytoplasmic membrane and vesicles on one pole of the stack have fringed membranes. Many filamentous fungi have Golgi bodies and they have also been reported in spheroplasts of the yeast *Schizosaccharomyces pombe* when new cell wall is being synthesized (cited by Marchant & Smith, 1968). Bouch (1965) reported continuity of the nuclear envelope with Golgi lamellae and endoplasmic reticulum in brown algae. Other references to algal Golgi bodies are given in Table 5.

Bacterial cell-wall lipopolysaccharide may be synthesized by an organized system in the cell envelope of Gram-negative organisms (Salton, 1968); otherwise there is no evidence for structures similar to the Golgi apparatus in bacteria.

Lomasomes and Woronin bodies

Lomasomes in fungi, algae and higher plants (reviewed by Calonge, Fielding & Byrde, 1969) resemble bacterial mesosomes (Pl. 4; and Bracker, 1967). They may be involved in fungal cross-wall or membrane synthesis (Brenner & Carroll, 1968).

Woronin bodies have been seen only in fungi (Pl. 5, and Bracker, 1967). They are refractile under phase-contrast microscopy and are usually close to, or apparently blocking, pores in hyphal septa (Brenner & Carroll, 1968). Brenner & Carroll suggest that they form from 'pouch-like membrane systems'.

Lysosomes and vacuoles

Lysosomes are cytoplasmic membrane-bounded structures containing hydrolytic enzymes (De Duve & Wattiaux, 1966; Dingle, 1968). Rat liver and kidney lysosomal enzymes are nearly all most active at below pH 7, and they hydrolyse virtually all important biological macro-molecules (Tappel, 1968). De Duve & Wattiaux (1966) have discussed the nomenclature of structures involved in intracellular digestion; these

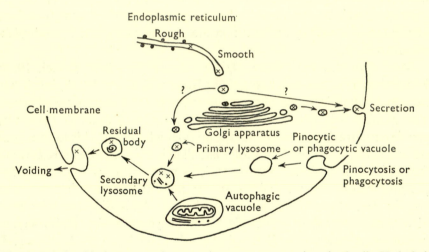

Fig. 4. Relationship between various membranous structures in animal cells. Hydrolytic enzymes (denoted by crosses) are given as an example of a product of protein synthesis at the rough endoplasmic reticulum. Secondary lysosomes are formed on fusion of primary lysosomes with autophagic and/or pinocytic or phagocytic vacuoles. Not to scale.

and other cytological terms are defined concisely by Roodyn (1967a). Material to be digested, from inside or outside the cell, is first sur-rounded by a membrane to form a vacuole (Fig. 4), which acquires hydrolytic enzymes by fusion of its membrane with that of a primary or a secondary lysosome, so becoming a secondary lysosome. Different types of secondary lysosomes may fuse (Dingle, 1968). Products of hydrolysis pass through the membrane (Cohn & Ehrenreich, 1969; Ehrenreich & Cohn, 1969). Resistant material accumulates and may be concentrated in 'residual bodies' or 'dense bodies' which may remain in the cell or be voided (Fig. 4). Healthy organisms have considerable control of the amount and probably some control of the nature of the material digested.

Aleurone grains in cottonseed have lysosomal enzymes (Yatsu & Jacks, 1968). There is some evidence that vacuoles and other particles in maize roots (Matile, 1966, 1968a) and aleurone grains from pea seeds (Matile, 1968b) contain acid hydrolases (see Wardrop, 1968; Gahan 1969). Aleurone grains may be involved in the mobilization of reserves for germination. Matile & Moor (1968) suggested that maize root-cell vesicles of different appearances in freeze-etched preparations are formed from the endoplasmic reticulum and Golgi apparatus. Some are secretory vesicles, others fuse to form the mature cell vacuole. Plant vacuoles can also fuse with, or form, smaller vesicles (Northcote, 1968a); their role in digestion is unknown.

Many protozoa ingest material by pinocytosis or phagocytosis and can obtain all nutrients for growth by digestion of this material in food vacuoles (Salt, 1968; Curds & Cockburn, 1968) which are analogous to lysosomes in function (Müller, 1967) and formation (Dembitzer, 1968). Autolysosomes also occur in protozoa, especially in cultures in the stationary phase (Elliott & Bak, 1964), in starved organisms (Brandes, Buetow, Bertini & Malkoff, 1965; Levy & Elliott, 1968) and in encysting amoebae (B. Bowers & E. D. Korn, personal communication; and Pl. 5, fig. 9). Fusion of autolysosomes with food vacuoles has not been reported. The biochemical properties of these protozoan organelles differ from those of mammals (Müller, Baudhuin & De Duve, 1966). Changes occur in such properties and in the buoyant density of lysosomes from starved cells or cells from aged cultures (R. Brightwell & D. Lloyd, unpublished observations).

Matile & Wiemken (1967) found high specific activities of hydrolytic enzymes in a 'vacuole' fraction from lysed *Saccharomyces cerevisiae* protoplasts; the proportion of the total activity in the fraction was not calculated. We have been unable to repeat this result with *Saccharomyces carlsbergensis*, but 30% of the total acid phosphatase and esterase activity sedimented in sucrose (Cartledge *et al.* 1969). Marchant & Smith (1968) discussed the yeast vacuole. Other fungi probably have lysosomes (Pitt, 1968) which may be involved in intracellular digestion (Thornton, 1967) and in the infection of higher plants (Pitt & Coombes, 1968).

Work on lysosomes of yeast, fungi, algae and protozoa has met difficulties of fractionation and also difficulties caused by extracellular hydrolases which are usually inducible. This is more confusing if such enzymes are secreted as shown in Fig. 4; there is some suggestive evidence for this in fungi (Calonge *et al.* 1969). De Duve & Wattiaux (1966) suggested that infolding of the bacterial cell membrane and the subsequent attachment of hydrolytic enzymes (Pollock, 1962) may have been an early stage in the evolution of lysosomes. It is true that many intracellular hydrolytic enzymes of bacteria have similar functions to the lysosomes of eukaryotes in hydrolysing only denatured, partially damaged or imperfect macromolecules. However, little is known about their precise location and regulation.

The plasma or cytoplasmic membrane

Permeability of the plasma membrane, which is a unit (double) membrane is discussed elsewhere in this Symposium (Mitchell, p. 121). Pinocytosis is another means by which material can enter cells. This process consists of invagination of the plasma membrane to form vacuoles which pass into the cell to form food or phagocytic vacuoles (Holter, 1965). Alternatively the vacuoles may fuse with the plasma membrane and in so doing release their contents into the cytoplasm (Fig. 4). In amoebae any part of the cytoplasmic membrane may invaginate, but only specialized parts in other organisms, e.g. ciliate protozoa (Holter, 1965; Elliott & Clemons, 1966). Energy (ATP) is required for the formation of pinocytic vacuoles and protein synthesis is also essential in mammalian cells (Cohn, 1966) and *Tetrahymena pyriformis* (R. Brightwell, unpublished observations). There are no reports of pinocytosis or phagocytosis in higher plants or prokaryotes.

Membranes of secretory vesicles fuse with plasma membranes in many eukaryotes including fungi (Calonge *et al.* 1969) and protozoa (Holter, 1965), but there is no evidence that prokaryotic membranes take part in such processes.

In all prokaryotes the plasma membrane is the dominant membranous structure of the cell. In the eubacteria, blue-green algae and actinomycetes, with few exceptions osmoregulation is poor and rigid walls of high tensile strength are developed in association with membranes (Hurst & Stubbs, 1969). These rigid walls are generally not present in the halophilic bacteria, mycoplasmas, rickettsias and bacterial L forms. The association between the cell wall and the plasma membrane in prokaryotes is discussed by D. J. Ellar in this Symposium (p. 167). Little is know about the formation of the specialized walls, pellicles and cytoskeleton in the eukaryotic microbes such as the Foraminifera.

Mesosomes

Improvements in fixation and embedding have revealed that invaginations of the plasma membrane (mesosomes) in prokaryotes are more common than once thought, but they are possibly not present in some chemolithotrophic bacteria such as *Ferrobacillus* (Remsen & Lundgren, 1966). Plasma membrane invagination may be limited to one or two mesosomes per cell which then appear as circular or oval vesicles containing convoluted membranes with similar dimensions to the plasma membrane (Pl. 6); the surrounding membrane is structurally separate from the internal membranes. Alternatively, the invaginations may be extensive and fill the cell with a series of stacked or coiled membranes as in *Azotobacter* (Gelman *et al*. 1967) and the methane-oxidizing organisms recently described by Whittenbury (1969; and Pl. 6). In many phototrophic bacteria vesicular chromatophores lacking internal membranes are formed by plasma membrane invagination and fill the cell.

Owing to the limitations of histochemistry and to the difficulty of isolating intact membranous structures from bacteria (Salton & Chapman, 1962; Salton, 1967), there is as yet no single function attributed to mesosomes which is not also attributed to the plasma membrane. In fact most mesosome and plasma-membrane preparations are admixed in preparations from cells fragmented by physical methods or made from protoplasts prepared and lysed by the more usual methods. Advantage has been taken of the observation that in some organisms vesicles resembling mesosomes largely lacking their internal membranes are extruded during the controlled plasmolysis of spheroplasts (Fitz-James, 1968).

It should however be pointed out that, in some organisms (*Streptomyces*, *Bacillus*), mesosomes may quickly lose much of their internal structures under conditions such as lack of O_2, presence of penicillin (Highton, 1969; Silva, 1967) and also during fixation. Ghosh & Murray (1968) described the isolation of a series of membrane fractions extruded by controlled lysis of spheroplasts of *Listeria monocytogenes* and subsequent separation by density-gradient centrifugation in polyglucose (Ficol). A fraction consisting mainly of tubules or vesicles was identified as mesosomes. There was little difference in chemical or enzymic composition as between this fraction and one identified as plasma membrane; together they contained the bulk of the cell phospholipid. However, the mesosome fraction from growing cells showed a marked uptake of ^{32}P, which suggests a high turnover rate in this fraction during rapid cell division, as compared with a slower uptake by the bulk of the plasma membrane. Other differences included a possible difference in distribution of NADH and Nitro blue tetrazolium (NBT) reductases and a much higher ATPase activity in the plasma membrane fraction; the latter enzyme is a marker for the plasma membrane (Cole & Hughes, 1964). Similar methods have been used for the fractionation of *Bacillus licheniformis* 6346 (Reaveley & Rogers, 1969). Marker enzymes usually used for the cytoplasmic membrane (NADH oxidase, succinate dehydrogenase) were low in the fraction identified as mesosomes, and the protein content was different. The cytochrome spectra of both fractions were remarkably similar although the cytochrome content was very low for such fractions. In contrast, Ferrandes, Chaix & Ryter (1966) found a high cytochrome concentration in mesosomes isolated from *Bacillus subtilis*.

The function of mesosomes has been reviewed by Salton (1967) and Gelman *et al.* (1967). Apart from the isolation studies already discussed little additional evidence other than that from electron microscope observations has since been obtained. By comparison with the function of eukaryotic membranous systems, various functions of mesosomes have been described.

(*a*) As mitochondrial equivalents. This is based largely on their apparently high tellurite and NBT reductase activities (van Iterson, 1965) as well as their greater elaboration in obligate aerobes as opposed to anaerobes (see Pl. 6, fig 1). Their possible role in oxidative phosphorylation was discussed by Gelman *et al.* (1967).

(*b*) Another function attributed to mesosomes has been as DNA and plasmid distribution and replication sites in cell division. Here their function may be likened both to the nuclear membrane and spindle of eukaryotes (Reaveley, 1968; Cuzin & Jacob, 1968).

(*c*) Mesosomes have been associated with local concentrations of enzymes such as penicillinase, with some permeases and with so-called periplasmic enzymes which are easily lost by some bacteria as a result of cold, heat, osmotic shock or vibration (Heppel, 1967). It is equally possible that these properties may be due to the plasma membrane plugs which penetrate the pores of cell walls (Hurst & Stubbs, 1969), which could pinch-off in a fashion analogous to the voiding of autophagosomes by eukaryotes (p. 310).

(*d*) Other suggested functions of mesosomes in cell division include those of new cell-wall and cross-wall formation (Vanderwinkel & Murray, 1962); new membrane synthesis (FitzJames, 1968); differentiation (Mandelstam, 1969).

CONCLUSION

In eukaryotes the gap between our knowledge of the morphology of cell components and a knowledge of biochemical function, although still large, is nevertheless rapidly closing so that an integrated cell model can now be constructed which may give guidance for further experimentation. This is due largely to the successful development of cell fractionation methods and the rational choice of organisms that are most amenable to this approach (see Hughes, 1970). In prokaryotes, however, the gap between structure and function is still wide and will remain so until better methods of cell fractionation, particularly of cell disintegration, are developed beyond the present empirical stage. Nevertheless the position is clear enough in prokaryotes and eukaryotes to show that we are concerned largely with the dynamic changes associated with growth and differentiation of membrane-enclosed systems in eukaryotes, and probably with what, for want of a better description, can be called membrane-associated systems in prokaryotes. We are therefore faced with the task of ascribing functions to phospholipid-protein associations for which the basic physics and chemistry are still in an elementary state. In discussing mitochondria Lloyd (1969) asked basic questions which apply not only to mitochondria, but also to all of the membranous systems described in the present article, as follows; (1) The manner of integration of specific metabolic systems in membrane structures. (2) The site of synthesis of individual membrane components and their intracellular translocation and reproduction. (3) The mechanism of regulation of the biosynthesis of individual components and whole organelles and their relation to growth and differentiation. (4) The nature of the interplay between the nucleus and extranuclear DNA in controlling membrane components.

Eukaryotes offer a wide range of organisms in which experiments on growth and differentiation can be controlled and well-defined cell components isolated in a reasonably intact condition. Prokaryotes, on the other hand, are particularly useful for studying genetic control, not so readily studied with eukaryotes. In addition, rapid enzymic changes can be induced in membrane-bound systems of prokaryotes, such as the electron-transport system of phototrophs and chemotrophs, permeases, and membrane-bound enzymes concerned with the metabolism of methane and other hydrocarbons. It is, therefore, a matter of regret that regulatory mechanisms such as those discussed by Wimpenny (1969) cannot yet be tackled in anything but an elementary fashion at a structural level, because we have no available methods for their precise dissection. Until more of the gaps in our knowledge of the organization of functions in prokaryotes are filled, the fascinating ideas now current about prokaryotic symbiotic associations in eukaryotic evolution (Sagan, 1967) must remain speculative.

We express our thanks especially to Drs Highton, Whittenbury and Bracker for allowing us to publish their micrographs and Drs Reaveley and Rogers for the discussion of their paper before publication. Thanks are also due to our Cardiff colleagues for material in Fig. 1 and for their useful discussions.

REFERENCES

ANDERSON, N. G. (1966). The development of zonal centrifuges. *Natn. Cancer Inst. Monogr.* **21**, 253.

AVERS, C. J. & FEDERMAN, M. (1968). Occurrence in yeast of cytoplasmic granules which resemble microbodies. *J. Cell Biol.* **37**, 555.

BAINTON, D. F. & FARQUHAR, M. G. (1966). Formation of two types of granules from opposite faces of the Golgi body at different times in the polymorphonuclear leucocyte life cycle. *J. Cell Biol.* **28**, 277.

BAUDHUIN, P., MÜLLER, M., POOLE, B. & DE DUVE, C. (1965). Non-mitochondrial oxidizing particles (microbodies) in rat liver and kidney and in *Tetrahymena pyriformis. Biochem. biophys. Res. Commun.* **20**, 53.

BEACHEY, E. H. & COLE, R. M. (1966). Cell wall replication in *Escherichia coli* studied by immunofluorescence and immuno electronmicroscopy. *J. Bacteriol.* **93**, 1245.

BEAMS, H. W. & KESSEL, R. G. (1968). The Golgi apparatus: structure and function. *Int. Rev. Cytol.* **23**, 209.

BOUCH, G. B. (1965). Fine structure and organelle associations in brown algae. *J. Cell Biol.* **26**, 523.

BOWEN, S. M. & GRIFFITHS, A. J. (1969). Changes in the levels and distribution of certain enzymes during encystment in *Hartmanella castellanii. J. gen. Microbiol.* **57** (Abstract).

BOWERS, B. & KORN, E. D. (1968). The fine structure of *Acanthamoeba castellanii.* I. The trophozoite. *J. Cell Biol.* **39**, 95.

BRACKER, C. E. (1967). Ultrastructure of fungi. *A. Rev. Pl. Pathol.* **5**, 343.

BRANDES, D., BUETOW, D. E., BERTINI, F. & MALKOFF, D. B. (1965). Role of lysosomes in cellular lytic processes. *Exp. Mol. Pathol.* **3**, 583.

BREIDENBACH, R. W., KAHN, A. & BEEVERS, H. (1968). Characterization of glyoxysomes from castor bean endosperm. *Pl. Physiol.* **43**, 705.

BRENNER, D. M. & CARROLL, G. C. (1968). Fine-structural correlates of growth in hyphae of *Ascodesmis sphaerospora. J. Bact.* **95**, 658.

BRIGHTWELL, R., LLOYD, D., TURNER, G. & VENABLES, S. E. (1968). Subcellular fractionation of *Tetrahymena pyriformis* by zonal centrifugation: changes in enzyme distributions during the growth cycle. *Biochem. J.* **109**, 42 P.

BROWN, R. M. (1969). Observations on the relationship of the Golgi apparatus to wall formation in the marine chrysophycean alga *Pleurochrysis scherffelii* Pringsheim. *J. Cell Biol.* **41**, 109.

CALONGE, F. D., FIELDING, A. H. & BYRDE, R. J. W. (1969). Multivesicular bodies in *Sclerotinia fructigena* and their possible relation to extracellular enzyme secretion. *J. gen. Microbiol.* **55**, 177.

CAMPBELL, P. N. & LAWFORD, G. R. (1968). The protein synthesizing activity of the endoplasmic reticulum in liver. In *Structure and Function of Endoplasmic Reticulum in Animal Cells*, p. 57. Ed. F. C. Gran. London and New York: Academic Press.

CARTLEDGE, T. G., BURNETT, J. K., BRIGHTWELL, R. & LLOYD, D. (1969). Subcellular fractionation of *Saccharomyces carlsbergensis* by zonal centrifugation. *Biochem. J.* (In Press.)

CHANTRENNE, H. & COURTOIS, C. (1954). Formation de catalase induite par l'oxygène chez la levure. *Biochim. biophys. Acta* 14, 397.

CHAPPELL, J. B. (1968). Mitochondrial transporting systems. *Br. med. Bull.* 24, 150.

CHAPPELL, J. B. & HANSFORD, R. (1969). Preparation of mitochondria from animal tissues and yeasts. In *Subcellular Components*, p. 43. Eds. G. D. Birnie and S. M. Fox. London: Butterworths.

COHN, Z. A. & EHRENREICH, B. A. (1969). The uptake, storage and intracellular hydrolysis of carbohydrates by macrophages. *J. exp. Med.* 129, 201.

COHN, Z. A. (1966). The regulation of pinocytosis in mouse macrophages. I. Metabolic requirements as defined by the use of inhibitors. *J. exp. Med.* 124, 557.

COLE, H. A. & HUGHES, D. E. (1964). The enzymic activity of the outer shell of *Lactobacillus arabinosus*. *J. gen. Microbiol.* 40, 81–95.

COOPER, R. A., JONES, M. G., VENABLES, S. E. & LLOYD, D. (1969). Subcellular Fraction of *Polytomella caeca* by zonal centrifugation. *Biochem. J.* 114, 65P

CURDS, C. R. & COCKBURN, A. (1968). Studies on the growth and feeding of *Tetrahymena pyriformis* in axenic and monoxenic culture. *J. gen. Microbiol.* 54, 343.

CUZIN, F. & JACOB, F. (1968). Existence chez *E. coli* K_{12} d'une unité genetique de transmission formée des differentes replicons. *Annls Inst. Pasteur, Paris* 112, 529.

DANIELS, E. W. (1964). Morphological changes in the Golgi apparatus in feeding and starving amoebae. *Z. Zellforsch. mikrosk. Anat.* 64, 38.

DAVIES, E. & MACLACHLAN, G. A. (1968). Effects of indolyl-acetic acid on the intracellular distribution of β-gluconase activities in the pea epicotyl. *Archs Biochem. biophys.* 128, 595.

DE DUVE, C. & BAUDHUIN, P. (1966). Peroxisomes (microbodies and related particles). *Physiol. Rev.* 46, 323.

DE DUVE, C. & WATTIAUX, R. (1966). Functions of lysosomes. *A. Rev. Physiol.* 28, 435.

DE DUVE, C., WATTIAUX, R. & BAUDHUIN, P. (1962). Distribution of enzymes between subcellular fractions in animal tissues. *Adv. Enzymol.* 24, 291.

DEMBITZER, H. M. (1968). Digestion and distribution of acid phosphatase in *Blepharisma*. *J. Cell Biol.* 37, 329.

DINGLE, J. T. (1968). Vacuoles, vesicles and lysosomes. *Br. med. Bull.* 24, 141.

DOLIN, M. I. (1961). Survey of microbial electron transport mechanisms. In *The Bacteria*, vol. 2, p. 319. Eds. I. C. Gunsalus and R. Y. Stanier. London and New York: Academic Press.

DUBUY, H. G., MATTERN, C. F. T. & RILEY, F. L. (1966). Comparison of the DNA's obtained from brain nuclei and mitochondria of mice and from the nuclei and kinetoplasts of *Leishmania enriettii*. *Biochim. biophys. Acta* 123, 298.

EHRENREICH, B. A. & COHN, Z. A. (1969). The fate of peptides pinocytosed by macrophages *in vitro*. *J. exp. Med.* 129, 227.

ELLIOTT, A. M. & BAK, I. J. (1964). The fate of mitochondria during ageing in *Tetrahymena pyriformis*. *J. Cell Biol.* 20, 113.

ELLIOTT, A. M. & CLEMONS, G. L. (1966). An ultrastructural study of ingestion and digestion in *Tetrahymena pyriformis*. *J. Protozool.* 13, 311.

FERRANDES, B. CHAIX, P. & RYTER, A. (1966). Localization des cytochromes de *Bacillus subtilis* dans les structures mésosomiques. *Comp. Rend.* 263, 1632.

FitzJames, P. C. (1968). The collection of mesosome vesicles extruded during Protoplasting. In *Microbial Protoplasts, Spheroplasts and L forms*. Ed. L. B. Gaze. Baltimore: Williams & Wilkins Co.

Gahan, P. B. (1969). Subcellular localization and function of hydrolases in differentiating plant cells. *Biochem. J.* **111**, 3 P.

Gallwitz, D. & Mueller, G. C. (1969). Histone synthesis *in vitro* by cytoplasmic microsomes from HeLa cells. *Science, N.Y.* **163**, 1351.

Gelman, N. S., Lukoyanova, M. A. & Ostrovskii, D. N. (1967). *Respiration and Phosphorylation of Bacteria*. New York: Plenum Press.

Ghosh, B. K. & Murray, R. G. E. (1968). Fractionation and characterization of the plasma and mesosome membrane of *Listeria monocytogenes. J. Bact.* **97**, 426.

Gomori, G. (1952). *Microscopic Histochemistry*. University Chicago Press.

Grimstone, A. V. (1966). Structure and function in protozoa. *A. Rev. Microbiol.* **20**, 131.

Grove, S. N., Bracker, C. E. & Morre, D. J. (1968). Cytomembrane differentiation in the endoplasmic reticulum-Golgi apparatus-vesicle complex. *Science, N.Y.* **161**, 171.

Hawker, L. E. & Gooday, M. A. (1967). Delimitation of the gametangia of *Rhizopus sexualis*: an electron microscope study of septum formation. *J. gen. Microbiol.* **49**, 371.

Hendler, R. W. (1965). The importance of membranes in protein biosynthesis. *Nature, Lond.* **207**, 1053.

Heppel, L. A. (1967). Selective release of enzymes from bacteria. *Science, N.Y.* **156**, 1451.

Hicks, R. M. (1966). The function of the Golgi complex in transitional epithelium: synthesis of the thick cell membrane. *J. Cell Biol.* **30**, 623.

Highton, P. G. (1969). An electron microscopic study of cell growth and mesosomal structure in *Bacillus licheniformis. J. Ultrastruct.* **26**, 130.

Hill, G. C. & White, D. C. (1968). Respiratory pigments of *Crithidia fasciculata. J. Bact.* **95**, 2151.

Holt, S. J. & O'Sullivan, D. G. (1958). Studies in enzyme cytochemistry. *Proc. Roy. Soc. Lond.* B **148**, 465.

Holter, H. (1965). Passage of particles through cell membranes. *Symp. Soc. gen. Microbiol.* **15**, 87.

Horne, R. W. & Muscatello, U. (1968). Effect of the tonicity of some negative staining solutions on the elementary structure of membrane bounded systems. *J. Ultrastruct. Res.* **25**, 73.

Hughes, D. E. (1970). Some studies on cyto-differentiation in microbes. Eds. W. Bartley and H. Kornberg.

Hughes, D. E., Wimpenny, J. W. T. & Lloyd, D. (1970). Methods of cell disintegration. In *Methods in Microbiology*. Eds. J. R. Norris and D. W. Ribbons. London and New York: Academic Press.

Hurst, A. & Stubbs, J. B. (1969). Electron microscopic study of membranes and walls of bacteria and changes occurring during growth initiation. *J. Bact.* **97**, 1466.

Van Iterson, W. (1965). Symposium on the fine structure and replication of bacteria and their parts. *Bact. Rev.* **29**, 299.

Jamieson, J. D. & Palade, G. E. (1968). Intracellular transport of secretory proteins in the pancreatic exocrine cell. *J. Cell Biol.* **39**, 580.

Kay, D. H. (1965). *Techniques for Electron Microscopy*. 2nd ed. Oxford: Blackwell.

Kisaki, T. & Tolbert, N. E. (1969). Glycolate and glyoxylate metabolism by isolated peroxisomes or chloroplasts. *Pl. Physiol., Lancaster* **44**, 242.

Kotelnikova, A. V. & Zvjagilskaja, R. A. (1966). In *Symposium on Mitochondrial Structure and Function*, p. 66. Acad. Sci. USSR, Moscow.

LEIGHTON, F., POOLE, B., BEAUFAY, H., BAUDHUIN, P., COFFEY, J. W., FOWLER, S. & DE DUVE, C. (1968). The large scale separation of peroxisomes, mitochondria and lysosomes from the livers of rats injected with Triton WR-1339. *J. Cell Biol.* **37**, 482.

LEVY, M. R. & ELLIOTT, A. M. (1968). Biochemical and ultrastructural changes in *Tetrahymena pyriformis* during starvation. *J. Protozool.* **15**, 208.

LIGHT, P. A., RAGAN, C. I., CLEGG, R. A. & GARLAND, P. B. (1968). Iron limited growth of *Torulopsis utilis*, and the reversible loss of mitochondrial energy conservation at site I and of sensitivity to rotenone and piericidin A. *FEBS letters* **1**, 4.

LLOYD, D. (1965). Respiratory control in mitochondria isolated from the colourless alga, *Prototheca zopfii*. *Biochim. biophys. Acta* **110**, 425.

LLOYD, D. (1969). The development of organelles concerned with energy production. *Symp. Soc. gen. Microbiol.* **19**, 299.

LLOYD, D. EVANS, D. A. & VENABLES, S. E. (1968). Propionate assimilation in the flagellate *Polytomella caeca*. An inducible mitachondrial enzyme system. *Biochem. J.* **109**, 897.

LLOYD, D. & GRIFFITHS, A. J. (1968). The isolation of mitochondria from the amoeba, *Hartmanella castellanii* Neff. *Expl Cell Res.* **51**, 291.

LLOYD, D. & HUGHES, D. E. (1970). Subcellular fractionation. In *Advances in Study of Metabolic Control Processes*, vol. 1. Eds. J. B. Chappell and P. B. Garland. New York: John Wiley & Sons.

LORENC, R., DUKWICZ, A. & KWIEK, S. (1968). Catalase activity of yeast grown in anaerobic conditions. *Acta microbiol. pol.* **17**, 309.

MANDELSTAM, J. (1969). Regulation of bacterial spore formation. *Symp. Soc. gen. Microbiol.* **19**, 377.

MARCHANT, R. & SMITH, D. G. (1968). Membranous structures in yeasts. *Biol. Rev.* **43**, 459.

MARR, A. G. (1960). Enzyme localization in bacteria. *A. Rev. Microbiol.* **14**, 241.

MATILE, P. (1966). Enzyme der Vakuolen aus Wurzelzellen von Maiskeimlingen. *Z. Naturf.* **216**, 871.

MATILE, P. (1968*a*). Aleurone vacuoles as lysosomes. *Z. Pflanzenphysiol.* **58**, 365.

MATILE, P. (1968*b*). Lysosomes of root tip cells in corn seedings. *Planta* **79**, 181.

MATILE, P. & MOOR, H. (1968). Vacuolation: origin and development of the lysosomal apparatus in root-tip cells. *Planta* **80**, 159.

MATILE, P. & WIEMKEN, A. (1967). The vacuole as the lysosome of the yeast cell. *Arch. Mikrobiol.* **56**, 148.

McCLURE, W. K., PARK, D. & ROBINSON, P. M. (1968). Apical organization in the somatic hyphae of fungi. *J. gen. Microbiol.* **50**, 177.

MIYAKE, Y., GAYLOR, J. L. & MASON, H. S. (1968). Properties of a submicrosomal particle containing P-450 and flavoprotein. *J. biol. Chem.* **243**, 5788.

MOLLENHAUER, H. H., EVANS, W. & KOGUT, C. (1968). Dictyosome structure in *Euglena gracilis*. *J. Cell Biol.* **37**, 579.

MOOR, H. & MÜHLELTHALER, K. (1963). Fine structure in frozen etched yeast cells. *J. Cell Biol.* **17**, 609.

MOURE, Y. (1968). Biochemical characterization of the components of the endoplasmic Reticulum in rat liver cell. In *Structure and Function of the Endoplasmic Reticulum in Animal Cells*, p. 1. Ed. F. C. Gran. London and New York: Academic Press.

MÜHLPFORDT, H. (1963). Über die Bedeutung und Feinstruktur des Blepharoplastein bei parasitischen Flagellaten. *Z. Tropenmed. Parasit.* **14**, 357.

MÜLLER, M. (1967). Digestion. In *Chemical Zoology*, vol. 1, p. 351. Ed. G. W. Kidder. London and New York: Academic Press.

MÜLLER, M., BAUDHUIN, P. & DE DUVE, C. (1966). Lysosomes in *Tetrahymena pyriformis*. I. Some properties and lysosomal localization of acid hydrolases. *J. Cell Physiol.* **68**, 165.

MÜLLER, M., HOGG, J. F. & DE DUVE, C. (1968). Distribution of tricarboxylic acid cycle and of glyoxylate cycle enzymes between mitochondria and peroxisomes in *Tetrahymena pyriformis*. *J. biol. Chem.* **243**, 5385.

MÜLLER, M. & MØLLER, K. M. (1967). Peroxisomes in *Acanthamoeba* sp. *J. Protozool.* **14** (Suppl.), 11.

NEUTRA, M. & LEBLOND, C. P. (1969). The Golgi apparatus. *Scient. Am.* **220**, 100.

NORTHCOTE, D. H. (1968a). The organization of the endoplastic reticulum, the Golgi bodies and microtubules during growth and cell division. In *Plant Cell Organelles*, p. 179. Ed. J. B. Pridham. London: Academic Press.

NORTHCOTE, D. H. (1968b). Structure and function of plant cell membranes. *Br. med. Bull.* **24**, 107.

OHNISHI, T., SOTTOCASA, G. & ERNSTER, L. (1966). Current approaches to the mechanism of energy-coupling in the respiratory chain. Studies with yeast mitochondria. *Bull. Soc. Chim. biol.* **48**, 1189.

ORRENIUS, S., GNOSSPELIUS, Y., DAS, M. L. & ERNSTER, L. (1968). In *Structure and Function of the Endoplasmic Reticulum in Animal Cells*, p. 81. Ed. F. C. Gran. London and New York: Academic Press.

PARSONS, D. F., WILLIAMS, G. R., THOMPSON, W., WILSON, D. & CHANCE, B. (1967). Improvements in the procedure for purification of outer and inner membrane. In *Mitochondrial Structure and Compartmentation*, p. 29. Eds. E. Quagliariello, S. Papa, E. C. Slater and T. M. Tager. Bari: Adriatica Editrice.

PITT, D. (1968). Histochemical demonstration of certain hydrolytic enzymes within cytoplasmic particles of *Botrytis cinerea*. *J. gen. Microbiol.* **52**, 67.

PITT, D. & COOMBES, C. (1968). The disruption of lysosome-like particles of *Solanum tuberosum* cells during infection by *Phytophthora erythroseptica*. *J. gen. Microbiol.* **53**, 197.

POLLOCK, M. R. (1962). Exoenzymes. In *The Bacteria*, vol. 4, p. 121. Eds. I. C. Gunsalus and R. Y. Stanier. London and New York: Academic Press.

RAGNOTTI, G., LAWFORD, G. R. & CAMPBELL, P. N. (1969). Biosynthesis of microsomal NADP-cytochrome-*c* reductase by membrane-bound polysomes from rat liver. *Biochem. J.* **112**, 139.

RAMBOURG, A., HERNANDEZ, W. & LEBLOND, C. P. (1969). Detection of complex carbohydrates in the Golgi apparatus of rat cells. *J. Cell Biol.* **40**, 395.

REAVELEY, D. A. & ROGERS, H. (1969). Some enzymic activities of the mesosomes and cytoplasmic membranes of *Bacillus licheniformis* 6346. *Biochem. J.* **112**, 1126.

REAVELEY, D. A. (1968). Isolation and characterization of cytoplasmic membranes and mesosomes of *Bacillus subtilis*. *Biochem. biophys. Res. Comm.* **30**, 649.

REDMAN, C. M. & SABATINI, D. D. (1966). Vectorial discharge of peptides released by puromycin from attached ribosomes. *Proc. natn. Acad. Sci. U.S.A.* **56**, 608.

REITH, A., SCHÜLER, B., VOGEL, W. & KLINGENBERG, M. (1967). Elektronmikroskopisch-autoradiographische untersuchungen zur Intramitochondrialen lokalisation ^{3}pH-Markieter Adeninnucleotide. *Histochemie* **11**, 33.

REMSEN, C. & LUNDGREN, D. G. (1966). Electron microscopy of the cell envelope of *Ferrobacillus ferroxidase* prepared by freeze-etching and chemical fixation techniques. *J. Bact.* **92**, 1765.

ROODYN, D. B. (1967a). *Enzyme Cytology*. London and New York: Academic Press.

ROODYN, D. B. (1967b). *Some Methods for the Isolation of Nuclei from Animal Cells in Subcellular Components*, p. 15. Eds. G. D. Birnie and S. M. Fox. London: Butterworth.

ROODYN, D. B. & WILKIE, D. (1968). *Biogenesis of Mitochondria*. London: Methuen.

SAGAN, L. (1967). On the origin of mitosing cells. *J. theor. Biol.* **14**, 225.

SALT, G. W. (1968). The feeding of *Amoeba proteus* on *Paramecium aurelia*. *J. Protozool.* **15**, 275.

SALTON, M. R. J. (1967). Structure and function of bacterial cell membranes. *A. Rev. Microbiol.* **21**, 417.

SALTON, M. R. J. (1968). Microbial cytology. In *Comprehensive Biochemistry*, vol. 23, p. 134. Eds. M. Florkin and E. H. Stotz. London: Elsevier Publ. Co.

SALTON, M. R. J. & CHAPMAN, J. (1962). Isolation of the membrane mesosome structure from *M. lysodeikticus*. *J. Ultrastruct. Res.* **6**, 489.

SCHLESSINGER, D. (1963). Protein synthesis by polyribosomes on protoplast membranes of *Bacillus megaterium*. *J. molec. Biol.* **7**, 569.

SILVA, M. T. (1967). Electron microscopic aspects of membrane alterations during bacterial cell lysis. *Expl Cell Res.* **46**, 74.

SJÖSTRAND, F. S. (1967). *Electron Microscopy of Cells and Tissues*. London and New York: Academic Press.

SOTTOCASA, G. L., KUYLENSTIERNA, B., ERNSTER, L. & ANDERS, B. (1967). In *Methods in Enzymology*, vol. 10, p. 448. Eds. R. W. Estabrook and M. E. Pullman. London and New York: Academic Press.

STEEVE, R. L. (1957). Electron microscopy of structural detail in frozen biological specimens. *Biophys. biochem. Cytol.* **3**, 45.

STEINERT, M. & VAN ASSEL, S. (1967). The loss of kinetoplastic DNA in two species of Trypanosomatidae treated with acriflavine. *J. Cell Biol.* **34**, 489.

SULEBELE, G. A. & REGE, D. V. (1968). The nature of the glucose effect on the induced synthesis of catalase in *Saccharomyces cerevisiae*. *Enzymologia* **35**, 321.

TANI, J. & HENDLER, R. W. (1964). On the cytological unit for protein synthesis *in vivo* in *E. coli*. I. Studies with spheroplasts of type K12. *Biochim. biophys. Acta* **80**, 279.

TAPPEL, A. L. (1968). Lysosomes. In *Comprehensive Biochemistry*, p. 77. Eds. M. Florkin and E. H. Stotz, London and Amsterdam: Elsevier Publ. Co.

THORNTON, R. M. (1967). The fine structure of Phycomyces. I. Autophagic vesicles. *J. Ultrastruct. Res.* **21**, 269.

TOLBERT, N. E., OESER, A., KISAKI, T., HAGEMAN, R. H. & YAMAZAKI, R. K. (1968). Peroxisomes from spinach leaves containing enzymes related to glycolate metabolism. *J. biol. Chem.* **243**, 5179.

TURNER, G. & LLOYD, D. (1968). The mitochondria of *Tetrahymena pyriformis* strain ST. *Fedn Eur. Biochem. Soc.*, *Prague* A 822.

VANDERWINKEL, E. & MURRAY, R. G. E. (1962). Organelles intracytoplasmiques bactériens et site d'activité oxidoreductase. *J. Ultrastruct. Res.* **7**, 185.

VICKERMAN, K. (1965). Polymorphism and mitochondrial activity in sleeping sickness Trypanosomes. *Nature, Lond.* **208**, 762.

WARDROP, A. B. (1968). Occurrence of structures with lysosome-like function in plant cells. *Nature, Lond.* **218**, 979.

WATSON, K. & SMITH, J. E. (1967). Oxidative phosphorylation and respiratory control in mitochondria from *Aspergillus niger*. *Biochem. J.* **104**, 332.

WHITTENBURY R. (1969). (In process of publication.)

WIMPENNY, J. W. T. (1969). Oxygen and carbon dioxide as regulators of microbial growth and metabolism. *Symp. Soc. Gen. Microbiol.* **19**, 161.

YATSU, L. Y. & JACKS, T. J. (1968). Association of lysosomal activity with aleurone grains in plant seeds. *Archs Biochem. Biophys.* **124**, 466.

EXPLANATION OF PLATES

PLATE 1

Electron micrographs of section of *Tetrahymena pyriformis*; osmic acid fixed; stained with lead citrate. Fig. 1. ($\times$ 5,500.) Showing nuclear membrane (*nm*), mitochondria (*m*), vacuoles (*v*), pellicle (*p*), cilia (*c*), cell membrane (*cm*). Fig. 2. ($\times$ 14,000.) Showing ribosomes (*r*), rough endoplasmic reticulum (*er*), mitochondria (*m*), food vacuole (*fv*), inner mitochondrial membrane (*im*), outer mitochondrial membrane (*om*).

PLATE 2

Electron micrographs of fractions obtained by zonal centrifugation of a homogenate of cells from a 7-day culture of *Tetrahymena pyriformis*.

Fig. 1. Granules of reserve material equilibrating at $\rho = 1\cdot14$. ($\times$ 7,500.)

Fig. 2. Multivesicular structures, vacuoles and other membranous structures from a fraction containing solution of $\rho = 1\cdot15$ with high activities of acid hydrolase. Note absence of mitochondria. ($\times$ 7,500.)

Fig. 3. Vacuoles (*v*) and mitochondria (*m*) from a fraction containing solution of $\rho = 1\cdot19$ and high activities of acid hydrolases. There are structures present which may be autophagic vacuoles (av). ($\times$ 6,700.)

Fig. 4. Mitochondria from a fraction containing solution of $\rho = 1\cdot22$. No obvious contamination by other organelles is visible. ($\times$ 4,500.)

PLATE 3

Section of the hypha of the oomycete *Pythium ultimum*. Hyphal wall (*cw*), cell membrane (*cm*), mitochondrion (*m*), vesicle (*ve*), vacuole (*v*), Golgi apparatus (*g*), nucleus (*n*), nucleolus (*nu*), nuclear envelope (*ne*). ($\times$ 36,000.)

Inset (lower left). Rough endoplasmic reticulum with attached and free ribosomes. The two-unit membranes and lumen can be seen. ($\times$ 48,000.) Fixed in glutaraldehyde, post-fixed in OsO_4.

PLATE 4

Fig. 1. A lomasome (*l*) between the cell membrane (*cm*) and hyphal wall (*w*) of *Pythium ultimum*. ($\times$ 56,000.) Fixed in $KMnO_4$.

Fig. 2. As Fig. 1, but fixed in glutaraldehyde; post-fixed in OsO_4.

PLATE 5

Fig. 1. Woronin (*w*) body in hypha of the discomycete *Ascodesmis nyricans* close to the pore (*p*) in a septum (*s*). The crystalline internal part of the Woronin body is bounded by a unit membrane. ($\times$ 125,000.) Fixed in $KMnO_4$.

Fig. 2. Autolysosomes in encysted amoeba *Hartmanella castellanii*. Partially degraded membranous structures from within the organism, including possibly membranes from primary lysosomes, are surrounded in these cases by unit membranes. ($\times$ 50,000.) Fixed in glutaraldehyde, post-fixed in OsO_4 stained with uranyl acetate and lead citrate. Reproduced by kind permission of Mr M. Stratford.

PLATE 6

Prokaryotic mesosomes

Sections of *Bacillus licheniformis*

Fig. 1. Showing mouth of mesosome invagination in plane of division. ($\times$ 80,000.)

Fig. 2. Transverse section. ($\times$ 100,000.)

Mesosomes in methane-oxidizing bacteria

Fig. 3. Section. ($\times$ 50,000).

Fig. 4. Negatively-stained, autolysed cells in which it is thought that the bundles of closely packed membranes consist of disc-shaped vesicles. ($\times$ 45,000.)

PLATE 1

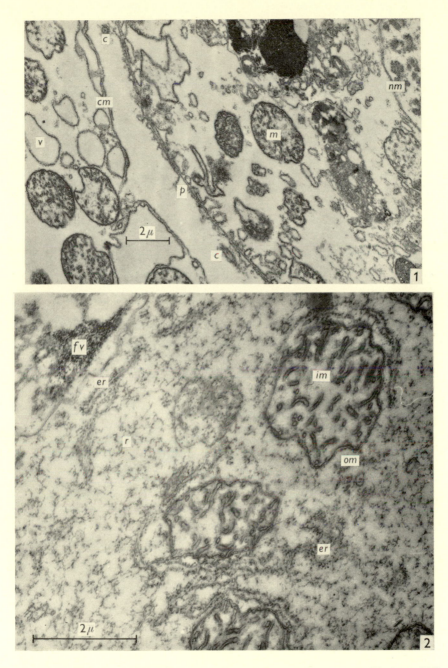

PLATE 2

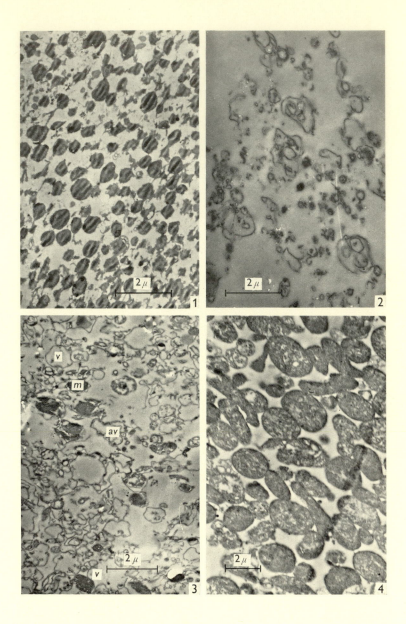

PLATE 3

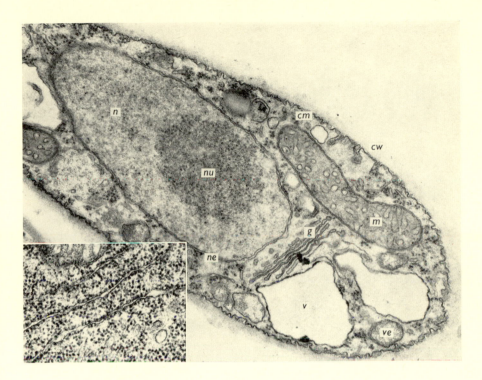

PLATE 4

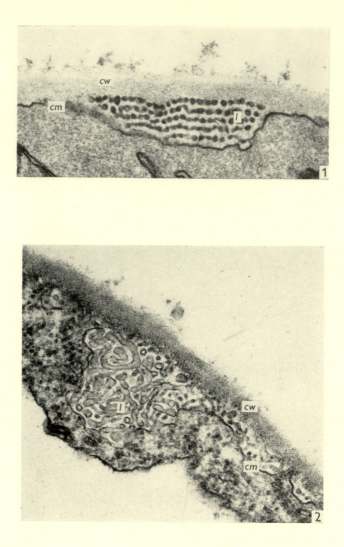

PLATE 5

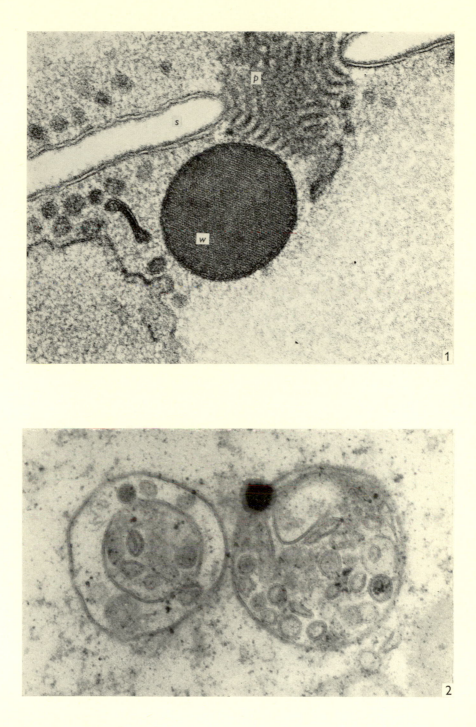

PLATE 6

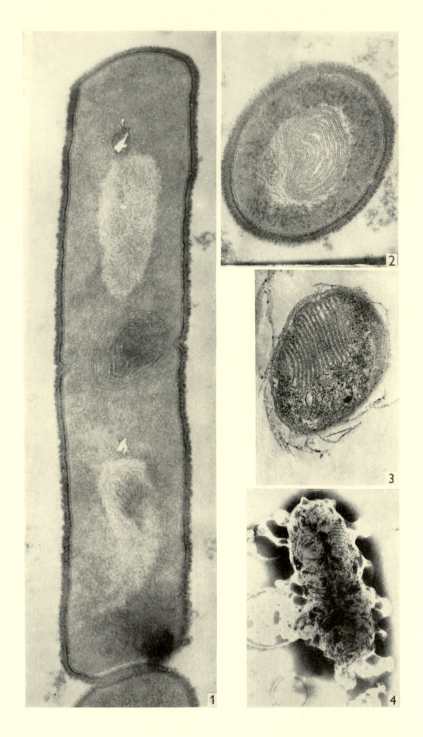

THE REQUIREMENTS OF A VIRUS

ALISON A. NEWTON

Department of Biochemistry, University of Cambridge

Viruses are obligate intracellular parasites. They are thus entirely dependent on a host cell if they are to multiply, but the extent of dependence varies from one virus to another. A virus that is dependent on a cell for several functions will have so many and varied types of interaction with it that one would expect that the virus could only grow in a limited range of hosts. On the other hand a virus that could specify many of its own requisites for growth might have more autonomy and grow in a wider range of cells.

Viruses are generally recognized by the alterations that they cause to the growth or morphology of a particular cell. It has thus been usual to classify viruses by the type of host that they normally infect, and to regard plant, animal and bacterial viruses (bacteriophage, phage) as being distinct classes of organism. Such an approach has many disadvantages; one of the more obvious is that some viruses, notably those pathogens of plants or animals which require an intermediate insect vector, can grow in at least two different types of cell. Furthermore, classification by host obscures similarities between viruses that infect widely different types of cell. As more viruses have been characterized and as our knowledge of the fundamental process involved in virus replication has increased it has become clear that viruses of similar chemical composition multiply in similar ways in widely different types of host; this may be seen most clearly with small RNA-containing viruses which are found in a wide variety of hosts (Montagnier, 1968). When viruses are grouped according to chemical composition (Table 1) it may be seen that most types of virus are found in a wide variety of hosts. There are still some groups of organisms for which no viruses have been described, but this is likely to be due either to lack of knowledge of the groups, or lack of a suitable detection system. Thus, amongst invertebrates only insects have a well documented list of viral pathogens. Examination of the viruses listed in Table 1 does show some unequal distribution of virus types amongst host groups. The possession of a lipoprotein envelope was thought to be limited to viruses growing in eukaryotic cells although the recent description of a phage containing lipid and associated with membrane-like material (Espejo &

Table 1. *Examples of some viruses infecting various host cells*

Nucleic acid component of virus	Shape of virus	Suggested group or family*	Host	Examples	Reference
Single-stranded RNA	Rod	Protovirus	Some plants	Tobacco mosaic virus	Rhodes & van Rooyen (1968)
	Flexible rod	Pachyvirus	Some plants	Tobacco rattle virus	Rhodes & van Rooyen (1968)
		Leptovirus	Some plants	Potato virus X	Rhodes & van Rooyen (1968)
	Isometric	Androphage	*E. coli*	Qβ,f2	Bradley (1967); Höfschneider & Hausen (1968)
			Pseudomonas	PP 7	Bradley (1967)
			Mammals	Foot-and-mouth disease, polio	Höfschneider & Hausen (1968); Rhodes & van Rooyen (1968)
		Picornavirus	Birds	Acute bee paralysis	Bellett (1968)
			Bees	Infectious pancreatic necrosis†	Wolf (1966)
			Fish	'Statolon' virus	Banks *et al.* (1968)
			Penicillium	Turnip yellow mosaic	Höfschneider & Hausen (1968)
			Cruciferae		
	Enveloped, polymorphic	Myxovirus	Mammals	Influenza	
			Birds	Fowl plague	
		Paramyxovirus	Mammals	Mumps, sendai	
			Birds	Newcastle disease	
		Thylaxovirus	Mice	Gross leukaemia virus	Rhodes & van Rooyen (1968)
			Birds	Rous sarcoma virus	
		Arbovirus	Mammals + arthropods	Western equine encephalitis	
				Yellow fever	
		—	Many plants	Tomato spotted wilt†	Hull (1970)
	Enveloped, bullet-shaped	Rhabdovirus	Mammals	Vesicular stomatitis	Rhodes & van Rooyen (1968)
			Birds	Flanders-Hart Park	Rhodes & van Rooyen (1968)
			Insects	Sigma virus of *Drosophila*	Bellett (1968)
			Fish	Egtved	Wolf (1966)
	Enveloped, bacilliform	—	Several plants + leafhopper	Lettuce necrotic yellows, plantain B	Hull (1970)

Nucleic acid	Shape	Name	Host	Virus / disease	Reference
Double-stranded RNA	Isometric	Reo virus	Mammals, Marsupials, Birds	Reovirus	Rhodes & van Rooyen (1968)
			Insects	Cytoplasmic polyhedrosis†	Vago & Bergoin (1968)
			Many plant species + insects	Wound tumour	Sinha (1968)
				Rice dwarf	
Single-stranded DNA	Rod	Ionophage	E. coli	fd, fl phage	Bradley (1967)
	Isometric	Picodna virus, microvirus	E. coli	φX174	
		Parvovirus	Rodents	Kilham rat virus	Mayor & Melnick (1968)
Double-stranded DNA	Rod	—	Insects	Polyhedrosis, granulosis	Vago & Bergoin (1968)
	Complex, tailed	Bacteriophages	Most bacterial species including E. coli	T phages	Adams (1959)
			Bacillus	SP phages	Bradley (1967)
			Brucella	—	
			Streptomyces	—	
			Vibrio	—	
		Cyanophage	Blue-green algae	LPP1	Bradley (1967)
	Isometric	Polyoma, papilloma	Mammals	Polyoma virus, papilloma virus	Rhodes & van Rooyen (1968)
		Adenovirus	Mammals	Adenovirus	Rhodes & van Rooyen (1968)
		Iridescent virus	Birds, Insects	Tipula iridescent virus	Bellett (1968)
			Fish	Lymphocystosis†	Bellett (1968)
			Amphibia	Frog virus 1†	Bellett (1968); Lunger (1966)
	Enveloped, isometric	Herpes virus	Mammals	Herpes simplex	
			Birds	Avian herpes	
			Amphibia	Lucke renal carcinoma	Rhodes & van Rooyen (1968)
			Fish	Fishpox†	
	Enveloped, brick-shaped	Poxvirus	Mammals, Birds	Vaccinia, Fowlpox	
		Vagoiavirus	Insects	Spindle disease†	Vago & Bergoin (1968)

* The nomenclature of viruses is under consideration by the International Committee for nomenclature of viruses. Latinized names have not been used here, but the more common names have been used where possible.

† These viruses have not yet been assigned with certainty to a particular group; the most probable position is indicated.

Canelo, 1968) may indicate that this limitation is not a real one. Until recently all known plant viruses have contained RNA, but a DNA containing virus infecting cauliflowers has now been described (Shepherd, Wakeman & Romanko, 1968).

It has been stated that the problems of virus multiplication are those of replication of the nucleic acid and protein components. Since the overall processes of replication of nucleic acids and translation of this information into amino acid sequences are similar in all organisms, one might anticipate that viruses of similar nucleic acid content would replicate in a similar way in different hosts. Furthermore, one might expect that many viruses could multiply in a wide range of hosts, but this is clearly not the case. Most viruses are specific for the type of cell that they infect; some indeed are very highly specific as shown by the limitation of polio virus replication to certain morphological cell types of primates (Holland, 1964), or of the ability to use phage infection as a means of typing salmonellas (Craigie & Felix, 1947). There have been many attempts to grow viruses in cells which differ markedly from the normal host cell, e.g. to grow animal viruses in bacteria (Abel & Trautner, 1964) and bacteriophages in plants (Sander, 1964), but very few successful results have been reported. Clearly it is not a phenomenon that occurs readily. The chances of it occurring frequently under natural circumstances seem small, and yet it is to the advantage of the virus to be able to spread to as wide a variety of hosts as possible. We do not yet understand all the factors that determine the ability of a virus to grow in one cell and not in another.

From the point of view of the virus, the cell is much more than a convenient incubation mixture for replication of viral nucleic acid and protein synthesis. It undoubtedly presents many problems to the invading virus. Firstly, the virus has to gain entry to the cell cytoplasm and this frequently involves penetrating a chemically-resistant and mechanically-strong cell wall. The cell membrane must be crossed, and in the case of viruses that multiply in the nuclei of eukaryotic cells, many membranes may have to be crossed. Once inside the cell the virus must resist any defence mechanisms that the host may possess. It must then accomplish replication of its own genetic material and to achieve this it must compete with the synthetic activities of the cell for precursor materials, energy and such enzymes as it does not specify or provide for itself. Having achieved this, the viral materials must be assembled, transported within the cell and finally be released from it if the virus is to continue its growth cycle. All these processes involve an interaction between viral and cell components. In solving the problems posed by infection, the virus has had to adapt to the conditions existing in different cells. There have been few detailed studies of the growth of one virus in a variety of hosts. We have little information which indicates why a virus can grow well in one system, and poorly or not at all in another. Examination of the stages of the growth cycle where specific interactions between virus and host are required may shed some light on this. Also, study of ways in which viruses are specialized for infection of various types of cell should reveal much about normal cell functions. Lysogenic phages and tumour viruses are undoubtedly dependent on the host cell to a greater degree than other viruses and will not be considered here.

Attachment and penetration

The major problem facing a virus is that of gaining entry to the cytoplasm of the cell. Many cells are protected by a mechanically strong barrier which may be of a thickness comparable to the size of the virus; this has frequently required the evolution of a complex mechanism for attachment and penetration. Plant viruses, however, have apparently not evolved such a mechanism and rely on damage to the cell wall to gain access to the cytoplasm: this may be produced by accidental damage or through the agency of an intermediate host acting as vector. In either case it is an inefficient and frequently non-specific process requiring the production of immense numbers of virus particles to maintain the life cycle.

There is no doubt that much of the specificity of a virus for a particular cell type is manifest during the phase of adsorption to the cell surface. Viruses are not motile and have a random chance of collision with a susceptible cell. The establishment of a more durable contact is all-important if the virus is to infect that cell; moreover, the virus will clearly be more efficient if it is only bound in this way to a cell in which it is capable of multiplying. Specific adsorption presupposes the formation of some specific bonds between virus and cell suface. The ability to interact is a property both of the virus and of the susceptible cell and may be genetically determined in each. A mutation which affects surface structures may render a cell resistant to some viruses but confer ability to adsorb others. Even minor modifications of chemical structure can have a profound effect on the properties of cell-wall material; physiological and environmental conditions also affect these properties. The ability of a virus to adsorb to a cell reflects the known changes that occur in the cell surface properties. Viruses frequently have precise requirements for particular ionic conditions if adsorption is to occur, indicating the importance of a particular charge or conformation of cell surface and virus in the adsorption process.

Adsorption may initially be a reversible phenomenon; viruses may be eluted from the cell by changes in environmental conditions. However, adsorption is frequently the first in a series of processes which culminate in penetration of the cell by the virus or its nucleic acid component. The reactions that accompany adsorption frequently trigger changes which result in irreversible modification of the virus structure (see Figs. 1–4).

There is little knowledge of the chemistry of this interaction for any one virus, although in most cases one or more of the protein components of the virus particle is known to be responsible for the attachment. Certainly the specificity for the susceptible cell resides in the protein coat of polio virus. Holland (1964) showed that whereas polio virus can only infect cells of primate origin in tissue culture, removal of the protein coat allows the purified viral RNA to infect a wide range of cells, even cells from birds. The ability of infectious nucleic acid to infect a wider host range than the parent virus is seen in many cases. Some animal viruses may exhibit phenotypic mixing, in which the nucleic acid component of one virus is enclosed in the outer layers of another: the host range is determined by the surface material.

Of all known viruses only the bacteriophages containing double-stranded DNA and the viruses which attack blue-green algae are known to possess a specialized organ for adsorption and penetration. It has been known for many years that phages attach tail first to susceptible bacteria; the details of this remarkably intricate viral organ have been reviewed by Kozloff (1968). T-even bacteriophages adsorb by the long tail fibres and probably secondarily by the shorter tail-plate pins. Adsorption to the bacterial surface results in a change in conformation of the fibres and the base plate which together with contraction of the tail sheath eventually result in injection of the bacterium with phage DNA (Fig. 1).

Attachment of the virus to the cell requires only passive participation by the cell. Killed cells or even isolated cell components may adsorb virus and frequently act as competitive inhibitors of adsorption to sensitive cells; indeed they may serve in this way to neutralize viral infectivity. This has made it possible to isolate cell-surface material and to examine the nature of the receptor material.

In the cell wall of *Escherichia coli* B there are several receptor substances for phages of the T series. Weidel (1968) showed that the receptor substance for each phage has a different chemical composition. The sites for attachment of T_2 and T_6 are in the outermost lipoprotein layer, while sites for T_3, T_4 and T_7 are in the rigid mucopeptide layer beneath this. Receptor material for T_5 contains lipopolysaccharide and lipoprotein which can be isolated in particulate form, and each receptor particle combines with one phage particle. None of the receptor substances has yet been fully purified and the mechanism of interaction with phage tail protein is unknown. Bayer (1968) observed that T phages adsorb to particular regions of the cell surface which are seen to be in close contact with the cell membrane after plasmolysis of the cell. He postulated that these areas are regions where phage DNA may more easily enter the bacterial cytoplasm. Since various T phages

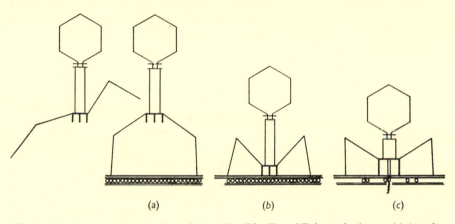

Fig. 1. Attachment and penetration of the cell wall by T_2 and T_4 bacteriophages. (*a*) Attachment of tail fibres to specific receptor sites. (*b*) Change in conformation of the long tail fibres, bringing short tail-plate pins into contact with the cell wall. (*c*) Contraction of the tail sheath. The inner tail tube penetrates the lipoprotein, lipopolysaccharide and mucopeptide layers but may not penetrate the membrane (Simon & Anderson, 1967 Copyright © Academic Press, London and New York).

compete for these same sites Bayer proposed that wall-membrane attachment sites form the 'bull's-eye' for the phage tail, surrounded by a mosaic of receptor sites for individual tail fibres. Bayer estimated the number of such sites to be about 100 per cell. The function of these sites in the uninfected cell is not known, but if the phage can only attach and inject successfully in these regions, it suggests that there is a remarkable specialization of virus for the bacterial surface.

The firm binding achieved by the attachment of both long and short tail fibres of the T phages is probably necessary to stabilize the virus particle during the contraction of the sheath that follows. Simon & Anderson (1967) showed that this contraction results in forcing the phage tail tube through the outer, softer, lipoprotein and lipopolysaccharide layers of the bacterial wall. They believe, however, that the tail tube does not penetrate the inner mucopeptide layer and that it is possible that a lysozyme associated with the phage tail is involved in local lysis of mucopeptide. However, mutants which lack lysozyme may also infect cells; the role of lysozyme in this process is not certain. It is not known whether the membrane is punctured by the tail; alternative possibilities are that the DNA is released into the space between wall and membrane, or is injected directly into the membrane channels described by Bayer (1968).

Although only certain phages have this specialized tail structure some of the simpler RNA- and DNA- containing phages have specialized requirements for attachment to host cell. Many of these phages are specific for the male bacterium and it has been shown that they are

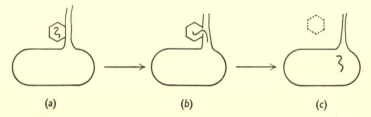

Fig. 2. Attachment of male-specific RNA bacteriophages and release of RNA. (*a*) Attachment of phage to F-pilus; (*b*) alteration of phage coat; penetration of pilus by RNA; (*c*) transfer of RNA down pilus into cell; release of coat protein (Silverman & Valentine, 1969).

adsorbed specifically to F-pili (see review by Hoffman-Berling, Kaernes & Knippers, 1966). When the pili are removed by shearing, the bacteria are rendered temporarily immune to infection, and bacteria which are normally resistant may be rendered susceptible by acquisition of the F factor. F⁻ (female) bacteria cannot be infected by whole virus, but their spheroplasts may be productively infected with infectious nucleic acid extracted from these viruses. Filamentous DNA phage, e.g. fd, adsorb end-on to the tips of the pili; shearing the virus gives an end fragment which may still adsorb; it is possible that a specialized protein component of the virus is involved in this attachment. The RNA phages adsorb laterally to the pili. It is not known how either of these types of virus release their nucleic acid, but coat protein does not enter the cell. It is therefore supposed that the nucleic acid is released into the pili and passes down a hypothetical channel in the centre of the pilus (Brinton & Beer, 1967). The RNA from phage f2 becomes transiently sensitive to ribonuclease and is thought to be within the pili; removal of pili in the absence of cations at this stage releases the RNA (see Fig. 2).

Other phages adsorb specifically to bacterial flagella and the adsorption of only one particle of phage is sufficient to immobilize the whole complement of flagella. Such phages may inject their DNA only after moving down the flagellum to its base (Schade, Adler & Ris, 1967; Raimondo, Lundh & Martinez, 1968).

Animal viruses do not have one specialized region of their surface capable of combining with the cell surface, but seem rather to have many combining groups. Many animal viruses agglutinate red blood cells which implies the existence of at least two combining groups. Recent evidence indicates that there is an average of 500 haemagglutinin groups on the surface of one influenza particle (Laver & Valentine, 1969). The brick-shaped vaccinia virus is able to adsorb so that either the long or the short axis is parallel to the cell surface.

The phenomenon of haemagglutination has been used as a model system for the study of the adsorption of viruses to cell surfaces (Hirst, 1965). Adsorption of many animal viruses to red cells and other animal cells is abolished by pre-treatment of the cells with neuraminidase, so that a common receptor substance for these viruses seems to be the N-acetyl-neuraminic acid side-chain of the cellular mucoprotein. The impetus to study of this class of compound came from a study of haemagglutination by influenza virus (Gottschalk, 1959). Viruses of the influenza group possess a specialized haemagglutinin which is a glycoprotein present in the viral surface (Laver & Valentine, 1969). The particles also contain a neuraminidase, but although both these proteins have been analysed the role of each in the attachment of virus to susceptible cells is not clear.

Many animal viruses, including those of the myxovirus group, are surrounded by a lipoprotein envelope. This envelope is derived partly from the host cell in which the virus has been grown, and it also contains virus-specified material. These viruses react with antiserum directed against viral protein, and also with antiserum against uninfected host cell. The host antigen of influenza virus is largely carbohydrate and is covalently linked to viral protein (Laver & Valentine, 1969). Viruses which possess such an envelope are seen to attach to the cells by alignment of the two lipoprotein layers, which may be followed by a fusion of the membranes. Membranes of animal cells show a natural tendency to fuse and may form five-layered leaflets, as in synaptic junctions: similar structures are formed in the process of virus adsorption. Such a process may occur frequently in the normal cell during pinocytosis where vesicles are pinched off from the plasmalemma. Possession of an envelope having many features of the normal cell membrane allows the virus to exploit this normal phenomenon in order to attach and possibly

to gain entry to the cell. Also the ability of the normal cell membran
to fuse may be enhanced by the effects of virus modifications of membrane structure (Roizman, 1962). Cells infected by many viruses fuse
to give giant multinucleate cells at a late stage of infection; fusion of
cells can also be achieved without productive infection by the addition
of very large numbers of virus particles to cell cultures. Virus particles
inactivated by ultraviolet irradiation are equally effective in producing
this type of cell fusion. This suggests that viral multiplication is unnecessary and that fusion is a consequence of the virus:cell interaction.
Sendai virus exhibits this cell-fusion ability to a remarkable degree, and
is now widely used as a tool for fusing cells of differing types (Harris,
Watkins, Ford & Schoefl, 1966). The mechanism of cell fusion is not
understood, but it is known that active cellular metabolism is required
for fusion to occur, and fusion does not occur at 4° but takes place
rapidly at 37°. It has been considered that viruses capable of inducing
cell fusion might possess some lipolytic activities; phospholipase activity
has not been detected in Sendai virus. Moreover, cells susceptible to
fusion possess lysolecithinase so fusion is unlikely to be a consequence
of the presence of lysolecithin in the virus particle (Elsbach, Holmes &
Choppin, 1969). Enveloped viruses frequently contain enzymes derived
from host membrane, e.g. adenosine triphosphatase, and these may
play some part in cell fusion.

The possession of an envelope is thus an advantage to the virus in
facilitating attachment of virus to cell membrane, but even for viruses
which normally have envelopes these may not always be essential for
infectivity. Herpes virus is found in both enveloped and non-enveloped
forms, both of which are thought to be infectious; however the form
which lacks envelopes is much less infective (Watson, 1968).

When the virus envelope has attached to the plasmalemma and
possibly fused with it, there are two ways in which it may enter the
cytoplasm (see Fig. 3). The fused membranes may break down (Fig. 3a),
so that the viral capsid enters the cytoplasm directly. This overcomes the
difficulty of passage of particles through membranes (see later) and has
been observed to occur with influenza and Sendai viruses (Morgan &
Rose, 1968; Morgan & Howe, 1968). With these viruses the internal
ribonucleic protein helix is found in the cytoplasm very shortly after
infection. Intermediate stages in the process can only be observed when
the cellular activity is restricted by chilling or unfavourable incubation
conditions, since integration of viral and cellular membranes occurs
very rapidly when the metabolic activity of the cells is restored. Attached
virus, but which is still outside the cell, is sensitive to neutralizing

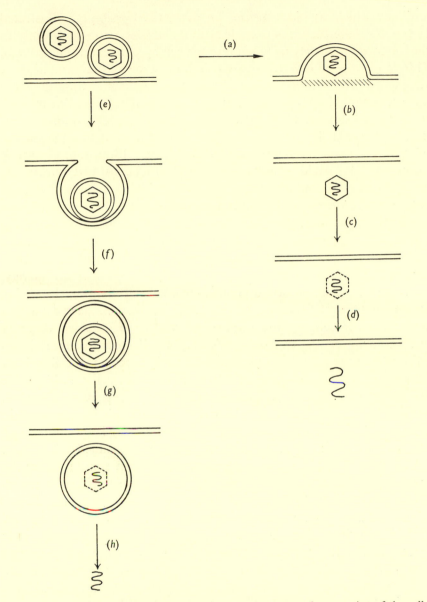

Fig. 3. Attachment of enveloped viruses to mammalian cells and penetration of the cells. (*a*) Fusion of viral envelope with cell membrane; dissolution of fused membranes. (*b*) Entry of viral nucleoprotein into cytoplasm. (*c*) Removal of viral proteins. (*d*) Release of viral nucleic acid. It is not known how or when this occurs.

Or (*e*) Attachment of viral envelope to cell membrane; invagination of cell membrane enclosing complete virus. (*f*) Virus enters cytoplasm by viropexis, enclosed in a phagocytic vesicle. (*g*) Breakdown of vesicle and viral membranes. Release of viral nucleoprotein into the cytoplasm. Much of the viral coat protein may have been removed at this stage. (*h*) Release of viral nucleic acid; it is not known how or when this occurs.

antibody, which can block fusion. An alternative possibility is illustrated in Fig. 3*e*; here the virus particle with its envelope attached firmly to the cell membrane may be taken into the cell by viropexis (Dales, 1965). This is closely similar to pinocytosis and involves the formation of vesicles by invagination (Holter, 1965); these vesicles enclose the virus and are released into the cytoplasm by pinching-off a piece of membrane from the cell surface. Virus particles have been observed within vesicles in the cytoplasm shortly after infection; it is of interest that the surface of these vesicles is frequently seen to be 'coated' (Simpson, Hauser & Dales, 1969). Coated vesicles are thought to be formed during the natural process of pinocytosis. Furthermore, it is possible that such vesicles are formed from specialized areas of cell membrane coated with mucopolysaccharide (Fawcett, 1965). If so, only limited areas of the cell surface may be capable of adsorbing virus particles at any one time.

It is not yet possible to decide between these two possibilities. Some viruses may be taken up by fusion and others by viropexis. Alternatively, the same virus may on occasion enter by either mechanism, depending on the physiological state of the cell. Different workers have examined penetration of cells by the same virus and have obtained different results. However, the two processes may not be very different; internalization of the cell membrane may be a rapid process in a cell actively engaged in pinocytosis, and the fusion of viral and cell membranes at the surface or in an intracellular location may be a matter of chance. It must be remembered that to detect the uptake of virus particles in sections of infected cells, it is necessary to use very large numbers of infecting particles: only a few of these can lead to productive infection.

The process of pinocytosis may be stimulated in some cells by the presence of certain cofactors, e.g. oleic acid, linoleic acid or N-acetyl-neuraminic acid, or by adsorption of protein with a particular charge (Cohn & Parks, 1967); acidic proteins are more effective than neutral proteins in stimulating pinocytosis, but basic proteins penetrate cells more rapidly. It would be of interest to know whether any viruses possess the correct charge or the cofactors necessary to stimulate the cell to engulf them, or whether the virus particles are merely passengers which rely on the natural activity of the cell. Cells will engulf inert substances, such as carbon particles, provided that these are initially attached to the cell surface. It may be relevant that the infectivity of many viruses and infectious nucleic acids is increased by the addition of the cationic polymer DEA–dextran. This may function by increasing the affinity of virus for cell surface, which is negatively charged at physiological pH values.

Most of the simpler animal viruses do not possess envelopes; other methods of attachment must be used. Most is known about the reaction between polio virus and its host cell (Holland, 1964). Material that acts as a receptor substance has been isolated from the cell membrane and

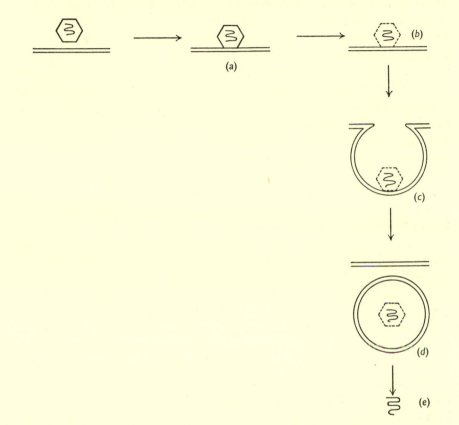

Fig. 4. Attachment of non-enveloped viruses to mammalian cells. (*a*) Attachment of viral coat protein to receptors on the cell surface. (*b*) Alteration in structure of viral coat. (*c*) Invagination of the cell membrane, enclosing the virus. (*d*) Enclosure of the virus in cytoplasmic vesicle. Possible removal of virus coat proteins. (*e*) Dissolution of vesicle and release of viral nucleic acid.

is thought to be a lipoprotein. This is only present in a few types of human and primate cells in brain, spinal cord and intestine. The absence of receptors from other cells is sufficient to explain the tissue tropism of this virus. Human and primate cells cultivated *in vitro* acquire in parallel the receptors for polio virus and sensitivity to infection. The receptor substance may be removed from cells by trypsin treatment but is slowly regenerated.

After adsorption to the cell surface several irreversible changes occur in the properties of the virus. It loses the ability to react with antiserum and becomes sensitive to proteolytic enzymes; virus eluted from the surface is no longer infectious but still contains infectious nucleic acid. These changes suggest that there is a change in conformation of the sub-units of protein coat, triggered by adsorption (Fig. 4); however, the RNA remains insensitive to ribonuclease. Adsorbed and modified virus is thought to enter the cell by viropexis as described for enveloped viruses. It is possible that plant viruses also may enter the cell cytoplasm by a similar method. The insect or other vector or mechanical damage may introduce the virus only into the space between wall and membrane, and it may then be taken up into vesicles (Cocking & Pojnar, 1968). It is not clear whether these engulfed particles are subsequently capable of promoting infection in that cell.

Penetration of cell membranes

To initiate infection the nucleic acid component of the virus must gain access to the cell cytoplasm. This means that a very large particle or macromolecule has to penetrate the cell membrane. This problem is of increasing importance in biological research, and involves the mechanism of transport of large molecules across organized membranes (Gross, 1967). We have little knowledge of how this is achieved for any virus.

The study of bacteriophage penetration by differential labelling of protein and nucleic acid components has shown that the major protein components of all phages remain outside the cell. Release of phage nucleic acid is triggered during the process of adsorption and it is the nucleic acid which must cross the membrane. It is possible that the tail tube of the T_4 phage is forced through the membrane during contraction of the tail sheath, and that DNA is either pushed or drawn directly into the cytoplasm. Leakiness of the bacterial membrane occurs temporarily after infection and infection at high multiplicity can lead to 'lysis from without'. However, it has been suggested that DNA is injected into the space between wall and membrane, and that the leakiness is a consequence of penetration of the membrane by the rigid DNA molecule. This process might occur by a process similar to pinocytosis, although one might then expect that purified DNA could infect spheroplasts. DNA from T phages can only infect spheroplasts when the DNA is still associated with some viral protein, although smaller DNA molecules from other phages such as ϕX174 or λ can infect spheroplasts. Alternatively, the phage may make use of some normal host mechanism. It is not known how transforming DNA or the DNA transferred during conjugation penetrates the cell membrane, but Bayer (1968) claimed to observe channels in the membrane through which the DNA might pass. Uptake of transforming DNA has been found to occur preferentially at the ends and middles of cells, possibly near the mesosomes (Javor & Tomasz, 1968). Uptake of the complete T_5 DNA requires active protein synthesis in the bacterium (Lanni, 1965); when this is inhibited only a small part of the DNA enters. Since conjugation only occurs when the bacteria possess F-pili, it seems likely that DNA transferred during conjugation passes out through the pili (Brinton, 1965), synthesis of DNA in either donor or recipient cell being necessary if transfer is to occur. Phages which adsorb to pili may make use of a similar mechanism, perhaps requiring some co-operation from the host cell, to achieve uptake of their nucleic acid component. The competition between the male-specific RNA- and DNA-containing phages at some stage subsequent to adsorption suggests that both may use the same channel for nucleic acid transport (Silverman & Valentine, 1969).

The viruses of eukaryotes do not release their nucleic acid before entry into the cell. The proposed mechanism whereby enveloped viruses fuse with the plasmalemma and release their nucleoprotein directly into the cytoplasm is attractive because it avoids the problem of membrane penetration. However, the mechanism of membrane dissolution is unclear. Viruses that enter the cytoplasm enclosed within a vesicle still have to escape from the vesicle through a membrane. Study of sections of infected cells shows that the nucleoprotein cores of virus particles are released into the cytoplasm but yields no information about the mechanism of dissolution of vesicles. The vesicles formed during the normal processes of pinocytosis and phagocytosis also change appearance; the membranes may disintegrate after entry into the cytoplasm. Many animal viruses replicate in the nucleus of the cell; this implies that the nucleic acid component of these viruses must enter the nucleus. The nuclear membrane is a double unit membrane and the potential difference across this membrane may be as great as that across the plasmalemma (Loewenstein, 1964) so that it presents a formidable barrier to diffusion. There are numerous large pores in the nuclear membrane, but these are not open to the cytoplasm and are filled with material of unknown nature. Feldherr & Harding (1964) showed that amoebae could take up colloidal gold particles and that these particles entered the nucleus through the nuclear pores, provided that they were of diameter less than 150 Å. Virus particles could well enter by a similar mechanism, although the cores of herpes virus and adenovirus appear to be larger than this. An alternative possibility was suggested by Dales (1962), who had evidence that the residual protein of adenovirus might be left in the cytoplasm and that only DNA entered the nucleus through the nuclear pores.

It has been argued that the lack of an unequivocal demonstration of virus growth within cytoplasmic organelles, such as chloroplasts and mitochondria, is due to the difficulty of passage of viral material through the complex limiting membranes. However, recent observation on the biogenesis of these organelles has shown that large protein molecules can penetrate mitochondrial membranes.

It is not only during initial infection of an organism that viruses have to cross cell membranes. Cell to cell transfer must occur if the virus is to multiply effectively in multicellular organisms. There is some evidence that virus particles or even viral nucleic acid may pass through cytoplasmic bridges formed by the transient fusion of cell membranes. Herpes virus can infect adjacent cells without entry into the surrounding culture fluid (Stoker, 1959). Many viruses of the herpes group are noninfectious in the extracellular state and *in vitro* these viruses can only be spread by the addition of infected cells to fresh susceptible cells. This suggests that direct passage from cell to cell may be the normal means of spread for these viruses. Some plant viruses have been observed in passage through plasmadesmata (Esau, Cronshaw & Hoefert, 1967; Davison, 1969). This is thought to be a pathway for natural spread of virus within the plant.

Infectious nucleic acid may be prepared from many of the simpler viruses and it is clear that this can enter cells without difficulty but with greatly decreased efficiency as compared with whole virus. The uptake of infectious nucleic acid can be increased by changing the physiological condition of the cell, for example by incubation in extremes of salt concentration or by addition of polycations.

Uncoating

It is clear that the interaction between bacteriophages and the receptors on the cell surface is sufficient to bring about release of the nucleic acid from the protein. With many animal viruses, however, penetration occurs some time before the nucleic acid becomes sensitive to nuclease digestion. Apparently intact virions can be seen in the cytoplasm either free or enclosed in vesicles. The changing appearance of the virus particles with time after infection indicates that disintegration of the viral particle may be a slow process. Joklik (1966) used vaccinia virus labelled in either lipid, protein or DNA components, and showed that uncoating of this virus occurred in two stages. Shortly after infection most of the phospholipid and about half the protein became acid-soluble. This process apparently took place within cytoplasmic vesicles and involved the participation of pre-existing host cell enzymes. Virus cores, which at this stage are resistant to deoxyribonuclease, are released from the vesicles into the cytoplasm and there is a lag before the DNA is released from the remaining protein material. The enzyme that brings about this second stage has not been idenified but it is apparently not present in normal cells, since synthesis of RNA and protein following infection are essential if the DNA is to be successfully uncoated. The discovery within the vaccinia particle of RNA polymerase (see Woodson, 1968) capable of transcribing the DNA of the core while it is still complexed with protein has suggested that the information for this enzyme is provided by the infecting virus itself.

Although the uncoating of other animal viruses has not been studied in such detail it is probable that host enzymes play an important part in this process. Phagocytic vesicles are thought to be closely associated with lysosomes which contain active proteases and other lytic enzymes. It is probable that viruses engulfed in phagocytic vesicles are exposed to these enzymes which may well divest the viruses of their coat proteins. It is of particular interest that polio virus, which is resistant to attack by proteolytic enzymes in the free state, becomes sensitive to them as a result of the interaction with the cell surface. Reo virus is concentrated in the lysosome fraction (Silverstein & Dales, 1968) and the protein is released while it is in this intracellular location. Activation of lysosomal enzymes has been observed during infection by several viruses (Allison, 1967); the 'free' activity of lysosomal enzymes increased markedly after infection of mouse cells by herpes virus, although the total activity of the enzymes was unaltered (A. Newton, unpublished data, 1963), indicating either an increase in lysosomal fragility or a breakdown of lysosomal particles.

Host restriction and the destruction of viral nucleic acid

Once viral nucleic acid has entered the cytoplasm it is separated from protective coat protein and may be susceptible to attack by degradative enzymes. All cells contain nucleolytic enzymes which may have a normal function in the regulation of nucleic acid metabolism, as in the degradation of mRNA or in the synthesis and repair of DNA (Shugar & Sierakowska, 1967). It is not known how the activity of these enzymes against host components is regulated; they might be capable of hydrolysing nucleic acids foreign to the cell. Other nucleolytic enzymes are

normally separated from their substrates by restriction to certain intracellular compartments, such as lysosomes. The possibility of exposure
of virus nucleic acid to these nucleases raises the question: how does
viral nucleic acid resist digestion? Reo virus RNA is double-stranded
and is not susceptible to attack by ribonuclease either *in vitro* or in the
lysosomes (Silverstein & Dales, 1968). Many viruses contain circular
DNA that is resistant to exonuclease attack. The DNA of T-even phages
is known to be glucosylated (see Cohen, 1968); this confers some resistance to deoxyribonuclease digestion in the host. Growth of these phages
in a mutant of *Escherichia coli* deficient in UDPG-pyrophosphorylase
gives progeny virus containing little glucose (see review, Stacey, 1965).
These deficient phages are unable to grow in *E. coli* B, where the DNA
is rapidly degraded, but can grow normally in shigellas, which do not
contain enzyme capable of hydrolysing the phage DNA (Fukusawa,
1964). This is one example of the phenomenon of host-controlled
modification, where the ability of a phage to infect a particular bacterial
strain is modified by the bacterial strain on which the phage has been
grown; the modification is not inherited in the normal Mendelian manner (reviews, Arber, 1965, 1968). An example of host-controlled modification that may be of general significance has been investigated in
detail by Arber (1968). Phages such as λ that have been grown on
E. coli K 12 only infect *E. coli* B with low efficiency, although they can
grow normally on some other strains of *E. coli*. Arber showed that it is
the DNA of the phage that has been modified by growth in the first
host, and that it is probable that such modification involves methylation
of a particular base or bases in the phage DNA. When such modified
DNA is injected into a new host it will be degraded unless it has been
modified in the correct position. The restricting host appears therefore
to possess an efficient system for checking the base composition of
invading DNA. It is possible that this system includes a specific nuclease
that may be closely linked with the modifying enzyme. Episomic elements are also known to be subject to this type of restriction; this may
prove to be a general mechanism for protecting the cell against foreign
nucleic acids. It is also possible that if the bacterial DNA itself is not
correctly methylated it may be treated as 'foreign' and hydrolysed in
the same manner (Lark, 1968). The type of restriction is affected by the
presence of lysogenic phage and episomes in the bacteria.

The nucleic acid of many animal and plant viruses is sensitive to
nuclease attack, after isolation. 'Rare' bases have not been demonstrated
in these viruses, and it is not clear how these viral nucleic acids can
survive within the host cell. It is possible that nucleases are highly

localized in the cell and that virus is not exposed to them. However, the probable dependence of viruses on lysosomal proteolytic enzymes for release from coat protein suggests that the nucleic acid will be exposed to lysosomal nucleases. Internal proteins of the virion may serve to protect the nucleic acid against degradation. There is no experimental evidence to suggest whether viral nucleic acid is completely free from protein within the cytoplasm, and the ability of vaccinia DNA to serve as a template while still associated with protein (Kates & McAuslan, 1967) suggests that complete freedom from protein may not be essential for expression of nucleic acid function.

Dependence on host cell systems

The virus is dependent on the host cell for multiplication; the extent of the dependence varies with the virus. The simplest viruses can only code for three proteins; these proteins are essential for production of infective progeny and are not present in normal cells. Smaller viruses exist such as tobacco necrosis satellite virus but these are not only dependent on the cell but on other viruses. Even the largest viruses which contain enough information to code for 500 to 900 proteins, which have a complex structure and contain RNA polymerase, other enzymes and coenzymes, are still dependent on the host cell for its complex machinery and for the supply of low molecular weight precursors.

The provision of energy and low molecular weight precursor molecules

By definition (Lwoff, 1957) viruses contain no enzymes involved in the generation of ATP. They must therefore either possess genetic information which will code for the production of new energy-generating systems or they must rely on pre-existing host systems for the supply of necessary high energy compounds. It is obvious that the small viruses could not code for such a complex enzyme system. While larger viruses might contain enough information for such a system there is no conclusive evidence for the production of a new enzyme concerned with energy production. In general, respiration of cells is little affected during the early stages of viral multiplication. In the terminal stages, when cell structures may show signs of damage, marked inhibition of energy-generating systems may be seen, but even this is not always apparent and some animal cells may show unchanged rates of glycolysis in spite of gross degenerative changes. There are, however, many reports of some alterations in activity of glycolysis or pentose-phosphate pathways following infection. It is difficult to know whether such changes reflect a specific viral effect and how much relatively non-specific effects such as permeability changes may alter the rates of enzymic activity in these metabolic systems. Slight changes in the intracellular concentrations of adenine nucleotides may, for example, have a significant effect on the overall rate of glycolysis.

Viruses may also be dependent on the host cell systems for the supply of low molecular weight precursors required for the synthesis of viral macromolecules. Many bacterial viruses contain bases in their nucleic acid that are not present in the nucleic acids of the host. In these cases one might anticipate that new enzyme systems must be present in the infected cell to catalyse the formation of the precursor nucleotides and this reasoning led to the discovery of several new enzymes produced after phage infection (Cohen, 1968). These include the enzymes required for the synthesis and glycosylation of the hydroxymethylcytosine found in the phage DNA. More surprising, however, is the finding that there is also an increase in activity of many enzymes already present in the host. Such an increase in activity might result from many causes, including stimulation of a pre-existing host enzyme, de-repression of host-controlled synthesis of the enzyme, stabilization of an unstable enzyme or production of new enzyme specified by the viral genome. In many cases it has been shown that the extra enzymic activity found after infection is due to the presence of an enzyme having different physical and antigenic properties to the normal host enzyme, and this enzyme may show different requirements for activating ions, altered Michaelis constant, etc. (see reviews by Cohen, 1968; Keir, 1968). If, also, the enzyme activity can be induced by infection of a mutant cell normally lacking the enzyme and similar enzymes can be induced in different cells by the same virus, it seems conclusive evidence that the information for the enzyme is provided by the virus and not by the host. In addition to the enzymes found in bacteria infected by T-even phages, such requirements have been satisfied in the case of thymidine kinase induction by several viruses, including vaccinia, herpes and pseudorabies. However, reports of a virus-stimulated increase in activity of other enzymes involved in nucleotide metabolism are not so conclusive. The increase in activity of thymidylate kinase following infection of cells by herpes and pseudorabies viruses may be due to stabilization of a usually unstable enzyme in the infected cell (Kaplan, 1967; Sheltawy & Newton, 1970).

The induction of a new enzyme in a cell already possessing that activity invites speculation about the necessity for such enzyme synthesis. Perhaps the activities of host enzymes are insufficient to provide for the synthesis of viral material. Another possibility is that the activity of the enzyme may vary with the physiological state of the cell. Thus, in a population of growing eukaryotic cells only some will be engaged in DNA synthesis, those not in the DNA synthetic phase may have very low activities of enzymes involved in deoxynucleotide synthesis. In whole multicellular organisms very few cells will be synthesizing DNA at the same time. The ability of a virus to induce these particular enzymes may allow it a degree of independence of the division status of the host which would ensure a larger population of cells giving productive infections. Moreover, several of the virus-induced enzymes have smaller Michaelis constants than the host enzyme. This suggests that the virus should be able to compete successfully for intermediates in the presence of the host synthetic system, especially when these systems are in different cellular compartments. Little is known about the intracellular distribution of these enzymes, and the sites of virus synthesis may be separated from the normal host sites. Such a situation might also require the synthesis of a new enzyme induced by the virus.

Synthesis of nucleic acids

Replication of viral nucleic acid is a fundamental process in the multiplication of virus and it is not surprising that enzymes involved in this process are usually specified by the invading virus. There is good evidence that new DNA polymerases are produced in cells after infection by many DNA viruses, including phage T_2, vaccinia and herpes virus (reviews: Sinsheimer, 1968; Keir, 1968). Host enzymes involved in nucleic acid synthesis are under strict control and may be firmly bound to their templates. There is evidence that proteins necessary for DNA synthesis have to be synthesized for the initiation of each round of replication in *Escherichia coli* (Lark & Lark, 1964). Dependence on these activities might limit the capacity of an invading virus to multiply. Nevertheless it is becoming increasingly clear that some viruses are dependent on the host, in varying degree, for the synthesis of viral nucleic acid. The role of the host in replication of viral DNA has been summarized by Sinsheimer (1968), who used mutants of *E. coli* having temperature-sensitive enzymes involved in DNA synthesis. Such techniques have confirmed that T-phages are independent of the host but that phages such as λ and ϕX174 are much more dependent. Replication of phage ϕX174 occurs in several stages, the first of which is the conversion of single-stranded parental DNA to the double-stranded replicative form. Sinsheimer showed that this is dependent on a host enzyme present before infection. Subsequent stages of the replication are also dependent on the host, although some necessary protein, possibly involved in initiation, is produced after infection.

Replication of RNA viruses involves the use of RNA as template to direct the synthesis of new viral RNA (review: Höfschneider & Hausen, 1968). Normal cells are not able to perform this reaction, so that, the production of a new enzyme would appear to be mandatory for these viruses. Such enzymes have now been found in animal cells infected with such viruses as polio, reo and influenza, plant cells infected with tobacco mosaic virus and *Escherichia coli* infected with phages f2, MS2, Qβ. The latter enzymes have been extensively purified and it has been found that the active enzyme dissociates into two sub-units (Eikhorn, Stockley & Spiegelman, 1968) neither of which is independently capable of synthesizing viral RNA. One component is present only in infected cells, but the lighter component is also found in uninfected cells. The function of this protein in the normal cells is unknown. A co-operation between host protein and virus protein in this way implies a particularly intimate relationship of virus and host. Although this type of host dependence has not yet been shown unequivocally for other viruses, it is probable that it may be more general. The use of temperature-sensitive mutants of host cells should prove of great value in such investigations. Useful information has also been obtained by the investigation of the action of rifamycin on virus multiplication. This compound combines with bacterial RNA polymerase and prevents initiation of RNA synthesis (di Mauro *et al.* 1969). Rifamycin inhibits production of *m*RNA at all stages of the growth of phage T_4 (Haselkorn, Vogel & Brown, 1969) and phage SPO1 (Geiduschek & Sklar, 1969), suggesting that host RNA polymerase is involved in the reaction, although the polymerase from cells infected with phage T_4 is not identical with that in uninfected *E. coli*. RNA polymerase from *E. coli* only binds poorly *in vitro* to phage T_4 unless a protein fraction σ is also present (Travers & Burgess, 1969). It has been suggested that the σ fraction determines the specificity of binding of polymerase to the DNA. Similar factors may be directed by the virus and there is evidence that such a factor may determine the ability of extracts of infected *E. coli* to transcribe phage T_4 DNA into 'late' *m*RNA *in vitro* (Crouch, Hall & Hager, 1969).

RNA polymerase of animal cells is not inhibited by rifampicin, a derivative of rifamycin, but rifampicin prevents multiplication of vaccinia and other pox viruses (Subak-Sharpe, Timbury & Williams, 1969). It has been suggested that RNA polymerase induced after infection is distinct from the normal host enzyme. Rifampicin was without action on a wide range of other viruses infecting animal cells. Unlike many RNA-containing viruses the growth of influenza virus and other myxoviruses of subgroup 1 is inhibited by actinomycin D (Barry, 1964).

Actinomycin D is effective only in the early stages of virus growth and has been shown to affect the appearance but not the *in vitro* activity of the RNA polymerase thought to be involved in replication of the viral RNA (Ho & Walters, 1966). It has been suggested that some cellular function is required during the early stages of virus infection, but the nature of this function is unknown.

Protein synthesizing apparatus

It was thought for many years that viruses were completely dependent on host machinery for the synthesis of viral proteins. The classical experiment of Brenner, Jacob & Meselson (1961) showed that phage *m*RNA combined with host cell ribosomes formed before infection, and that these ribosomes with attached RNA were the sites of protein synthesis. The formation of 'viral polysomes' by interaction of viral *m*RNA with host cell ribosomes has been observed in many types of virus-infected cell, and a similar type of interaction seems to occur whether the viral *m*RNA has been transcribed from viral DNA or whether it is viral RNA itself acting as messenger.

Many viruses cause a rapid inhibition of host RNA synthesis; synthesis of ribosomal RNA in particular seems to be depressed following infection. The synthesis of virus-encoded ribosomal RNA has never been demonstrated and appears to be unlikely. The virus therefore must rely on host ribosomes for protein synthesis. Depletion of host ribosomes by alteration of the physiological state of cells before infection may be expected to alter the rate of production and final yield of virus. Many viruses cause a rapid inhibition of synthesis of host proteins (Levinthal, Hosoda & Shrub, 1967); this is the result of degradation of polysomes following infection (Penman, Scherrer, Becker & Darnell, 1963). The mechanism leading to degradation of polysomes is not known, but it does not seem to be a consequence of the inhibition of host RNA synthesis, because in animal cells it occurs more rapidly than degradation of polysomes following treatment with actinomycin; Willems & Penman (1966) suggested that it may result from a change in the interaction between ribosomes and *m*RNA.

Synthesis of certain RNA phage proteins *in vitro* by using cell-free extracts from uninfected bacteria shows that the host system is sufficient to synthesize viral proteins (Fraenkel-Conrat & Weissman, 1968), and in particular that the mechanism of initiation is the same. However, other attempts have been less successful and it seems probable that more complex viruses require some modification of host systems before synthesis of viral proteins can occur. Several such modifications have been detected in infected cells, including production of new *t*RNAs, new amino acid activating enzymes and modification of existing *t*RNAs by methylation or other changes (review: Subak-Sharpe, 1968). Other alterations may become apparent when more is known of the function of the various transfer factors.

Subak-Sharpe (1968) proposed that one possible reason for production of new *t*RNA molecules might be a scarcity of certain *t*RNAs within the host. A virus of base composition differing widely from that of its host would require alternative *t*RNA molecules and would be at a disadvantage when these molecules were not present.

Although protein synthesis in the normal and infected prokaryotic cell seems to occur by the same mechanism, there are some indications that virus-directed protein synthesis is not identical with the protein synthesis observed in normal eukaryotic cells. Interferon is a protein active in preventing the growth in animal cells of both RNA- and DNA- containing viruses, and it is probable that its action is to prevent translation of viral *m*RNA into viral protein (Joklik, 1968; Marcus & Salb, 1966). These workers have shown that ribosomes from interferon-treated cells are inactive in protein synthesis when viral RNA is used as messenger, but are fully active with host *m*RNA; viral *m*RNA is unable to combine *in vitro* with the ribosomes of interferon-treated cells. It has been proposed that interferon promotes the synthesis within the cell of a secondary protein that either combines with the ribosomes (Marcus & Salb, 1966) or with the viral *m*RNA (Joklik, 1968) to prevent interaction of the two. There is some indication that reaction with the smaller ribosomal sub-unit is prevented, so that viral polysomes cannot form in interferon-treated cells (Levy & Carter, 1968). This indicates that there must be some difference between host *m*RNA and viral *m*RNA. However, we still do not know enough about the interactions of ribosomes and *m*RNA in eukaryotes to be precise about the action of interferon. It has been proposed that the initiating codons for viral protein synthesis differ from those of eukaryotes (Noll, 1966) but this is not yet proven. Several authors have reported that at least some of the *m*RNA present in eukaryotes is associated with protein (review: Hadjiolov, 1968), but the possible mechanism of interaction between protein-bound RNA and ribosomes is obscure. However, it is very likely that viral *m*RNA is produced in a form different from this, and so the interaction with ribosomes might well be different.

One important difference between prokaryotes and eukaryotes is the ability to use polycistronic *m*RNA. While it is evident that bacterial systems use such templates, there is an increasing amount of information to suggest that eukaryotes do not normally do so. Many small RNA viruses use viral RNA as template for the synthesis of viral proteins. Within cells infected by polio virus polyribosomes have been detected of a size which could contain all the virus RNA (Penman *et al.* 1963); this is sufficient to code for 8 to 10 proteins. Jacobson & Baltimore (1969) observed very large proteins in cells after infection by polio virus; these large proteins subsequently formed the smaller viral coat-proteins. This has suggested the possibility that polio virus RNA does not contain a termination codon recognizable by the ribosomes of eukaryotes, but is translated as a continuous message. The virus protein produced as a large molecule must have a structure that can be divided in specific regions, but nothing is known about this mechanism of subdivision. If a host enzyme is required to achieve this hydrolysis, the protein sequence must be such that hydrolysis occurs only at specific locations recognized by the host enzyme. The RNA of other viruses of eukaryotes such as reovirus and influenza virus is effectively discontinuous so that only relatively small lengths of RNA are transcribed from them (Watanabe & Graham, 1968; Duesberg, 1968). Plant viruses such as tobacco rattle virus seem to occur as two distinct particles, each containing part of the RNA genome (Bancroft, 1968). This may represent an alternative adaptation to the requirements for protein synthesis in eukaryotes.

The nucleolus is now known to be the site of synthesis and processing of ribosomal RNA in eukaryotes (review: Perry, 1968). Many viruses inhibit the synthesis of ribosomes in the host and it is noticeable that an early change in nucleolar structure is detectable in many virus infections. Bernhard & Granboulan (1968) showed that various viruses caused first an enlargement of the nucleolus followed by dispersion of nucleolar material. Some of the changes of nucleolar structure seen are reminiscent of the changes produced by actinomycin treatment, which at low concentrations is known preferentially to inhibit rRNA synthesis. The nucleolus may have other functions besides ribosome synthesis; Harris, Sidebottom, Grace & Bramwell (1969), for example, obtained evidence that nucleolar activity was essential for transfer of RNA from nucleus to cytoplasm. It is not clear whether this activity was a consequence of nucleolar involvement in ribosomal synthesis, the ribosomes then serving to transport other RNA into the cytoplasm, or whether a distinct nucleolar function was involved. In view of the drastic alteration in nucleolar structure resulting from some virus infections one may speculate about the likely effect of viruses on the transport of RNA from the nucleus. Inhibition of transfer of host RNA to the cytoplasm is known to occur shortly after infection by vaccinia virus

(Salzman, Shatkin & Sebring, 1964). However, the mechanism of transfer of viral mRNA synthesized in the nucleus, from nucleus to cytoplasm, is unknown. Do viruses growing in the nucleus rely on cellular
machinery to transport RNA, or is this another virus-directed activity?

In eukaryotes protein synthesis is thought to occur mainly or exclusively in the cytoplasm. Darnell (1968b) and others have claimed that
mature ribosomes are absent from the nucleus and hence that protein
synthesis within the nucleus is impossible. Isolated nuclei do incorporate
amino acids and show requirements for the incorporation which differ
from cytoplasmic protein synthesis, but these nuclei may not have been
free from cytoplasmic contamination. The site of synthesis of proteins
normally assembled within the nucleus, e.g. ribosomal proteins and
histones, is not known. Many viruses are known to be assembled within
the nucleus and evidence has been accumulating that the protein components of these viruses are synthesized in the cytoplasm and transported
into the nucleus for assembly into mature virus particles (Ben-Porat,
Shimono & Kaplan, 1969; Olshevsky, Levitt & Becker, 1967). Electron
microscopy of cells infected with herpes virus and treated with ferritin-
labelled antibody has shown that viral protein is distributed over all
the membranes of the cell with the exception of mitochondrial membranes. Again, the mechanism of transport of these viral proteins within
the cell is unknown. It is apparent that under certain circumstances
large quantities of protein can enter the nucleus from the cytoplasm of
normal cells (Gurdon & Weir, 1969).

Interaction with host cell membranes and other specialized sites

As in other branches of cell biology, the attention of virologists is
now being drawn to the importance of membranes in the spatial
organization and control of biosynthetic processes. It is becoming clear
that the growth of many viruses involves interaction with membrane
structures of the host. The attachment of bacterial DNA to a membrane
site is believed to exert some control over the initiation of DNA synthesis (Lark, 1966). DNA polymerase activity and newly formed DNA
are associated with membrane material (Ganesan & Lederburg, 1965;
Smith & Hanawalt, 1967). There is clear evidence (Sinsheimer, 1968)
that replication of phage ϕX174 involves attachment of the parental
replicative form to a specific bacterial site, and that the number of such
sites varies from one to four according to the history of the cells.
Starved cells have only one site, whereas log-phase cells may show up
to 4 sites (Stone, 1967). It has been postulated that these sites are those
involved in normal cellular DNA replication. The replicative form of

phage ϕX174 can be isolated from cells attached to a structure that appears to be derived from the membrane; it is not covalently bonded to this and may be released by detergents (Knippers & Sinsheimer, 1968).

The growth of some phages is decreased by ultraviolet irradiation of the host cells before infection; this has suggested that only a limited number of 'phage-synthesizing centres' may exist. Schachtele, Anderson & Rogers (1968) used canavanine, an analogue of arginine that causes unusual effects on bacterial membranes and inhibits RNA polymerase, to study the capacity of cells to support phage T_4 replication: their data suggest that there are 2 to 8 specialized centres in each bacterium of *Escherichia coli*.

If viral DNA displaces host DNA from some site essential for its replication, one might anticipate that synthesis of host DNA would be halted. Such a phenomenon is observed during replication of phage ϕX174 (Sinsheimer, 1968). Many T-phages cause degradation of host DNA, but it is not known whether this host DNA must be displaced from cellular sites before degradation can start.

The existence of replication sites for DNA on the nuclear membrane of animal cells has been shown by Comings & Kakefuda (1968); it will be interesting to see whether DNA viruses that multiply within the nucleus exhibit any interactions with these sites. The association of other viruses with specialized regions or organelles within a cell may indicate a requirement for a particular synthetic function of the cell. Tobacco rattle virus is found strikingly arrayed along the outer mitochondrial membranes of infected plants (Harrison & Roberts, 1968); the reason for this is not known. Reovirus is associated with the spindle fibres of dividing cells, and the spindle fibres become coated with viral proteins. However, replication is not dependent on spindle formation since colchicine, which prevents polymerization of spindle protein, has no effect on the yield of virus (Dales, 1963).

Many animal viruses are replicated in association with large membranous structures in the animal cell. The whole replication complex of polio virus including RNA polymerase, viral polysomes and completed virus particles are associated with these structures which have been called 'factories' (see Darnell, 1968a). Treatment of the complex with detergent releases the components from the sites. Neither the function or origin of these membranes is known. They may serve to isolate and concentrate viral assembly processes within the cytoplasm of infected cells, or they may serve some function in the assembly processes of the virus. There is some evidence (Simon, 1969) that phage T_4 is associated with the bacterial membrane during assembly. It is probable that concentration of protein sub-units, together with restriction to a two-dimensional space might lead to more efficient assembly of the complex viral structure.

Many viruses of eukaryotes are surrounded by a lipoprotein envelope and this envelope is usually derived from host membrane. Viruses that have such envelopes are released from the cell by a process of budding through a cell membrane. Myxoviruses are an example of such viruses that are not detectable as entities within the cell; viral nucleoprotein is assembled beneath the plasmalemma and the virus particle is formed by pinching-off a piece of cell membrane to enclose the nucleoprotein material (see Compans, Holmes, Dales & Choppin, 1966). Other enveloped viruses of the arbovirus group are formed in an essentially similar way, but bud into vesicles within the cell cytoplasm (Hackett, Zee, Schaffer & Talens, 1968). There is considerable evidence that the phospholipid composition of myxoviruses is characteristic of the cell in which the virus was grown (Kates, Allison, Tyrell & James, 1962). Furthermore, cellular lipids synthesized before infection are incorporated into the envelopes of myxoviruses (Schafer, 1959), an arbovirus (Pfefferkorn & Hunter, 1962), and even the lipid-containing bacteriophage PM_2 (Espejo & Canelo, 1968). However the cell membrane is not unchanged by infection. Changes in appearance of cell membranes in the region of virus assembly have been seen, and frequently the membranes appear thickened at these sites (Compans *et al.* 1966). The membrane acquires the ability to react with antiviral serum; considerable areas of cell membrane may be altered in this way (Nii, Morgan, Rose & Hsu, 1968). The membranes of cells infected by many viruses thus seem to be a mosaic of virus and host cell material, including phospholipid and viral protein. This change in the composition of the surface may have a profound effect on the properties of the cell, as for example its ability to fuse with other cells. Transformation of cells by tumour viruses is associated with a change in surface properties and the acquisition of new antigens characteristic of the transforming virus. Many of the features that distinguish tumour cells from the normal may be due to altered membrane properties; indeed it has been suggested that the alterations in membrane structure are the basis of the change from a normal cell to a tumour cell (Sachs, 1968).

The properties of the cell membrane may also have an influence on the development of the virus. The myxovirus SV5 is released in large numbers by budding from the surface of monkey kidney cells, and the cells show little apparent damage as a result of virus growth (Klenk & Choppin, 1969). However, the same virus grows only poorly in hamster kidney cells, but causes considerable cell damage and death. The virus particles are only formed inefficiently at the surface of hamster cells, and the cells fuse after infection. The differences have been correlated with a difference in the chemical compositions of the host cell membranes. A mutant of herpes virus studied by Spring, Roizman & Schwartz (1968) grows normally in hamster cells and acquires its envelope by budding through the nuclear membrane; however, this mutant cannot mature in dog kidney cells and the failure to mature has been correlated with inability to be enveloped by the membranes of the dog kidney cells. Thus it seems likely that the interaction of virus with cell membranes may have important consequences both for virus and for host cells.

Bacterial viruses, in contrast to those of eukaryotes, are usually released from the cell after lysis of the cell wall. However, this process involves the production of lytic enzymes specified by the phage; mutants of various phages have been obtained which lack the necessary lytic enzyme and these are not released from cell. Enzymes such as lysozyme, lipase and polysaccharide depolymerase have all been described in cells at a late stage of phage infection. The filamentous DNA-containing phages however do not cause lysis of the host cell and are released from the infected cell over long periods; growth and cell division is not inhibited by the phage (review: Hoffman–Berling et al. 1966). Phage fd particles are rapidly transferred to the outer medium but the mechanism of their egress is not understood. It is of interest that mutants of this phage that cannot be released kill the host cell, perhaps due to accumulation of phage material at some essential bacterial site. Such a system may prove useful in the study of transport through bacterial membranes.

Conclusion

A few types of interaction between viruses and their host cells have been discussed. These interactions may be highly specific and require the specialization of a virus for a particular host cell. If the virus is to multiply successfully in a cell that cell must possess certain features required by that particular virus. Viruses can only introduce a few modifications to the normal patterns of activity in the cell, and virus multiplication requires more co-operation with the cell than has at times been believed. That the host cell is frequently killed as a result of virus growth is perhaps incidental; the viruses which have adapted most successfully to the host are those that do not kill the cell but continue to multiply in a growing and dividing host cell system.

Viruses of prokaryotes and eukaryotes have in general similar demands to make of the metabolic machinery of the host. The major differences between the growth of viruses in these cells concern prob-

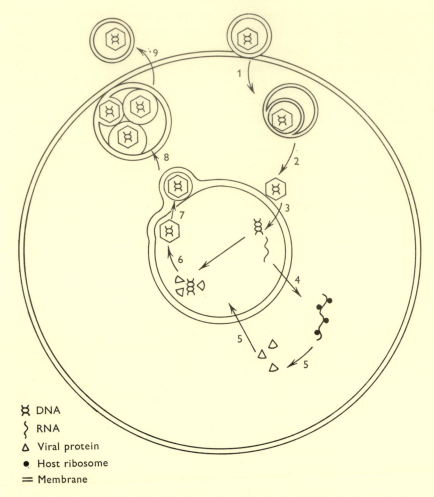

DNA
RNA
△ Viral protein
● Host ribosome
= Membrane

Fig. 5. Representation of the growth of a virus (e.g. herpes virus) in the nucleus of a eukaryotic cell. The stages involving transport within the cell are emphasized. (1) Attachment of the
virus to the cell membrane and penetration through the cell membrane. (2) Release from
phagocytic vesicle into cytoplasm. (3) Release of viral nucleic acid from coat protein, and
transport into the nucleus. (4) Transfer of viral *m*RNA from nucleus into cytoplasm;
Attachment of *m*RNA to host cell ribosomes in cytoplasm. (5) Synthesis of viral proteins
and transfer from cytoplasm to nucleus. (6) Assembly of viral particles in nucleus. (7) Transfer into perinuclear space by 'blebbing' through cell membrane. (8) Transfer through cytoplasm enclosed in vesicles or tubules; several virus particles may be enclosed in the same
vesicle. (9) Release from the cell, possibly by fusion of the vesicle membrane with the cell
wall.

lems of transport and organization within the cell. Both are specialized
to gain entry to a particular type of cell; the differences here reflect the
different organizations of the cell wall or cell membrane structure. However, inside a eukaryotic cell there are major problems for assembly of

viruses due to the compartmentation of the host; this problem is accentuated for viruses that grow in the nucleus (illustrated in Fig. 5).

It is difficult and probably foolish at this stage to make any further generalizations about the growth of viruses in different types of cell. It must be emphasized that most of our knowledge of the intracellular events occurring during replication of viruses has been obtained by study of infected *Escherichia coli* or infected mammalian and chicken cells. With this in mind it seems that the larger viruses, particularly those containing DNA, are more independent of the host cell for synthesis of nucleic acid and its precursors than are smaller viruses. But the components of the smaller viruses have been synthesized *in vitro* and these viruses have been assembled from their components *in vitro* so that little co-operation from cellular structures may be needed in these latter stages of small virus growth. Although assembly of phage T_4 from preformed components has now been achieved in cell lysates (Edgar & Wood, 1966) it seems probable that assembly of the larger, more complex viruses requires organized processes within the cell.

REFERENCES

ABEL, P. & TRAUTNER, T. A. (1964). Formation of an animal virus within a bacterium. *Z. VererbLehre* **95**, 66.

ADAMS, M. H. (1959). *Bacteriophages*. New York: Interscience Publishers Inc.

ALLISON, A. (1967). Lysosomes in Virus Infected Cells. In *Perspectives in Virology*, vol. 5, p. 29. Ed. M. Pollard. London and New York: Academic Press.

ARBER, W. (1965). Host controlled modification of bacteriophage. *A. Rev. Microbiol.* **19**, 365.

ARBER, W. (1968). Host Controlled Restriction and Modification of Bacteriophage. In *The Molecular Biology of Viruses*, p. 295. 18*th Symp. Soc. gen. Microbiol.* Eds. L. V. Crawford and M. G. P. Stoker. Cambridge University Press.

BANCROFT, J. B. (1968). Plant Viruses: Defectiveness and Dependence. In *The Molecular Biology of Viruses*, p. 229. 18*th Symp. Soc. gen. Microbiol.* Eds. L. V. Crawford and M. G. P. Stoker. Cambridge University Press.

BANKS, G. T., BUCK, K. W., CHAIN, E. B., HIMMELWEIT, F., MARKS, J. E., TYLER, J. M., HOLLINGS, M., LAST, F. T. & STONE, O. M. (1968). Viruses in fungi and interferon stimulation. *Nature, Lond.* **218**, 542.

BARRY, R. D. (1964). Effect of Inhibitors of Nucleic Acid Synthesis on the Production of Myxoviruses. In *Cellular Biology of Myxoviruses*, p. 51. Eds. G. E. Wostenholme and J. Knight. London: J. & A. Churchill & Sons.

BAYER, M. E. (1968). Adsorption of bacteriophages to adhesions between wall and membrane of *Escherichia coli*. *J. Virol.* **2**, 346.

BELLETT, A. J. D. (1968). The iridescent virus group. *Adv. Virus Res.* **13**, 225.

BEN-PORAT, T., SHIMONO, H. & KAPLAN, S. A. (1969). Synthesis of proteins in cells infected with herpes virus. II. Flow of structural proteins from cytoplasm to nucleus. *Virology* **37**, 56.

BERNHARD, W. & GRANBOULAN, N. (1968). Electron Microscopy of the Nucleolus in Vertebrate Cells. In *The Nucleus*, p. 81. Eds. A. J. Dalton and F. Haguenau. London and New York: Academic Press.

BRADLEY, D. E. (1967). Ultrastructure of bacteriophages and bacteriocins. *Bact. Rev.* **31**, 230.

BRENNER, S., JACOB, F. & MESELSON, M. (1961). An unstable intermediate carrying information from genes to ribosomes for protein synthesis. *Nature, Lond.* **190**, 576.

BRINTON, C. (1965). Structure, function synthesis and genetic control of bacterial pili. *Trans. N.Y. Acad. Sci.* **27**, 1003.

BRINTON, C. C. & BEER, H. (1967). The Interactions of Male Specific Bacteriophages with F-pili. In *The Molecular Biology of Viruses*, p. 251. Eds. J. S. Colter and W. Paranchych. London and New York: Academic Press.

COCKING, E. C. & POJNAR, E. (1968). A study of the infection of tomato fruit by tobacco mosaic virus. *Phytopath. Z.* **2**, 317.

COHEN, S. (1968). *Virus Induced Enzymes.* New York: Columbia University Press.

COHN, Z. A. & PARKS, C. (1967). The regulation of pinocytosis in mouse macrophages. II. Factors inducing vesicle formation. *J. exp. Med.* **125**, 213.

COMINGS, D. E. & KAKEFUDA, T. (1968). Initiation of deoxyribonucleic acid replication at the nuclear membrane in human cells. *J. molec. Biol.* **33**, 225.

COMPANS, R. W., HOLMES, K. V., DALES, S. & CHOPPIN, P. W. (1966). An electron microscopic study of moderate and virulent virus cell interactions of the parainfluenza virus SV 5. *Virology* **30**, 411.

CRAIGIE, J. & FELIX, A. (1947). Typing of typhoid bacilli with V 1-bacteriophages. *Lancet*, **252**, 823.

CROUCH, R. J., HALL, B. D. & HAGER, G. (1969). Control of gene transcription in T-even bacteriophage. Alteration in RNA polymerase accompanying phage infection. *Nature, Lond.* **223**, 476.

DALES, S. (1962). An electron microscopic study of the early association between two mammalian viruses and their hosts. *J. cell Biol.* **13**, 303.

DALES, S. (1963). Association between the spindle apparatus and Reo virus. *Proc. natn. Acad. Sci. U.S.A.* **50**, 268.

DALES, S. (1965). Penetration of animal viruses into cells. *Prog. med. Virol.* **7**, 1.

DARNELL, J. E. (1968a). Considerations on Virus Controlled Functions. In *Molecular Biology of Viruses*, p. 149. *18th Symp. Soc. gen. Microbiol.* Eds. L. V. Crawford and M. G. P. Stoker. Cambridge University Press.

DARNELL, J. E. (1968b). Ribonucleic acid from animal cells. *Bact. Rev.* **32**, 262.

DAVISON, E. M. (1969). Cell to cell movement of tobacco ringspot virus. *Virology* **37**, 694.

DI MAURO, E., SNYDER, L., MARINO, P., LAMBERTI, A., COPPO, A. & TOCCHINI-VALENTINI, G. P. (1969). Rifampicin sensitivity of the components of DNA-dependent RNA polymerase. *Nature, Lond.* **222**, 533.

DUESBERG, P. H. (1968). The RNAs of influenza virus. *Proc. natn. Acad. Sci. U.S.A.* **59**, 930.

EDGAR, R. S. & WOOD, W. B. (1966). Morphogenesis of bacteriophage T_4 in extracts of mutant-infected cells. *Proc. natn. Acad. Sci. U.S.A.* **55**, 498.

EIKHORN, T. S., STOCKLEY, D. J. & SPIEGELMAN, S. (1968). Direct participation of a host protein in the replication of viral RNA in vitro. *Proc. natn. Acad. Sci. U.S.A.* **59**, 506.

ELSBACH, P., HOLMES, K. V. & CHOPPIN, P. W. (1969). Metabolism of lecithin and virus induced cell fusion. *Proc. Soc. exp. Biol. Med.* **130**, 903.

ESAU, K., CRONSHAW, J. & HOEFERT, L. L. (1967). Relation of beet yellows virus to the phloem and to movement in the sieve tube. *J. Cell Biol.* **32**, 71.

ESPEJO, R. T. & CANELO, E. S. (1968). Origin of phospholipid in bacteriophage PM 2. *J. Virol.* **2**, 1235.

FAWCETT, D. W. (1965). Surface specialization of absorbing cells. *J. Histochem. Cytochem.* **13**, 75.

FELDHERR, C. M. & HARDING, C. V. (1964). The permeability characteristics of the nuclear envelope at interphase. *Protoplasmatologia* **5**, 35.

FRAENKEL-CONRAT, H. & WEISSMAN, C. (1968). In Vitro Synthesis of Viral Components. In *Molecular Basis of Virology*, p. 209. Ed. H. Fraenkel-Conrat. New York: Reinhold Book Corporation.

FUKUSAWA, T. (1964). The course of infection with abnormal bacteriophage T_4 containing non-glucosylated DNA in *Escherichia coli* strains. *J. molec. Biol.* **9**, 525.

GANESAN, A. T. & LEDERBURG, G. J. (1965). A cell membrane bound fraction of bacterial DNA. *Biochem. biophys. Res. Commun.* **18**, 824.

GEIDUSCHEK, E. P. & SKLAR, J. (1969). Continual requirement for a host RNA polymerase component in bacteriophage development. *Nature, Lond.* **221**, 833.

GOTTSCHALK, A. (1959). Chemistry of Virus Receptors. In *The Viruses*, vol. 3, p. 31. Eds. F. M. Burnet and W. M. Stanley. London and New York: Academic Press.

GROSS, L. (1967). Active membranes for active transport. *J. theor. Biol.* **15**, 298.

GURDON, J. B. & WEIR, R. S. (1969). Cytoplasmic proteins and the control of nuclear activity in early amphibian development. *Biochem. J.* **114**, 52.

HACKETT, A. J., ZEE, Y. C., SCHAFFER, L. & TALENS, L. (1968). Electron microscopic study of the morphogenesis of vesicular stomatitis virus. *J. Virol.* **2**, 1154.

HADJIOLOV, A. A. (1968). Ribonucleic acids and information transfer in animal cells. *Prog. Nucleic Acid Res.* **7**, 196.

HARRIS, H., SIDEBOTTOM, E., GRACE, D. M. & BRAMWELL, M. E. (1969). The expression of genetic information: a study with hybrid animal cells. *J. Cell Science* **4**, 499.

HARRIS, H., WATKINS, J., FORD, C. E. & SCHOEFL, G. I. (1966). Artificial heterokaryons of animal cells from different species. *J. Cell Science* **1**, 1.

HARRISON, B. D. & ROBERTS, I. M. (1968). Association of tobacco rattle virus with mitochondria. *J. gen. Virol.* **3**, 121.

HASELKORN, R., VOGEL, M. & BROWN, R. D. (1969). Conservation of rifamycin sensitivity during T_4 development. *Nature, Lond.* **221**, 836.

HIRST, G. K. (1965). Cell Virus Attachment and the Action of Antibodies on Viruses. In *Viral and Rickettsial Infections of Man*. 4th edn, p. 216. Eds. F. L. Horsfall and I. Tamm. London: Pitman Medical Publishing Co. Ltd.

HO, P. K. & WALTERS, P. (1966). Influenza virus induced ribonucleotidyl transferase and the effect of actinomycin D on its formation. *Biochemistry*, **5**, 231.

HOFFMAN-BERLING, H., KAERNES, H. & KNIPPERS, R. (1966). The small bacteriophages. *Adv. Vir. Res.* **12**, 329.

HÖFSCHNEIDER, P. H. & HAUSEN, P. (1968). The Small RNA Viruses of Plants, Animals and Bacteria; c. The replication cycle. In *The Molecular Basis of Virology* p. 169. Ed. H. Fraenkel-Conrat. New York: Reinhold Book Corporation.

HOLLAND, J. J. (1964). Enterovirus entrance into specific host cells and subsequent alterations of cell protein and nucleic acid synthesis. *Bact. Rev.* **28**, 3.

HOLTER, H. (1965). Passage of Particles and Macromolecules Through Cell Membranes. In *Function and Structure in Micro-organisms*, p. 89. 15th *Symp. Soc. gen. Microbiol.* Eds. M. R. Pollock and M. H. Richmond. Cambridge University Press.

HULL, R. (1970). Large RNA Plant-infecting Viruses. In *The Biology of Large RNA viruses*. London and New York: Academic Press. (In Press.)

JACOBSON, M. F. & BALTIMORE, D. (1969). Polypeptide cleavage in the formation of polio virus proteins. *Proc. natn. Acad. Sci. U.S.A.* **61**, 77.

JAVOR, G. T. & TOMASZ, A. (1968). An autoradiographic study of genetic transformation. *Proc. natn. Acad. Sci. U.S.A.* **60**, 1216.

JOKLIK, W. K. (1966). The pox viruses. *Bact. Rev.* **30**, 33.

JOKLIK, W. K. (1968). Studies on the Mechanism of Action of Interferon. In *Interferon*, p. 111. Eds. G. E. W. Wolstenholme and M. O'Connor. London: J. & A. Churchill Ltd.

KAPLAN, A. S. (1967). Studies on the Control of the Infective Process in Cells Infected with Pseudorabies Virus. In *Molecular Biology of Viruses*, p. 527. Eds. J. S. Colter and W. Paranchych. London and New York: Academic Press.

KATES, M., ALLISON, A. C., TYRELL, D. A. & JAMES, A. T. (1962). Origin of lipids of influenza virus. *Cold Spring Harb. Symp. quant. Biol.* **27**, 293.

KATES, J. R. & McAUSLAN, B. R. (1967). Pox virus DNA-dependent RNA polymerase. *Proc. natn. Acad. Sci. U.S.A.* **58**, 134.

KEIR, H. M. (1968). Virus Induced Enzymes in Mammalian Cells Infected with DNA Viruses. In *The Molecular Biology of Viruses*, p. 67. 18*th Symp. Soc. gen. Microbiol.* Eds. L. V. Crawford and M. G. P. Stoker. Cambridge University Press.

KLENK, H. D. & CHOPPIN, P. W. (1969). Lipids of plasma membranes of monkey and hamster kidney cells and para influenza virions grown in these cells. *Virology* **38**, 255.

KNIPPERS, R. & SINSHEIMER, R. L. (1968). Process of infection with bacteriophage ϕX174. XX. Attachment of the parental DNA of bacteriophage ϕX174 to a fast sedimenting cell component. *J. molec. Biol.* **34**, 17.

KOZLOFF, L. M. (1968). Biochemistry of the T-even Bacteriophages of *Escherichia coli*. In *Molecular Basis of Virology*, p. 435. Ed. H. Fraenkel-Conrat. New York: Reinhold Book Corporation.

LANNI, Y. T. (1965). DNA transfer from phage T_5 to host cells: dependence on intercurrent protein synthesis. *Proc. natn. Acad. Sci. U.S.A.* **53**, 969.

LARK, C. (1968). Effect of methionine analogues ethionine and norleucine on DNA synthesis in *Escherichia coli* 15 T⁻. *J. molec. Biol.* **31**, 401.

LARK, C. & LARK, K. G. (1964). Evidence for two distinct aspects of the mechanism regulating chromosome replication in *Escherichia coli*. *J. molec. Biol.* **10**, 120.

LARK, K. G. (1966). Regulation of chromosome replication and segregation in bacteria. *Bact. Rev.* **30**, 3.

LAVER, W. G. & VALENTINE, R. C. (1969). Morphology of the isolated haemagglutinin and neuraminidase sub-units of influenza virus. *Virology*, **38**, 105.

LEVINTHAL, C., HOSODA, J. & SHRUB, D. (1967). The Control of Protein Synthesis after Phage Infection. In *The Molecular Biology of Viruses*, p. 71. Eds. J. S. Colter & W. Paranchych. London and New York: Academic Press.

LEVY, H. B. & CARTER, W. A. (1968). Molecular basis of action of interferon. *J. molec. Biol.* **31**, 561.

LOEWENSTEIN, W. R. (1964). Permeability of the nuclear membrane as determined with electrical methods. *Protoplasmatologia* **5**, 26.

LUNGER, P. D. (1966). Amphibia related viruses. *Adv. Virus Res.* **12**, 1.

LWOFF, A. (1957). The concept of virus. *J. gen. Microbiol.* **17**, 239.

MARCUS, P. I. & SALB, J. M. (1966). Molecular basis of interferon action: inhibition of viral RNA translation. *Virology* **30**, 502.

MAYOR, H. D. & MELNICK, J. L. (1966). Small deoxyribonucleic acid containing viruses (picodna virus group). *Nature, Lond.* **210**, 331.

MONTAGNIER, L. (1968). The Replication of Viral RNA. In *Molecular Biology of Viruses*, p. 125. 18*th Symp. Soc. gen. Microbiol.* Eds. L. V. Crawford and M. G. P. Stoker. Cambridge University Press.

MORGAN, C. & HOWE, C. (1968). Structure and development of viruses as seen in the electron microscope. IV. Entry of para-influenza I (Sendai) virus. *J. Virol.* **2**, 1122.

MORGAN, C. & ROSE, H. M. (1968). Structure and development of viruses as seen in the electron microscope. VIII. Entry of influenza virus. *J. Virol.* **2**, 925.

NII, S., MORGAN, C., ROSE, H. M. & HSU, K. C. (1968). Electron microscopy of herpes simplex virus. IV. Studies with ferritin-conjugated antibodies. *J. Virol.* **2**, 1172.

NOLL, H. (1966). Chain initiation and control of protein synthesis. *Science, N.Y.* **151**, 1241.

OLSHEVSKY, U., LEVITT, J. & BECKER, Y. (1967). Studies on the synthesis of herpes simplex virions. *Virology* **33**, 323.

PENMAN, S., SCHERRER, K., BECKER, Y. & DARNELL, J. E. (1963). Polyribosomes in normal and polio virus infected Hela cells and their relationship to messenger RNA. *Proc. natn. Acad. Sci. U.S.A.* **49**, 654.

PERRY, R. P. (1968). The nucleolus and the synthesis of ribosomes. *Prog. Nucleic Acid Res.* **6**, 220.

PFEFFERKORN, E. R. & HUNTER, H. S. (1963). The source of ribonucleic acid and phospholipid of Sindbis virus. *Virology* **20**, 446.

RAIMONDO, L. M., LUNDH, N. P. & MARTINEZ, R. J. (1968). Primary adsorption site of phage PBSI: the flagellum of *Bacillus subtilis*. *J. Virol.* **2**, 256.

RHODES, A. J. & VAN ROOYEN, C. E. (1968). *Textbook of Virology*. 5th edn. Baltimore: The Williams and Wilkins Co.

ROIZMAN, B. (1962). Polykaryocytosis. *Cold Spring Harb. Symp. quant. Biol.* **27**, 327.

SACHS, L. (1968). An Analysis of the Mechanism of Neoplastic Cell Transformation by Polyoma Virus, Hydrocarbons and X Irradiation. In *Current Topics in Developmental Biology*, vol. 2, p. 129. Ed. A. A. Moscona. London and New York: Academic Press.

SALZMAN, N. P., SHATKIN, A. J. & SEBRING, E. D. (1964). The synthesis of DNA-like RNA in the cytoplasm of Hela cells infected by vaccinia virus. *J. molec. Biol.* **8**, 405.

SANDER, E. (1964). Evidence for the synthesis of a DNA phage in leaves of tobacco plants. *Virology* **24**, 245.

SCHACHTELE, C. F., ANDERSON, D. L. & ROGERS, P. (1968). Mechanism of canavanine death in *Escherichia coli*. II. Membrane bound canavanyl protein and nuclear disruption. *J. molec. Biol.* **33**, 861.

SCHADE, S. Z., ADLER, J. & RIS, H. (1967). How bacteriophage χ attacks motile bacteria. *J. Virol.* **1**, 599.

SCHAFER, W. (1959). Some Observations Concerning the Reproduction of RNA Containing Animal Viruses. In *Virus Growth and Variation*, p. 61. *9th Symp. Soc. gen. Microbiol.* Eds. A. Isaacs and B. W. Lacey. Cambridge University Press.

SHELTAWY, M. & NEWTON, A. A. (1969). Thymidylate kinase activity in cells infected with herpes virus. *J. gen. Microbiol.* (In Press.)

SHEPHERD, R. J., WAKEMAN, R. J. & ROMANKO, R. R. (1968). DNA in cauliflower mosaic virus. *Virology*, **36**, 150.

SHUGAR, D. & SIERAKOWSKA, H. (1967). Mammalian nucleolytic enzymes and their localization. *Prog. Nucleic Acid. Res.* **7**, 369.

SILVERMAN, P. M. & VALENTINE, R. C. (1969). The RNA injection step of bacteriophage infection. *J. gen. Virol.* **4**, 111.

SILVERSTEIN, S. C. & DALES, S. (1968). The penetration of reovirus RNA and initiation of its genetic function in L strain fibroblasts. *J. cell Biol.* **36**, 197.

SIMON, L. D. (1969). The infection of *Escherichia coli* by T_2 and T_4 bacteriophages as seen in the electron microscope. III. Membrane associated intracellular bacteriophages. *Virology* **38**, 203.

SIMON, L. D. & ANDERSON, T. F. (1967). The infection of *Escherichia coli* by T_2 and T_4 bacteriophages as seen in the electron microscope. I. Attachment and penetration. *Virology* **32**, 279.

SIMPSON, R. W., HAUSER, R. E. & DALES, S. (1969). Viropexis of vesicular stomatitis virus by L cells. *Virology* **37**, 285.

SINHA, R. C. (1968). Recent work on leafhopper-transmitted viruses. *Adv. Virus Res.* **13**, 181.

SINSHEIMER, R. L. (1968). The Replication of Viral DNA. In *The Molecular Biology of Viruses*, p. 101. 18*th Symp. Soc. gen. Microbiol.* Eds. L. V. Crawford and M. G. P. Stoker. Cambridge University Press.

SMITH, D. W. & HANAWALT, D. C. (1967). Properties of the growing point region in the bacterial chromosome. *Biochim. biophys. Acta* **148**, 519.

SPRING, S. S., ROIZMAN, B. & SCHWARTZ, J. (1968). Herpes simplex products in productive and abortive infection. II. Electron microscopic and immunologic evidence for failure of viral envelopment as a cause of abortive infection. *J. Virol.* **2**, 384.

STACEY, K. A. (1965). Intracellular modification of nucleic acids. *Br. med. Bulletin* **21**, 211.

STOKER, M. G. P. (1959). Growth Studies with Herpes Virus. In *Virus Growth and Variation*, p. 142. 9*th Symp. Soc. gen. Microbiol.* Eds. A. Isaacs and B. W. Lacey. Cambridge University Press.

STONE, A. B. (1967). Some factors which influence the replication of the replicative form of bacteriophage ϕX174. *Biochem. biophys. Res. Commun.* **26**, 247.

SUBAK-SHARPE, H. (1968). Virus Induced Changes in Translation Mechanism. In *The Molecular Biology of Viruses*, p. 47. 18*th Symp. Soc. gen. Microbiol.* Eds. L. V. Crawford and M. G. P. Stoker. Cambridge University Press.

SUBAK-SHARPE, J. H., TIMBURY, M. C. & WILLIAMS, J. F. (1969). Rifampicin inhibits the growth of some mammalian viruses. *Nature, Lond.* **222**, 341.

TRAVERS, A. A. & BURGESS, R. R. (1969). Cyclic re-use of the RNA polymerase sigma factor. *Nature, Lond.* **222**, 537.

VAGO, C. & BERGOIN, M. (1968). Viruses in invertebrates. *Adv. Virus Res.* **13**, 248.

WATANABE, Y. & GRAHAM, A. F. (1968). Structural units of Reovirus ribonucleic acid and their possible functional significance. *J. Virol.* **1**, 665.

WATSON, D. H. (1968). The Structure of Animal Viruses in Relation to their Biological Functions. In *The Molecular Biology of Viruses*, p. 207. 18*th Symp. Soc. gen. Microbiol.* Eds. L. V. Crawford and M. G. P. Stoker. Cambridge University Press.

WEIDEL, W. (1958). Bacterial viruses (with particular reference to adsorption penetration). *A. Rev. Microbiol.* **12**, 27.

WILLEMS, M. & PENMAN, S. (1966). Mechanism of host cell protein synthesis inhibition by polio virus. *Virology* **30**, 355.

WOLF, K. (1966). The fish viruses. *Adv. Virus Res.* **12**, 36.

WOODSON, B. (1968). Recent progress in pox virus research. *Bact. Rev.* **32**, 127.

THE ORGANIZATION OF DNA IN EUKARYOTIC CHROMOSOMES

ROBIN HOLLIDAY

National Institute for Medical Research,
Mill Hill, London N.W.7

It would be surprising if the characteristic appearance of individual chromosomes during meiotic and mitotic division in eukaryotic organisms did not reflect an underlying order in their molecular organization. It is generally agreed that there is a dearth of unequivocal information about the nature of this order. Observations with the electron microscope have all too often revealed an apparently chaotic situation: a tangled mass of threads or fibres usually of much greater width than double-stranded DNA, and often a good deal of extraneous non-fibrous material. Greater success has been achieved with chromosomes in particular configurations, for instance, the synaptinemal complex involving the paired chromosomes at prophase of meiosis (for review, see Moses, 1968), or the lampbrush stage of meiosis and the nucleolar DNA in amphibian oocytes (Miller, 1965; Miller & Beatty, 1969). In general, however, the electron microscope has not contributed as much information as might have been expected towards solving the problem of the organization of DNA in eukaryotic chromosomes, and in this review the emphasis will be placed on other sources of information.

The chromosome is a linear structure

The compact rod-like appearance of the metaphase chromosome is the result of the coiling of a much longer thread. Particular treatment of such chromosomes causes them partially to uncoil and convincingly reveals the underlying linearity (e.g. Ohnuki, 1968). The thread-like structure is clearly visible during the early stages of meiotic division (leptotene, zygotene and pachytene). At pachytene the chromosome has a characteristic beaded appearance, with densely-staining regions alternating with lightly-staining ones. The linear structure of the chromosome is also apparent in the cells of the salivary glands and other specialized tissues in Dipteran insects. Here the giant chromosomes are visible in cells which are not undergoing division, and again they reveal a chromomeric structure, in this case the chromomeres appearing as bands. It is well established that the giant chromosomes are polytene in structure, that is, they are formed as a result of a large number of divisions of the basic chromosome thread without concomitant separation of the individual strands (Beermann & Pelling, 1965). In *Drosophila*, the total length of the salivary gland chromosomes is about one-fifth that which would be expected if the DNA in individual strands were completely extended. (Some confusion may arise over the use of the term 'strand'. The cytologist refers to single-stranded or multi-stranded chromosomes, meaning chromosomes containing one DNA duplex or many. It is hoped that in each case the context will distinguish this usage from description of DNA as single or double stranded.)

Only fairly recently has Kleinschmidt's spreading technique been successfully applied to eukaryotic DNA. Solari (1967), using sea urchin DNA, detected single strands up to 100 μm long with no evidence of discontinuities or branching in the structure. Wolstenholme, Dawid & Ristow (1968) studied DNA molecules from *Chironomus* chromosomes and detected pieces up to 240 μm in length. Treatment with detergent, pronase or RNAse did not significantly decrease their size, whereas DNAse decreased them to very small fragments (< 1 μm). A similar result has been obtained with yeast by Stevens (cited by Williamson, Moustacchi & Fennell, 1969). *Saccharomyces cerevisiae* has at least 17 chromosomes (Mortimer & Hawthorne, 1969) and the haploid cell contains 3×10^{-14} g. DNA. Allowing for the fact that the DNA replicates early in the cell cycle (Williamson, 1965), each chromosome would have about 10^{-15} g., 300 μm DNA. Stephens detected pieces up to 100 μm long, which represents a substantial part of one chromosome arm and must contain about 200 genes. These later studies have not confirmed the earlier report of Hotta & Bessel (1965) that some of the DNA from boar and wheat consisted of small closed circles.

The linear structure of the genetic material is also revealed by genetic analysis. Classical studies on the linkage groups of *Drosophila*, maize, mouse, *Neurospora* and many other organisms demonstrate that the genes are arranged in a linear sequence along each chromosome. In fungi, the use of techniques allowing the selection of rare recombinants has carried genetic analysis to the fine-structure level. There are numerous examples of single gene maps in which the order of mutant sites is linear, and none where the map is branched or circular. Moreover, the length of a single gene may be greater than the distance from one end of the gene to a mutation in another gene of apparently unrelated function. The *y* (yellow conidia) and *ad*$_8$ (adenine requirement) mutants in *Aspergillus nidulans* (Pritchard, 1955) provide an example of this. In the same organism the *ad*$_9$ and *paba*$_1$ genes are probably adjacent (see

Whitehouse & Hastings, 1965). Finally, the use of closely-linked markers on each side of a gene which has been mapped at the fine-structure level indicates that the gene is integrated linearly in the chromosome structure (e.g. Pritchard, 1955; Siddiqi, 1962; Case & Giles, 1958; Murray, 1963). It is difficult to obtain comparable data with more complex eukaryotes, but the evidence strongly suggests that the same situation applies in *Drosophila* (Chovnick, 1966) and maize (Nelson, 1962) as in fungi.

Neither the genetic data nor the direct examination of chromosomal DNA provides any support for the idea that there is other material inserted at intervals—between genes—along the length of the chromosome; not is there any evidence for branched or circular genetic structures in the chromosome. Moreover Wolstenholme *et al.* (1968) calculated that the pieces of DNA they isolated from *Chironomus* must have extended across several chromomere regions. One must conclude that the chromomere merely represents a denser folding of the DNA than in other parts of the chromosome.

The DNA content of eukaryotic nuclei

It was first demonstrated by Boivin and the Vendrelys (for review, see Vendrely, 1955) that mammalian cells from different tissues contain the same amount of DNA, and furthermore that sperm cells contain half this amount. In several species of fungi vegetative diploid cells contain twice the amount of DNA contained by haploid ones (see Fincham & Day, 1965). It is also well known that DNA is metabolically stable. These observations strongly suggest that most, if not all, the DNA has a genetic function. The difficulty arises when the actual amounts of DNA per nucleus in different species are considered. The DNA content of cell nuclei of a large number of species has now been measured: a selected sample of this information is given in Fig. 1.

The *Escherichia coli* cell probably contains between 1000 and 2000 enzymes; add to this a large number of repressors, operator and promotor regions, the structural proteins required for membranes, ribosomes, etc., as well as the ribosomal and transfer RNA, and one reaches a figure of perhaps 5000 genes. If each gene has on average 600 base pairs then the total DNA content would be 3×10^6 base pairs, a molecular weight of 2×10^9 daltons and a length of $1000 \, \mu m$ (1 mm). Cairns (1963) found the length of the *Escherichia coli* genome to be $1400 \, \mu m$ (other bacteria appear to have somewhat shorter genomes; see Kleinschmidt, 1967). In other words the DNA content is roughly what one might expect. It is not unreasonable that *Neurospora* or *Aspergillus*, with their complex cellular structure, capacity for cellular

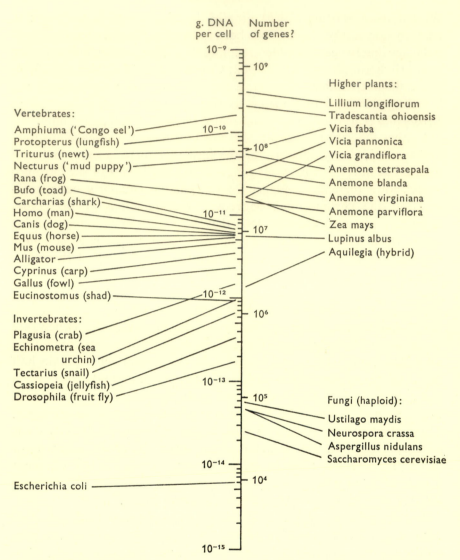

Fig. 1. The DNA content per cell in various eukaryotes. Also indicated is the corresponding number of gene copies per cell, assuming each gene has an average of 600 base-pairs. Sources: vertebrates (Mirsky & Ris, 1951; Vialli, 1957); invertebrates (Mirsky & Ris, 1951; Kurnick & Herskowitz, 1952), higher plants (McLeish & Sunderland, 1961; Rothfels, Sexsmith, Heimburger & Krause, 1966; Martin, 1968), fungi (Fincham & Day, 1965).

differentiation, an elaborate mechanism for sexual reproduction and meiotic division, should have about 10 times as much DNA as a bacterium. Nor is it surprising that so complex an organism as the fruit fly, *Drosophila melanogaster*, should have about twice as much DNA as a fungus. Moving up the evolutionary scale it is perhaps somewhat surprising that mammals should have about 40 times as much DNA as *Drosophila*. This might be explicable if a large amount of DNA were necessary to code for the complexities of the higher nervous system: but this seems unlikely since the DNA contents per cell of man and mouse are very similar. That DNA content cannot be related to the organism's position on the evolutionary scale is convincingly shown by the enormous DNA value for certain amphibia and the primitive lung-fish *Protopterus*, which have up to 30 times the mammalian amount. Even more surprising results have come from studies on the amount of DNA in higher plants. McLeish & Sunderland (1961) measured the DNA content of diploid nuclei in ten species and found a 60-fold difference between the lupin (*Lupinus albus*) and the lily (*Lilium longi-florum*), the others being of intermediate value. Later it was shown that even species within the same genus may have very different amounts of DNA; for instance, Rees, Cameron, Hazarika & Jones (1966) showed that *Vicia faba* had seven times as much DNA as *V. sativa*, although their chromosome number was the same, and *Lathyrus* species showed a two- to three-fold variation in DNA content. In further studies on *Vicia*, Martin (1968) measured the DNA content of 12 species and noted that they fell into three groups on the basis of their DNA content, the amounts approximating to a 1:2:4 ratio. Rothfels, Sexsmith, Heimburger & Krause (1966) examined 22 species in six related genera of the Ranunculaceae. The variation in DNA content was 40-fold, the values appearing to fall in a series 1:8:12:16:20:24:40. Variation in DNA content between closely related species is not restricted to higher plants. In one group of the Hemipteran bugs, closely related species had a two-fold variation, and in one case a probable tetraploid species, *Thyanta calceata*, had the same DNA content as closely related diploid *Thyanta* species (Schrader & Hughes-Schrader, 1956; Hughes-Schrader & Schrader, 1956). Callan (1967) cited the example of two *Mesostomum* (flatworm) species with the same chromosome number but an 11-fold difference in DNA content per nucleus.

Two major conclusions can be drawn from such studies. First, if the chromosomes consist of a linear array of single copies of individual genes, then many organisms have far more DNA than would be required to specify the structure and regulate the synthesis of RNA and protein molecules. Second, enormous changes in DNA content per genome can evolve quite rapidly.

There appear to be three basic solutions to this problem. (1) Many of the genes are duplicated longitudinally to form a tandem array of copies within a single stranded (unineme) chromosome structure. (2) The total DNA in the chromosome is duplicated many times to produce a multi-stranded (polyneme) chromosome consisting of several identical copies of the basic genetic information. (3) Only a proportion of the DNA has any genetic function and the rest—which in many cases must be the major part—has some other role to play.

Longitudinal multiplicity of genetic information

Detailed studies of the giant lampbrush chromosomes in the oocytes of the newt *Triturus* and the polytene chromosomes in Diptera provide the main evidence for the 'master and slave' hypothesis of chromosome organization which has been suggested by Callan (1967). This preserves the concept of the single-stranded chromosome and suggests that many of the genes are multiplied many times within the linear structure to give 'genes' consisting of a master template and a number of serially repeated slave templates. The lampbrush chromosome consists of a backbone of chromomeres containing most of the DNA and numerous closed loops which extend from the chromomeres into the nucleoplasm. The pattern of chromomeres and loops is constant for each chromosome and appears to be genetically determined. Stretching of the chromosome leads to the splitting of the basal chromomere into two parts held together by loops. This shows that the longitudinal strands do not pass directly along the backbone of the chromosome but rather from chromomere regions into loop and back, then to the next chromomere region, loop, and so on. Treatment of the chromosome with DNAase leads to the splitting of the loops and the disintegration of the chromosome (for a full description of the lampbrush chromosome see Callan & Lloyd, 1960). Two studies suggest that the DNA in the loops consists of a single double-helix. Gall (1963) examined the kinetics of loop and chromosome breakage following treatment with pancreatic DNAase, which causes single-strand breaks in DNA. He found that the breakage of the loop followed 2-hit kinetics, whereas the breakage of the chromosome followed roughly 4-hit kinetics. Since the chromosomes in the oocyte are replicated and each chromomeric region has two loops, this result suggested that the DNA in each loop contained a single DNA molecule, and also that there were no further DNA strands running along the backbone of the chromosome. Miller (1965) studied the structure of the loops by electron microscopy after digesting away the RNA and protein. He found that the width of the remaining material was of the order of 30 Å, which shows that there could not be more than two molecules of DNA and that there was probably only one.

The giant salivary gland chromosomes of the related subspecies *Chironomus thummi thummi* and *Chironomus thummi piger* have a very similar pattern of chromomeric bands along their length. However, the DNA content of the nuclei (whether in salivary gland or other cells) differs by 27%. Keyl (1965) found the explanation for this difference by measuring the DNA content of individual homologous bands in the two subspecies. He found that certain *thummi* bands had 2, 4, 8 or 16 times as much DNA as the corresponding *piger* bands. This showed that certain regions of the chromosome had undergone rounds of replication and the additional DNA was permanently integrated into the chromosome structure.

Callan's hypothesis that genetic information is serially repeated is based largely on these observations. It receives additional support from several hybridization studies on the DNA of the nucleolar organizer region in *Drosophila* and *Xenopus* (Ritossa, Atwood & Spiegelman, 1966; Wallace & Birnstiel, 1966; Brown & Weber, 1968). These show that ribosomal RNA hybridizes with the DNA of the nucleolar organizer, and furthermore that this DNA must contain a very large number of copies of the ribosomal genes. In the case of *Xenopus* there are 450 serially repeated copies of the genes coding for 28s and 16s ribosomal RNA, alternating with regions of non-hybridizable DNA. This structure for nucleolar DNA has been brilliantly confirmed by the electron microscope studies of Miller & Beatty (1969), which show regions of DNA synthesizing RNA alternating with regions of inactive DNA.

The idea of multiple copies of each gene poses considerable problems with regard to the origin and the evolution of mutations of genes. To avoid this difficulty Callan suggested that one of the many copies has the role of master template, and that once during the life cycle the slaves are matched against the master. Any differences in base sequence which may have arisen in the various slave copies during the many intervening mitotic divisions are ironed out by this process. This idea of matching, which is not a new one, will be discussed in a later section along with the problem of recombination between such chromosomes. Callan believes that the matching process occurs at the diplotene stage in meiosis, when the lampbrush chromosome configuration appears. There is evidence (Gall & Callan, 1962) that the loops are not static but that the DNA is being continually spun out from one of the chromomeric regions and wound in at the other. This process is believed to allow both sequential transcription of slave copies and also their matching against the master strand.

The experiments of Taylor, Woods & Hughes (1957) on the mechanism of chromosome replication in *Vicia* are frequently cited as evidence for a single-stranded chromosome. These experiments are so well known that they need not be described here. They show that the replication of the chromosome is semi-conservative, just as is that of DNA. In further studies Taylor (1958) detected and analysed sister strand exchanges (occasional breakage and reunion of the products of chromosome replication), and deduced from this analysis that the structure which replicates semi-conservatively consists of two sub-units which are not identical, in that only like sub-units will rejoin with each other. Thus, with regard to replication and this type of recombination, the chromosome behaves as if it contained a single double-helix of DNA. Unfortunately, as will be discussed, additional information from the same type of experiment suggests that this interpretation may be too simple.

Quite different experiments also suggest that the chromosome may be single stranded. Evidence has been presented by Sparrow & Evans (1961) that there is a direct relationship between the DNA content per cell and the radiosensitivity of the organism. Plants with high DNA values are far more sensitive to radiation than those with low values. Since ionizing radiation is known to break DNA and chromosomes, the result can be simply explained on this basis. Clearly the longer the DNA in a chromosome, the more likely it is to be broken by a given dose of radiation. If organisms with large amounts of DNA have multi-stranded chromosomes, one might expect them to be more radiation-resistant, as a break in one of many strands would be unlikely to break the chromosome and might well be rapidly repaired.

Additional support for the hypothesis of longitudinal multiplicity comes from the study of meiosis in hybrids between *Allium cepa* and *A. fistulosum*, which have a 1·5-fold difference in DNA content (Jones & Rees, 1968). When the pairing pattern of homologous chromosomes of different size is examined at pachytene of meiosis, it can be seen that various regions in the longer chromosome are looped out and unpaired, which is what would be expected if certain regions of these chromosomes had additional material inserted longitudinally into their structure.

Lateral multiplicity of genetic information

Many cytologists have supported the idea that the chromosomes do not each contain a single copy of the genetic material, but a number of parallel strands each presumably containing the full complement of genetic information (for review, see Wolff, 1969). Much of the evidence for this view comes simply from observation with the light microscope of stained or specially-treated chromosomes. Frequently the products of chromosome replication have been seen to have a double structure; recent very clear examples have been provided by Vosa (1968), Trosko & Wolff (1965) and Martin (1968). Since the subchromatid structures appear to represent an exact longitudinal division of chromosomal material, it is almost impossible to explain them in terms of a particular pattern of folding of a single fibre. This is not true of various reports of multiple strands detected by electron microscopy (e.g. Wolfe & Hewitt, 1966) since it is possible that one long fibre could fold back on itself many times to give an apparently multistranded structure in certain regions of the chromosome.

The experiments of Beermann & Pelling (1965) have proved what had long been suspected, namely, that the giant chromosomes in Diptera are multistranded or polytene in structure. An original single strand replicates many times without nuclear division to produce a chromosome with many hundreds of lateral copies. Apart from salivary glands, such chromosomes are found in several other tissues of dipteran insects, as well as in protozoa (Alonso & Perez Silva, 1965) and in the suspensor cell in the ovule of higher plants (Nagl, 1969).

If certain cells can replicate their DNA without separating the products into separate cells, it is possible to imagine the evolution of a mechanism for replicating such a chromosome. Once this had been established in the germ line, there might in different cases be a selective advantage for an increase or a decrease in the number of strands. Thus drastic changes in DNA content might occur by fairly simple steps if the change involved simply a doubling or halving of the total or a proportion of the total number of strands. This would provide an explanation for the wide difference in DNA content between closely related species and also for the observation that these DNA values fall into an arithmetic or geometric series.

Evidence that the DNA content may be halved under particular conditions is provided by experiments with developing sea urchin eggs (Bibring, cited by Mazia, 1960). Treatment of the fertilized egg with mercaptoethanol prevents cleavage and DNA synthesis; removal of the block at the time the control embryos undergo their second cleavage, results in the formation of four nuclei, each with the full complement of chromosomes but with half the normal amount of DNA. It is almost impossible to understand how this might occur if each chromosome contained a single long DNA molecule.

The experiments of Taylor and his colleagues showed that the chromosome replicates semi-conservatively, but the conclusion that this type of replication represents the division of a single DNA structure has been challenged by La Cour & Pelc (1958), Peacock (1963, 1965) and, more recently, by Deaven & Stubblefield (1969) and Darlington & Haque (1969). By using the same autoradiographic techniques, they found that exceptions to semi-conservative replication were quite common. When a chromosome replicates once in the presence of a tritiated thymidine, both daughter chromosomes are labelled. If replication is semi-conservative, the next time the chromosome divides the label should segregate, to form labelled and unlabelled granddaughter chromosomes. However, it is frequently observed that the products of a labelled daughter are *both* labelled although, according to Taylor, the labelled daughter should have only one polynucleotide chain containing tritium. There is no plausible explanation for this *iso*labelling unless the chromosome consists of more than one double-helix of DNA. Indeed, the finding of Deaven & Stubblefield (1969) that *iso*labelling can be detected five or more chromosome divisions after labelling implies that many strands must be present. On this interpretation, the frequently observed semi-conservative replication of chromosomes is explained by supposing that the products of replication of each individual DNA molecule normally segregate to different daughter chromosomes, but that occasionally this rule is broken and both products of replication of a single DNA molecule are included in *one* daughter chromosome.

Taylor's (1958) interpretation of the observed pattern of sister strand exchanges has also been questioned. Heddle (1969) pointed out that the results obtained by different investigators are inconsistent, and that the observations as a whole are not incompatible with a multistranded chromosome structure.

A metabolic or structural function for DNA?

The problem of large amounts of DNA per genome might be explained if only a proportion of the DNA had a genetic function and the rest was redundant or played some other role in the cell. Metabolic DNA has now been shown to exist. In the amphibian oocyte, not only are there multiple copies of DNA coding for ribosomal RNA in the nucleolus, but there are also large numbers of additional nucleoli (each containing the same type of DNA) free in the nucleoplasm. This extrachromosomal DNA accounts for some 70 % of the total DNA in the cell (Perskowska, Macgregor & Birnstiel, 1968). Another possible example of metabolic DNA is the synthesis of the DNA in the puff regions of the giant chromosomes in the dipteran *Sciara* (Crouse & Keyl, 1967). However, the Boivin–Vendrely law of the constancy of DNA in the cells of various tissues strongly argues against this being a widespread phenomenon in differentiated cells; and since it has frequently been shown that all the DNA (as measured in the interphase nucleus) is accounted for by the total DNA content of the metaphase chromosomes, any large body of metabolic DNA would have to be part and parcel of the chromosomes and invariable in amount.

Detailed hybridization studies have shown that many eukaryotic organisms contain a fraction of DNA with repetitive base sequences (Britten & Kohne, 1968). Detailed studies have been made with rodent DNA by Flamm, Walker & McCallum (1969) and Walker, McLaren & Flamm (1969). The homogeneous rapidly-annealing fraction has a different density from the bulk of the nuclear DNA, which is heterogeneous and anneals extremely slowly after denaturation. The separate strands of the homogeneous fraction can be separated, as their density is different, and this makes it possible to show that they contain some degree of internal homology. The satellite DNA is probably not transcribed since no RNA capable of hybridizing with it has been detected. A further peculiarity is the fact that the satellite fraction from various related rodent species show no obvious cross-species homology. This suggests that this fraction may be 'non-genetic' and not subject to the usual rules governing the evolution of proteins. Walker *et al.* (1969) suggested that this DNA—about 10% of the total—may have a 'housekeeping' role to play in the structure, organization or behaviour of chromosomes. The repeated base sequences would allow hydrogen bonding between different parts of the chromosomes and may provide a basis for their three-dimensional structure, for instance in the folding of certain regions into chromomeres. Less likely, it might be necessary for the general coiling of metaphase chromosomes. Holliday (1968) suggested that a particular sequence of bases scattered along the length of chromosomes might be a substrate for a fibrillar pairing protein which is responsible for bringing homologues into close apposition at prophase of meiosis, and which forms part of the synaptinemal complex. It was also suggested that other base sequences might be the specific substrate for an endonuclease which carries out the first step in genetic recombination. Gierer (1966) suggested that regions of internal homology in DNA would allow the formation of hydrogen-bonded loops (analogous to those in *t*RNA). He suggested that the formation of such loops from linear DNA might be a mechanism for the control of transcription, in other words a mechanism for shutting off a gene's activity. This simple and ingenious idea does not appear to have received the attention it deserves.

In summary, there appear to be several possible functions for DNA which does not code for messenger RNA and protein synthesis. Direct information about such functions is negligible and one can only guess at the total amount of DNA which may be involved. It would be very profitable to extend the studies of Britten & Kohne (1968) and Walker *et al.* (1969) to a group of related organisms which have large differences in cellular DNA content, to see whether the rapidly-annealing component was a constant fraction of the total DNA or whether this fraction varied enormously from species to species.

The problem of gene mutation

Genetic experiments with many species over several decades have proved that the gene is a stable structure capable of replication without alteration. Mutations are occasional random changes in the structure which are transmitted to all replicas of the altered gene. There are numerous examples of gene mutations being induced with single-hit kinetics by radiation or chemical treatment. If each gene consisted of multiple copies in a cell, to obtain a mutation all copies would have to be altered and in many cases the particular alteration would have to be identical in all the copies. With multiple genes different mutations would gradually accumulate in the various copies and it is impossible to see how evolution could occur, or how genetics could exist as a science, since the phenotypes of mutations in a family of genes would be so ill-defined as to be unrecognizable in breeding experiments. (In passing, it is interesting to note that Fincham (1967) suggested that the properties of certain unusual unstable genes in higher plants, which often have a graded series of phenotypes, might be explained on the basis of Callan's 'master and slave' hypothesis.)

The traditional view of the geneticist that the gene consists of a single template has, however, already been undermined, since each gene copy consists of a short stretch of DNA which, of course, contains *two* complementary copies of the genetic information. It would be expected that the chemical change producing a mutation would occur in one strand, and therefore of the two daughter cells only one should carry the mutation. Although this does frequently happen, there are now numerous examples of mutation being transferred to both daughter cells; in other words, in these cases a mutation must occur in both strands of the DNA. This has been clearly shown in *Escherichia coli* (see Bridges & Munson, 1968) and among eukaryotes evidence comes from *Drosophila* (e.g. Muller, Carlson & Schalet, 1961; Altenburg & Browning, 1961; Inagaki & Nakao, 1966), *Ustilago* (Holliday, 1962), *Schizosaccharomyces* (Nasim & Clarke, 1965; Nasim & Auerbach, 1967),

Saccharomyces (Ito, Yamasaki & Matsudaira, 1962; Resnick, 1969) and *Paramecium* (Kimball, 1964). The most direct demonstration comes from the extensive pedigree analyses made with *Schizosaccharomyces* by Haefner (1967). He irradiated haploid cells with ultraviolet (u.v.) radiation and then isolated their descendants by micromanipulation. He found several instances of an induced mutation being present in *both* viable daughter cells. In discussing the mechanisms of whole-gene mutation after u.v. treatment, Holliday (1962) suggested that the repair mechanism might be responsible for both strands being altered. U.v.-irradiation produces intrastrand pyrimidine dimers which are largely eliminated by an excision repair mechanism. The explanation suggested for mutation was that occasionally, by misrepair, rather than the dimer being excised, the bases opposite the dimer were removed. When the single strand gap was subsequently filled by repair synthesis on the intact strand, the distortion in the template caused by the dimer would lead to the insertion of incorrect bases. As a consequence of this mis-repair two altered strands would be produced from a single u.v. hit. It was also suggested that any mismatched or non-complementary base pairs in heteroduplex DNA might be corrected by a repair mechanism, since if this were so, the phenomenon of gene conversion during genetic recombination could be neatly explained (Holliday, 1962; Whitehouse, 1963).

The significance of these speculations in a discussion of chromosome structure is that they may relate to the problem of mutation in chromosomes which contain multiple copies of all or some of the genes. In Callan's model of chromosome organization the slave copies are responsible for transcription, whereas the master is responsible for genetic homogeneity and continuity. Once in every generation the slave genes are matched in turn against the master, the matching process involving the formation of single strands and the successive annealing of slave strands against the master. It is then necessary to invoke a special unidirectional repair mechanism which will result always in the removal of the mismatched base in the slave strand. The molecular manipulations which would be necessary to match a hundred or more slave copies against the master would of course be of formidable complexity.

By preserving the idea that the chromosome contains one enormously long strand of DNA, Callan attempts to avoid the difficulties which any multistranded model introduces. However, to explain mutation he invokes a mechanism which could readily be used to explain mutation in a multistranded chromosome also. One of the parallel strands would be a master, and the others matched against it once during the life cycle, presumably during or after meiosis.

Another possibility, previously discussed (Holliday, 1962) is that there may be a tendency for DNA molecules to unravel and re-anneal with homologous sister strands. For a chromosome consisting of four parallel DNA molecules there would have to be two rounds of un-ravelling and annealing: the idea being that a mutation in one molecule would either be eliminated, or spread to all the molecules, by the action

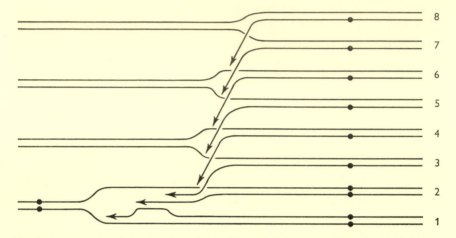

Fig. 2. A scheme for the replication of a multistranded chromosome from a single master template, based on Jehle's (1965) model of DNA replication. The arrow indicates the growing point of each new poly-nucleotide chain. A mutation in the master molecule, as on the left, is transmitted to all replicas, as on the right. All the strands may become mutant if repair of mis-matched bases occurs (see text). Replication is semi-conservative if even-numbered strands segregate from odd-numbered ones. Exceptions to this will account for occasional replication which is not semi-conservative (see text).

of the repair mechanism which recognizes and corrects mismatched bases. As in the lampbrush model, this process would have to occur at least once during the life cycle, and also, the correction of mismatched bases cannot be arbitrary but must follow definite rules.

Only one attempt appears to have been made to find a relation-ship between DNA value and mutation rate. Using five plants with different amounts of DNA per nucleus, Sparrow, Baetcke, Shaver & Pond (1968) measured somatic mutation rates after irradiation (dominant petal colour to recessive). They found that the higher the DNA value the greater the mutation rate. This surprising result argues against any master strand hypothesis, and perhaps suggests that a mutation in any one of a number of copies of the gene can lead to loss of function of all of them. Unfortunately the system used does not exclude the possibility that the 'mutations' are chromosome deletions or mitotic cross-overs.

Although it is generally assumed that both strands of a DNA duplex transmit genetic information, this view has not gone unchallenged. Several experiments with bacteria and phages suggests that only one of the two templates transmits information to both newly-synthesized strands (e.g. Baricelli & Del Zuppo, 1968; Kubitschek, 1964; and see Jehle & Parke, 1967). Stahl (1969) has argued the case for Jehle's (1964) model for replication of DNA in which only one parent acts as template for the synthesis of a new strand which is in turn template for the fourth. The two new strands anneal with the old and semi-conservative replication is preserved. This scheme can be extended to a polyneme chromosome in which one DNA strand acts as master template; this is shown in Fig. 2. A mutation in the master would be transmitted to all daughter molecules. Again, it is necessary to invoke a unidirectional correction mechanism to make all the strands uniform for base sequence.

Genetic recombination

Recombination is a precise process involving the breakage and reunion of DNA molecules at homologous points. If recombination were not exact then deletion or duplication of genetic material would occur in the recombinants. One simple demonstration that this cannot be the case is the fact that 50 or so recombination events in the yeast zygote (see Mortimer & Hawthorne, 1969) do not result in the production of a high proportion of non-viable products of meiosis. Single deletions in haploid cells would in many cases be lethal, yet the observed non-viability is of the order of 10%: this means that less than 0·5% of the exchanges could be unequal. The precision of recombination analysis was pointed out some years ago by Pontecorvo & Roper (1956). Although most of the fine structure mapping in eukaryotes has been done with fungi, all the indications are that recombination at the fine structure level also occurs in more complex organisms (see Pontecorvo, 1958; Chovnick, 1966).

If the chromosomes contain a single DNA molecule with a linear array of single genes, it is possible to devise models which account for most, if not all, of the extensive fungal recombination data (reviews: Holliday, 1968; Whitehouse, 1969). However, as in the case of mutation, complications arise if there are multiple copies of genetic material. In the case of longitudinal multiplicity, the primary difficulty lies in the fact that for any one gene represented many times, there would be multiple pairing segments or regions of homology. Recombination would frequently generate unequal numbers of gene copies. The situation would essentially be an unstable one, with all the individuals in the population ultimately having different numbers of slave gene copies. Another difficulty, pointed out by Whitehouse (1967), is that genes of different function can be adjacent to each other on the recombination maps, which should not be observed if genes are serially repeated. To explain this Whitehouse suggested that a special mechanism restricts recombination to the master gene. His model cannot be elaborated here: it demands several assumptions, the validity of which the reader should perhaps judge for himself.

At first sight it seems impossible to account for the precision of recombination if the chromosomes have a polyneme structure. If the breaks did not occur at precisely the same points in all the molecules then the recombinant molecules would not be identical—some would contain duplications and some deletions. However, this difficulty is not a real one (see Holliday, 1968): the use of simple wire models makes it possible to see that recombination of polyneme chromosomes could occur without heterogeneity in structure, or entangling, of the recombinant strands. In meiotic prophase a special mechanism ensures the close pairing of homologous chromosomes. The visible manifestation of this is the synaptinemal complex (see Moses, 1968), but the mechanism of the process is unknown. The first step in recombination could be initiated at a short specific base sequence (the recombinator sequence— scattered at irregular intervals along the length of the chromosome) which serves as substrate for the action of a specific endonuclease. This enzyme produces cuts in all the molecules at the same point and in strands of the same polarity. After this, the steps in recombination are essentially the same as in the model outlined elsewhere (Holliday, 1964, 1968); only one other point will be mentioned here. To obtain a complete cross-over obviously all the strands have to be broken and rejoined. Although the first single-strand breaks must be at a defined point, the breaks in the strands of the other polarity, after the formation of heteroduplex or hybrid DNA regions, need not be at a defined point or between corresponding base pairs in all strands. The hydrogen bonding between complementary base pairs ensures a continuity in recombinant chromosome structure and single-strand gaps or overlaps could be filled or trimmed by a repair mechanism.

Recombination in polyneme chromosomes appears to present fewer problems than does mutation. It must be realized that most of the detailed information about recombination in eukaryotes comes from studies with fungi, and there is as yet no reason to believe that these organisms have chromosomes of this structural complexity. There is a real need for information about recombination at the fine structure level in an organism with a high DNA content.

Conclusions

The genetic evidence that a particular gene is represented only once in a chromosome has frequently been used as an argument against the idea that the chromosome has a complex organization in which gene copies are represented many times. If, however, *special mechanisms* exist which make multiple copies of a gene behave as a single unit then the geneticists' argument breaks down. It is important to realize that these special mechanisms have to be invoked whether the multiplicity of genetic information is longitudinal *or* lateral.

The evidence for both longitudinal and lateral multiplicity seems very strong. Clearly the fact that the nucleolar organizer region of the chromosome region contains numerous serially repeated copies of the genes coding for ribosomal RNA, means that there must be a mechanism for the evolution and maintenance of such a region. And if one region exists, why not many others in the genome? It may be that *Triturus* with its very high DNA content per genome has organized most of its DNA in this way. The hybridization studies with rodent DNA have provided strong evidence that repetitive base sequences do not occur in most of this DNA. It would be very interesting to apply the same techniques to *Triturus* or its relatives to see whether this proportion was very much smaller.

Much of the evidence for polyneme chromosomes comes from studies of higher plants, and it may well be that this group of organisms has evolved a special mechanism for the replication and maintenance of such chromosomes. Certainly, once such a mechanism had evolved it is relatively easy to imagine that the number of strands in a chromosome could change by rather simple evolutionary steps, and this would account for the enormous differences in DNA content between related species much more easily than would numerous changes in longitudinal multiplicity.

It may therefore be a mistake to attempt to interpret the available data in terms of a single model of chromosome organization. It seems easier to imagine that a selective advantage in having many copies of certain genes has resulted in the evolution of different means of achieving this end, together with special mechanisms for ensuring the stability and possibility of further evolution of such complex chromosomes.

The nature of these special mechanisms poses a challenging problem for geneticists in the future, and one that is perhaps not likely to be solved by concentrating on microbial systems.

It is quite evident that at the moment there are more questions to be asked about the organization of DNA in eukaryotic chromosomes than there are answers available. Enough is now known to show that the answers will not be obtained by electron microscopy, nucleic acid chemistry or genetics alone, but only by a combination of all three approaches. Moreover, there is a real need to select certain suitable species from different groups of organisms and to concentrate research on these. As we have seen, information from diverse eukaryotic species can be very conflicting and this frequently leads to confusion and disagreements in this very difficult field.

REFERENCES

ALONSO, P. & PEREZ SILVA, J. (1965). Giant chromosomes in protozoa. *Nature, Lond.* **205**, 313.

ALTENBURG, E. & BROWNING, L. S. (1961). The relatively high frequency of whole body mutations compared with fractionals induced by X-rays in Drosophila sperm. *Genetics, Princeton* **46**, 203.

BARICELLI, N. A. & DEL ZUPPO, G. (1968). Genotypic reversion by methylene blue: the orientation of guanine hydroxymethyl cytosine at mutated sites in r11 mutants of phage T 4. *Molec. gen. Genetics* **101**, 51.

BEERMANN, W. & PELLING, C. (1965). H³ Thymidinmarkierung einzelner Chromatiden in Riesen chromosomen. *Chromosoma* **16**, 1.

BRIDGES, B. A. & MUNSON, R. J. (1968). Genetic Radiation Damage and Its Repair in *Escherichia coli*. In *Current Topics in Radiation Research*, vol. 4, p. 95. Eds. M. Ebert and A. Howard. Amsterdam: North Holland.

BRITTEN, R. J. & KOHNE, D. E. (1968). Repeated sequences in DNA. *Science, N.Y.* **161**, 529.

BROWN, D. D. & WEBER, C. S. (1968). Gene linkage by RNA-DNA hybridization. II. Arrangement of the redundant gene sequences for 28 s and 18 s ribosomal RNA. *J. molec. Biol.* **34**, 681.

CAIRNS, J. (1963). The chromosome of *Escherichia coli*. *Cold Spring Harb. Symp. quant. Biol.* **28**, 43.

CALLAN, H. G. (1967). The organization of genetic units in chromosomes. *J. Cell Sci.* **2**, 1.

CALLAN, H. G. & LLOYD, L. (1960). Lampbrush chromosomes of crested newts *Triturus cristatus* (Laurenti). *Phil. Trans. Roy. Soc. Lond.* B **243**, 135.

CASE, M. E. & GILES, N. H. (1958). Recombination mechanisms at the *pan-2 locus* in *Neurospora crassa*. *Cold Spring Harb. Symp. quant. Biol.* **23**, 119.

CHOVNICK, A. (1966). Genetic organization in higher organisms. *Proc. Roy. Soc. Lond.* B **164**, 198.

CROUSE, H. V. & KEYL, H. G. (1967). Extra replications in the 'DNA puffs' of *Sciara coprophila*. *Chromosoma* **25**, 357.

DARLINGTON, C. D. & HAQUE, A. (1969). The replication and division of polynemic chromosomes. *Heredity, Lond.* **24**, 273.

DEAVEN, L. L. & STUBBLEFIELD, E. (1969). Segregation of chromosomal DNA in Chinese hamster fibroblasts *in vitro*. *Expl Cell Res.* **55**, 132.

FINCHAM, J. R. S. (1967). Mutable genes in the light of Callan's hypothesis of serially repeated gene copies. *Nature, Lond.* **215**, 864.

FINCHAM, J. R. S. & DAY, P. R. (1965). *Fungal Genetics*. Oxford: Blackwell.

FLAMM, W. G., WALKER, P. M. B. & McCALLUM, M. (1969). Some properties of the single strands isolated from the DNA of the nuclear satellite of the mouse *Mus musculus*. *J. molec. Biol.* **40**, 423.

GALL, J. G. (1963). Kinetics of deoxyribonuclease action on chromosomes. *Nature, Lond.* **198**, 36.

GALL, J. G. & CALLAN, H. G. (1962). [H³]uridine incorporation in lampbrush chromosomes. *Proc. natn. Acad. Sci. U.S.A.* **48**, 562.

GIERER, A. (1966). Model for DNA and protein interactions and the function of the operator. *Nature, Lond.* **212**, 1480.

HAEFNER, K. (1967). Concerning the mechanism of ultraviolet mutagenesis. A micromanipulatory pedigree analysis in *Schizosaccharomyces pombe*. *Genetics, Princeton* **57**, 169.

HEDDLE, J. A. (1969). Influence of false twins on the ratios of twin and single sister chromatid exchanges. *J. theor. Biol.* **22**, 151.

HOLLIDAY, R. (1962). Mutation and replication in *Ustilago maydis*. *Genet. Res.* **3**, 472.

HOLLIDAY, R. (1964). A mechanism for gene conversion in fungi. *Genet. Res.* **5**, 282.

HOLLIDAY, R. (1968). Genetic Recombination in Fungi. In *Replication and Recombination of Genetic Material*, p. 157. Eds. W. J. Peacock and R. D. Brock. Canberra: Australian Academy of Science.

HOTTA, Y. & BESSEL, A. (1965). Molecular size and circularity of DNA in cells of mammals and higher plants. *Proc. natn. Acad. Sci. U.S.A.* **53**, 356.

HUGHES-SCHRADER, S. & SCHRADER, F. (1956). Polyteny as a factor in the chromosomal evolution of the Penlatomini. *Chromosoma* **8**, 135.

INAGAKI, E. & NAKAO, Y. (1966). Comparison of frequency patterns between whole body and fractional mutations induced by X-rays in *Drosophila melanogaster*. *Mut. Res.* **3**, 268.

ITO, T., YAMASAKI, T. & MATSUDAIRA, Y. (1962). Studies on the genetic multiplicity of a gene in yeast cells. I. Characteristics of the induced mutations by subcritical temperature and ultraviolet light. *Jap. J. Genet.* **37**, 276.

JEHLE, H. (1965). Replication of double-strand nucleic acids. *Proc. natn. Acad. Sci. U.S.A.* **53**, 1451.

JEHLE, H. & PARKE, W. C. (1967). Nucleic Acid Replication and Transcription. In *Structural Chemistry and Molecular Biology*, p. 399. Eds. A. Rich and N. Davidson. San Francisco: W. H. Freeman.

JONES, R. N. & REES, H. (1968). Nuclear DNA variation in Allium. *Heredity, Lond.* **23**, 591.

KEYL, H.-G. (1965). Duplikationen von Untereinheiten der Chromosomalen DNS während der Evolution von *Chironomus thummi*. *Chromosoma* **17**, 139.

KIMBALL, R. F. (1964). The distribution of X-ray induced mutations to chromosomal strands in *Paramecium aurelia*. *Mut. Res.* **1**, 129.

KLEINSCHMIDT, A. K. (1967). Structural Aspects of the Genetic Apparatus of Viruses and Cells. In *Molecular Genetics*, pt. 2, p. 47. Ed. J. H. Taylor. London and New York: Academic Press.

KUBITSCHEK, H. E. (1964). Mutation without segregation. *Proc. natn. Acad. Sci. U.S.A.* **52**, 1374.

KURNICK, N. B. & HERSKOWITZ, I. H. (1952). The estimation of polyteny in Drosophila salivary gland nuclei based on determination of desoxyribonucleic acid content. *J. cell. comp. Physiol.* **39**, 281.

LA COUR, L. F. & PELC, S. R. (1958). Effect of colchicine on the utilization of labelled thymidine during chromosomal reproduction. *Nature, Lond.* **182**, 506.

MCLEISH, J. & SUNDERLAND, N. (1961). Measurement of deoxyribosenucleic acid (DNA) in higher plants by Feulgen photometry and chemical methods. *Expl Cell Res.* **24**, 527.

MARTIN, P. G. (1968). Differences in Chromosome Size Between Related Plant Species. In *Replication and Recombination of Genetic Material*, p. 93. Eds. W. J. Peacock and R. D. Brock. Canberra: Australian Academy of Science.

MAZIA, D. (1960). The analysis of cell reproduction. *Ann. N.Y. Acad. Sci.* **90**, 455.

MILLER, O. L. (1965). Fine structure of lampbrush chromosomes. *Natn. Cancer Inst. Monogr.* **18**, 79.

MILLER, O. L. & BEATTY, B. R. (1969). Visualization of nucleolar genes. *Science, N.Y.* **164**, 955.

MIRSKY, A. E. & RIS, H. (1951). The desoxyribonucleic acid content of animal cells and its evolutionary significance. *J. gen. Physiol.* **34**, 451.

MORTIMER, R. K. & HAWTHORNE, D. C. (1969). Yeast Genetics. In *The Biology of Yeasts*. Eds. A. H. Rose and J. S. Harrison. London and New York: Academic Press.

MOSES, J. M. (1968). Synaptinemal complex. *Am. Rev. Genet.* **2**, 363.

MULLER, H. J., CARLSON, E. A. & SCHALET, A. (1961). Mutation by alteration of the already existing gene. *Genetics, Princeton* **46**, 213.

MURRAY, N. E. (1963). Polarized recombination and fine structure within the *me*-2 gene of *Neurospora crassa*. *Genetics, Princeton* **48**, 1163.

NAGL, W. (1969). Banded polytene chromosomes in the legume *Phaseolus vulgaris*. *Nature, Lond.* **221**, 70.

NASIM, A. & AUERBACH, C. (1967). The origin of complete and mosaic mutants from mutagenic treatment of single cells. *Mut. Res.* **4**, 1.

NASIM, A. & CLARKE, C. H. (1965). Nitrous acid induced mosaicism in *Schizosaccharomyces pombe*. *Mt. Res.* **2**, 395.

NELSON, O. E. (1962). The waxy locus in maize. I. Intra locus recombination frequency estimates by pollen and by conventional analysis. *Genetics, Princeton* **47**, 737.

OHNUKI, Y. (1968). Structure of chromosomes. I. Morphological studies of the spiral structure of human chromosomes. *Chromosoma* **25**, 402.

PEACOCK, W. J. (1963). Chromosome duplication and structure as determined by autoradiography. *Proc. natn. Acad. Sci. U.S.A.* **49**, 793.

PEACOCK, W. J. (1965). Chromosome replication. *Natn. Cancer Inst. Monogr.* **18**, 101.

PERSKOWSKA, E., MACGREGOR, H. C. & BIRNSTIEL, M. L. (1968). Gene amplification in the oocyte nucleus of mutant and wild type *Xenopus laevis*. *Nature, Lond.* **217**, 649.

PONTECORVO, G. (1958). *Trends in Genetic Analysis*. London: Oxford University Press.

PONTECORVO, G. & ROPER, J. A. (1956). The resolving power of genetic analysis. *Nature, Lond.* **178**, 83.

PRITCHARD, R. H. (1955). The linear arrangement of a series of alleles of *Aspergillus nidulans*. *Heredity, Lond.* **9**, 343.

REES, H., CAMERON, F. M., HAZARIKA, M. H. & JONES, G. H. (1966). Nuclear variation between diploid Angiosperms. *Nature, Lond.* **211**, 828.

RESNICK, M. A. (1969). Induction of mutations in *Saccharomyces cerevisiae* by ultraviolet light. *Mut. Res.* **7**, 315.

RITOSSA, F. M., ATWOOD, R. C. & SPIEGELMAN, S. (1966). A molecular explanation of the bobbed mutants of Drosophila as partial deficiencies of 'ribosomal' DNA. *Genetics, Princeton* **54**, 819.

ROTHFELS, K., SEXSMITH, E., HEIMBURGER, M. & KRAUSE, M. O. (1966). Chromosome size and DNA content of species of anemone L and related genera (Ranunculaceae). *Chromosoma* **20**, 54.

SCHRADER, F. & HUGHES-SCHRADER, S. (1956). Polyploidy and fragmentation in the chromosomal evolution of various species of *Thyanta* (Hemiptera). *Chromosoma* **7**, 469.

SIDDIQI, O. H. (1962). The fine genetic structure of the *paba*-1 region of *Aspergillus nidulans*. *Genet. Res.* **3**, 69.

SOLARI, A. J. (1967). Electron microscopy of native DNA in sea urchin cells. *J. Ultrastruct. Res.* **17**, 421.

SPARROW, A. H., BAETCKE, K. P., SHAVER, D. L. & POND, V. (1968). The relationship of mutation rate per roentgen to DNA content per chromosome and to interphase chromosome volume. *Genetics, Princeton* **59**, 65.

SPARROW, A. H. & EVANS, H. J. (1961). Nuclear factors affecting radio-sensitivity. I. The influence of nuclear size and structure, chromosome complement, and DNA content. *Brookhaven Symp. Biol.* **14**, 76.

STAHL, F. W. (1969). *The Mechanics of Inheritance*, 2nd edn. New Jersey: Prentice Hall.

TAYLOR, J. H. (1958). Sister chromatid exchanges in tritium-labeled chromosomes. *Genetics, Princeton* **43**, 515.

TAYLOR, J. H., WOODS, P. S. & HUGHES, W. L. (1957). The organization and duplication of chromosomes as revealed by autoradiographic studies using tritium labeled thymidine. *Proc. natn. Acad. Sci. U.S.A.* **43**, 122.

TROSKO, J. E. & WOLFF, S. (1965). Strandedness of *Vicia faba* chromosomes as revealed by enzyme digestion studies. *J. Cell Biol.* **26**, 125.

VENDRELY, R. (1955). The Deoxyribonucleic Acid Content of the Nucleus. In *The Nucleic Acids*, vol. 11, p. 155. Eds. E. Chargaff and J. N. Davidson. London and New York: Academic Press.

VIALLI, M. (1957). Volume et contenu en ADN par noyau. *Expl Cell Res.* (Suppl.) **4**, 284.

VOSA, C. G. (1968). A method to reveal subchromatids in somatic chromosomes. *Caryologia* **21**, 381.

WALKER, P. M. B., McLAREN, A. & FLAMM, W. G. (1969). The problem of highly repetitive DNA in higher organisms. In *Molecular Cytology*. Ed. Lima de Faria. Amsterdam: North Holland.

WALLACE, H. & BIRNSTIEL, M. L. (1966). Ribosomal cistrons and the nucleolar organizer. *Biochem. biophys. Acta* **114**, 296.

WHITEHOUSE, H. L. K. (1963). A theory of crossing over by means of hybrid deoxyribonucleic acid. *Nature, Lond.* **199**, 1034.

WHITEHOUSE, H. L. K. (1967). Cycloid model for the chromosome. *J. Cell Sci.* **2**, 9.

WHITEHOUSE, H. L. K. (1969). *Towards an Understanding of Heredity*, 2nd edn. London: Arnold.

WHITEHOUSE, K. L. H. & HASTINGS, P. J. (1965). The analysis of genetic recombination on the polaron hybrid DNA model. *Genet. Res.* **6**, 27.

WILLIAMSON, D. H. (1965). The timing of deoxyribonucleic acid synthesis in the cell cycle of *Saccharomyces cereviseae*. *J. Cell Biol.* **25**, 517.

WILLIAMSON, D. H., MOUSTACCHI, E. & FENNELL, D. (1969). In preparation.

WOLFE, S. L. & HEWITT, G. M. (1966). The strandedness of meiotic chromosomes from *Oncopeltus*. *J. Cell Biol.* **31**, 31.

WOLFF, S. (1969). Strandedness of chromosomes. *Intn. Rev. Cytol.*, vol. 25, p. 279. Eds. G. C. Bourne and J. F. Danielli. London and New York: Academic Press.

WOLSTENHOLME, D. R., DAWID, I. B. & RISTOW, H. (1968). An electron microscope study of DNA molecules from *Chironomus tentans* and *Chironomus thummi*. *Genetics, Princeton* **60**, 759.

REPRODUCTION OF MITOCHONDRIA AND CHLOROPLASTS

D. WILKIE

Department of Botany and Microbiology, University College, London, England

For many years the most controversial aspect about mitochondria and chloroplasts has been their auto-reproductive capacity. Previous claims, based mainly on microscopic and genetic evidence, that the organelles were self-perpetuating have been put on more solid foundations in recent years by the discovery that intrinsic DNA occurs in these structures and that mitochondria and plastids incorporate amino acids into their protein *in vitro* by means of an intrinsic protein-synthesizing system which comprises ribosomes, transfer ribonucleic acids (*t*RNA) and appropriate enzymes. The extent of the genetic information encoded in organelle DNA presumably defines the extent of the self-reproducing capacity of these units; but it has yet to be shown that any part of organelle DNA codes for a definitive protein such as an enzyme. This could only be done by showing the correlation between a point-mutation in the DNA and an alteration in a specific protein. Much information has now been accumulated about the physical and biochemical nature of organelle DNA and of the protein-synthesizing mechanism which, together with the analysis of cytoplasmically inherited mutations affecting the system, allow some conclusions to be drawn about organelle autonomy. At the same time this should define the extent to which organelle biogenesis is under the control of the cell, that is, on the degree of dependence on genetic information carried in the nucleus.

MITOCHONDRIA

Mitochondria are of primary importance among the membrane-bound structures in the cytoplasm of eukaryotes. Apart from a few facultative anaerobic eukaryotes in which the organelle can be dispensed with in certain circumstances, mitochondrial function is an essential feature of energy metabolism. The same basic structure of a smooth outer membrane surrounding an inner membrane with many infoldings or cristae is seen in all organisms, but the extent of pleiomorphic changes seen by phase-contrast microscopy in living cells indicates considerable flexibility of the system. What follows about the biochemical features to be discussed is not intended to be an exhaustive review of the subject. General findings will be presented; for details of technical procedures and literature the reader is referred to the reviews of Borst & Kroon (1969), Kroon (1969), Rabinowitz (1968a) and Roodyn & Wilkie (1969).

Deoxyribonucleic acids (DNA)

Mitochondrial DNA (MDNA) has been isolated from many organisms extending from fungi to man. Electron microscopy of MDNA of animals including insects, gasteropods, amphibia, birds and mammals reveals a homogeneous population of twisted circular molecules consisting of two covalently-closed strands of average contour length varying from 4·45 μm for sea urchin eggs to 5·6 μm for guinea-pig liver cells. Complexes of circles may arise by the joining of individual molecules in various ways. The contour lengths of the basic units of animal MDNA are all of the order of about 5 μm, but it is important to establish the validity of the reported differences in contour lengths before attempting to assess the amount of genetic information carried in each case. Wolstenholme & Dawid (1968) made a detailed study of the MDNA of urodele and anuran Amphibia and concluded that the greater contour length of anuran MDNA (5·66 μm. as against 5·09 μm) was real, and that in preparations containing mixtures of the two types each could be distinguished clearly in the electron microscope. Also, the difference in length correlated with a difference in buoyant density in CsCl gradient centrifugation, urodele MDNA being 1·695 and anuran 1·702 g./cm.[3]. It was concluded these differences reflected evolutionary diversification.

Preparations of MDNA of higher plants have revealed linear molecules ranging up to 62 μm (Wolstenholme & Gross, 1968) corresponding to a molecular weight of 10^8 daltons. It is possible that these molecules are comprised of repeating sequences of a smaller basic unit.

In electron microscope analysis of MDNA of the yeast *Saccharomyces cerevisiae* linear molecules up to 20 μm. and circles with contour lengths ranging from 0·5 μm. to 10 μm. have been seen (Avers, 1967; Shapiro, Grossman, Marmur & Kleinschmidt, 1968). Borst and associates reported that MDNA of *S. carlsbergensis* appeared as heterogeneous linear duplex molecules of various lengths up to 18 μm., but from a more detailed study (Hollenberg, Borst, Thuring & Van Bruggen, 1969) concluded that the heterogeneity seen was due to the MDNA having been degraded during its isolation and that the undegraded molecules consisted of 26 μm circles. Furthermore, in re-naturation studies in which the MDNA re-annealed with second order kinetics without evidence of heterogeneity, they concluded that the content of genetic information in the yeast MDNA was equivalent to its molecular size. This conclusion had already been reached for chick liver MDNA (Borst, Ruttenberg & Kroon, 1967).

The homogeneity of MDNA from various organisms indicates that the mitochondrial genome is represented by a duplex monomer, circular in most if not all species. Differences in contour lengths of about 5 μm in animals and several times this value in yeasts and probably in higher plants, have been interpreted in terms of evolutionary processes, with the mitochondria of more primitive organisms retaining more of their genetic information, and more evolved types showing greater control by the nucleus. In the case of a 5 μm molecule, which, as suggested by Borst, may be the minimum size for mitochondrial function, it can be calculated that the information is encoded in some 15,000 base pairs which could specify about 30 polypeptides each of molecular weight of 20,000. The yeast mitochondrial genome could specify about five times this number of proteins, but this is still short of the amount required to make a mitochondrion complete with ribosomes (see Roodyn & Wilkie, 1968).

MDNA re-natures rapidly to give a product which bands at native density in CsCl gradients. Dawid & Wolstenholme (1968) studied the hybridization of related and unrelated MDNAs. The formation of common high molecular weight complexes during the joint re-annealing of mixtures of DNAs gives a measure of the extent of base-sequence homology. The validity of such tests for homology can be seen from the work of Britten & Waring (1965) in which unrelated DNAs from nuclei of different organisms were shown to form separate complexes during re-annealing. Dawid & Wolstenholme (1968) detected no homologies between nuclear and mitochondrial DNA either in the case of *Xenopus laevis* or of *Rana pipiens*. Homologies were found between the nuclear DNAs of *X. laevis* and mouse, and between the MDNAs of *X. laevis* and chick, but none between the MDNAs of *X. laevis* and yeast. The precise nature of the homologous regions are unknown. In the case of nuclear DNA + mitochondrial DNA hybridization, in which no re-annealing has been detected, it has been pointed out by Borst *et al.* (1967) that even though there were homologous sequences in the nuclear DNA these would comprise a small percentage of the total sites available and would not be detectable in the banding.

Experiments by Parsons & Simpson (1967) with isolated mitochondria from rat liver and by Wintersberger (1968) with yeast mitochondria have shown that the mitochondria can incorporate labelled deoxyribo-nucleoside triphosphates into their DNA by a system which involves MDNA polymerase; the DNA synthesized had the same buoyant density as the parent molecule. Reich & Luck (1966), who used [^{15}N]-labelled DNA precursors, concluded that MDNA replication in *Neuro-spora* is semi-conservative. These findings show that mitochondria have an intrinsic system for reduplicating their DNA, but on the other hand may be dependent on nuclear information for the specification of the replication enzymes.

Ribonucleic acids (RNA)

RNA in mitochondria is present in ribosomes, *t*RNA and presumably messenger RNA (*m*RNA). Although contamination with cytoplasmic RNA is a problem it is possible to distinguish the RNA components of mitochondria from those of the cytoplasm on the basis of sedimentation properties. RNA molecules with sedimentation coefficients of 23s and 16s have been identified from mitochondria of *Neurospora* sp., yeast and mammalian cells and are considered to be ribosomal. A 4s species of RNA from yeast mitochondria was shown by Wintersberger (1966) to accept amino acids in a manner characteristic of cytoplasmic *t*RNA. More detailed analyses of mitochondrial ribosomes of *Neurospora* sp. have been made by Küntzel & Noll (1967) and Küntzel (1969). A comparison with the cytoplasmic ribosomes of the *Neurospora* sp. showed s values of 77s for the cytoplasmic units and 73s for the mitochondrial units. Furthermore, the mitochondrial ribosomes dissociated at 0·1 mM-magnesium and the cytoplasmic ribosomes at 0·001 mM-magnesium. Ribosomal sub-units of the two systems also differed in sedimentation properties, with 50s and 37s for mitochondria, and 60s and 37s for cytoplasm. A difference in base composition between the respective RNAs of the two ribosomal types was also detected (see also Rifkin, Wood & Luck, 1967). The evidence from *Neurospora* sp. justifies the more general conclusion that mitochondria contain ribosomes smaller than those of the cytoplasm and that their RNA components are like those of bacteria in sedimentation properties.

Isolated mitochondria can incorporate nucleotides into RNA and possess a DNA-dependent RNA polymerase (Wintersberger, 1966). This finding also provides evidence that MDNA functions genetically and that transcription takes place in the organelle. More direct evidence of transcription is available from the DNA + RNA hybridization studies of Fukuhara (1967) and Wintersberger & Wiehhauser (1968) with a yeast, and of Suyama (1967*a*, *b*) with a *Tetrahymena* sp. The evidence shows that mitochondrial RNA (MRNA) hybridized with MDNA, whereas cytoplasmic ribosomal RNA hybridized only with nuclear DNA and not with mitochondrial DNA to any significant extent. Wintersberger (1966) found that the 16s component of the MRNA consistently hybridized more efficiently with MDNA than did the 23s MRNA, but both RNAs hybridized to some extent with nuclear DNA; hybridization of MDNA with 4s MRNA was also seen (Table 1).

Table 1. *Hybridization by RNA components with mitochondrial DNA (MDNA) derived from wild-type cells or from a* petite *mutant of Saccharomyces cerevisiae; from E. Wintersberger and G. Viehhauser (1968)*

DNA on filter	[³²P]RNA	RNA bound to filter*	% RNA/DNA
Wild type MDNA	mit. 23s	1508	3·92
Wild type MDNA	mit. 16s	1833	4·73
Wild type MDNA	mit. 4s	1291	3·64
Mutant MDNA	mit. 23s	105	0·27
Mutant MDNA	mit. 16s	172	0·45
Mutant MDNA	mit. 4s	306	0·86

* C.p.m. corrected for background adsorption on to blank filters.

In the studies of Wood & Luck (1969) with a *Neurospora* sp., MRNAs of 23s and 19s were obtained. These MRNAs were complementary to 6·1% and 2·8% of the MDNA, respectively, while cytoplasmic RNAs of 28s and 18s were complementary to 0·67% and 0·33% of the nuclear DNA, respectively. Wood & Luck calculated from the re-naturation kinetics of the MDNA that the sequence length molecular weight was more than 66×10^6 and therefore that the genes specifying the 25s and 19s MRNAs were repeated at least four times in the MDNA; in other words, the MDNA of the *Neurospora* sp. contained redundant sequences. These deductions are based, of course, on the complementarity figures and have a bearing on attempts to assess the genetic potential of MDNA from the overall amount of genetic material.

Transfer ribonucleic acids (tRNA)

Wintersberger (1965) found that yeast mitochondria contain *t*RNA and amino-acyl-*t*RNA synthetases. Similarly in *Neurospora* sp. mitochondria Barnett, Brown & Epler (1967) showed the presence of *t*RNA and activating enzymes capable of reacting with several different amino acids. They found that aspartyl-, leucyl- and phenylalanyl-*t*RNA synthetases could not acylate the *Neurospora* cytoplasmic *t*RNAs. Chromatographic analysis has further emphasized the differences between mitochondrial and cytoplasmic *t*RNAs of *Neurospora* (Brown

& Novelli, 1968), *Tetrahymena* (Suyama, 1967*a*, *b*) and rat liver
(Fournier & Simpson, 1968). Buck & Nass (1969) in chromatographic
studies of rat liver found that the mitochondria possessed unique species
of leucyl-, tyrosyl-, aspartyl-, valyl- and seryl- *t*RNA and that cytoplasmic activating enzymes were unable to acylate species of *t*RNA
which were exclusively mitochondrial.

Protein synthesis

Although the conditions governing the incorporation of labelled
amino acids into isolated mitochondria are well known (see Roodyn &
Wilkie, 1968) it has yet to be demonstrated that the product of incorporation is a definable protein. The highest specific activity after incorporation of labelled amino acids is found in the so-called structural protein,
and none of the main soluble enzymes is radioactive. Roodyn (1965)
fractionated, by differential centrifugation, rat liver mitochondria
labelled *in vitro* with [^{14}C]valine (Fig. 1). The most radioactive protein
was in a fraction which sedimented after one hour at 100,000 g and was
rich in RNA, phospholipid and succinoxidase. This corresponded to
the membrane fractions isolated by other investigators. Negligible
radioactivity was found in the soluble fraction. Neupert, Brdiczka &
Sebald (1968) studied incorporation of labelled amino acids into isolated
rat liver mitochondria and separated the outer and inner membranes of
the organelle after the incubation. By using appropriate enzyme markers
they calculated that the outer membrane had a radioactivity less than
5% of that of the inner membrane. These results indicated that the
mitochondrial inner membrane was the site of accumulation of radioactivity or, in other words, that the proteins synthesized in the organelle,
presumably on the mitochondrial ribosomes, were incorporated into the
inner membrane. The corollary is that the outer membrane and soluble
enzymes of the matrix were synthesized on the cytoplasmic ribosomes.
In this connection Kadenbach (1968) followed the transfer of proteins
from microsomes into mitochondria by using rat liver pulse-labelled
with [^{14}C]leucine and concluded that cytochrome *c* was completely
synthesized together with the haem group at the microsomes and
transported into the mitochondria. The mechanism of transport across
the mitochondrial membrane is unknown. These findings agree with
those of Cadavid & Campbell (1967) on the kinetics of incorporation of
[^{14}C]lysine into cytochrome *c* from microsomes and mitochondria,
respectively, in intact cells. The indications are that cytochrome *c*
and probably other mitochondrial enzymes are synthesized in the
cytoplasm and transported through the mitochondrial membrane.

Evidence of cytoplasmic synthesis of the proteins of mitochondrial ribosomes has been presented by Küntzel (1969) and will be further discussed below. A significant feature of mitochondrial protein synthesis is that the chain initiator is of the bacterial type as indicated by the findings of Smith & Marcker (1968) that the mitochondria contain formylmethionyl-*t*RNA.

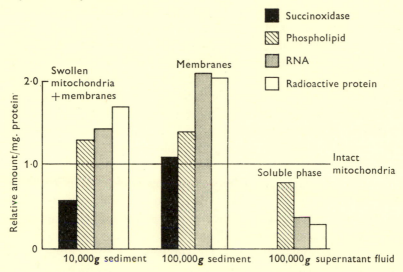

Fig. 1. Sub-fractionation of mitochondria labelled *in vitro* with [^{14}C]valine. Radioactivity data of intact mitochondria taken from Roodyn, Reis & Work (1961); see Roodyn & Wilkie (1968).

Use of inhibitors of protein synthesis. One of the most important findings in recent years is the inhibition by the antibacterial antibiotic chloramphenicol of amino acid incorporation into the mitochondrial fractions of a wide range of organisms (e.g. Mager, 1960; Kroon, 1964; Wheeldon & Lehninger, 1966) including yeast (Wintersberger, 1965). Chloramphenicol inhibits amino acid incorporation into protein in systems using bacterial ribosomes (see Vasquez, 1966) but does not inhibit incorporation of amino acids *in vitro* in systems with cytoplasmic ribosomes of animals (e.g. Von Ehrenstein & Lipmann, 1961) or yeast (Bretthauer *et al.* 1963). With intact cells of yeast the effect of chloramphenicol is to inhibit the synthesis of the mitochondrial membrane-bound cytochromes *a* and *b* but not cytochrome *c* (Clark-Walker & Linnane, 1967). It is claimed that cycloheximide, a known inhibitor of cytoplasmic ribosomes of eukaryotes, inhibits the synthesis of cytochrome *c* but not cytochrome *a* in yeast cells (Yu, Lukins & Linnane, 1968; but see Rabinowitz, 1968*b*). The main difficulty in studying the effects of cycloheximide on intact yeast cells is the extreme sensitivity of the cytoplasmic ribosomes to this antibiotic. In view of the obvious dependence of mitochondrial growth on protein synthesis in the cytoplasm, any conclusions about effects on cytochrome *a* synthesis must be equivocal. It is known for example that in yeasts cytochrome *a* synthesis is regulated by cytochrome *c* (Ycas, 1956; Reilly & Sherman, 1965) so that any inhibition of synthesis of cytochrome *c* would be expected to have an effect on the amount of cytochrome *a* synthesized. However, mutant yeast strains (from this laboratory) with known cycloheximide-resistant cytoplasmic ribosomes (Cooper, Banthorpe & Wilkie, 1967) were able to grow on non-fermentable glycerol medium and synthesized all their cytochromes in the presence of comparatively high concentrations of cycloheximide (Wilkie, 1968), indicating that the mitochondrial ribosomes were insensitive to its action (assuming the *in vivo* mitochondria were permeable to cycloheximide in these cases).

This ability to distinguish between cytoplasmic and mitochondrial ribosomes by means of cycloheximide, which inhibits the cytoplasmic ribosomes but not the mitochondrial ribosomes, and chloramphenicol which acts conversely, was used by Küntzel (1969) to investigate the synthesis of ribosomal proteins. He showed that mitochondrial ribosomes but not cytoplasmic ribosomes from a *Neurospora* sp. were resistant to cycloheximide and sensitive to chloramphenicol in cell-free systems; but the synthesis *in vivo* of the proteins of both ribosome types was inhibited to the same extent by cycloheximide, and unaffected by chloramphenicol. These results indicate that the proteins of both types of ribosomes are synthesized in the cytoplasm. The biochemical aspects of the biogenesis of mitochondria outlined here clearly show that the organelle has a unique protein-synthesizing system with bacteria-like properties. From the results of nucleic acid hybridization experiments it would seem that at least some of the RNA components of the system are coded for by the DNA of the mitochondrion.

Genetic aspects

Nearly all available information about the genetics of mitochondria comes from micro-organisms and most of this from a single species of yeast. The cytoplasmically inherited respiratory deficiency known as *petite* in *Saccharomyces cerevisiae* provided early evidence of the genetic autonomy of mitochondria. Since its discovery and analysis by Ephrussi and Slonimski some twenty years ago the condition has been studied intensively. Experiments show the following results.

(*a*) The mutation is apparently irreversible, indicating the loss of information, hence the general use of the notation ρ to designate the cytoplasmic factor, with $\rho-$ (rho minus) for *petite*, $\rho+$ for normal.

(*b*) In crosses between $\rho-$ and $\rho+$, the resulting zygotes and their vegetative diploid and sexual haploid progeny are $\rho+$ except in certain cases where the $\rho-$ condition is dominant and apparently excludes or suppresses the normal ρ factor in the zygote.

(*c*) The *petite* mutation can be specifically induced with high frequency by mutagens such as acriflavine, ethidium bromide and ultraviolet radiation, and by non-mutagenic compounds including phenylethyl alcohol, nalidixic acid and some antibiotics (see below). The common factor in the activity of these agents appears to be an effect on the replication of DNA either directly or indirectly.

(*d*) In crosses between different $\rho-$ mutants only stable $\rho-$ diploids are produced, whatever the way in which the respective $\rho-$ mutants were produced. $\rho-$ Mutants are thus incapable of undergoing genetic recombination to make a good ρ factor, indicating that all $\rho-$ mutants are similarly deficient since a mechanism of genetic recombination between ρ factors seems to be available (Thomas & Wilkie, 1968*a*).

A series of nuclear gene mutations has been identified and each results in the *petite* phenotype although the cells are still $\rho+$. The most likely interpretation of these results is that nuclear genes control the transcription of the ρ factor, perhaps by coding for the appropriate enzymes. Other nuclear genes have been discovered which give $\rho-$ cells when they mutate, leading to the conclusion that nuclear gene action is necessary for ρ factor replication. These genetical findings are clarified by the biochemical analysis of *petite* mutants.

Petite mutants have multiple deficiencies in mitochondrial enzymes of the inner membrane assembly, including deficiency of cytochromes *a* and *b*; cytochrome *c*, on the other hand, is synthesized in abundance. Since these enzymes of the respiratory chain are absent, *petite* mutants cannot utilize non-fermentable substrates such as glycerol. Because *Saccharomyces cerevisiae* is a facultative anaerobe, the respiratory incapacity does not affect growth to any extent when fermentable sugars are available.

It has been shown that the *petite* mutation is associated with an alteration, usually extensive, in MDNA; this will now be discussed. The buoyant density of MDNA of $\rho+$ cells is of the order of 1·686 g./cm.³ while that of $\rho-$ strains is consistently lower, a density of 1·671 g./cm.³ having been reported in some cases. This low value corresponds to the density of a polymer which contains mainly deoxyadenylate and deoxythymidylate, whereas the wild-type MDNA would have about 17 mole % guanine + cytosine (GC) content. Alterations in DNA which are detectable as changes in buoyant density must involve many base-pair alterations. This leads to the conclusion that MDNA of $\rho-$ cells is largely or entirely nonsensical. A further conclusion is that part at least of the MDNA is the genetic factor ρ. In a particular *petite* strain investigated by Borst & associates (see Hollenberg *et al.* 1969) the MDNA had a buoyant density 10 mg./cm.³ lower than the wild-type strain from which it was derived. In the electron microscope the mutant DNA appeared as numerous duplex circles of 1 μm. contour length; the significance of this is not clear. How the *petite* mutation arises, which it does with a high spontaneous rate, is an acute problem in view of the large number of copies of the mitochondrial genome in respiratory adapted cells. Slonimski and co-workers (see Rabinowitz, 1968a) have claimed to have found as much as 18% of total cell DNA in the mito-chondria of some strains. How the system can be over-run with grossly aberrant mutant genomes is difficult to understand. Wintersberger (1967) found that the species of MRNA from the mitochondria of the wild-type yeast were undetectable in the corresponding *petite* mutant which synthesized unspecified MRNA in small amount compared to wild type. Also the MDNA of the *petite* mutant showed little or no capability of hybridizing with MRNA of the wild type (Table 1). These results support the hypothesis that MDNA carries genetic information specifying MRNA.

The original statement made by Clark-Walker & Linnane (1967) that the inhibition by antibacterial antibiotics of mitochondrial protein synthesis in yeast cells is reversible requires qualification. In a detailed analysis of several strains of *Saccharomyces cerevisiae*, Williamson, Maroudas & Wilkie (1969) found that chloramphenicol and erythro-mycin induced the *petite* mutation; 100% induction usually resulted among daughter cells when cultures were maintained for more than ten cell generations in the presence of the antibiotics. Only mutants with drug-resistant mitochondria which synthesized protein *in vivo* and *in vitro* in the presence of inhibitor (see below) were not affected in this respect. An explanation of these findings may be that a protein (or proteins) involved in MDNA replication is synthesized on mitochondrial ribosomes, and that in the presence of antibiotic this synthesis is greatly decreased or halted. The protein is ultimately diluted out on continued cell division, leading to a breakdown in MDNA replication and thereby resulting in the *petite* condition.

The findings of Clark-Walker & Linnane (1967) about the *in vivo* inhibition by chloramphenicol were extended by these authors to include tetracycline and erythromycin. The strain of yeast used in these tests was diploid and inherently resistant to 2 mg. chloramphenicol/ml. as defined by the ability of the organism to grow on non-fermentable solid medium at this concentration of antibiotic. When subjected to tetrad analysis, the segregation patterns indicated that this yeast strain carried two nuclear genes each conferring resistance to about 1 mg. chloramphenicol/ml. but having an additive effect together (D. Wilkie, unpublished data). This was the first indication that the response of mitochondria to antibacterial antibiotics was genetically controlled. Several haploid strains of *Saccharomyces cerevisiae* (of this laboratory) were then tested and the degree of mitochondrial inhibition by certain antibiotics was found to depend upon the strain used. Spontaneous resistant-mutants were isolated from strains which showed sensitivity to one or other of chloramphenicol, erythromycin or tetracycline; these have been the subject of concentrated study. In the genetic analysis it was found that some of the mutants resistant to erythromycin (E^R) when crossed to a sensitive strain (E^S) showed non-nuclear cytoplasmic inheritance of the resistance, and segregation of resistance and sensitivity among the cells of diploid clones from individual zygotes. On the assumption that the genetic factor for resistance was located in the MDNA, the *petite* mutation was induced by acriflavine in the E^R strain. In the cross between $\rho-$ E^R and $\rho+$ E^S which, as expected, yielded p+ zygotes, resistance was not transmitted. In the reciprocal cross $\rho+$ $E^R \times \rho-$ E^S all progeny inherited resistance (Thomas & Wilkie, 1968b; Linnane, Saunders, Gingold & Lukins, 1968). The hypothesis of MDNA location of resistance was substantiated by these results since, simultaneously with the MDNA being rendered ineffective, there was loss of the resistance factor. MDNA association of this kind was found to apply to resistance to other antibiotics, in particular to spiramycin and the aminoglycoside, paromomycin (Thomas & Wilkie, 1968a). When crosses were made between these different resistant strains, some of the zygotes gave recombinant clones. Thus in the cross $E^R \times P^R$ (paromomycin resistant), some clones were comprised of double-resistant cells ($E^R P^R$) while others were of double-sensitive cells ($E^S P^S$). The mechanism of the recombination process giving new mitochondrial genomes is not clear, but presumably involves genetic crossing-over. Apparent anomalies in the system in which the $\rho-$ strain transmits resistance to some extent in certain cases (see Linnane *et al.* 1968) can be explained in terms of recombination if it be postulated that MDNA of $\rho-$ cells may not be entirely nonsensical (Thomas, 1969; Thomas &

Wilkie, 1969). This might be expected in those *petite* strains which show a comparatively small alteration in buoyant density of their MDNA as compared with the normal.

Amino acid analogues

In view of the reports of differences in the activating enzymes of cytoplasm and mitochondria, it seemed reasonable to expect that these differences would be reflected in different responses to the inhibitory effects of amino acid analogues (Wilkie, 1969). To test this, yeast strains were grown on fermentable sugar-containing (YES) media and non-fermentable glycerol-containing (YEG) media and containing one or other of the analogues canavanine, *p*-fluorophenylalanine (*p*FP) or ethionine, which are the analogues of arginine, phenylalanine and methionine, respectively. With about half of the 27 strains used, the mitochondrial system was more sensitive than the cytoplasm with canavanine or *p*FP, but ethionine did not show this selective inhibition to any great extent. In some strains there was a 50-fold difference in this respect. Thus, in one strain growth proceeded on sugar-containing medium (YES) in the presence of 5 mg. *p*FP/ml. while on glycerol medium (YEG) there was no growth at 0·1 mg. *p*FP/ml. In these tests the corresponding amino acid was supplied in the medium with the analogue. In strains showing mitochondrial sensitivity cells grown to stationary phase in YES medium in the presence of analogue synthesized considerably smaller amounts of cytochromes *a* and *b* than did the controls. Spontaneous resistant-mutants were picked from the YES and from the YEG series and tested for cross-resistance. Some of the results obtained with canavanine are given in Table 2, which show that it was possible to pick up resistance which affected either cytoplasm or mito-chondria. Genetic analysis has shown so far that different nuclear genes control the respective resistances. If it can be shown that the mechanism of resistance in the mitochondrion is an alteration in the *t*RNA-activating enzyme system, genetic analysis should give the location of the genetic information involved.

Experiments wtih [^{14}C]*p*FP have shown the incorporation of the analogue into the proteins of these yeasts (Rhodes & Wilkie, 1969). For example, by using a strain with resistant cytoplasm and sensitive mito-chondria, and fractionation of the mitochondria after a period of growth in presence of [^{14}C]*p*FP, it may be possible to determine the site of synthesis of various mitochondrial proteins; the proteins with the most radioactivity would be those that were synthesized in this organelle.

In summarizing the work on yeast mitochondria it is apparent that an intricate system for the perpetuation and evolution of the organelle is in operation, the details of which are open to analysis by using an approach which combines genetics and biochemistry. Although there are features of yeast mitochondria which distinguish them from other mitochondria, such as their repressibility in anaerobic conditions, the main characteristics of the protein-synthesizing apparatus, structure and biochemistry show such striking similarities with other mitochondria that extrapolation to other organisms is probably valid.

Table 2. *Canavanine-sensitivity of yeast strains and spontaneous mutants on fermentable sugar medium (YES) and non-fermentable glycerol medium (YEG) from Wilkie (1969)*

	Canavanine (mg./ml.)								
	YEG series				YES series				
Strain	0·05	0·1	0·15	0·25	0·1	0·15	0·25	0·5	1·0
	Arbitrary units of growth								
D6 *ar*	4	3	2	1	4	4	2	1	0
D10	5	5	3	2	5	4	3	1	0
−R–S	5	5	5	4	5	5	5	5	4
D15	5	5	4	3	5	4	3	0	0
D18 *ar*	0	0	0	0	4	4	4	2	0
−R–S	0	0	0	0	4	4	4	4	4
−R–G	5	4	3	1	5	5	4	2	0
D74 *ar*	4	3	2	1	5	4	3	2	0
−R–S	4	3	2	0	5	5	4	4	4
D75	5	5	5	4	5	5	5	3	1
A7F	0	0	0	0	2	1	0	0	0
−R–G	5	4	2	1	5	4	2	1	0
A433	0	0	0	0	2	1	0	0	0
−R–S	5	5	4	3	5	5	4	2	0
−R–G	5	4	2	1	5	4	2	1	0
B9	1	1	0	0	4	4	3	1	0
−R–S	0	0	0	0	4	4	4	4	4
−R–G	5	4	3	2	5	4	4	2	0
B11	5	5	5	3	5	5	4	3	1
41	2	1	0	0	5	4	3	2	0
−R–G	5	5	3	1	5	5	4	1	0

R–S, R–G, resistant mutants from YES and YEG series respectively. *ar*, requirement for arginine. 0 to 5, arbitrary units of growth from no growth to full growth. All untreated controls allocated 5.

CHLOROPLASTS

Nearly all that has been described above in relation to mitochondria applies to chloroplasts. Chloroplasts possess DNA distinct from nuclear DNA (see review by Kroon, 1969) and are capable of synthesizing protein *in vitro*. Chloroplast fractions from cells of higher plants and algae contain ribosomes of the 70s type as compared to the 80s type of the cytoplasm. The literature is well documented in the article by Hoober & Blobel (1969) in which the characteristics of the chloroplastic and cytoplasmic ribosomes of the alga *Chlamydomonas reinhardii* are compared. The chloroplast ribosomes have a sedimentation value of 68s while those in the cytoplasmic matrix had 80s. These compare with the previous findings of Sager & Hamilton (1967) of 69s and 79s, respectively, for this organism. The two types of ribosomes were also reported by both groups of investigators to vary in regard to the concentration of $MgCl_2$ required for their dissociation. Hoober & Blobel (1969) also showed, in an extension of the findings of many other investigators with protein inhibitors (again giving a review of the literature), that chloramphenicol and cycloheximide together were required to inhibit completely amino acid incorporation into protein by intact cells, each antibiotic inhibiting maximally about one-half of the total incorporation. Also, chloramphenicol affected the 68 s ribosomes but not the 80s, and conversely for cycloheximide. Electron microscope studies of chloroplast ribosomes indicate that these units are smaller than cytoplasmic ribosomes; this further supports the above findings and justifies the conclusion that the chloroplast ribosomes are similar to bacterial ribosomes.

As in mitochondria, DNA-RNA hybridization has shown that the RNA components of chloroplast ribosomes are complementary to chloroplast DNA (Scott & Smillie, 1967; Tewari & Wildman, 1968). From these experiments it has been estimated that the information content of chloroplast DNA is between 4 and 8×10^8 daltons, assuming that chloroplast DNA is made of repeating units from a basic single strand with no redundancy.

Quantitative re-naturation experiments by Wells & Birnstiel (1967) put the total information content of higher plant chloroplast DNA at 80×10^6 daltons, which corresponds to about 120,000 base pairs.

The genetic function of chloroplast DNA is manifest in the non-Mendelian inheritance of certain chloroplast mutations in higher plants and in the extensive studies of Sager and her collaborators with *Chlamydomonas* (see Wilkie, 1964). Nuclear involvement in chloroplast biogenesis is apparent in the numerous chloroplast mutants whose origin can be traced to changes in chromosomal genes (see Wilkie, 1964). As in mitochondria, it is clear that the chloroplast is under the control of an intricate system which integrates genetic information from nucleus and organelle. An attempt to represent the system diagrammatically in the case of mitochondria is given in Fig. 2.

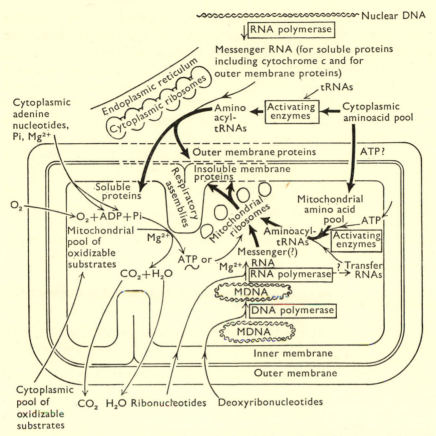

Fig. 2. Probable pathways for synthesis of mitochondrial proteins. Bold arrows: amino acid incorporation; other arrows: related processes; ⟋⟍⟋⟍, site of membrane formation. (From Roodyn & Wilkie, 1968.)

CONCLUSIONS

A complete answer to the initial question about the precise role of organelle DNA, which would give an accurate measure of the degree of autonomy of organelles, is still outstanding. Indeed, from the reports of wide differences in the amounts of mitochondrial DNA in different organisms, a general answer may be difficult to give, even if the measurements apply to a single basic genome. The available evidence suggests that organelle DNA mainly specifies the RNA components of the protein-synthesizing system. In doing this, the coding capacity even of a 5μm molecule would not be exhausted, assuming that it does not have many repeating sequences. It is likely that the code for some of the proteins of the membranes of the organelles is also carried in the organelle DNA.

The most striking feature of the growth and reproduction of organelles is the degree of similarity of certain aspects of this with the growth and reproduction of bacteria. This gives considerable support to the old hypothesis that the organelles were originally bacteria which formed a symbiotic relationship with various cell types and were subsequently integrated into the cell system. The evolutionary progress of the integration can be thought of as a progressive acquisition by the nucleus of genetic information from organelles. Perhaps in the case of animal mitochondria the furthest development in this process has been reached.

REFERENCES

AVERS, C. J. (1967). Heterogeneous length distribution of circular DNA filaments from yeast mitochondria. *Proc. natn. Acad. Sci. U.S.A.* **58**, 620.

BARNETT, W. E., BROWN, D. H. & EPLER, J. L. (1967). Mitochondrial specific aminoacyl RNA synthetases. *Proc. natn. Acad. Sci. U.S.A.* **57**, 1775.

BORST, P. & KROON, A. M. (1969). Mitochondrial DNA: physico-chemical properties; replication and genetic function. *Int. Rev. Cytol.* (In Press.)

BORST, P., RUTTENBERG, G. J. C. M. & KROON, A. M. (1967). Mitochondrial DNA. I. Preparation and properties of mitochondrial DNA from chick liver. *Biochim. biophys. Acta* **149**, 140.

BRETTHAUER, R. K., MARCUS, L., CHALOUPKA, J., HALVORSON, H. O. & BOCK, R. M. (1963). Amino acid incorporation into protein by cell free extracts of yeast. *Biochemistry, N.Y.* **2**, 1079.

BRITTEN, R. J. & WARING, M. (1965). Renaturation of the DNA of higher organisms. *Yb. Carnegie Instn Wash.* **64**, 316.

BROWN, D. H. & NOVELLI, (1968). Chromatographic differences between the cytoplasmic and mitochondrial *t*RNAs of *Neurospora crassa*. *Biochem. biophys. Res. Commun.* **31**, 262.

BUCK, C.A. & NASS, M. M. K. (1969). Studies on mitochondrial *t*RNA from animal cells. I. A comparison of mitochondrial and cytoplasmic *t*RNA and aminoacyl-*t*RNA synthetases. *J. molec. Biol.* **41**, 67.

CADAVID, N. F. G. & CAMPBELL, P. N. (1967). Incorporation, *in vivo*, of ^{14}C-lysine into cytochome *c* of rat liver sub-cellular fraction. *Biochem. J.* **102**, 39 P.

CLARK-WALKER, G. D. & LINNANE, A. W. (1967). The biogenesis of mitochondria in *Saccharomyces cerevisiae*. *J. Cell Biol.* **34**, 1.

COOPER, D., BANTHORPE, D. V. & WILKIE, D. (1967). Modified ribosomes conferring resistance to cycloheximide in *Saccharomyces cerevisiae*. *J. molec. Biol.* **26**, 347.

DAWID, I. B. & WOLSTENHOLME, D. R. (1968). Renaturation and hybridization studies of mitochondrial DNA. *Biophys. J.* **8**, 65.

FOURNIER, M. J. & SIMPSON, M. V. (1968). The Occurrence of Amino Acid Activating Enzymes and *s*RNA in Mitochondria. In *Biochemical Aspects of the Biogenesis of Mitochondria*, p. 227. Eds. E. L. Slater, J. M. Tager, S. Papa and E. Quagliariello. Bari, Italy: Adriatica Editrice.

FUKUHARA, H. (1967). Informational role of mitochondrial DNA studied by hybridization with different classes of RNA in yeast. *Proc. natn. Acad. Sci. U.S.A.* **58**, 1065.

HOLLENBERG, C. P., BORST, P., THURING, R. W. J. & VAN BRUGGEN, E. P. J. (1969). Size, structure and genetic complexity of yeast mitochondrial DNA. *Biochim. biophys. Acta*. (In Press.)

HOOBER, J. K. & BLOBEL, G. (1969). Characterization of the chloroplastic and cytoplasmic ribosomes of *Chlamydomonas reinhardii*. *J. molec. Biol.* **41**, 121.

KADENBACH, B. (1968). Transfer of Proteins from Microsomes into Mitochondria. Biosynthesis of Cytochrome *c*. In *Biochemical Aspects of the Biogenesis of Mitochondria*, p. 415. Eds. E. L. Slater, J. M. Tager, S. Papa and E. Quagliariello. Bari, Italy: Adriatica Editrice.

KROON, A. M. (1964). Inhibitors of mitochondrial protein synthesis. *Biochim. biophys. Acta* **76**, 165.

KROON, A. M. (1969). DNA and RNA from Mitochondria and Chloroplasts: Their Contribution to the Biogenesis of These Organelles. In *Handbook of Molecular Biology* (In Press.)

KÜNTZEL, H. (1969). Proteins of mitochondrial and cytoplasmic ribosomes from *Neurospora crassa*. *Nature, Lond.* **222**, 142.

KÜNTZEL, H. & NOLL, H. (1967). Mitochondrial and cytoplasmic polysomes from *Neurospora crassa*. *Nature, Lond.* **215**, 1340.

LINNANE, A. W., SAUNDERS, G. W., GINGOLD, E. B. & LUKINS, H. B. (1968). The biogenesis of mitochondria. V. Cytoplasmic inheritance of erythromycin resistance in *Saccharomyces cerevisiae*. *Proc. natn. Acad. Sci. U.S.A.* **59**, 903.

MAGER, J. (1960). Chloramphenicol and chlortetracycline inhibition of amino acid incorporation into proteins in a cell-free system from *T. pyriformis*. *Biochim. biophys. Acta* **38**, 150.

NEUPERT, W., BRDICZKA, D. & SEBALD, W. (1968). Incorporation of Amino Acids into the Outer and Inner Membrane of Isolated Rat-Liver Mitchondria. In *Biochemical Aspects of the Biogenesis of Mitochondria*, p. 395. Eds. E.L. Slater, J. M. Tager, S. Papa and E. Quagliariello. Bari, Italy: Adriatica Editrice.

PARSONS, P. & SIMPSON, M. V. (1967). Biosynthesis of DNA by isolated mitochondria: incorporation of thymidine triphosphate-2-C¹⁴. *Science, N.Y.* **155**, 91.

RABINOWITZ, M. (1968*a*). Extranuclear DNA. *Bull. Soc. Chim. biol.* **50**, 311.

RABINOWITZ, M. (1968*b*). Addenda. In *Biochemical Aspects of the Biogenesis of Mitochondria*, p. 437. Bari, Italy: Adriatica Editrice.

REICH, E. & LUCK, D. J. L. (1966). Replication and inheritance of mitochondrial DNA. *Proc. natn. Acad. Sci. U.S.A.* **55**, 1600.

REILLY, C. & SHERMAN, F. (1965). Glucose repression of cytochrome *a* synthesis in cytochrome-deficient mutants of yeast. *Biochim. biophys. Acta* **95**, 640.

RHODES, M. & WILKIE, D. (1969). In preparation.

RIFKIN, M. R., WOOD, D. D. & LUCK, D. J. L. (1967). Ribosomal RNA and ribosomes from mitochondria of *Neurospora crassa*. *Proc. natn. Acad. Sci. U.S.A.* **58**, 1025.

ROODYN, D. B. (1965). Further studies affecting amino acid incorporation into protein by isolated mitochondria. *Biochem. J.* **97**, 782.

ROODYN, D. B., REIS, P. J. & WORK, T. S. (1961). Protein synthesis in mitochondria: requirements for the incorporation of radioactive amino acids into mitochondrial protein. *Biochem. J.* **80**, 9.

ROODYN, D. B. & WILKIE, D. (1969). *The Biogenesis of Mitochondria*. London: Methuen & Co. Ltd.

SAGER, R. & HAMILTON, J. (1967). Cytoplasmic and chloroplast ribosomes of Chlamydomonas: ultracentrifuge characterization. *Science, N.Y.* **157**, 709.

SCOTT, N. S. & SMILLIE, R. M. (1967). Evidence for the direction of chloroplast ribosomal RNA synthesis by chloroplast DNA. *Biochem. biophys. Res. Commun.* **28**, 598.

SHAPIRO, L., GROSSMAN, L. I., MARMUR, J. & KLEINSCHMIDT, A. K. (1968). Physical studies on the structure of yeast mitochondrial DNA. *J. molec. Biol.* **33**, 907.

SMITH, A. E. & MARCKER, K. A. (1968). N-formylmethionyl transfer RNA in mitochondria from yeast and rat liver. *J. molec. Biol.* **38**, 241.

SUYAMA, Y. (1967*a*). The origins of mitochondrial ribonucleic acids in *Tetrahymena pyriformis*. *Biochemistry. N.Y.* **6**, 2829.

SUYAMA, Y. (1967*b*). Leucyl-*t*RNA and leucyl-*t*RNA synthetase in mitochondria of *Tetrahymena pyriformis*. *Biochem. biophys. Res. Commun.* **28**, 746.

TEWARI, K. K. & WILDMAN, S. G. (1968). Function of chloroplast DNA. I. Hybridization studies involving nuclear and chloroplast DNA with RNA from cytoplasmic (80 s) and chloroplast (70 s) ribosomes. *Proc. natn. Acad. Sci. U.S.A.* **59**, 569.

THOMAS, D. Y. (1969). Ph.D. thesis. University of London.

THOMAS, D. Y. & WILKIE, D. (1968*a*). Recombination of mitochondrial drug-resistance factors in *Saccharomyces cerevisiae*. *Biochem. biophys. Res. Commun.* **30**, 368.

THOMAS, D. Y. & WILKIE, D. (1968*b*). Inhibition of mitochondrial synthesis in yeast by erythromycin: cytoplasmic and nuclear factors controlling resistance. *Genet. Res.* **11**, 33.

THOMAS, D. Y. & WILKIE, D. (1969). Genetical and biochemical aspects of mitochondrial drug-resistance in *Saccharomyces cerevisiae*. In preparation.

VASQUEZ, D. (1966). Binding of chloramphenicol to ribosomes. The effect of a number of antibiotics. *Biochim. biophys. Acta* **114**, 277.

VON EHRENSTEIN, G. & LIPMANN, F. (1961). Experiments on haemoglobin biosynthesis: *Proc. natn. Acad. Sci. U.S.A.* **47**, 941.

WELLS, R. & BIRNSTIEL, M. L. (1967). A rapidly renaturing deoxyribonucleic acid component associated with chloroplast preparations. *Biochem. J.* **105**, 53 P.

WHEELDON, L. W. & LEHNINGER, A. L. (1966). Energy linked synthesis and decay of membrane proteins in isolated rat-liver mitochondria. *Biochem.* **5**, 3533.

WILKIE, D. (1964). *The Cytoplasm in Heredity*. London. Methuen & Co. Ltd.

WILKIE, D. (1968). Mutants of the Respiratory System of Yeast in Problems of Mitochondrial Biogenesis. In *Biochemical Aspects of the Biogenesis of Mitochondria*, p. 457. Eds. E. L. Slater, J. M. Tager, S. Papa and E. Quagliariello. Bari, Italy: Adriatica Editrice.

WILKIE, D. (1969). Selective inhibition of mitochondrial synthesis in *Saccharomyces cerevisiae* by canavanine. *J. molec. Biol.* (In Press.)

WILLIAMSON, D. H., MAROUDAS, N. M. & WILKIE, D. (1969). Induction of cytoplasmic respiratory deficiency in yeast by antibiotics. In preparation.

WINTERSBERGER, E. (1965). Proteinsynthese in isolierten Hefe-Mitochondrion *Biochem. Z.* **341**, 409.

WINTERSBERGER, E. (1966). In *Regulation of Metabolic Processes*, p. 439. Eds. J. M. Tager, S. Papa, E. Quagliariello & E. L. Slater. Amsterdam: Elsevier.

WINTERSBERGER, E. (1967). A distinct class of ribosomal RNA components in yeast mitochondria as revealed by gradient centrifugation and by DNA-RNA-hybridization. *Hoppe Seyler's Z. physiol. Chem.* **348**, 1701.

WINTERSBERGER, E. (1968). Synthesis of DNA is Isolated Yeast Mitochondria. In *Biochemical Aspects of the Biogenesis of Mitochondria*, p. 189. Eds. E. L. Slater, J. M. Tager, S. Papa and E. Quagliariello. Bari, Italy: Adriatica Editrice.

WINTERSBERGER, E. & WIEHHAUSER, G. (1968). Function of mitochondrial DNA in yeast. *Nature, Lond.* **220**, 699.

WOLSTENHOLME, D. R. & DAWID, I. B. (1968). A size difference between mitochondrial DNA molecules of urodele and anuran Amphibia. *J. Cell Biol.* **39**, 222.

WOLSTENHOLME, D. R. & GROSS, N. J. (1968). The form and size of mitochondrial DNA of the red bean, *Phaseolus vulgaris*. *Proc. natn. Acad. Sci. U.S.A.* **61**, 245.

Wood, D. D. & Luck, D. J. L. (1969). Hybridization of mitochondrial ribosomal RNA. *J. molec. Biol.* **41,** 211.

Ycas, M. (1956). Formation of haemoglobin and the cytochromes by yeast in the presence of antimycin A. *Expl Cell Res.* **11,** 1.

Yu, R., Lukins, H. B. & Linnane, A. W. (1968). Selective *in vivo* Action of Cycloheximide on the Synthesis of Soluble Mitochondrial Proteins. In *Biochemical Aspects of the Biogenesis of Mitochondria*, p. 359. Eds. E. L. Slater, J. M. Tager, S. Papa and E. Quagliariello. Bari, Italy: Adriatica Editrice.

THE ROAD TO DIPLOIDY WITH EMPHASIS ON A DETOUR

JOHN R. RAPER AND ABRAHAM S. FLEXER

Department of Biology, Harvard University,
Cambridge, Massachusetts 02138, U.S.A.

It has been said that much of human inquiry is little more than the elaborate elucidation of the obvious. We fear that this judgement can justly be applied to much of what we shall examine here, because practically all of the basic facts that we can consider in connection with diploidy have long been accepted by biologists generally. We do hope, however, that a fresh examination of these long-known fundamental facts in a context somewhat different from their usual exposition may suggest unexpected relationships or reveal unrecognized evolutionary trends.

DIPLOIDY AS A WAY OF LIFE

By all available criteria diploidy is an ancient and highly successful condition. Its origins are not only difficult to discern, but the relevant schemes that could account for diploidy are not readily subject to experimental verification. Our examination of diploidy is accordingly restricted to: (a) a survey of the distribution of diploidy and of alternatives to diploidy among extant taxa, (b) a search for correlations between this distribution and the characteristics of existing groups, (c) an attempt, through deductive reasoning, to read evolutionary meaning into the observed distribution and correlations.

Diploidy is, of course, a condition possible only—or at least recognizable only—in eukaryotic organisms. Between prokaryotes and eukaryotes there is a structural and developmental gulf that Stanier, Doudoroff & Adelberg (1963) aptly characterized as the greatest evolutionary discontinuity in the biological world. With this, happily, we need not here be concerned. By whatever course of events eukaryotes originated—the postulate of Margulis (1968) is perhaps as feasible as any available alternative—the early course of eukaryotic evolution seems most likely to have passed through sexually reproducing forms that were somatically or vegetatively haploid. Many extant species, members of groups that are commonly considered to be among the most primitive eukaryotes, still retain this basic organization. If evolutionary achievement is judged by the dual characters of structural complexity and biomass, however, there is a totally compelling correlation between somatic or vegetative diploidy and biological success. In very practical terms, the restriction of haploidy to simple and primitive forms and the universal presence of diploidy in the more highly advanced forms argues strongly that the diploid condition confers significant biological advantages.

It is perhaps surprising that only a modest range of answers can be given to the simple question 'What are the advantages of diploidy?' The answers relate, with various minor modifications, to the simple fact that the diploid individual, population, or species has a greater genetic endowment than would a corresponding haploid form. This difference in genetic endowment leads to several consequences, some of which confer obvious advantages, others of which are of questionable significance. The epitome of the obvious, of course, is the fact that, cell for cell, the diploid organism has twice the amount of genetic information, but the relevance of this is somewhat difficult to assess. In certain cases there is evidence that both of the duplicated components of the diploid genome may be functioning alternatively or alternately, as in somatic mosaicism of the human female for characters carried by the X chromosome (Stern, 1968). And what about the phenotypic differences recently described between genetically identical armadillo quadruplets (Storrs & Williams, 1968)?

It has been suggested (Commoner, 1964) that DNA, in addition to its function as a template for the assembly of proteins, may also regulate cellular metabolism by controlling the concentration of free nucleotides. If such speculations may be extended, the additional DNA contained in the diploid genome may be viewed as the basis for a more subtle control of cellular activity and possibly also of morphogenetic control. In any event, the evident superiority of the diploid does not seem to lie in the greater productivity by the doubled genome, for in many other cases, such as in the heterokaryons of many fungi, alleles in substantial minority still provide adequate quantities of their specific gene products (Davis, 1966).

At the population level, the differences in genotypic capacity between haploid and diploid organisms are quite impressive. For paired alleles at n loci, a haploid species could contain 2^n distinct genotypes, whereas the corresponding diploid species could generate 3^n genotypes. The phenotypic capacity of a haploid is equal to the genotypic capacity. The number of possible phenotypes in a diploid species would lie between the haploid and diploid values for genotypic capacity and would be determined by the proportions of dominant, partially dominant or nondominant allelic pairs. The significance of these differences between haploidy and diploidy, although substantial, may not be critical. One possible point of diploid superiority might lie in what has been termed the Baldwin effect, the phenotypic accommodation of individuals to conditions that are considerably suboptimal for the population during the period required for genetic adjustment within the population (Grant, 1963). The diploid organism, with its dual set of genes, may very well have available a wider range of accommodative responses to unfavourable conditions.

A further feature of the diploid, genetic buffering, may constitute a critical advantage in certain cases. The masking of a wide variety of selectively neutral or even deleterious genetic characters tremendously increases the plasticity and potential variability of diploid species. Such plasticity and potential variability permit rapid adjustments of gene frequencies in diploid populations under rapidly changing selective pressures. Micro-organisms have evolved, within the confines of haploid or even asexual life cycles, a variety of mechanisms to ensure at least a functional approximation to the level of genetic plasticity conferred by diploidy. In some asexual forms, tremendous populations of vegetative cells or propagules comprise vast arrays of genotypes that arise through mutations—genotypes, that, through rapid vegetative multiplication, are collectively competent quickly to take advantage of a wide variety of ecological situations. Genetic plasticity, however, is vastly increased in haploid sexually-reproducing forms that have available the benefits of crossing-over and assortment. The more complex of the volvocine series of green algae have achieved a degree of genetic versatility which, though decidedly less effective than that enjoyed by diploid organisms, may account for the evolution of the series. In *Volvox* spp., for example, all four meiotic products remain associated through many rounds of vegetative division to form a many-celled colony. Papazian (1954) pointed out that after a few sexual cycles Mendelian segregation among the descendants of a single ancestral zygote generates as many gentoypes as would a diploid or tetraploid species. Moreover, the continued association of the four meiotic products in a single colony provides a degree of genetic buffering: pre-adaptive genes and gene-combinations may be retained in the population without being subjected to the ruthless selection that operates on strictly haploid species. This combination of plasticity and genetic buffering may provide a sufficient selective advantage to rationalize the evolution of the volvocine series. One might also advance similar arguments for the convergent evolution of colonial forms among the eubacteria, the blue-green algae and several other algal groups. Genetic buffering, although a significant aspect of diploidy, is thus not necessarily of universal benefit to micro-organisms that produce tremendous numbers of asexual propagules or colonial forms, but it appears to be essential in higher organisms. The benefits of genetic buffering would thus seem more likely to be correlated with the duration of the sexual cycle and the efficiency of asexual reproductive processes. Self-fertility, in conjunction with haploidy, very short sexual cycles, and prolific asexual reproduction appears to be the successful solution for primitive forms in aquatic habitats. The only disadvantage to haploidy under such circumstances, if indeed it is a disadvantage, might well be a limitation on the evolutionary competence to achieve greater structural complexity—a limitation not solved by colonial organization. Certainly, genetic buffering would appear to be imperative in those more highly evolved forms that are self-sterile but cross-fertile, with long sexual cycles and without asexual reproduction.

HAPLOIDY VERSUS DIPLOIDY IN LIFE CYCLES

These various consequences of the greater genetic plasticity of the vegetative or somatic diploid must at least add up to a small selective advantage that could account for the universal diploidy of the most highly evolved forms. The diploid life cycle, typical of the animal kingdom, however, might well be considered simply as the terminal point of a continuous sequence. The other extreme in this sequence would be the haploid life cycle, with its single nuclear generation in the diploid. Between these limits, at least theoretically, there is an opportunity for continuously altered proportions of haploid and diploid phases. Three basic types of life cycle are thus immediately apparent (Fig. 1):

(a) *Haploid cycle*, in which the entire life history is passed in the haploid except for the diploid zygote nucleus.

(b) *Diploid cycle*, in which the entire life history is passed in the diploid except for the immediate products of meiosis, the gametes.

(c) *Haplo-diploid cycle*, in which there is an alternation of haploid and diploid vegetative phases. The distinction between diploid and haplo-diploid cycles sometimes requires an arbitrary definition of the haploid phase. This is nicely illustrated in the myxomycetes and in many yeasts, in which the immediate products of meiosis, motile swarm cells and ascospores, respectively, are competent to undergo sexual fusions. They are equally competent, however, to divide mitotically to generate clones of unicells, the individual members of which also can behave as gametes. The life cycle in such cases is here considered haplo-diploid; a haploid phase is considered to occur wherever the immediate products of meiosis are competent to multiply vegetatively.

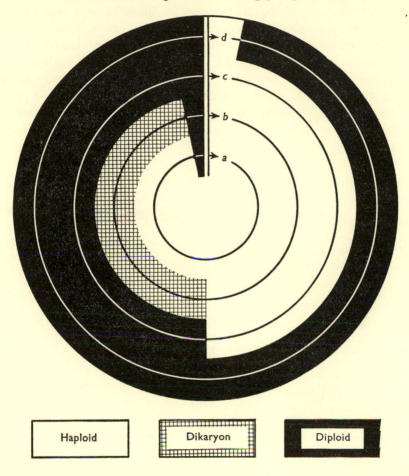

Nuclear phase

Fig. 1. Schematic comparison of basic life cycles. In each cycle, changes in nuclear phase are indicated, in clockwise sequence, by changes in shading. The double vertical line at the top of the diagram represents meiosis, and each of the two adjacent sectors represents a single nuclear generation (revised from Raper, 1954, fig. 2. Copyright 1954 by the American Association for the Advancement of Science).

(*d*) *Haplo-dikaryotic cycle.* A fourth basic life cycle is found only among higher fungi and includes a stable association of paired haploid nuclei, the dikaryophase, that is propagated vegetatively. The dikaryophase is the substitute for a vegetative diplophase and is, in effect, the genetic and physiological equivalent of the diploid. This fourth type of life cycle proceeds through haploid and dikaryotic phases of varying proportions prior to a single nuclear generation in the diploid and thence to meiosis.

An examination of the distribution of these various types of life cycles in the major groups of organisms (Table 1) reveals certain rather striking developments and correlations. The most conspicuous of these is the fact that the entire animal kingdom, with the exception of the most primitive and simplest group, the protozoa, are exclusively diploid organisms—as indeed are most of the protozoa as well. Haploid protozoa are largely confined to the flagellates; there are only a few forms in which there is an alternation of generations, and these are restricted to the Foraminifera (Grell, 1967).

Table 1. *Distribution of life cycles in eukaryotic organisms*

Organisms	Haploid	Haplo-diploid	Haplo-dikaryotic	Diploid
Protozoa	X	X	.	X
Higher animals	.	.	.	X
Myxomycetes	.	X	.	.
Algae				
Green	X	X	.	.
Red	X	X	.	.
Brown	.	X	.	X
Diatoms	.	.	.	X
Golden	X	?	.	.
Fungi				
Uniflagellate water moulds	X	X	.	X (rare)
Biflagellate water moulds	?	.	.	X
Zygomycetes	X	.	.	.
Hemiascomycetes	X	X	.	.
Euascomycetes	.	.	X	.
Basidiomycetes	.	.	X	.
Embryophyta				
Mosses and liverworts	.	X	.	.
Ferns, etc.	.	X	.	.
Seed plants	.	X	(—effectively→	X)

The situation in the plant kingdom is, of course, totally different. Among the algae, haploid, haplo-diploid and diploid forms are present, as they are also in the simpler fungi. Between the lower and higher fungi, however, there is a discontinuity as evidenced by the dikaryon in the latter, of which more later. Among higher plants, the Embryophyta comprising mosses, liverworts, ferns and seed plants, the life cycle is exclusively haplo-diploid, with a progressive decrease in the extent and size of the haploid phase in the series from the mosses to the higher plants. The end-point in this progression, as represented by the flowering plants, is effectively diploid, although a distinct haploid gametophytic generation is present but reduced to microscopic proportions (three cells in the male and eight cells in the female). The basic life cycles of flowering plants and of animals thus differ only in detail, but the evolutionary histories in the two cases appear to be vastly different. Diploidy apparently became fixed very early in the evolutionary history of the animal kingdom. Plants by contrast have approached effective, though not actual, diploidy in a long evolutionary history of increasing primacy of the diploid phase in a cycle that features an alternation of generations. One possible correlation of significance of this progression will be considered later.

One of the more intriguing observations in all of this is the occurrence of the dikaryon in the higher fungi. Dikaryosis, a monopoly of the higher fungi, the Euascomycetes and Basidiomycetes, is truly a substitute for diploidy and differs from it primarily in the spatial separation of the two genomes in a genetically balanced system. The dikaryon thus represents a temporary interruption in the transition from the haploid to the diploid. As such, the dikaryon deserves somewhat special consideration, to be detailed later, particularly in reference to the process of dikaryotization.

Beyond these very striking aspects of the distribution of haploidy and diploidy in the various groups of organisms, attention may be called to a number of additional points of interest. In the simpler organisms, the protozoa and the various groups of algae and fungi, there is a curious mixture of forms that are haploid, forms that are diploid, and forms in which there is an alternation of haploid and diploid generations. This complete range of cycles occurs in at least two groups, the protozoa and the uniflagellate water moulds. It is perhaps not surprising that all three types of cycles are found in as large and phylogenetically diverse a group as the protozoa, but it is difficult to rationalize in the uniflagellate water moulds—a group that appears to be phylogenetically homogeneous. Several of the groups of lower plants have two of the three major types of cycles, often in closely related taxa. The underlying bases of this admixture of life cycles are obscure, and it can only be suggested that each of the three types of cycle endows some slight but critical selective advantage to forms that occupy an almost infinite range of subtly differing environments. The problem is neatly posed by the haploid and haplo-diploid yeasts, both of which forms share what to us appears to be a single highly specialized and very uniform ecological niche.

Another interesting point is the strong correlation in plants of the dominance of the diploid phase with increasing structural complexity. Among lower plants having more or less equal haploid and diploid generations, such as the algae *Ulva* and *Ectocarpus* and the fungus *Allomyces*, there is essentially no difference in the structural complexity of the two generations. The larger brown algae, however, are either haplo-diploid with the haploid phase reduced to microscopic proportions, e.g. *Laminaria*, or diploid, e.g. *Fucus*. In these cases, a high degree of structural complexity is achieved in the diploid as compared to that in the diploid phase of the more equally balanced forms such as *Ectocarpus*. This correlation of the dominance of the diploid with increasing structural complexity is even more plainly evident in the Embryophyta. In the mosses and liverworts, the haploid phase is the persistent vegetative structure, with the diploid sporophytic phase being a transient structure parasitic on the haploid, but structurally far more complex than the vegetative haploid. Among the ferns and their relations, the haploid phase is greatly reduced, with the independent diploid phase achieving significantly greater structural complexity than the diploid phase of the mosses and liverworts. The end-point in this series is of course reached in the flowering plants, in which the haploid phase is reduced to microscopic proportions with the diploid phase a vegetative structure of great complexity.

This strong correlation between the dominance of the diplophase and the attainment of structural complexity as well as the distribution of diploidy throughout the biological world leads to the suspicion that diploidy is necessary, but in itself not sufficient, for the extensive differentiation of tissues. Valid comparisons here are necessarily restricted to the plant kingdom, and the leafy liverwort is immediately recognized as the most highly differentiated vegetative haploid structure known; but compare the low degree of differentiation in this organism with the many highly differentiated tissues of higher plants and animals. The basis for the correlation between high differentiative competence and diploidy is obscure; again it can only be surmised that the redundancy inherent in the double genetic endowment of the diploid permits a more intricate organization. The occurrence of relatively simple diploid organisms, however, such as diatoms and water moulds of both the uniflagellate and biflagellate series, warns against the easy acceptance of oversimplified explanations. The meaning of the phrase 'extensive differentiation of tissue' must also be considered at two temporal levels: the ontogenetic and the phylogenetic. The argument advanced here is that diploidy may confer the *phylogenetic* capacity to evolve differentiated tissues but does not address itself to the issue at the ontogenetic level. As a case in point, consider the recent demonstration that suitably pampered pollen grains of tobacco will form well-developed, though undersized, leafy haploid plants that flower profusely but which do not set seed (Nitsch & Nitsch, 1969).

The predominance of diploidy in the animal kingdom and of an alternation of haploid-diploid generations in the plant kingdom calls to mind certain other generally valid distinctions between plants and animals and the question whether further correlations of significance might be found. Consider, for example, the matter of motility. Animals are characteristically motile at least at some time in their development and in a large majority of cases have at least limited competence to seek out the desirable things of life, such as food, a favourable habitat and a mate. By contrast, plants are characteristically sedentary and with few exceptions are initially distributed totally by chance and ultimately by the good fortune of deposition in favourable locations. Under these circumstances, the retention of a functionally motile or at least mobile male haploid phase, e.g. pollen in higher forms, would seem to have tremendous survival value. A similar selective advantage may be ascribed to the role played after fertilization by those elements of the female gametophyte, e.g. the polar nuclei, which participate in the formation of the endosperm as an ideal milieu for the development of the embryo. In the absence of motility, it is difficult to conceive of any mechanism by which the immediate cellular products of meiosis, i.e. the gametes, could possibly be effective except in universal self-fertility.

These considerations lead to the conclusion that certain advantages accrue to different types of life cycle which emphasize the haploid phase in some forms and the diploid in others. It is clear, however, that the achievement of structural sophistication depends upon the presence— and presumably the interaction—of two distinct genomes. This association of distinct genomes, whether alike or dissimilar, is achieved either by their union in a diplophase or by their association in separate haploid nuclei in heterokaryosis.

DIKARYOSIS

Heterokaryosis, the association of distinct genomes in a system capable of vegetative propagation, occurs only in the fungi and represents an alternative to diploidy. Two types of heterokaryons may be distinguished, both types frequently occurring in the same species (Raper & San Antonio, 1954). (a) *Indefinite heterokaryons* are those in which the ratio of component nuclear types is indefinite and subject to change; heterokaryons of this type are basically vegetative. (*b*) *Dikaryons* are genetically balanced heterokaryons, each containing two haploid genomes in a ratio of 1:1; heterokaryons of this highly specialized type constitute an integral phase of the sexual cycle of all higher fungi and, in the Basidiomycetes, are capable of indefinite vegetative propagation. The dikaryon, as mentioned before, is the genetic and physiological equivalent of the diploid and differs from a diplophase primarily in the competence of the haploid nuclei to act independently. Nuclear fusion is ultimately required before meiosis can occur, but in the peculiar biology of the higher fungi the retention of the two separate genomes in a relationship of confederation has obviously had sufficient selective advantage to become fixed in these forms (Raper, 1966, 1968). Because it represents a sustained interrupted stage in the process of diploidization and because it constitutes the only functional alternative to diploidy, the dikaryon deserves careful examination, both intrinsically and as a possible vehicle for analyses of the evolution and operation of diploidy.

The dikaryon constitutes an integral stage in the sexual cycle of two groups of fungi, the Euascomycetes and the Basidiomycetes, both of which groups have haplo-dikaryotic life cycles. Beyond these basic similarities, however, the details of the life cycles and the characteristics of the dikaryons are quite different. In the Ascomycetes the vegetative phase is haploid throughout, the dikaryotic condition being initiated by the fusion of differentiated sexual organs. Nuclei from both mates then become associated in pairs within the female organ (gametangium), and the process of dikaryotization in Ascomycetes is thus synonymous with sexual fusion. Propagation of the dikaryon occurs through the synchronous division of the paired nuclei and their movement into filamentous processes that grow out of the fertilized gametangium. These filamentous processes, the ascogenous hyphae, collectively comprise the dikaryon, which is confined within a sheath of sterile haploid cells from which it derives all nutrients. Although the dikaryon may become quite extensive in some forms, it is an obligate parasite on the mycelium of the maternal haploid parent. All attempts to culture the dikaryon independently of the maternal mycelium have failed, and dikaryons of ascomycetes accordingly have limited utility as experimental objects. This largely explains the prevailing uncertainty about the significance of the dikaryon in the biology of the Ascomycetes.

The dikaryophase of the higher Basidiomycetes, by contrast, is readily culturable in many species and has been the subject of intensive study for the better part of a century. It is this type of dikaryon that is of interest here. An appreciation of the establishment, maintenance and special features of the dikaryon, however, requires a brief description of the life cycle and biology of a representative member of this group of fungi. A very common wood-rotting mushroom, *Schizophyllum commune*, typifies a majority of the higher Basidiomycetes with respect to the life cycle and to the nature and biology of the dikaryon. With minor differences, what follows in this article applies to numerous other forms,

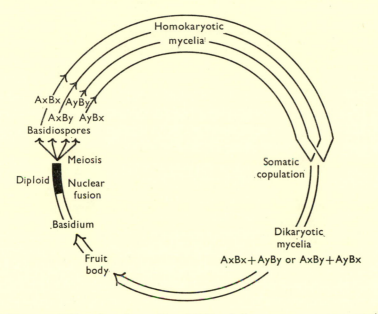

Fig. 2. Diagram of life cycle and sexuality of *Schizophyllum commune*, a representative of the higher Basidiomycetes (revised from Raper & Miles, 1958). —, Homokaryotic; =, dikaryotic; ■, diploid.

such as *Coprinus lagopus, C. fimetarius, C. radiatus, C. macrorhizus, Collybia velutipes, Pleurotus ostreatus, P. spondoleucus, Lentinus edodes, Polystictus versicolor*, etc. (Quintanilha & Pinto-Lopes, 1950; Esser, 1967). In all of these forms, the ultimate products of sexual reproduction and the immediate products of meiosis are basidiospores (Fig. 2). Each basidiospore germinates to produce haploid vegetative mycelium constituted of filamentous hyphae that are characteristically made up of uninucleate cells and have simple septa (i.e. septa lacking external appendages). A haploid plant of this sort contains nuclei of a single genotype only and is properly termed a homokaryon. After an indefinite period of vegetative growth, homokaryotic mycelia give rise to dikaryotic mycelia. In about 10% of the species examined, individual homokaryotic mycelia are self-fertile and, in isolation, each may become converted to a dikaryon. In a majority of species, all of those listed above included, individual homokaryotic mycelia are self-sterile but cross-fertile with mycelia of compatible mating types. Dikaryotization in these forms occurs by the fusion of undifferentiated vegetative cells of the two mates and the reciprocal exchange of nuclei. This droll process is the essential sexual interaction of the higher Basidiomycetes.

Whatever the way in which it is established the dikaryotic mycelium is characteristically made up of binucleate cells with an external appendage, the clamp connection, at each septum. Capable of an existence that is independent of the original homokaryon(s), the dikaryon may grow vegetatively for an indefinite period of time; it is this capacity that makes these dikaryons such favourable experimental objects. Fruiting bodies are eventually borne on dikaryotic mycelia, and here, in the basidia, occur the final stages of the sexual cycle: karyogamy followed immediately by meiosis.

THE BULLER PHENOMENON

Dissemination of the higher fungi with few exceptions is through dispersal by wind and the chance deposition of propagules on suitable substrates. Two basidiospores deposited close together upon a newly decorticated apple branch, for example, would germinate to form homokaryons which, with very high probability, would become mutually dikaryotized upon contact. The resultant dikaryon would continue to grow and would eventually produce fruiting bodies. A second pair of spores falling upon the substrate might very well repeat this history, if neither of the two new homokaryons came in contact with the dikaryon before they met each other. Here probably lies the *raison d'être* of the dikaryon in these forms: at the first contact of an established dikaryon with a homokaryon, the homokaryon is dikaryotized by nuclei from the dikaryon. Dikaryotization of the homokaryon under these circumstances was first described by Buller in 1931 and was later termed the Buller Phenomenon in his honour (Quintanilha, 1937); the phenomenon is also often referred to as a dikaryon-monokaryon, or di-mon, mating (Papazian, 1950). The origin of the dikaryotizing nuclei differs with the genetic relationships of the three homokaryotic strains involved. The following basic relationships are commonly recognized (Papazian, 1950).

(*a*) *Legitimate*

(i) *Compatible:* both component strains of the dikaryon are compatible with the homokaryon; nuclei of either strain or of both strains may serve to dikaryotize.

(ii) *Hemi-compatible:* only one of the component strains of the dikaryon is compatible with the homokaryon; nuclei of the compatible strain dikaryotize.

(*b*) *Illegitimate*

Neither strain of the dikaryon is compatible with the homokaryon; nuclei recombinant for factors required for compatibility originate within the dikaryon and serve as the agents of dikaryotization.

To these perhaps should be added yet another major category:

(*c*) *Invasive*

Nuclei of both types present in the dikaryon invade the homokaryon and replace the resident nuclei. Whether this is a legitimate or an illegitimate process is a moot question, as it occurs with about equal frequency in legitimate and illegitimate matings; moot or not, it is certainly sneaky.

The result of any di-mon interaction is the conversion of a homokaryon into a dikaryon. Di-mon matings are thus *bona fide* sexual interactions.

The extent to which dikaryotization in Nature occurs through di-mon matings as opposed to homokaryotic matings is impossible to assess with accuracy. On the basis of available information it seems likely that a major fraction of dikaryons are established in di-mon interactions. Buller (1931) first recognized the possible implications of the phenomenon in Nature and spoke of it in terms of 'social organization'. Others have since considered the probable role of the Buller Phenomenon in the distribution, sexuality and evolutionary history of the higher basidiomycetes (Raper, 1968; Raper & Flexer, 1967). Several aspects of the di-mon interaction and of its product, the derived dikaryon, are significant in a consideration of the dikaryon as an alternative to diploidy.

Mosaicism

Multiple inoculations of a substrate—be it over a period of time in wood-rotting forms such as *Schizophyllum commune* or simultaneously by the exposure of pre-inoculated substrates to conditions favourable for growth of coprophilous forms such as *Coprinus* species—result in physiologically unified dikaryons constituted of many independent accretions of diverse origins. The product is a genetic mosaic, all parts of which may produce fruiting bodies and liberate basidiospores. The entire integrated dikaryotic mycelium may be very extensive and is, in fact, a mosaic population capable of harbouring enormous variability. A very few experimental studies have provided relevant information

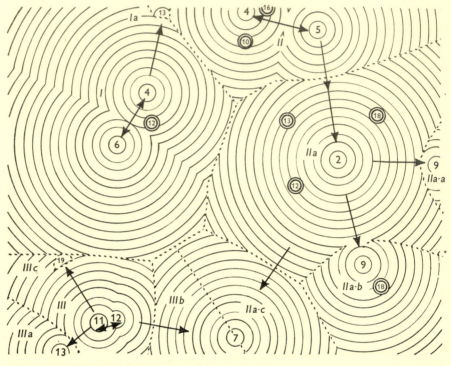

Fig. 3. Model of the development of a mosaic dikaryotic mycelium. Encircled arabic numerals specify the times, expressed in arbitrary units after exposure of the substrate, and locations of randomized spore-falls; concentric circles represent vegetative growth from spores on free substrate; double arrows represent homokaryon-homokaryon interactions to establish initial dikaryons *I*, *II* and *III*; single arrows indicate di-mon interactions to establish derived dikaryons *Ia*, *IIa*, etc.; genotypically distinct dikaryons are delimited by broken lines; spores that fell on established dikaryons are encircled with double heavy lines and are assumed to play no role in the development of the mosaic (Raper, 1966).

about the detailed genetic structure of such naturally occurring di-karyons. In the most extensive of these studies Burnett & Partington (1957) used mating-type factors as 'tell-tales' in a survey of several species of gregarious mushrooms; they found a distribution of factors in two different species that strongly suggested the contiguous sharing of factors in distinct genomes, as would be expected if clusters had indeed been formed by successive di-mon interactions.

A simple model (Raper, 1966) based on known characteristics of di-mon interactions permits some understanding of the overall composite structure of a mosaic dikaryon and serves as a base for a number of predictions (Fig. 3). With a high probability, contiguous portions of the dikaryon would share genetic characters in haploid-genome units. With an equally high probability, contiguous portions should differ genetically by single genome units. Finally, certain haploid genotypes should predominate as repeating components in the entire dikaryotic mosaic. These predictions can be tested in simple and direct, if laborious, experiments. Such experiments would require naturally occurring dikaryotic mosaics of considerable extent, but these are seen from time to time. For example, *Schizophyllum commune* occurs as a wound parasite on apple trees in epidemic outbreaks and became very common in the Upper Galilee of Israel on trees damaged by shelling in the Six-Day War of 1967. Work now in progress seeks to analyse some of these populations, each consisting of a hundred or so fruiting bodies taken from an individual tree, in terms of a variety of parameters relating to population biology, such as (*a*) the distribution of shared and dissimilar genetic factors and (*b*) the growth rates of dikaryons cultured directly from fruiting bodies and homokaryotic progeny derived therefrom (G. Simchen, personal communication). Easy rationalization of any observed distribution of haploid genotypes within a mosaic dikaryon, however, is effectively thwarted by two other phenomena which occur regularly in di-mon interactions. These reveal further potential for plasticity of the dikaryophase of Basidiomycetes and deserve brief comment.

Internuclear selection

In a compatible di-mon mating, both nuclear types of the dikaryon are compatible with the homokaryon, and the natural expectation demands that in a replicated sample of such a mating, the two components of the dikaryon would serve as dikaryotizing agents with equal frequency. However, such is rarely the case. The first test of this expectation showed for each specific compatible di-mon mating a predilection for dikaryotization by one particular component of the dikaryon. In a limited sample this bias was attributed to a selective advantage enjoyed by one of the mating-type factors (Quintanilha, 1939). Later work confirmed the fact of very strong internuclear selection (Papazian, 1950), but different workers have found convincing evidence to attribute the effect to different genetic bases. (*a*) Crowe (1963) concluded that the same mating-type factor implicated by Quintanilha, that is the *A*-factor, is responsible for the phenomenon. (*b*) Ellingboe & Raper (1962*b*) ascribed the effect to the alternate mating-type factor, that is, the B-factor. (c) Kimura (1958) asserted that other genes in the background genomes mediate the selection. Whatever may be the basic cause of the preferential selection of one nuclear type over the other in any particular instance, the effect is often appreciable; the preference in many cases is 25:1 or greater. Such strong selection of certain haploid genotypes should also have a marked impact on the formation and composition of the mosaic dikaryon; this selection might account for a considerable departure from a random distribution in the mosaic of its component haploid genotypes.

Somatic recombination

Another feature common to di-mon matings also contributes to the complexity of the final mosaic dikaryon. That dikaryotization of the homokaryon occurred in illegitimate di-mon matings, in which neither haploid component was compatible with the homokaryon, was an important part of Buller's original description of the phenomenon. Various schemes were offered to explain this unexpected result (cf. Raper, 1966, p. 61, for review); in 1939 Quintanilha reported the results of work that clearly showed somatic recombination between the genomes of the two dikaryotic components to be involved. Papazian (1950) confirmed Quintanilha's basic findings and further found that recombinant nuclei often served as the dikaryotizing agents for homokaryotic members of fully compatible di-mon matings. Crowe (1960) made the first detailed analysis of this process and found 13 cases of dikaryotization by recombinant nuclei in a sample of 76 compatible di-mon matings. Her work revealed recombination to involve crossing-over as well as independent assortment; from this she concluded that somatic recombination here occurred through 'precocious meiosis' of rare diploid nuclei in the dikaryon.

More recent work (see Raper, 1966, p. 183 for review) has shown the underlying process, or processes, of recombination to be extraordinarily difficult to resolve. Several different interactions certainly occur simultaneously whenever a homokaryon comes into contact with a dikaryon with which it is incompatible. Moreover, what *appears* to happen depends in large measure on the means of examination, particularly on the selective procedures used to recover the products of the recombinational process. Occasional recombinant nuclei have been recovered from 'unmated' dikaryons, i.e. isolated dikaryons (Parag, 1962b); the homokaryon may thus sometimes serve only as a selecting agent for rare recombinant compatible nuclei in the illegitimate di-mon mating. However, it has been clearly established that in most cases the homokaryon is no mere neutral selecting agent: genetic markers of all *three* original haploid genomes are frequently recovered among the recombinant genotypes. On the basis of more or less equally convincing evidence two underlying mechanisms have been advanced to account for genetic recombination in these vegetative systems: (*a*) parasexuality (*sensu* Pontecorvo, 1956) involving haploidization either of triploid nuclei or of two successively formed diploids in the dikaryon and with the homokaryon, respectively (Swiezynski, 1962, 1963; Prud'homme, 1963, 1965); (*b*) a specific transfer of mating-type factors from both components of the dikaryon into the background genome of the homokaryon (Ellingboe & Raper, 1962*a*; Ellingboe, 1963). Again, as with internuclear selection, the underlying bases of somatic recombination are not clear, but it obviously constitutes an important source of variability in the establishment of the mosaic dikaryon. This variability derives from the interchange of genetic characters between all components of the mosaic dikaryon without the restrictions that would be imposed by assortment of static haploid genomes. All in all, the stable vegetative dikaryon is not only the physiological and genetic equivalent of a diplophase, it is a far more plastic and adaptable consortium of two genomes than is the diploid; this arrangement quite clearly has served the higher fungi well. So well, in fact, that they seem to have evolved an effective means of protecting it.

DIPLOIDY IN HIGHER BASIDIOMYCETES

Until very recently the dikaryon was the only condition known in the higher fungi in which a persistent association of two genomes was maintained in a vegetative system. The recovery of aneuploid isolates in some of the earlier studies of somatic recombination suggested the existence of at least transient diploid nuclei in vegetative systems (Gans & Prud'homme, 1958; Prud'homme & Gans, 1958; Middleton, 1964). Within the past decade several workers have reported relatively stable diploids to originate in various types of di-mon matings (Swiezynski, 1962, 1963; Prud'homme, 1963, 1965), in incompatible matings between homokaryons (Casselton, 1965; Parag & Nachman, 1966; Mills & Ellingboe, 1969; Day & Roberts, 1969), and in compatible matings between homokaryons (Koltin & Raper, 1968). In all but the latter case diploidy appears to be quite sporadic, statistically predictable in large samples but highly improbable at the level of individual nuclei. In general, these diploids are primarily recovered by nutritional or other selective means; morphologically they closely mimic the heterokaryons to which they correspond *vis-à-vis* the factors that control mating. Because of their sporadic appearance and their close mimicry of heterokaryotic mycelial types, such diploids are interpreted as selectively favoured rare diploid nuclei which arise continuously in heterokaryons and which ordinarily undergo rapid haploidization. Diploids of this type are of considerable experimental interest in (*a*) their interactions with homokaryons, (*b*) their interactions with other diploids, (*c*) the opportunities they provide for analyses of the comparative effectiveness of complementation between genes in *cis* and *trans* arrangements in diploids and dikaryons (Day & Roberts, 1969).

A single report of the regular and uniform formation of diploids in matings between homokaryotic strains suggests a more basic phenomenon than the cases listed above (Koltin & Raper, 1968). This phenomenon was first observed in a mating between two fully compatible strains, each of which carried several mutations that had been induced by X-irradiation. The result of this mating was not the expected dikaryon but a mycelium that appeared to be homokaryotic, i.e. uninucleate cells and no clamp connections. This exceptional mycelium was also prototrophic whereas the original mates were both auxotrophic. Furthermore, this mycelium produced fruiting bodies; an initial analysis of its progeny revealed all known markers of both original homokaryons. When either of the original mates was out-crossed with a wild strain and the progeny were mated with the alternate original mate, dikaryons resulted in half the cases and uninucleate-celled mycelia in the other half. Further analysis confirmed the hypothesis that a recessive gene, *dik*, homozygous by chance in the exceptional mating, determined uniform and rapid diploidization when present in both mates, regardless of the mating types of the two partners and of any other known genetic factors. Homozygosity for the dominant allele, *dik*$^+$, or heterozygosity led to normal dikaryotization when the two mates were compatible and to the specific common-factor heterokaryons when the two mates were incompatible (see below). The two strains carrying *dik* may have had a common ancestor in which a single mutation of the *dik* locus occurred. It is important to note, however, that only the dominant allele, *dik*$^+$, is known in strains collected from Nature.

If, as seems very likely from the facts and relationships considered above, the Buller Phenomenon has a special significance in the biology of the higher fungi, highly efficient selection against diploids of all these types is readily rationalized. The various types of diploids are quite stable in isolation, and they form dikaryons and other types of heterokaryons in interactions with homokaryons. In these haploid-diploid systems, the diploid nuclei are commonly very unstable; rampant aneuploidy is the usual result. Whether the aneuploids arise as sectors from vegetative systems or segregate as monosporous progeny in meiosis, aneuploids are decidedly less fit than are their haploid counterparts. These considerations thus point to several reasonable conclusions. (*a*) As a balanced genetic system of two associated genomes, the dikaryophase is a true alternative to diploidy in that it has the basic features of complementation, somatic recombination and buffering that underlie the biological effectiveness of the diploid. (*b*) Beyond these features shared with the diploid, the dikaryophase has the further competence to participate in *all* of the interactions of which its component strains are separately capable. (*c*) The dikaryophase is thus an exceptionally versatile system combining features of diploidy and haploidy which has proved crucial to the survival of the higher fungi. (*d*) The basis for the development or retention of the dikaryon appears to lie in the evolution of genetic devices to protect and maintain it, rather than in any inability to 'manage' the diploid state. That only a single specific genetic factor to assure dikaryosis rather than diploidy has to date been discovered in no way argues against additional, and probably less simple, genetic systems which contribute toward the stabilization of a basic advantageous feature.

As mentioned above, the dikaryon is something of a biological oddity and an evolutionary cul-de-sac, although a highly successful one, and may be considered as a sustained interruption of the process of diploidization. If this view be admissible, the process of dikaryotization, or dikaryosis, which occurs in several successive stages and is subject to considerable dissection by genetic means, might well be viewed as diploidization in slow motion. Dikaryosis is under precise genetic control, as are the determination of mating competence among homokaryons and the characteristics of the mycelial products of these interactions. An understanding of the dikaryon and of dikaryosis is thus necessarily inextricably interrelated with the details of the underlying incompatibility system which regulates these different aspects of sexuality of the higher fungi.

SEXUAL MORPHOGENESIS

The mating system of a typical form exemplified here by *Schizophyllum commune* stands in stark contrast to that of most other organisms. The worldwide population of this species consists of about 42,000 distinct mating types, and individuals of these are inter-fertile in > 98% of all possible combinations. This already high degree of outbreeding efficiency is increased to 100% by the option afforded by the Buller Phenomenon, because any homokaryon may be dikaryotized, fertilized, by any dikaryon. This degree of opportunism is only one of the several unusual manifestations of a genetic system, designated tetrapolar or bifactoral incompatibility, which regulates a morphogenetic progression unique to the higher Basidiomycetes. The control of this morphogenetic progression is achieved by two clearly separable parts of a regulatory system: the *regulating component* and the *regulated component*. The regulating component comprises four distinct loci, each with an extensive series of multiple alleles. The four loci are linked in two pairs, designated the *A* and *B* incompatibility factors. The *A* and *B* factors serve to 'turn on' or to 'turn off' two distinct but interrelated morphogenetic sequences that transform two compatible haploid plants into a uniform dikaryon. The regulated component consists of a large but unknown number of genes scattered throughout the genome and which mediate inidividual events in these sequences.

The morphogenetic sequence begins with an interaction of two homokaryons, each of which contains nuclei of a single genotype that includes a specific *A* factor and a specific *B* factor. When two such mycelia meet, contiguous cells of the two fuse as a matter of course to establish binucleate fusion cells. When the nuclei brought together in the fusion cells carry different *A* and different *B* factors, there ensues a sequence of events that converts both mates into a fertile heterokaryon, the dikaryon. This sequence comprises six distinct stages (Raper & Raper, 1968) (Fig. 4).

(1) *Nuclear migration.* The nuclei of the fusion cells pass into the filaments of both mates and migrate (presumably through successively disrupted septa) throughout both pre-established mycelia.

(2) *Nuclear pairing.* When the migrating nuclei reach the apical cells of either mycelium, a pairing occurs between a migrant nucleus and the resident nucleus in each apical cell.

(3) *Formation of hook-cell.* Before nuclear division, a short lateral outgrowth, directed basipetally, forms on the apical cell in the region of the paired nuclei.

(4) *Conjugate division.* The two paired nuclei then divide synchronously, one nucleus dividing along the axis of the cell, the other in such a way as to deposit the basipetal daughter in the lateral hook-cell.

(5) *Septation of apical cell and hook-cell.* A septum is then laid down across the dividing cell immediately behind the hook-cell; simultaneously a septum forms across the base of the hook-cell. The temporarily

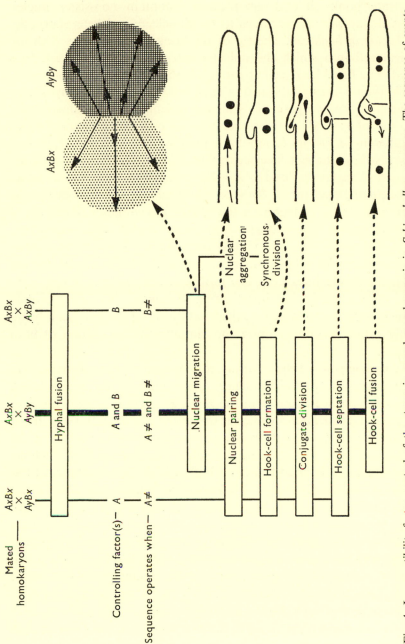

Fig. 4. Incompatibility-factor control of the stages in sexual morphogenesis in *Schizophyllum commune*. The sequence of events leading to the establishment of the dikaryon in a compatible mating ($A \neq B \neq$) is traced by the heavy central line. The partial sequence evoked by the A and B factors separately in $A = B \neq$ and $A \neq B =$ heterokaryons are traced by the thinner lines to left and right. The events comprising the morphogenetic progression are diagrammed on the right (from Raper & Raper, 1968).

uninucleate hook-cell and subapical cell contain non-sister nuclei, whereas the apical cell continues to be binucleate and heterokaryotic.

(6) *Fusion of hook-cell.* The tip of the hook-cell then fuses with the subapical cell into which the nucleus entrapped in the hook-cell is released. The fused hook-cell apparatus is the clamp connection.

A dikaryon is thus established in each cell of the growing mycelium, and nuclei of the two parental types are indefinitely maintained at an exact ratio of 1:1. All of these events occur only when the two mates have different A factors (symbolized by $A \neq$) and different B factors, $A \neq B \neq$. None of the sequence occurs when A and B factors are the same in the two mates (symbolized by $A = B =$); both homokaryotic mates remain essentially unchanged. Only certain events of the sequence occur whenever either the A factor or the B factor is common. When the A factor is the same in both mates ($A = B \neq$) only nuclear migration occurs; when the B factor is the same in both mates ($A \neq B =$) little or no nuclear migration ensues, but all subsequent stages proceed normally except the final fusion of the hook-cell. The morphogenetic process is thus constituted of two partial but interlocking sequences that are separately regulated by the A and B factors, the final fusion of the hook-cell being the only event dependent upon the simultaneous operation of both sequences.

The characteristics of the resulting common-factor heterokaryons are determined by the partial sequences operating in these interactions. Only the B sequence operates in the $A = B \neq$ interaction; this involves rapid heterokaryotization by nuclear migration and a continuing alteration in the stability of septal structures. The disruption of septa and the lack of control over nuclear pairing result in the formation of a heterokaryon with cells of varied length and of variable nuclear content (0 to > 20 per cell). The ratio of the two constituent nuclear types also varies widely in different $A = B \neq$ heterokaryons, from about 1:1 to thousands:1. In *Schizophyllum commune*, but not in all tetrapolar species, $A = B \neq$ heterokaryons are metabolically subnormal and morphologically aberrant, with irregularly formed and branched hyphae and little aerial growth. The result of this interaction is quite different from that of the alternate common-factor interaction. Only the A sequence operates in the $A \neq B =$ interaction; there is little or no nuclear migration; heterokaryosis is accordingly restricted to the immediate vicinity of the original fusion cells. Because there is no extension of heterokaryosis by nuclear migration, vegetative propagation of the $A \neq B =$ heterokaryon from the fusion cells usually requires nutritional forcing. In such heterokaryons nuclear pairing and conjugate division provide each apical cell with a nucleus of each of the two constituent types, but the failure of the last stage of the morphogenetic progression leaves a nucleus permanently trapped in each hook-cell, and the subapical cell remains uninucleate. The $A \neq B =$ heterokaryon is thus made up predominantly of uninucleate cells except for the binucleate terminal cells; the overall nuclear ratio is approximately 1:1. The common-factor heterokaryons, $A = B \neq$ and $A \neq B =$, typically produce no fruiting bodies and are thus sexually infertile.

The regulated component

The roles of the many genes of the regulated component of the system can only be deduced from the aberrations caused by mutations. This component of the system is considered below with reference to disruptive modifications of the normal progression.

The regulating component

The regulating component of this system comprises two incompatibility factors, A and B, that assort independently. All available information indicates that A and B factors have a common basic structure, and a common pattern of interaction obtains among members of either series of factors. Each factor, A or B, is constituted of two linked loci, α and β, each with an extensive series of multiple alleles. Linkage between $A\alpha$ and $A\beta$ is generally far looser than between $B\alpha$ and $B\beta$, but the linkage in both cases is extremely variable: 1 to 40 morgans for $A\alpha$ and $A\beta$, < 0.1 to 8 morgans for $B\alpha$ and $B\beta$. The high degree of variability in recombination between αs and βs within each factor is controlled by genes that are external to the factors and specific for each factor (Simchen, 1967; Stamberg, 1968).

A given combination of $A\alpha$ and $A\beta$ alleles determines a unique A factor phenotype that to date can be recognized only on the basis of mating responses. A factors need differ in only one of their α or β components to be fully compatible; for example A $\alpha1-\beta1$ is equally compatible with A $\alpha1-\beta2$, A $\alpha2-\beta1$ and A $\alpha2-\beta2$. The numbers of alleles identified at the $A\alpha$ and $A\beta$ loci are 9 and 26, respectively, and the calculated numbers of alleles in the world population are nine of $A\alpha$ and about 50 of $A\beta$ (Raper, Baxter & Ellingboe, 1960). The A factor thus has about 450 fully compatible phenotypes. The number of alleles at the component loci of the B factor appears to be considerably smaller and the total number of B factors is 90 to 95 (Koltin, Raper & Simchen, 1967). Because each specific combination of an A factor and a B factor constitutes a unique mating type, there are almost 42,000 distinct mating types in the worldwide population of this species.

MODIFICATIONS OF SEXUAL MORPHOGENESIS

The regulated component

The expression of the several stages of sexual morphogenesis, determined by the proper combination of unlike incompatibility factors, is under the further control of numerous genes distributed throughout the genome

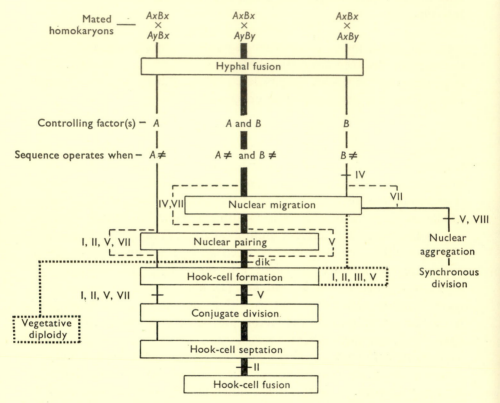

Fig. 5. Modifications of sexual morphogenesis by modifying mutations of the several phenotypic classes. A cross bar represents the blocking of all subsequent events in the series; a dashed line around an event indicates the bypassing of that event; a dotted line to an event indicates the induction of that event (revised from Raper & Raper, 1966).

which collectively comprise the regulated component of the morphogenetic sequence (Raper & Raper, 1964, 1966). Mutations in these genes disrupt the morphogenetic progression in a perplexing variety of ways (Fig. 5). A couple of dozen such 'modifying mutations' have now been characterized. They occupy many distinct loci, all segregate normally, and with few possible exceptions they assort independently of the *A* and *B* factors. Eight different categories of modifying mutations, types I to VIII, are defined by criteria such as: (a) dominance or recessiveness to their wild-type alleles, (b) the nature of interactions with other modifying mutations, (c) specific effects on the morphogenetic sequence, (d) whether only the A sequence is affected, only the B sequence is affected, or both sequences are affected. None of the modifying mutations is expressed in homokaryons that are otherwise wild type; they are expressed only under conditions in which part or all of the morphogenetic progression is 'turned on', i.e. in the $A \neq B \neq$, $A = B \neq$ and $A \neq B =$ heterokaryons (or, as described below, in their mutant-homokaryotic mimics). That many of the modifying mutations have pronounced effects upon events of both the A and the B sequences indicates an intricate relationship among the biochemical processes that underlie the several events of the two partial sequences.

The regulating component

Although no mutation in an incompatibility locus has ever been recovered from Nature, numerous mutations in the basic incompatibility loci have been induced, recovered and characterized during recent years (Day, 1963; Parag, 1962a; Raper, Boyd & Raper, 1965). Two classes of mutations have been studied: primary mutations induced in wild-type alleles, and secondary mutations induced in primary mutant alleles. Primary mutations have been induced in three of the four incompatibility loci; without exception to date, each of these mutations has resulted in the loss of the discriminatory function of self-recognition and self-incompatibility of the factor in which the mutant locus lies. A mutant factor in a homokaryon thus mimics the effects of paired different normal factors of the same series in a heterokaryon (Fig. 6). A homokaryon carrying a mutant B factor, in which the B sequence operates, is a phenotypic mimic of the $A = B \neq$ heterokaryon; an A factor mutant, with the A sequence operating, is a mimic of the $A \neq B =$ heterokaryon; and a homokaryon carrying a mutant A factor and a mutant B factor, in which both sequences operate, is a close mimic of the dikaryon—complete with binucleate cells, clamp connections and the competence to produce fertile fruiting bodies.

Among the primary mutations of the basic incompatibility loci, several in the $B\beta$ locus have been the objects of considerable study. In addition to the distinctive macroscopic phenotype of homokaryons carrying these mutations, i.e. mimics of the $A = B \neq$ heterokaryon, these mutations are of interest because they confer universal compatibility vis-à-vis the B factor, and they are accordingly self-fertile. Morphological aberrations extend to the ultrastructural level, and the septa of the mutant strains are frequently partial, disrupted or dissolved as in the $A = B \neq$ heterokaryon (Giesy & Day, 1965; Jersild, Mishkin & Nieder-pruem, 1966; Koltin & Flexer, 1969). Two different categories of mutations restore these $B\beta$ mutants to morphological normality: the secondary mutations in the $B\beta$ locus and modifying mutations of types IV, VI and VIII discussed above.

Several distinct classes of secondary mutations induced in mutant alleles of the $B\beta$ locus have been recovered and characterized (Raper *et al.* 1965; Raper & Raudaskoski, 1968; Raudaskoski, 1970; Raper & Raper, 1966, unpublished observations). All of these secondary

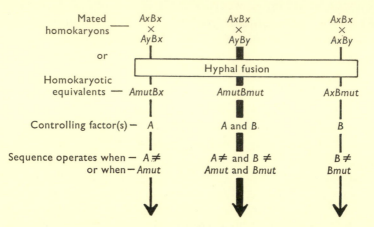

Fig. 6. Primary mutations at the basic incompatibility loci mimic in the homokaryon the effects of different, wild-type factors in the heterokaryon (see Fig. 4 above).

mutations restore normal homokaryotic morphology; otherwise they differ in numerous details in their interactions with variously constituted wild-type B factors. Some appear to be totally sterile; others are fertile with most wild B factors but not with their progenitors, that is, the factors in which they were induced; still others are fertile with all wild B factors, including their progenitors. The interaction between a strain carrying such a secondary mutation in the $B\beta$ locus and a strain carrying its progenitor is not wholly normal in that the secondary mutant donates but does not accept dikaryotizing nuclei. Such mutant loci, however, have new specificities and are the nearest approach yet achieved in the laboratory to the generation of new normal alleles at the incompatibility locus. It is uncertain whether this finding has any bearing on the origin of new alleles in the extensive series of alleles which characterize these loci in Nature.

Modifying mutations, malfunctions of the regulated component that restore normal morphology in $B\beta$-mutant strains but are located outside the B factor, are even more frequently encountered than secondary $B\beta$ mutations. These mutations do not affect the pattern of interactions of the mutant B factor, i.e. it remains compatible with all B factors. Again, as with the secondary $B\beta$ mutations, these modifying mutations restore normal homokaryotic morphology to $B\beta$-mutant strains; the distinctions between types depend upon the details of the interactions between mutant and wild-type strains. For example, types IV and VI interact unilaterally with wild-type strains, whereas type VIII interacts reciprocally with wild strains. In this latter case, the process of interaction is normal, but the pattern of interaction is not: two strains which carry both the $B\beta$ mutation and a type VIII modifier are reciprocally interfertile when they have different A factors. Of these three types, only mutations of type IV have been examined in any detail, but this study has yielded several findings of interest: (a) many loci are involved, and they are widely scattered, (b) each mutation is recessive to its wild allele; (c) independent mutations complement in certain pairings, but not in others, (d) noncomplementing mutations, however, often segregate independently (Y. Koltin, unpublished observations; C. Dubovoy, unpublished observations), (e) ultrastructural morphology of the septa of $B\beta$–VI doubly-mutant strains is indistinguishable from that of the wild-type homokaryon (A. S. Flexer, unpublished observations), (f) the mycelial proteins of the doubly-mutant strain are like those of the wild-type homokaryon rather than those of the $B\beta$-mutant strain (Wang & Raper, 1969a, 1970).

Much less is known about the biochemical aspects than of the genetic aspects of the morphogenetic progression, but some preliminary findings are relevant to our present concern. A comparison of the soluble proteins of the homokaryon with those of the dikaryon by serological and electrophoretic techniques have revealed no constant or significant differences between the protein spectra of two highly isogenic compatible homokaryons, i.e. homokaryons with identical genotypes except for the alleles at all four incompatibility loci. These spectra, however, were quite different from that of the dikaryon formed by the mating of the two isogenic homokaryons (Raper & Esser, 1961; Wang & Raper, 1969). Certain proteins are present in the homokaryons but are lacking in the dikaryon and *vice versa*, and the differences relate, not to the specific incompatibility factors present, but rather to the operation or inactivity of the morphogenetic sequence. Mutations which affect the system, both in the incompatibility loci and in a few modifying loci, have been introduced into genomes isogenic with the homokaryons and the dikaryon mentioned above, and the protein spectra of these fit the expectation that the protein spectra can be correlated with the operation of all, or parts of, or none of the morphogenetic sequence (Wang & Raper, 1970). Another study has correlated the activity of a specific enzyme, R-glucanase, the substrate of which is a component of the cell wall, with the operation of the *B* sequence (Wessels & Niederpruem, 1967; Wessels, 1969).

GENETIC REGULATION OF SEXUAL MORPHOGENESIS

The mode of operation of the incompatibility factors of *Schizophyllum commune* and other forms with the same pattern of sexuality has long been the subject of speculation. The intensive work of the last decade has made available such a wealth of new information that the formulation of models to explain the mode of action of the four basic incompatibility genes becomes more or less a compulsion. Even so, there is still much information needed to construct a model that describes how the system *does* work; present efforts are restricted to models of how the system *might* work. One thing appears certain: the morphogenetic sequence in *S. commune* and its genetic control must be much more complex and less direct than are the model regulatory systems in prokaryotes. In comparison with these latter systems, however, several generalities may be stated which may eventually form the basis of a fully acceptable model. (*a*) The two genes of each incompatibility factor can be likened to master regulating genes for one of the two interlocking sequences of events that together constitute the process of dikaryosis. (*b*) When only a single factor of either series, *A* or *B*, is present, whether in a homokaryon or in a common-factor heterokaryon, the corresponding morphogenetic sequence is inoperative. (*c*) When two different factors of either series are present, the corresponding sequence is operative and a common-factor heterokaryon results. The presence of different factors of both series results in the operation of both sequences and the establishment of the dikaryon. It is impossible on the basis of available information to distinguish between two different modes of action: the repression by a single factor of a constitutive system *or* the activation by interacting different factors (or by a single mutant factor) of a constitutive system. The former was earlier favoured as more nearly accounting for all known features of the system (Raper, 1966; Raper & Raper, 1968); but there are enough uncertainties to make the consideration of the alternatives attractive. (*d*) Whatever may prove to be the means by which the *A* and *B* factors regulate their respective sequences, the fact that none of the modifying mutations is expressed in the absence of morphogenetic activity very strongly suggests that the very large and still unknown number of such loci represent the structural genes that underlie the morphogenetic process and are directly responsible for the several events of the progression. (e) The numerous modifying mutations which block or otherwise alter specific stages of the morphogenetic sequence (those types affecting nuclear migration are best known) indicate a very complex link of control between the incompatibility factors and their target-genes, whether direct and multiradiate or indirect and multistage.

CONCLUSIONS

The degree of differentiation achieved by most organisms appears to be directly related to the organization of their genetic material. This relationship is reflected in an indistinctly graded series ranging from prokaryotes to haploids to haplo-dikaryons to diploids. Already enough is known to reject the simple notion that regulatory systems in prokaryotes can be borrowed directly to serve as models of regulatory systems underlying morphogenetic phenomena in eukaryotes. The course of differentiation from gene(s) to final morphological expression is known in no case and may be expected to be incredibly complex in even the simplest case. This was recognized in reference to sexual morphogenesis in *Schizophyllum commune* described above (Raper & Raper, 1968): 'Sexual morphogenesis in the higher Basidiomycetes is admittedly a complex business, and it would seem that a complete understanding of the system at the biochemical level is a very distant if not hopeless goal. Yet, as a morphogenetic process in an eukaryotic organism, it is ideally simple, as it involves only the transition from a uninucleate to a binucleate state. In other cases, less complex genetic systems may underlie grander morphological developments. This, however, appears unlikely, as it is only reasonable to expect a direct relationship between the magnitude of morphological change and the complexity of its genetic control.'

There is another point to ponder here. When a very simple alteration in a structurally simple dikaryotic organism, such as that considered above, is considered *in toto*, it proves to be extraordinarily complicated. The same basic pattern of control may well underlie comparably simple morphogenetic processes in evolutionarily more advanced diploid organisms. Every other feature common to eukaryotes, however, increases in complexity from primitive to advanced organisms, as additional refinements become 'standard equipment'. The innate complexity of the system regulating a morphogenetic progression may increase, not only with the complexity of the process itself as suggested above, but it may very well also increase with the overall complexity of the organism in which it occurs. Herein may lie the real superiority of diploidy: in the subtle control that, through interaction of two redundant genomes in intimate association, can be exerted on the minutiae of the differentiative processes. The problem here, however, is that of experimental analysis. All biologists are experimentalists of one sort or another, and nothing can be accepted without some sort of experimental verification. To those interested in a demonstration of the critical advantage of diploidy over an association of the same component genomes maintained in spatial separation, the higher Basidiomycetes are the one group of organisms which provides the two requisite genetically-balanced systems.

The work in this laboratory on sexuality and sexual morphogenesis in *Schizophyllum commune* was supported by grants from the National Institutes of Health of the U.S. Public Health Service, the National Science Foundation and the U.S. Atomic Energy Commission.

REFERENCES

BULLER, A. H. R. (1931). *Researches on Fungi*. IV. London: Longmans, Green & Co.

BURNETT, J. H. & PARTINGTON, M. (1957). Spatial distribution of fungal mating type factors. *Proc. R. phys. Soc. Edinb.* **26**, 61.

CASSELTON, L. A. (1965). The production and behaviour of diploid strains of *Coprinus lagopus*. *Genet. Res.* **6**, 190.

COMMONER, B. (1964). DNA and the chemistry of inheritance. *Am. Scient.* **52**, 365.

CROWE, L. K. (1960). The exchange of genes between nuclei of a dikaryon. *Heredity, Lond.* **15**, 397.

CROWE, L. K. (1963). Competition between compatible nuclei in the establishment of a dikaryon in *Schizophyllum commune*. *Heredity, Lond.* **18**, 525.

DAVIS, R. H. (1966). Heterokaryosis. In *The Fungi*, vol. 2, pp. 567–88. Eds. G. C. Ainsworth and A. S. Sussman. New York: Academic Press.

DAY, P. R. (1963). Mutations affecting the *A* mating type locus in *Coprinus lagopus*. *Genet. Res.* **4**, 55.

DAY, P. R. & ROBERTS, C. F. (1969). Complementation in dikaryons and diploids of *Coprinus lagopus*. *Genetics, Princeton* **62**, 265.

ELLINGBOE, A. H. (1963). Illegitimacy and specific factor transfer in *Schizophyllum commune*. *Proc. natn. Acad. Sci. U.S.A.* **49**, 286.

ELLINGBOE, A. H. & RAPER, J. R. (1962*a*). Somatic recombination in *Schizophyllum commune*. *Genetics, Princeton* **47**, 85.

ELLINGBOE, A. H. & RAPER, J. R. (1962*b*). The Buller Phenomenon in *Schizophyllum commune*: nuclear selection in fully compatible dikaryotic-homokaryotic matings. *Am. J. Bot.* **49**, 454.

ESSER, K. (1967). Die Verbreitung der Incompatibilität bei Thallophyten. *Handb. Pflanzenphysiol*, vol. 18, pp. 321–43. Eds. J. Straub and H. F. Linskens. Berlin: Springer-Verlag.

GANS, M. & PRUD'HOMME, N. (1958). Échanges nucléaire chez Basidiomycète *Coprinus fimetarius. C. r. hebd. Séanc. Acad. Sci., Paris* 247, 1895.

GIESY, R. M. & DAY, P. R. (1965). The septal pores of *Coprinus lagopus* (Fr.) *sensu* Buller in relation to nuclear migration. *Am. J. Bot.* 52, 287.

GRANT, V. (1963). *The Origin of Adaptations.* New York: Columbia University Press.

GRELL, K. G. (1967). Sexual Reproduction in Protozoa. In *Research in Proto-zoology.* Ed. T.-T. Chen. vol. 2, pp. 148–213. Oxford: Pergamon Press.

JERSILD, R., MISHKIN, S. & NIEDERPRUEM, D. P. (1966). Origin and ultrastructure of complex septa in *Schizophyllum commune* development. *Arch. Mikrobiol.* 57, 20.

KIMURA, K. (1958). Diploidization in the Hymenomycetes. II. Nuclear behaviour in the Buller Phenomenon. *Biol. J. Okayama Univ.* 4, 1.

KOLTIN, Y. & FLEXER, A. S. (1969). Alterations in nuclear distribution in *B*-mutant strains of *Schizophyllum commune. J. Cell Sci.* 4, 739.

KOLTIN, Y. & RAPER, J. R. (1968). Dikaryosis: Genetic determination in *Schizo-phyllum. Science, N.Y.* 160, 85.

KOLTIN, Y., RAPER, J. R. & SIMCHEN, G. (1967). Genetics of the incompatibility factor of *Schizophyllum commune*: The *B* factor. *Proc. natn. Acad. Sci. U.S.A.* 57, 55.

MARGULIS, L. (1968). Evolutionary criteria in Thallophytes: a radical alternative. *Science, N.Y.* 161, 1020.

MIDDLETON, R. B. (1964). Sexual and somatic recombination in common-*AB* heterokaryons of *Schizophyllum commune. Genetics, Princeton* 50, 701.

MILLS, D. I. & ELLINGBOE, A. H. (1969). A common-*AB* diploid of *Schizophyllum commune. Genetics, Princeton* 62, 271.

NITSCH, J. P. & NITSCH, C. (1969). Haploid plants from pollen grains. *Science, N.Y.* 163, 85.

PAPAZIAN, H. P. (1950). Physiology of the incompatibility factors in *Schizophyllum commune. Bot. Gaz.* 112, 143.

PAPAZIAN, H. P. (1954). A theoretical aspect of the genetics of Volvox. *Am. Nat.* 88, 172.

PARAG, Y. (1962*a*). Mutations in the *B* incompatibility factor in *Schizophyllum commune. Proc. natn. Acad. Sci. U.S.A.* 48, 743.

PARAG, Y. (1962*b*). Somatic recombination in dikaryons of *Schizophyllum commune. Heredity, Lond.* 17, 305.

PARAG, Y. & NACHMAN, B. (1966). Diploidy in the tetrapolar heterothallic Basidio-mycete *Schizophyllum commune. Heredity, Lond.* 21, 151.

PONTECORVO, G. (1956). The parasexual cycle in fungi. *A. Rev. Microbiol.* 10, 393.

PRUD'HOMME, N. (1963). Recombinaisons chromosomiques extrabasidiales chez un Basidiomycete *Coprinus radiatus. Annls de Génét.* 4, 63.

PRUD'HOMME, N. (1965). Somatic Recombination in the Basidiomycete *Coprinus radiatus.* In *Incompatibility in Fungi,* pp. 48–52. Eds. K. Esser and J. R. Raper. Berlin: Springer-Verlag.

PRUD'HOMME, N. & GANS, M. (1958). Formation de noyaux partiellement diploides au cours du phénomène de Buller. *C. r. hebd. Séanc. Acad. Sci., Paris* 247, 2419.

QUINTANILHA, A. (1937). Contribution à l'étude génétique du phénomène de Buller. *C. r. hebd. Séanc. Acad. Sci. Paris* 205, 745.

QUINTANILHA, A. (1939). Étude génétique du phénomène de Buller. *Bol. Soc. broteriana* 13, 425.

QUINTANILHA, A. & PINTO-LOPES, J. (1950). Aperçu sur l'état actuel de nos con-naissances concernant en 'conduite sexuelle' des espèces d'Hymenomycetes. *Bol. Soc. broteriana* 24, 115.

RAPER, C. A. & RAPER, J. R. (1964). Mutations affecting heterokaryosis in *Schizophyllum commune*. *Am. J. bot.* **51**, 503.

RAPER, C. A. & RAPER, J. R. (1966). Mutations modifying sexual morphogenesis in *Schizophyllum*. *Genetics, Princeton* **54**, 1151.

RAPER, J. R. (1954). Life Cycles, Sexuality, and Sexual Mechanisms in the Fungi. In *Sex in Micro-organisms*, pp. 42–81. Eds. D. H. Wenrich, I. F. Lewis and J. R. Raper. Washington: American Association for the Advancement of Science.

RAPER, J. R. (1966). *Genetics of Sexuality in Higher Fungi*. New York: Ronald.

RAPER, J. R. (1968). On the Evolution of Fungi. In *The Fungi*, vol. 3, pp. 677–93. Eds. G. C. Ainsworth and A. S. Sussman. London and New York: Academic Press.

RAPER, J. R., BAXTER, M. G. & ELLINGBOE, A. H. (1960). The genetic structure of the incompatibility factors of *Schizophyllum commune*: the *A* factor. *Proc. natn. Acad. Sci. U.S.A.* **46**, 833.

RAPER, J. R., BOYD, D. H. & RAPER, C. A. (1965). Primary and secondary mutations at the incompatibility loci in *Schizophyllum*. *Proc. natn. Acad. Sci. U.S.A.* **53**, 1324.

RAPER, J. R. & ESSER, K. (1961). Antigenic differences due to the incompatibility factors in *Schizophyllum commune*. *Z. VererbLehre* **92**, 439.

RAPER, J. R. & FLEXER, A. S. (1970). Mating Systems and Evolution of the Basidiomycetes. Ed. R. H. Petersen. In *International Symposium on the Evolution of the Higher Basidiomycetes*. (In Press.)

RAPER, J. R. & MILES, P. G. (1958). The genetics of *Schizophyllum commune*. *Genetics, Princeton* **43**, 530.

RAPER, J. R. & RAPER, C. A. (1968). Genetic regulation of sexual morphogenesis in *Schizophyllum commune*. *J. Elisha Mitchell scient. Soc.* **84**, 267.

RAPER, J. R. & RAUDASKOSKI, M. (1968). Secondary mutations at the Bβ locus of *Schizophyllum*. *Heredity, Lond.* **23**, 109.

RAPER, J. R. & SAN ANTONIO, J. P. (1954). Heterokaryotic mutagenesis in Hymenomycetes. I. Heterokaryosis in *Schizophyllum commune*. *Am. J. Bot.* **41**, 69.

RAUDASKOSKI, M. (1970). A new secondary βB mutation in *Schizophyllum* revealing functional differences in wild $B\beta$ alleles. (In Press.)

SIMCHEN, G. (1967). Genetic control of recombination and the incompatibility system in *Schizophyllum commune*. *Genet. Res.* **9**, 195.

STAMBERG, J. (1968). Two independent gene systems controlling recombination in *Schizophyllum commune*. *Molec. gen. Genet.* **102**, 221.

STANIER, R. Y., DOUDOROFF, M. & ADELBERG, E. A. (1963). *The Microbial World*, 2nd ed. Englewood Cliffs, N.J.: Prentice Hall.

STERN, C. (1968). *Genetic Mosaics and Other Essays*. Harvard University Press.

STORRS, E. E. & WILLIAMS, R. J. (1968). A study of monozygous quadruplet armadillos in relation to mammalian inheritance. *Proc. natn. Acad. Sci. U.S.A.* **60**, 910.

SWIEZYNSKI, K. M. (1962). Analysis of an incompatible di-mon mating in *Coprinus lagopus*. *Acta Soc. Bot. Pol.* **31**, 169.

SWIEZYNSKI, K. M. (1963). Somatic recombination of two linkage groups in *Coprinus lagopus*. *Genet. Pol.* **4**, 21.

WANG, C. S. & RAPER, J. R. (1969). Protein specificity and sexual morphogenesis in *Schizophyllum commune*. *J. Bacteriol.* **99**, 291.

WANG, C. S. & RAPER, J. R. (1970). Isozyme patterns and sexual morphogenesis in *Schizophyllum*.

WESSELS, J. G. H. (1969). Biochemistry of sexual morphogenesis in *Schizophyllum commune*: effects of mutations affecting the incompatibility system on cell-wall metabolism. *J. Bacteriol.* **98**, 697.

WESSELS, J. G. H. & NIEDERPRUEM, D. P. (1967). Role of a cell-wall glucan-degrading enzyme in mating of *Schizophyllum commune*. *J. Bacteriol.* **94**, 1594.

REPRINT* OF ARTICLE (1957) BY THE LATE E. C. DOUGHERTY

Neologisms needed for structures of primitive organisms;
E. C. Dougherty (*Department of Physiology, School of Medicine, University of California, Berkeley, California*)

The new terminology of this and the next two abstracts derives from studies designed to clarify the phylogeny of primitive organisms. The neologisms are not necessarily meant to replace terminology in current use, but to provide workers with a precise (and phylogenetic rational) nomenclature.

Although some have declined to term 'nuclei' the dumb-bell- or rod-shaped organelles occurring in bacteria and staining in a manner indicating DNA content, each of these structures in the Monera (= bacteria + blue-green algae) seem clearly homologous with a nucleus in higher organisms, even though the moneran organelle appears structurally simpler. I feel that *nucleus* can reasonably serve as a collective term for both structures, but that new terms are needed by comparative morphologists wishing to distinguish the moneran nucleus from the more organized type of nucleus found in all higher organisms, both primitive and advanced. For the moneran nucleus I propose **prokaryon** (from πρό, *before*; and κάρυόν, *kernel*—plural: **prokarya**), and for that of higher organisms, **eukaryon** (from εὐ, *well*; and κάρυόν—plural: **eukarya**). From these derive: (1) the adjectives **prokaryous** and **eukaryous** meaning, respectively, 'pertaining to prokarya and eukarya', and (2) the nouns **prokaryosis** and **eukaryosis** and their corresponding adjectives, **prokaryotic** and **eukaryotic** (from prokaryon, etc.; and -ωσιζ, a suffix signifying *condition*), denoting, respectively, 'the condition of possessing prokarya or eukarya'.

* Reprinted from J. Protozool (1957), **4** (Suppl.), p. 14.

GLOSSARY

The descriptions are intended to be informative rather than definitive. Words printed in capital letters, following a term to be explained, are synonyms; words in lower case are partial synonyms. Words in *italic* are explained separately under their own heading.

AXENIC CULTURE, PURE CULTURE. Defined by Dougherty (1953) as 'pertaining to growth of organisms of a single species in the absence of living organisms or living cells of any other species'.

BASAL GRANULE, blepharoplast, kinetosome, centriole. Cylindrical organelles of eukaryotic cells from which the axial filaments of cilia, flagella etc., arise. Have the same fine structure as *centrioles*, from which they sometimes arise; may also be self-perpetuating. In spermatogenesis a centriole generates the sperm tail to which it is related as a basal granule to its cilium.

CENTRIOLE, basal granule, kinetosome (see Stanier page 32). Like axial cores of cilia, they have nine peripheral microtubular components, but each has three sub-units, and the central microtubules found in cilia are absent. Cells usually have two centrioles, lying at right angles to each other in the centrosome; at division, each pair of centrioles generates another pair and they form the poles of the mitotic spindle. They may be concerned with the formation of asters and spindles, although they appear to be absent from higher plants which have lost the ability to form flagella (cilia).

CENTROMERE, SPINDLE ATTACHMENT, KINETOCHORE. Clear area seen in many eukaryotic chromosomes; appears to be last part of a chromosome to divide, and point at which divisive force of mitosis acts.

CHLOROPLASTS. The photosynthetic organelles of eukaryotic cells, bounded by a double membrane and having chlorophyll-containing lamellae (membranes) embedded in a protein matrix.

CHROMOMERES. Deeply staining granules, giving a beaded appearance to chromosomes, especially at prophase of meiosis, and giving a banded appearance to salivary-gland chromosomes of dipteran larvae; composed of densely compacted strands of DNA; may be units of replication.

CHROMONEMA (pl. CHROMONEMETA). The chromosome thread.

CHROMOSOME. 'One of the constituent genophores in a eukaryotic nucleus' (Stanier, page 32). Strongly basophilic rods into which the *eukaryon* (eukaryotic nucleus) resolves during division; probably always at least two per nucleus.

CILIUM, FLAGELLUM, sperm tail (see Stanier, page 32). No clear distinction can be made between flagella and cilia. Stanier suggests that the word flagellum be reserved for the flagellum of prokaryotes, and that the word cilium be used for flagella and cilia of eukaryotes.

CISTERNAE. Spaces enclosed between paired membranes of the endoplasmic reticulum.

CYANELLE. A blue-green alga living as an endosymbiont.

DIKARYON. Fungal mycelium having 'cells' containing two nuclei, usually of different mating type.

DIPLOID. Applied to the number of chromosomes present in a fertilized zygote and in cells derived from it by mitosis. Also applied to organisms having the diploid number of chromosomes.

ENDOCYTOSIS. Intake by cells of particles which do not enter by diffusion; *phagocytosis, pinocytosis* and *viropexis*.

ENDOPLASMIC RETICULUM. 'The irregular network of unit membranes which traverses the cytoplasmic region of a eukaryotic cell, often bearing ribosomes on its surface' (Stanier, page 31). The membranes form a system of interconnecting cavities (cisternae). The endoplasmic reticulum takes two forms, 'smooth' and 'rough' (granular), the latter having ribosomes closely adhering to its outer surface.

EUKARYON. Dougherty's term (see page 433) for the type of nucleus which has a nuclear membrane, forms true chromosomes and is found in all organisms except bacteria, actinomycetes and blue-green algae. EUKARYOSIS, EUKARYOTIC; the condition of possessing eukarya.

EXOCYTOSIS. Discharge by cells of particles which do not diffuse through the wall; reversed pinocytosis.

GENOPHORE. Ris (1961) stated that it was incorrect to use the word chromosome for the structure corresponding to a linkage group in prokaryotes and viruses; he proposed the term 'genophore' for the 'physical entity corresponding to a linkage group'. In eukaryotes the genophore is a *chromosome*, in viruses it is a single or double strand of RNA or DNA. The DNA in mitochondria and chloroplasts also comes within the definition, as do bacterial plasmids and episomes.

GOLGI APPARATUS, GOLGI BODY, DICTYOSOME (see Stanier, page 31). A localized membranous system in eukaryotic cells, in which osmium and silver are selectively deposited under certain conditions; it differs from the endoplasmic reticulum in having a more definite organization, in being localized within the cell, and often, in having slightly thicker membranes; there may be as many as twenty Golgi bodies per cell.

GRANA. Granules in which the chlorophyll of some plants is localized within the chloroplasts, consisting of highly organized stacks of membranous lamellae (see *thylakoids*); the lamellae protrude into, and branch in, the stroma, interconnecting with the lamellae of other grana. In some simpler plants the lamellae do not form grana.

HAPLOID. Refers to the number of chromosomes found in gametes and, by extension, to any organism which has one set of chromosomes in each nucleus.

INTERPHASE NUCLEUS. The nucleus when not dividing; originally, the 'resting' nucleus which may form between first and second division of meiosis.

KINETOPLAST. Unusual nucleoprotein organelle of trypanosomes, near base of flagellum; fine structure and staining reactions show similarity to mitochondria.

LAMPBRUSH CHROMOSOMES. Enlarged chromosomes, up to 800 μm long, of weak staining reaction, seen in early meiosis in some vertebrates; are banded into *chromomeres*, like salivary gland chromosomes, but are not *polytene*. The chromomeres show 'puffing', which is partial unravelling to reveal a pair of long loops of DNA double-helix embedded in RNA and protein; RNA synthesis takes place in the loops. Favourable material for the cytochemical study of chromosome function.

LATENCY, CRYPTICITY. Term used of several enzymes in *lysosomes*, and of catalase in *peroxisomes*, meaning relative inaccessibility by enzyme substrate; probably attributable to encapsulation of enzymes in membranes. For another usage see Vogel *et al.* pp. 113 *et seq.*

LOMASOMES, BOUNDARY BODIES. Electron microscopy reveals several unknown bodies in fungal hyphae. Lomasome ('border body') is a term for sponge-like structures, flattened against the inner wall of the hypha, and having their inner boundary delimited by the invaginated plasma membrane; their contents may be granular, vesicular, or both; occur in many fungi, and similar bodies occur in algae and higher plants.

LYSOSOMES, CYTOSOMES. Cytoplasmic particles in animal cells, containing several 'acid' hydrolytic enzymes in high concentration, the enzymes showing *latency*. Variable in form and content. Distribution in cells coincides with that of acid phosphatase. Act in intracellular digestion of cell components and of foreign matter taken in by phagocytosis and pinocytosis. Before lysosomes take part in digestion they have homogeneous or granular contents and are called primary lysosomes; when they have fused with phagosomes or vesicles containing engulfed matter they are called secondary lysosomes (phago-lysosomes). In plant cells spherosomes may be the functional equivalent of lysosomes.

MDNA. Mitochondrial DNA.

MEMBRANES. See Stanier, pages 31 and 32.

MICROBODIES, peroxisomes. 'Microbodies' is a morphological term applied to cytoplasmic, membrane-bound bodies (diam. 0·5 μm) in tubule cells of kidney, and to similar bodies in liver cells which have a 'crystalloid' microtubular core. Liver microbodies contain urate oxidase, D-amino acid oxidase, L-α-hydroxy acid oxidase and a high concentration of catalase; kidney microbodies and bodies from the protozoan *Tetrahymena* have similar but not identical biochemical properties. de Duve (1965) proposed 'peroxisome' as a general biochemical term for subcellular bodies functionally related to liver microbodies. Peroxisomes probably function in extra-mitochondrial oxidations. Catalase is 'latent' in intact peroxisomes; the other enzymes are not latent, probably because the peroxisome membrane is highly permeable.

MICROSOMES, MICROSOMAL FRACTION. Operational term for fraction which sediments more slowly than mitochondria; usually includes ribosomes and fragments of endoplasmic reticulum.

MICROTUBULES, fibres, fibrils. 'Rigid, hollow filaments with diameters ranging from 15 to 25 nm, which occur in many regions of the cytoplasm of the eukaryotic cell. They are particularly conspicuous (on account of their regular arrangements) in cilia and in the mitotic apparatus' (Stanier, page 32). Circular in section, unbranched, and may be almost as long as the cell; are common in irregularly-shaped cells. As neurotubules they extend the length of the axon or dendrite in nerve cells. Possible functions: contractile organelles, conduits, intracellular framework.

MITOCHONDRIA. Cytoplasmic organelles of eukaryotes, bounded by two membranes, the inner invaginated to form cristae; the seat of the tricarboxylic acid cycle; succinate dehydrogenase and cytochrome oxidase are localized in mitochondria. Contain DNA, and enzymes for replication, transcription and translation. Multiply by growth and division and may also arise from other parts of the cell. They contain insufficient DNA to code for all mitochondrial proteins, and nuclear genes act in the formation of mitochondria. Animal cells usually contain several hundred mitochondria.

MONOXENIC CULTURE. Refers to growth of organisms of one species in the presence of living organisms (or cells) of one other species (Dougherty, 1953). A xenic culture is one in which the primary species is associated with unidentified organisms.

NUCLEOLUS AND NUCLEOLAR ORGANIZER. Nucleoli are deeply staining bodies, rich in RNA, seen in resting nuclei with the light microscope. They lack a bounding membrane. Each nucleolus forms at a specific region of a chromosome, called the nucleolar organizer. In amphibian oocytes the nucleolar organizer multiplies to give about 1,000 detached organizers and associated nucleoli; genes in the organizers code for ribosomal RNA precursors, which are synthesized in large amounts, thereby permitting cytological detection of the ribosomal RNA genes. Oocyte nuclei are large in Amphibia (diam. up to 1 mm.), permitting rapid isolation and treatment of nucleoli before they denature.

NUCLEUS. 'A localized cellular region which contains most (or all) of the DNA. In eukaryotes, the nuclear DNA is carried on more than one *genophore* (chromosome); in prokaryotes, it is generally contained in a single genophore' (Stanier, page 32).

PEROXISOMES. See *microbodies*.

PHAGOCYTOSIS, endocytosis. Engulfment of extracellular particle(s) by cells, e.g. by amoebae, leucocytes and macrophages; the cell surface invaginates to form a membrane-bound cytoplasmic vacuole (*phagosome*) containing the engulfed particles. Ingested particles are digested by enzymes of *lysosomes*.

PHAGOSOME. Phagocytic or pinocytic vacuoles (vesicles) before they fuse with lysosomes. Plant peroxidase is taken up by phagosomes and may be used as a cytochemical marker.

PINOCYTOSIS, endocytosis. 'Cell drinking': uptake of liquid by cells in a manner similar to *phagocytosis*; the surface membrane of the cell invaginates, often forming deep tubular canals, and then closes to form pinocytic vesicles.

PLASMID. Introduced by Lederberg (1952) 'as a generic term for any extrachromosomal hereditary determinant. The plasmid itself may be genetically simple or complex...' Lederberg applied the word quite generally, e.g. to viruses, plasmagenes and symbiotic organisms. The term is in general use for non-chromosomal DNA replicons in bacteria, but should not be restricted to this usage. The word is usefully and appropriately applied to self-dependent organelles of eukaryotes, such as *mitochondria*, rather than to their DNA, for which other more specific words may be used (*genophore*, *replicon*).

PLASTID. General term for certain organelles in the cytoplasm of green plants. Chloroplasts are green photosynthetic plastids; leucoplasts are colourless plastids predominantly storing starch, oil or proteins; chromoplasts are coloured plastids, in for example petals and fruits, which lack chlorophyll but often contain considerable amounts of carotenoids. In some cases, at least, one type of plasmid may change into another.

POLYNEME. Pertaining to multistranded chromosomes.

POLYTENE CHROMOSOMES, GIANT CHROMOSOMES. Long thick chromosomes (up to $\frac{1}{4}$ mm. long) made up of thousands of strands, formed by reduplication of chromatids without subsequent separation and contraction. Found in giant cells in salivary glands etc. of dipteran larvae, they show characteristic sequences of bands (*chromomeres*) which usually correspond with genes; different banded regions 'puff' at specific stages of development: the puffs (Balbiani rings) are uncoiled regions of the DNA histone strands showing intense RNA synthesis, indicative of differential gene activation.

PROKARYON. Dougherty's term (see page 433) for nuclei of bacteria, actinomycetes and blue-green algae. PROKARYOTIC: the condition of possessing prokarya.

PYRENOIDS. Finely granular bodies, more electron dense than the ground substance (stroma), embedded in, or projecting from, the chloroplasts of algae and some liverworts; the granular material is proteinaceous; starch is often present as plates or grains near the pyrenoid surface. *Thylakoids* penetrate the pyrenoids in some species.

REPLICON. Jacob & Brenner's (1963) word for a unit of independent replication such as a bacterial chromosome, phage chromosome, or episome (bacterial *plasmid*); according to their model, replication would be governed by two specific and indirectly matching elements of the replicon: a structural gene coding for an 'initiator' molecule, and an operator or 'replicator' on which the initiator would act to initiate replication of the DNA attached to the replicator; the initiator would be similar to a repressor molecule in being cytoplasmic and able to act in the trans position but it would differ in that its effect would be to activate rather than repress.

SYMBIOSIS, MUTUALISM. The term is now commonly reserved for mutually beneficial associations, but as first used it includes parasitism, mutualism and commensalism, in which the host respectively suffers, benefits and is not affected.

TEKTINS. General term proposed by Mazia & Ruby (1968) for a class of similar proteins each of which forms different eukaryotic structures such as microtubules, microfilaments and membranes. Tektins show general resemblances to each other for example in amino-acid composition, and the molecules of each tektin specifically aggregate together.

THYLAKOIDS. The photochemical reactions of photosynthesis take place in lamellae (membranes) which are joined to form closed sacs called thyla-koids. In prokaryotes the thylakoids are of various shapes and are usually free in the cytoplasm. In eukaryotes the thylakoids are in chloroplasts and usually have the form of more or less flattened discs, which in higher plants are formed into dense stacks called *grana*.

VIROPEXIS. Process by which virus particles adsorbed to the cell surface are taken into the cytoplasm.

WORONIN BODIES, METACHROMATIC BODIES. Buller's (1933) term for elongate-oval bodies first described by Woronin in hyphae of ascomy-cetous fungi; under phase-contrast they are small, highly refractile, mobile bodies, up to 12 in number, randomly dispersed in apical cells but associ-ated with the perforate cross walls (septa) in older hyphae; under the electron microscope they are dense and membrane-bound with a lattice (crystalloid) structure. They plug septal pores adjacent to injured cells; may be composed largely of ergosterol.

REFERENCES

BULLER, A. H. R. (1933). *Researches on Fungi* 5, 127. University of Toronto Press.

DOUGHERTY, E. C. (1953). *Parasitology* 42, 259.

DE DUVE, C. (1965). *J. Cell Biol.* 27, 25A.

JACOB, F. & BRENNER, S. (1963). *C. r. hebd. Séanc. Acad. Sci., Paris* 256, 298.

LEDERBERG, J. (1952). *Physiol. Rev.* 32, 403.

MAZIA, D. & RUBY, A. (1968). *Proc. natn. Acad. Sci. U.S.A.* 61, 1005.

MOORE, R. T. & MCALEAR, J. H. (1961). *Mycologia,* 53, 194.

RIS, H. (1961). *Can. J. Cytol.* 3, 95.

STANIER, R. Y. 1970. This Symposium, p. 1.

INDEX

prokaryotes (*cont.*)

eukaryotes, 6, 28–9; energy transduction in, 144–52; evolution of, 112, 116–17; genetic code in, 45, 49–51; genetic flux in, 270–2; lysine synthesis in, 107–15; mesosomes of, 15, 22, 313–14, 337; movements through membranes of, 136, 137, 144; ornithine synthesis in, 115–16; photosynthetic apparatus in, 222–9; plasmids in, 249–77; recombination of genetic material in, 279–83, 288; ribosomes of, 55–8, (function of), 67–70, (protein of), 54–7, (sensitive to chloramphenicol), 70, 78, 241, 387; *r*RNAs of, 58–64, 78–9; structure and function in, 315; *see also* Actinomycetes, bacteria, blue-green algae

pro-lamellar bodies, in chloroplasts and blue-green algae, 229, 231–2

prophage P1 plasmid, 254

Propionibacterium, 65

protein synthesis: in bacteria, 307; in cell-free systems, 68; in chloroplasts, 393, 394; in cytoplasm, 5–6, 348, 386–7; inhibited by virus infection, 345; initiation of, 66, 68–9, 97; in mitochondria, 299, 386–8, 395

proteins: of homokaryons and dikaryons of *Schizophyllum*, 427–8, iron-containing, in photosynthetic bacteria, 212; of microtubules, 18; of ribosomes, 60, 61, 64–7, 97, 241; stimulation of pinocytosis by, 334; stored and concentrated in Golgi apparatus, 308, 309; structural self-assembling (tektins), 19; surface self-assembling, of *B. brevis*, 168; of virus coat 328, 333; of viruses, 336, 345

Proteus, 65

Proteus mirabilis, R-factor in, 256

protonmotive force, 145–6, 147–8

protons: coupling between metabolism and transport by, 149; symport and antiport reactions coupled by, 136, 141–2, 148; translocation of, in membranes, 127–9, 131, 144–52; uncoupling agents as conductors of, 132–3, 141

Protopterus, amount of DNA per cell in, 362, 363

Prototheca zopfii, mitochondria of, 300

protovirus, 324

protozoa: anaerobic spp. of, 25; autolysosomes in, 311; endoplasmic reticulum in, 307; endosymbionts in, 12, 14; Golgi apparatus in, 309; life cycles in, 406; lysine not synthesized in, 111, 112; mitochondria of, 300; polytene chromosomes in, 367; *r*RNAs of, 78–9

Prymnesium, Golgi apparatus in formation of surface scales of, 10, 12

Pseudomonas, FP-factor in, 254

Pseudomonas aeruginosa, genes for tryptophan metabolism in, 288

Pseudomonas fluorescens, lysine synthesis in, 110

pseudopodia, 16, 20, 28

pseudorabies virus, enzymes in cells infected by, 342

Psilotum, chloroplasts of, 235

Pteridophytes, chloroplasts of, 235

puromycin, effect of, on syntheses of: glycoprotein, 182; glycosaminoglycans, 186; *r*RNA, 97

purple photosynthetic bacteria, 15, 22; cytochromes of, 211; evolution of hydrogen by, 216; nitrogen-fixing systems from, 215; photosynthetic pigments of, 206; *see also* Athiorhodaceae, Thiorhodaceae

pyrenoids, in chloroplasts, 231, 232, 233, 235

Pyrrophyta, photosynthetic pigments of, 29

pyruvate: as electron donor for nitrogen fixation, 216; from fixation of CO_2, 210, 213; porter system for, 141

pyruvate oxidase, in mitochondria, 299

pyruvate synthase, 210, 213; absent from blue-green algae, 127

Pythium ultimum: Golgi apparatus in, 308; lysine synthesis in, 110

quinones, in photosynthetic electron transport systems, 206, 212

radiation: intrastrand pyrimidine dimers produced in DNA by, 371; mutagenic, 389; sensitivity to, varies with DNA content per cell, 366, 372

Rana pipiens, nuclear and mitochondrial DNA of, 383

Ranunculaceae, amount of DNA per cell of different spp. of, 363

rat, liver cells of: mitochondria of, 137, 147, 386; *r*RNA precursors in, compared with HeLa cells, 98, 101

recombination of genetic material, 373–5; in antibiotic-resistant strains of *Saccharomyces*, 391; in di-mon mating, 416–17; dual effect of, 290–1; in evolution, 281–8; and grouping of genes, 288–90, 292; in prokaryotes, 279–80

red blood cells (of mammals): agglutination of, by viruses, 331; membrane of, 19; porter systems in, 132, 136, 142; shape of, 168

reo virus, 325; concentrated in lysosome fraction of cell, 339; enzymes in cells infected by, 344; RNA of, 340, 347; on spindle fibres of dividing cells, 349

replicons, 250, 270–1; speciation in bacteria reflects restrictions on transmissibility of, 272